Ionic Liquids in Lipid Processing and Analysis

Ionic Liquids in Lipid Processing and Analysis

Opportunities and Challenges

Edited by

Xuebing Xu
Wilmar Global Research and Development Center, Shanghai, China

Zheng Guo
Department of Engineering, Aarhus University, Aarhus, Denmark

Ling-Zhi Cheong
Wilmar Global Research and Development Center, Shanghai, China; Department of Food Science, School of Marine Science, Ningbo University, China

ELSEVIER

AMSTERDAM • BOSTON • HEIDELBERG • LONDON
NEW YORK • OXFORD • PARIS • SAN DIEGO
SAN FRANCISCO • SINGAPORE • SYDNEY • TOKYO

Academic Press is an imprint of Elsevier

AOCS PRESS

Academic Press and AOCS Press
Academic Press is an imprint of Elsevier
125 London Wall, London EC2Y 5AS, UK
525 B Street, Suite 1800, San Diego, CA 92101-4495, USA
50 Hampshire Street, 5th Floor, Cambridge, MA 02139, USA
The Boulevard, Langford Lane, Kidlington, Oxford OX5 1GB, UK

Published in cooperation with American Oil Chemists' Society www.aocs.org

Director, Content Development: Janet Brown

British Library Cataloguing-in-Publication Data
A catalogue record for this book is available from the British Library

Library of Congress Cataloging-in-Publication Data
A catalog record for this book is available from the Library of Congress

ISBN: 978-1-63067-047-4

For information on all Academic Press publications
visit our website at https://www.elsevier.com/

Publisher: Nikki Levy
Acquisition Editor: Nancy Maragioglio
Editorial Project Manager: Karen Miller
Production Project Manager: Caroline Johnson
Designer: Ines Cruz

Typeset by TNQ Books and Journals
www.tnq.co.in

Contents

5. Rational Design of Ionic Liquids for Lipid Processing

Zheng Guo

6. Synthesis and Properties of Lipid-Inspired Ionic Liquids

Arsalan Mirjafari, Richard A. O'Brien, James H. Davis, Jr.

7. New Opportunities from Ionic Liquid for Chemical and Biochemical Processes of Lipids

Bianca Pérez, Wei Wei

List of Contributors

Kenneth Benjamin South Dakota School of Mines and Technology, Department of Chemical and Biological Engineering, Rapid City, SD, United States

Claire C. Berton-Carabin Wageningen University, Food Process Engineering Group, Wageningen, The Netherlands

Manat Chaijan Walailak University, School of Agricultural Technology, Department of Agro-Industry, Nakhon Si Thammarat, Thailand

Ling-Zhi Cheong Wilmar Global Research and Development Center, Shanghai, China; Department of Food Science, School of Marine Science, Ningbo University, China

Michael J. Cooney University of Hawaii at Manoa, Hawaii Natural Energy Institute, Honolulu, HI, United States

João A.P. Coutinho CICECO, University of Aveiro, Department of Chemistry, Aveiro, Portugal

James H. Davis, Jr. University of South Alabama, Department of Chemistry, Mobile, Alabama, United States

Mia Falkeborg Aarhus University, Department of Engineering, Aarhus, Denmark; Danish Technological Institute, Center for Food Technology, Aarhus, Denmark

Vicente Gotor Fernández University of Oviedo, Department of Organic and Inorganic Chemistry, Oviedo, Spain

Zheng Guo Department of Engineering, Aarhus University, Aarhus, Denmark

Diego O. Hartmann Universidade Nova de Lisboa, Instituto de Tecnologia Química e Biológica António Xavier, Oeiras, Portugal

Derya Kahaveci Yeditepe University, Department of Food Engineering, Istanbul, Turkey

Jingbo Li Aarhus University, Department of Engineering, Aarhus, Denmark

Arsalan Mirjafari Florida Gulf Coast University, Department of Chemistry and Physics, Fort Myers, Florida, United States

Mohd Firdaus Mohd Yusoff Aarhus University, Department of Engineering, Aarhus, Denmark; Universiti Kebangsaan Malaysia, UKM, Fakulti Sains dan Teknologi, Pusat Pengajian Sains Kimia dan Teknologi Makanan, Bangi, Selangor, Malaysia

Richard A. O'Brien University of South Alabama, Department of Chemistry, Mobile, Alabama, United States

Worawan Panpipat Walailak University, School of Agricultural Technology, Department of Agro-Industry, Nakhon Si Thammarat, Thailand

Caroline Emilie Paul Delft University of Technology, Department of Biotechnology, Delft, The Netherlands

Cristina Silva Pereira Universidade Nova de Lisboa, Instituto de Tecnologia Química e Biológica António Xavier, Oeiras, Portugal

Bianca Pérez Aarhus University, Department of Engineering, Aarhus, Denmark

Bethala Lakshmi Anu Prabhavathi Devi CSIR–Indian Institute of Chemical Technology, Center for Lipid Research, Hyderabad, India

Sónia P.M. Ventura CICECO, University of Aveiro, Department of Chemistry, Aveiro, Portugal

Tangadanchu Vijai Kumar Reddy CSIR–Indian Institute of Chemical Technology, Center for Lipid Research, Hyderabad, India

Wei Wei Aarhus University, Department of Engineering, Aarhus, Denmark

Xuebing Xu Wilmar Global Research and Development Center, Shanghai, China

Hua Zhao Savannah State University, Department of Chemistry and Forensic Science, Savannah, GA, United States

Nanjing Zhong Guangdong Pharmaceutical University, School of Food Science, Zhongshan, China

Preface

Ionic liquids (ILs) provide tremendous opportunities for designing new processes and improving or revolutionizing existing chemical/biochemical processes and production technology. There have been significant advances in the last decade in various fields related to ILs and although a list of books on ILs has been published summarizing the progress, no such lists exist in the field of lipids, even as book chapters. This book reviews the state-of-the-art research on the science and technology development in ILs focused on lipid processing and analysis. A detailed introduction of the book is presented as the first chapter.

We believe this work is the first collection about the applications of ILs in lipids processing and analysis. We hope this book covers most of the relevant topics expected by the scientific researchers and industrial engineers working in the lipids community. It is our great expectation that this book could stimulate more interest and discussion, or attract more attention to the tremendous opportunity and challenges from ILs as a newly emerging technology for lipids processing.

Finally, we would like to thank all contributors for their magnificent work in the collection of the latest research publications and their devotion to presenting accurate and detailed scientific information. The assistance from the AOCS Press is greatly appreciated, with special thanks to Janet and Lori.

Xuebing Xu
Zheng Guo
Ling-Zhi Cheong

Chapter 1

Are Ionic Liquids Ready for Lipids Processing?

An Introduction to the Book

Xuebing Xu
Wilmar Global Research and Development Center, Shanghai, China

Zheng Guo
Department of Engineering, Aarhus University, Aarhus, Denmark

Ling-Zhi Cheong
Wilmar Global Research and Development Center, Shanghai, China; Department of Food Science, School of Marine Science, Ningbo University, China

Room temperature ionic liquids (ILs) are neoteric (novel) green solvents with the potential to revolutionize some existing chemical/biochemical processes. Room temperature ILs are also evolving ionic functional materials, which expands applications in diverse fields.

1.1 IONIC LIQUIDS, NOT JUST A NEOTERIC SOLVENT: EXPANDING IN DIVERSE FIELDS

Evolving from high temperature molten salts, ILs can be dated back to the 19th century (Walden, P., 1914. Bull. Acad. Imper. Sci. (St. Petersburg), 1800). However, from conception to birth, ILs underwent a long, dreary, and unattractive gestation. A tremendous boost to this area occurred when air- and water-stable 1-ethyl-3-methylimidazolium tetrafluoroborate (EMIM BF_4) was synthesized and used in 1992 at the US Air Force Academy in Colorado Springs, Colorado (Chem. Commun. 1992, 965–967). Thereafter, a series of corresponding compounds was manufactured. Better properties conferred to this kind of novel solvent make ILs of practical interest and also link them to green chemistry.

Simply speaking, like normal salts, ILs are comprised entirely of ions; however, unlike the solid state of salts, the asymmetric cations and/or anions of

Ionic Liquids in Lipid Processing and Analysis. http://dx.doi.org/10.1016/B978-1-63067-047-4.00001-5

ILs result in them being liquid at room temperature or below 100°C. The discovery of ILs was regarded as breaking new ground by the chemical community and was accompanied by the introduction of the term "neoteric solvents" (Chem. Rev. 1999, 99, 2071–2083), indicating a class of novel solvents that offers a huge potential for industrial applications.

What features make ILs an innovative approach to green chemistry? What are the significant characteristics that distinguish them from conventional solvents? Depending on their composition, some ILs can possess specific properties. The following bullet points highlight the general key features of ILs. Overall, it is these properties that make ILs potentially attractive for many applications:

- Negligible vapor pressure that makes them nonevaporative, meaning no volatiles are discharged into atmosphere when employing ILs as media.
- Incredibly broad range of liquidity (from −96 to 300°C), providing ample leeway for specific reactions or design of unique processing conditions.
- Strong solvating power to allow solubilization of inorganic, organic, and polymeric materials. This property may permit some reactions or processes that are virtually impossible in traditional solvent systems to proceed in an IL system.
- Tailorablity of properties to meet specific requirements. The polarity and hydrophilicity/lipophilicity can be readily adjusted by judicious selection of cation, anion, and substituents. Accordingly, ILs are referred to as "designer solvents." If some functional groups are introduced, the ILs can be designed for specific tasks, yielding so-called "task-specific ILs."
- Easy to buy and simple to prepare because ILs exist as paired ions and are readily synthesized by simple metathesis (involving the swapping of reacting chemical structures).
- Availability of an extended family of ILs for selected purposes, which includes tetraammonium, tetraphosphonium, imidazolium, and pyridinium. In the past decades, the range of available anions and cations has expanded enormously. It is estimated that, if binary and ternary mixtures are included, there are approximately one trillion (10^{18}) accessible ILs. In contrast, useful conventional organic solvents number only around 300.

However, over two decades have passed since the first modern generation of ILs was manufactured in 1992 (Chem. Commun. 1992, 965–967). IL is no longer just a solvent to mediate chemical reactions/processes. Its applications have been greatly expanded to diverse fields (Scheme 1.1). It ranges from catalysis chemistry to physical chemistry, from electrochemistry to analytical chemistry, from separation to biomass conversion, from being biological excipients to industrial additives. The applications of ILs in lipid processing linked a few dimensions of property specificities pertaining to different types of ILs.

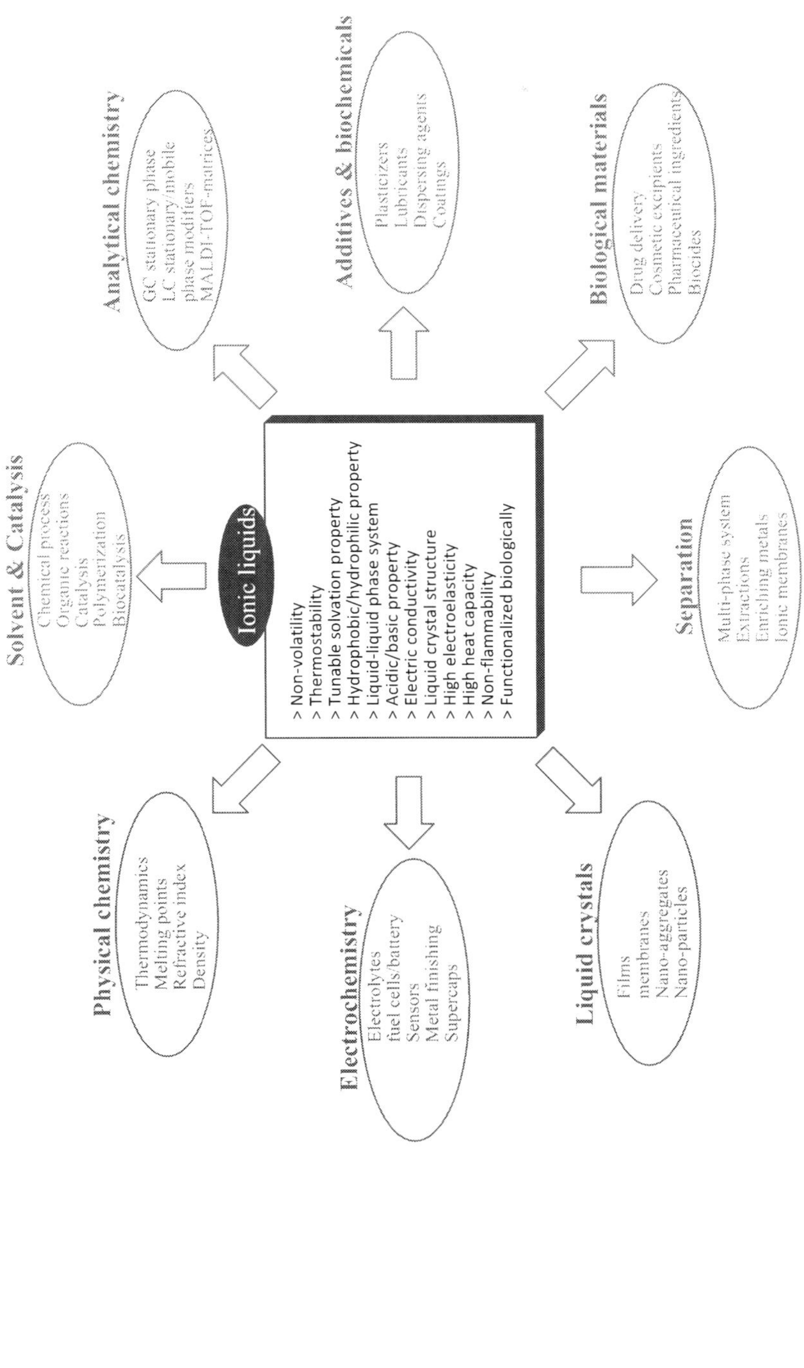

SCHEME 1.1 Matrix of IL application fields (the red (light gray in print versions) highlighting is partly involved for lipid processing in this book).

1.2 EVOLUTION OF IONIC LIQUID CONCEPT

As illustrated in Scheme 1.2, from being solvent replacements to biological materials/ingredients, ILs have tremendously evolved, which, according to Rogers's definition (New J. Chem. 2007, 31, 1429–1436), undergoes three generations. The first generation of ILs is purchasing solvent replacement for better reaction, catalysis, or green chemistry, where the accessible physical property set (e.g., non- or low volatility, thermal stability, or large liquid ranges) is achievable with many ILs. The second generation of ILs originates from growing interest in the materials applications of ILs, which utilize novel tunable physical and chemical property sets for applications as energetic materials, lubricants, analytical matrix, and functional ionic materials. The third generation of ILs is the targets endowed with biological property and functions or biological toxicity. They can be new ionic pharmaceutical ingredients or cosmetic formulation excipients.

The evolution or flexibility of the IL concept reflected the growing interests in the IL area; as the number of researchers in a variety of disciplines joined the ranks of the IL community, it was inevitable that new directions would emerge and new applications would be found. Progress will also be realized by taking advantage of the unique properties, namely the inherent modular nature of ILs, which gives a greater level of control over the physical, chemical, and biological properties of salts than is possible for molecular compounds. It can be anticipated that a natural outgrowth of the intense scrutiny of ILs by many different disciplines will ultimately lead to a much greater fundamental understanding of the interplay of strong and weak interactions in an IL-mediated system. Logically as a consequence this gives an opportunity to advance the technology in the relevant disciplines, including lipid processing and ionic lipid materials.

1.3 OPPORTUNITIES OFFERED BY IONIC LIQUIDS FOR LIPID PROCESSING

Since the birth of the modern "IL" concept, tremendous progress has been achieved. The distinct hallmarks are represented by (1) the growing number of commercial ILs; (2) the enlarging database of physical/chemical properties of ILs; (3) the increasing understanding and new knowledge of the structure–activity relationship of ILs, solvation, and interaction between ILs and substrates; and (4) emerging theoretical characterization and modeling. This constitutes the practical and theoretical basis, which thus definitely contributes to the development of new knowledge and advancing the technology in lipid processing.

Categorized by the molecular characteristics, lipids can be grouped into eight classes: fatty acyls, glycerolipids, glycerophospholipids, sphingolipids, saccharolipids, polyketides, sterol lipids, and prenol lipids (J Lipid Res. April 2009, 50, S9–S14). It is unclear how many different types of molecular

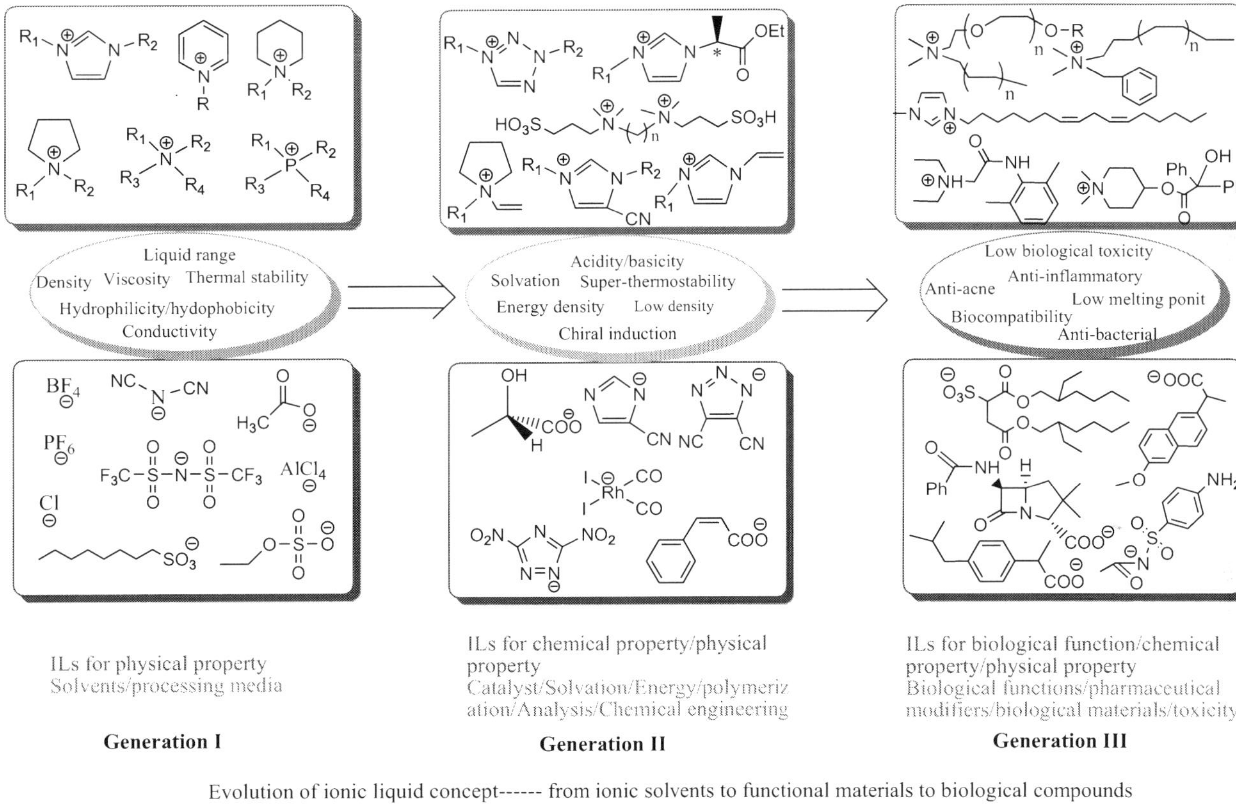

SCHEME 1.2 Evolution of ionic liquid concept for property and function.

species are in plant and microbial lipids sources. Recent progress in lipidomics reveals a remarkable diversity of lipids in the human body, in which the total number may be as many as 200,000 (Nat. Rev. Drug Discov. 2005, 4 (7), 594–610). It is evident that the conventional solvents, with numbers limited to 300, are insufficient to design variable processes that could match the structurally diverse lipids. However, the varied cation/anion structured ILs and combinations could yield one trillion accessible possibilities of ILs, which could provide enormous possibilities for designing and manufacturing ILs with the desired structure for a task-specific lipid process.

The opportunities, in relation to lipid/lipidic bioactive compounds processes, may include the following:

- Create or develop completely new reactions or processes by using newly-developed novel functionalized ILs or IL-type materials for innovative applications including lipid processing.
- Enable those reactions that do not occur in conventional solvents by creating compatible systems or enhancing solubility of substrates.
- Revolutionize or significantly improve currently existing processes or production lines by designing best appropriate ILs as processing media or processing aids.
- Develop “greener” processes to minimize discharge of volatile organic compounds and contribute to sustainability by design of degradable ILs using natural building blocks.
- Improve efficiency and selectivity of biocatalysis and biotransformation by developing/designing enzyme-benign/substrate-compatible IL systems.
- Improve extraction efficiency and recovery of bioactive compounds by designing/developing precise/effective multiphase IL extraction systems.
- Design/develop new IL-type pharmaceutical excipients/ingredients or deliver cargos for lipodic drugs, DNA/RNA, or protein delivery.
- Deconstruct/fractionate/regenerate lignocellulosic biomass for bioenergy and biochemical production by designing strong-solvation ILs.
- Improve lipid separation/analysis/identification by developing IL-based advanced GC, HPLC, HPLC-MS analysis techniques, specifically for lipidomic analysis.

1.4 WHAT IS NEW IN THIS BOOK?

This book reviews state-of-the-art progress with respect to the applications of ILs in lipid processing and relevant areas from different perspectives.

Part 1: Chemistry and Physics of Ionic Liquids for Lipid Processing

- Modeling and rational design of ionic liquids for lipid processing
- Synthesis of lipid-inspired ionic liquids
- Toxicity of ionic liquids

Part 2: Biocatalysis and Biotransformation in Ionic Liquids: Lipase Production, Purification, and Activation in Ionic Liquids

Part 3: Application of Ionic Liquids in Lipid Extraction and Recovery

Part 4: New Opportunity of Fats and Oils Modification and Processing in Ionic Liquids

Part 5: Ionic Liquids in the Synthesis of Antioxidant-Targeted Compounds

Part 6: Ionic Liquids in the Synthesis of Sugar/Carbohydrate and Lipid Conjugates

Part 7: Ionic Liquids in the Production of Biodiesels and Other Oleochemicals

Part 8: Ionic Liquids in Lipid Analysis

Chapter Biocatalysis and Biotransformations in Ionic Liquids elaborates on the successful applications of ILs in developing/enabling improved biocatalysis and biotransformation. The correlation between the molecular structures and their activities that influence enzyme activity/selectivity are discussed in detail, and involve most types of hydrolases. The chapter also discusses a specific type of ILs, deep eutectic solvents (DES), and reports on the latest progress of its application in biocatalysis.

Chapter Lipase Production and Purification From Fermentation Broth Using Ionic Liquids reviews the application of ILs in the downstream processing of biotechnological products, and the advantages, disadvantages, and potential of using ILs, especially in lipase production and purification.

Chapter Lipase Activation and Stability Enhancement in Ionic Liquids discusses the key factors that influence lipase activity and stability in ILs, as well as the methods and approaches to enhance lipase activity and stability in ILs. The recent application of DES for lipase activation is also reviewed.

Chapter Rational Design of Ionic Liquids for Lipid Processing summarizes the latest advances in the measurement of the physical and chemical properties of ILs and the available database of thermodynamic property datapoints. The rationality or preconditions and the essential steps toward a priori design of ILs are intensively discussed. The predictions of the physical/chemical properties of ILs including empirical (quantitative structure—property relationships and linear free energy descriptors) and experiment-independent modeling ab initio molecular dynamic modeling and conductor like screening model for real solvents (COSMO-RS) modeling are highlighted. Two case studies demonstrate how modeling helps achieve IL design for task-specific applications.

Chapter Synthesis and Properties of Lipid-Inspired Ionic Liquids reports on the idea of and progress in developing a new class of soft biomaterials: lipid-inspired ILs (LIILs), specifically in fatty acid-derived lipidic ILs and thioether-functionalized LIILs, with respect to methodology, synthetic chemistry, and characterization. The LIILs, having similar structure to natural lipids, are anticipating a great application potential in the biological and chemical industries.

Chapter New Opportunities From Ionic Liquid for Chemical and Biochemical Processes of Lipids reviews the most recent progress of the new opportunities presented by ILs for chemical/biochemical processes of lipids and lipodic pharmaceutical ingredients. The review mainly covers synthesis and characterization of active pharmaceutical ingredients (API), liquefaction of drugs with ILs, solvation of API with ILs, and IL-based microemulsion for drug delivery.

Chapter Ionic Liquids in Acylglycerol Synthesis and Modification highlights the progress and results, mainly in the Aarhus Group, in the exploration of the industry application opportunity of ILs in oil and fats modification, especially in design and development of IL-based systems for enzymatic production of partial glycerides (monoglycerides and diglycerides). The IL-containing reaction system and main reaction factors governing reaction selectivity and equilibrium shifting are intensively discussed. The COSMO-RS assisted reaction design and theoretical understanding are also discussed.

Chapter Ionic Liquids in Lipid Extraction and Recovery discloses the application potential of ILs as emerging solvents for extraction and recovery of lipid/bio-oils from natural sources. The focus is given to the discussion of unit operations with respect to extraction and recovery, and interplay between ILs and organic solvents. The molecular interactions and behaviors based ab initio modeling are also delineated.

Chapter Ionic Liquids in the Synthesis of Antioxidant Targeted Compounds summarizes related research regarding the applications of ILs in the extraction of natural antioxidants, and ILs as reaction media for lipophilization of natural antioxidants. Specifically for IL-mediated modification of phenolic compounds, the nature of ILs and other key parameters that influence enzyme activity and reaction progress and selectivity are intensively discussed.

Chapter Ionic Liquids in the Synthesis of Sugar/Carbohydrate and Lipids Conjugates reviews the latest progress in biocatalytic synthesis of sugar ester or carbohydrate-lipid conjugate by using ILs as an unconventional medium. The technical advantages offered by the unique property of ILs, such as increasing solubility of sugars and creating a compatible system for hydrophilic/hydrophobic substrates leading to enhanced conversion and selectivity, are described. The application of ILs in chemical synthesis of polysaccharide fatty acid esters is also reviewed.

Chapter Ionic Liquids in the Production of Biodiesel and Other Oleochemicals reviews the applications of ILs in production of biodiesels and biochemicals as both reaction media and functionalized acidic/basic catalysts. The technological advantages belonging to IL-based catalysts such as better compatibility, improved catalytic efficiency, and reusability are intensively discussed. The applications of ILs in mediating synthesis of biolubricants, polymers, and plasticizers are reviewed.

Chapter Ionic Liquids for Lipid Processing and Analysis: Opportunities and Challenges addresses an important factor limiting the application of

ILs—biological toxicity—focusing on its past (experimental assessment of IL toxicity), present (understanding the molecular basis of IL toxicity), and future (moving beyond toxicity toward potential new applications).

Last but not least, Chapter Ionic Liquids in Lipid Analysis discusses the applications of ILs in analytical chemistry including their applications in gas chromatography, high-performance liquid chromatography, matrix-assisted laser desorption ionization time-of-flight mass spectra and capillary electrophoresis, as well as in enrichment of n-3 polyunsaturated fatty acids/esters. Specifically, the applications of ILs in lipid analysis/separation are highlighted.

We believe this piece of work is the first collection about the applications of ILs in lipids processing. We hope this book covers most of the relevant topics expected by the scientific researchers and industrial engineers in the lipids community. It is our great expectation that this book could stimulate more interest and discussions, or attract more attention to the tremendous opportunity and challenges from ILs as a newly emerging technology for lipids processing.

Finally, we would like to thank all contributors for their magnificent work in the collection of the latest research publications and their devotion to presenting accurate and detailed scientific information. The assistance from the AOCS Press is greatly appreciated, with special thanks to Janet Brown and Lori Stewart.

Chapter 2

Biocatalysis and Biotransformation in Ionic Liquids

Caroline Emilie Paul
Delft University of Technology, Department of Biotechnology, Delft, The Netherlands

Vicente Gotor Fernández
University of Oviedo, Department of Organic and Inorganic Chemistry, Oviedo, Spain

LIST OF ABBREVIATIONS

Ac Acetyl
ABTS 2,2′-Azino-bis-(3-ethylbenzothiazoline-6-sulfonic acid) diammonium salt
ADH Alcohol dehydrogenase
ADH-A Alcohol dehydrogenase from *Rhodococcus ruber*
ADH-T Alcohol dehydrogenase from *Thermoanaerobium* sp.
ADH*Re* Alcohol dehydrogenase from *Rhodococcus erythropolys*
BAL Benzaldehyde lyase
Bu Butyl
BVMO Baeyer–Villiger monooxygenase
CAL-A *Candida antarctica* lipase type A
CAL-B *Candida antarctica* lipase type B
CIL(s) Chiral ionic liquid(s)
CLEA Cross-linked enzyme aggregates
C*p*ADH Alcohol dehydrogenase from *Candida parapsilosis*
CPO Chloroperoxidase
CRL *Candida rugosa* lipase
DESs Deep eutectic solvents
DHP Dehydropolymer
DKR Dynamic kinetic resolution
DMF Dimethylformamide
DMSO Dimethyl sulfoxide
E Enantioselectivity
EC Enzyme Commission
EH Epoxide hydrolase
EPA Eicosapentanoic acid
eq Equivalents
ER Ene reductase
Et Ethyl
Et_3N Triethylamine
FDH Formate dehydrogenase
GDH Glucose dehydrogenase
GI Glucose isomerase
***Hb*HNL** *Hevea brasiliensis* hydroxynitrile lyase
HLADH Alcohol dehydrogenase from horse liver
HNL Hydroxynitrile lyase
HRP Horseradish peroxidase
IL Ionic liquid

Ionic Liquids in Lipid Processing and Analysis. http://dx.doi.org/10.1016/B978-1-63067-047-4.00002-7

KPi Phosphate buffer
***Lb*ADH** Alcohol dehydrogenase from *Lactobacillus brevis*
***Lk*ADH** Alcohol dehydrogenase from *Lactobacillus kefir*
MDH Morphine dehydrogenase
Me Methyl
Me_2CO_3 Dimethyl carbonate
MeHNL Hydroxynitrile lyase from *Manihot esculenta*
MTBE Methyl *tert*-butyl ether
NAD(P)H β-nicotinamide adenine dinucleotide (phosphate), reduced form
NOX NAD(P)H oxidase
Oct Octyl
OYE Old Yellow Enzyme
***Pa*HNL** *Prunus amigdalus* hydroxynitrile lyase
PAMO Phenylacetone monooxygenase
PGA Penicillin G acylase
Ph Phenyl
PPL Porcine pancreas lipase
Pr Propyl
^{i}Pr Isopropyl
***Ras*ADH** Alcohol dehydrogenase from *Ralstonia* sp.
RML *Rhizomucor miehei* lipase
ROL *Rhizopus oryzae* lipase
rt Room temperature
RTIL Room-temperature ionic liquid
***Sy*ADH** Alcohol dehydrogenase from *Sphingobium yanoikuyae*
TADH Alcohol dehydrogenase from *Thermus* sp.
TBS *tert*-Butyldimethylsilyl ether
TEMPO (2,2,6,6-Tetramethyl-1-piperidinyl)oxidanyl
***Tes*ADH** Alcohol dehydrogenase from *Thermoanaerobacter ethanolicus*
ThDP Thiamine pyrophosphate
THF Tetrahydrofuran
TLL *Thermomyces lanuginosus* lipase
TON Turnover number
TSIL Task-specific ionic liquid
v/v Ratio volume/volume
VOC Volatile organic compound
UHP Urea-hydrogen peroxide
w/w Ratio weight/weight
YADH Yeast alcohol dehydrogenase

LIST OF ABBREVIATIONS FOR IONIC LIQUIDS

Cations

[amim] 1-Allyl-3-methylimidazolium
[bdmim] 1-Butyl-2,3-dimethylimidazolium
[bmim] 1-Butyl-3-methylimidazolium
[bmp] Butylmethylpyrrolidinium
[C_2OHmim] 1-(2-Hydroxyethyl)-3-methylimidazolium
[C_3OHmim] 1-(3-Hydroxypropyl)-3-methylimidazolium
[Ch] Choline
[ea] Ethylammonium
[emim] 1-Ethyl-3-methylimidazolium
[Et_3NH] Triethylammonium
[Et_3NMe] Triethylmethylammonium
[hmim] 1-Hexyl-3-methylimidazolium
[4-mbp] 4-Methyl-*N*-butylpyridinium
[$MeBu_3$P] 2-Methoxyethyl(tri-*n*-butyl) phosphonium
[mmim] 1,3-Dimethylimidazolium
[mtoa] Methyltrioctylammonium
[MTEOA] Tris(2-hydroxyethyl) methylammonium
[omim] 1-Octyl-3-methylimidazolium
[pmim] 1-Propyl-3-methylimidazolium
[pemim] 1-Pentyl-3-methylimidazolium
[$PrNH_3$] Propylammonium
[tea] Triethylammonium

Anions

[Arg] Arginate
[AOT] 1,4-Bis(2-ethylhexyl)-sulfosuccinate(docusate)
[BF_4] Tetrafluoroborate
[Br] Bromide
[$(CF_3SO_2)_2$N]/[NTf_2] Bis((trifluoromethyl)sulfonyl)amide/bistriflate

[CF_3SO_3] Trifluoromethanesulfonate
[Cl] Chloride
[$EtSO_4$] Ethylsulfate
[HSO_4] Hydrogensulfate
[$MeCO_2$] Acetate
[$MeSO_3$] Methylsulfonate
[$MeSO_4$] Methylsulfate
[Me_2PO_4] Dimethylphosphate
[Ms] Mesylate
[NO_3] Nitrate
[$OctSO_4$] Octyl sulfate
[OTf] Triflate
[PF_6] Hexafluorophosphate
[$PhCO_2$] Benzoate
[Pro] Prolinate
[Sac] Saccharinate

LIST OF ABBREVIATIONS FOR DEEP EUTECTIC SOLVENT

Hydrogen Bond Acceptors (HBAs)

ChAc Choline acetate
ChCl Choline chloride
Eac Ethylammonium chloride

Hydrogen Bond Donors (HBDs)

Acet Acetamide
EG Ethylene glycol
Gly Glycerol
MA Malonic acid
Ox Oxalic acid
U Urea

2.1 INTRODUCTION TO IONIC LIQUIDS FOR BIOCATALYSIS: AIM AND SCOPE

The application of biocatalytic methods for synthetic purposes plays an important role in synthetic chemistry since the discovery that the action of enzymes goes beyond performing hydrolytic reactions, and that biocatalysts can act with excellent levels of activity and selectivity in organic solvents as well as in neoteric systems. Thus, enzyme-catalyzed processes can provide benefits to existing methods for accessing bulk and fine chemicals in a selective and nonselective fashion under mild reaction conditions. Over the years, the advances in immobilization and protein engineering techniques have provided access to robust biocatalysts, which have found applications in various industrial sectors with the availability of different classes of enzymes, mainly hydrolases, oxidoreductases, and more recently, transferases. These advances have allowed the development of medium engineering studies, finding ionic liquids (ILs) to be valuable reaction media for the production of target chemicals. ILs are reusable solvents, which, depending on their nature, can stabilize an enzyme or inclusively improve its catalytical properties.

The aim of this chapter is to provide an update of relevant transformations carried out in ILs for the synthesis of valuable products, hence we have divided this contribution according to the classes of enzymes used for biotransformations developed in ILs. First, an introduction regarding the most common ILs used in combination with enzymes, and the main advantages of using ILs with synthetic applicability will be discussed. Later, selected reactions using a variety of enzymes will be presented, starting from the landmarks and onto interesting reactions recently developed. The use of mainly hydrolases and oxidoreductases will be presented, also paying attention to the lesser developed lyase or isomerase-catalyzed reactions in ILs. Nonsolvent applications of these neoteric solvents have been also included to provide a better understanding about the potential of ILs in organic chemistry. Finally, the state of the art regarding the use of enzymes in deep eutectic solvents (DESs) will be explained, a quite promising emerging type of media that can fill the gap between the application of these solvents not only in academia but also hopefully in industrial biotechnology in the near future.

2.1.1 Structural Properties

As described more in depth in other chapters of this book, ILs are salts consisting of a mixture of cations and anions with melting points near room temperature, although they are arbitrarily defined as salts with a melting point below 100°C (Wasserscheid and Welton, 2008; Hallett and Welton, 2011; Domínguez de María, 2012). In general, most ILs used are often liquid at room temperature, known as room-temperature ILs (RTILs). Commonly used ILs in biocatalysis consist of an organic cation, such as pyridine or imidazole derivatives, and an anion, generally inorganic but can also be organic, such as tetrafluoroborate (BF_4^-), bis(trifluoromethane)sulfonimide (NTf_2^-), and so on (Fig. 2.1).

Dialkylimidazolium cations with tetrafluoroborate or hexafluorophosphate anions are the most classic ILs used in biocatalysis, more specifically 1-alkyl-3-methylimidazolium salts and derivatives. Their main characteristics include (1) a negligible vapor pressure, hence emitting less volatile organic compounds (VOCs) than organic solvents (Earle et al., 2006; Ludwig and Kragl, 2007; Ahrenberg et al., 2014), (2) a high thermal stability (Kosmulski et al., 2004), (3) nonflammability, and (4) a high solvation capacity. More importantly, the physicochemical properties of ILs, such as their melting temperature, polarity, and hydrophobicity, can be fine-tuned by simply changing the structure or nature of the cation or anion, leading to a myriad of possibilities for new solvents (Wasserscheid and Welton, 2008; Chiappe and Rajamani, 2011; Passos et al., 2014). In this manner, ILs can be tailored to be immiscible with water or organic solvents of low or high polarity, thus facilitating conventional extraction methods for product separation and purification. Low melting points, as another example, can be achieved by incorporating bulky asymmetric cations, which disturb the stacking ability, together with weakly coordinating anions.

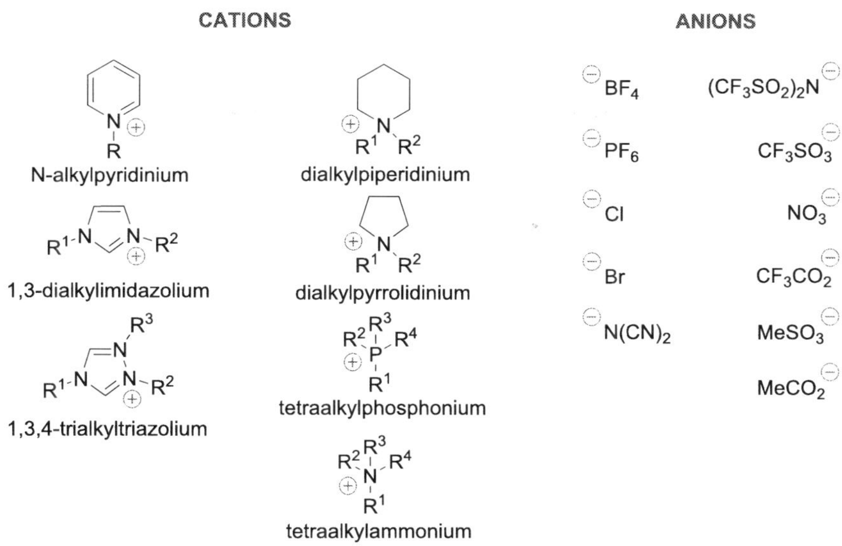

FIGURE 2.1 Examples of most commonly used cations and anions forming ionic liquids used in biocatalysis.

2.1.2 Classes of Ionic Liquids

The potential for tailoring the size, shape, and functionality of ILs offers opportunities that are unobtainable with conventional organic solvents. Changes in the type of ion, substitution, and composition produce new IL systems, each with a unique set of properties that can be used for a wide range of applications. The enormous potential of ILs lies within their intrinsic feature to be fine-tuned, as mentioned above, and thus to be tailored for a designated application leading to task-specific ILs (TSILs). Examples of TSILs include protic ILs, chiral ILs, multifunctional ILs, and supported ILs, designed for a large array of applications such as catalysis, synthesis, analytics, and so on. Recently, ILs made from natural products such as choline chloride, forming a third generation toward more bio-based ILs, and DESs, have been developed and advantageously used in biocatalysis (Domínguez de María, 2012).

ILs in biotransformations can be used as either an additive, a cosolvent, or a second hydrophobic phase, depending on its polarity. Certain ILs with a melting point of 50−100°C can also be used to coat enzymes. Other types of ILs, called kosmotropic ILs, have strongly hydrated ions that increase the water-network structure and lead to a more hydrated environment for the enzyme, whereas chaotropic ILs contain weakly hydrated ions, thus decreasing the water structure (Yang, 2009).

2.1.3 Benefits of Ionic Liquids in Biotransformations

ILs have been increasingly used in homogeneous and heterogeneous catalysis, and biocatalysis. Catalytic reactions in ILs have been investigated for more than 20 years; however, it is only during 2005 through 2015 that there has been an increase in their use in a wide range of catalytic and stoichiometric reactions as well as in many other applications (Madeira Lau et al., 2000; Sheldon, 2001; Sheldon et al., 2002; Kragl et al., 2002; van Rantwijk et al., 2003; Domínguez de María, 2008). Following the pioneering studies of Klibanov and coworkers (Zaks and Klibanov, 1988; Klibanov, 2001), the use of hydrolytic enzymes in anhydrous organic media became a valuable addition to the synthetic repertoire, especially for reactions that could not be performed in aqueous solution. Based on this work, Sheldon and coworkers reported the first example of a free enzyme, the lipase isoform B from *Candida antarctica* (CAL-B), in an IL to catalyze alcoholysis, ammonolysis, aminolysis, and perhydrolysis reactions (Madeira Lau et al., 2000). Thereafter, more processes were described in which enzymes demonstrated a high activity and selectivity level in ILs (Husum et al., 2001; Zhao, 2005; Moon et al., 2006; Lou et al., 2004). The use of ILs in biotransformations is still expanding to other enzyme classes such as oxidoreductases and lyases (Eckstein et al., 2002, 2004; Lou et al., 2005; Domínguez de María and Maugeri, 2011; Zhang et al., 2012). ILs can also be used to immobilize (Sate et al., 2007; Nakashima et al., 2009), activate (Zhao, 2005; Miyako et al., 2003), or stabilize enzymes (Lozano et al., 2001a,b; Kaftzik et al., 2002; De Diego et al., 2004; Lozano et al., 2005; De Diego et al., 2005; Zhao et al., 2006a; Lai et al., 2011; Attri and Choi, 2013; Feher et al., 2007), to improve the enantioselectivity (Schofer et al., 2001; Kim et al., 2001; Zhao and Song, 2007) and catalytic efficiency (Zhao et al., 2006b; Yang et al., 2009; Goldfeder and Fishman, 2014), and to minimize the formation of by-products (Sheldon, 2010).

The first three examples of biotransformations in ILs were reported in 2000 (Madeira Lau et al., 2000; Cull et al., 2000; Erbeldinger et al., 2000). Since then, many reviews, a book chapter, and an entire book have been written on the subject (Sheldon et al., 2002; van Rantwijk et al., 2003; Lou et al., 2004; Kragl et al., 2001; Park and Kazlauskas, 2003; van Rantwijk and Sheldon, 2007; Harjani et al., 2007; Gorke et al., 2010; Moniruzzaman et al., 2010; Yang and Pan, 2005; Klembt et al., 2008; Garcia et al., 2004; Cantone et al., 2007). ILs have indeed been shown to influence the performance of enzymes, such as its activity, and stability (Sheldon et al., 2002; Kragl et al., 2002; van Rantwijk and Sheldon, 2007). Various enzymes were found to be active in ILs, including proteases, lipases, esterases, glycosidases, and oxidoreductases (Klembt et al., 2008). Challenges when carrying out biotransformations in ILs include (1) the possible toxicity on microorganisms (when using whole cells) (Ranke et al., 2007), (2) their expensive synthesis, and (3) the requirement for more efficient methods for their reuse including product isolation.

Because of the malleability of ILs, enzymatic reactions carried in ILs can be developed as a one- or two-phase system. Therefore, depending on the desired system, biotransformations can be performed with whole cells in a two-phase IL—water system, with free or immobilized enzymes in a biphasic system (water phase and hydrophobic IL phase) or single IL phase, or alternatively in a mixture of IL and water phase (Dreyer and Kragl, 2008; Oppermann et al., 2011).

In biocatalysis, the solubility of substrates and products sometimes requires the addition of a cosolvent. The use of VOCs in industrial processes is often problematic due to their toxicity and flammability. ILs represent an alternative class of nonaqueous solvents because they do not exhibit a vapor pressure and have the potential to be reused. Additionally, ILs are able to dissolve a wide range of organic, inorganic, and organometallic compounds. A good example of sparingly soluble substrates are carbohydrates and nucleosides, with which ILs greatly improve the biocatalytic system (Liu et al., 2005; Zhao et al., 2008). The solubility of gases such as H_2, CO, and O_2 in ILs is generally good, making them attractive solvents for catalytic hydrogenations, carbonylations, hydroformylations, and aerobic oxidations (Reetz et al., 2002). To increase the solubility of hydrophobic substrates and improve the biocatalytic yield by minimizing the substrate or product inhibition, ILs can be used (1) as cosolvents with the aqueous medium, (2) as a second phase in a biphasic system, or (3) alone as nonaqueous solvents (eg, with lipases).

Whole-cell biocatalysis in ILs is also an important area (Gangu et al., 2009; Wood et al., 2011; Fan et al., 2014). Research has been done on hydrophobic ILs for the efficient asymmetric reduction of prochiral ketones using whole cells (Bräutigam et al., 2009). Using a biphasic IL:water reaction system to improve the oxygenase-catalyzed biotransformation with whole cells was also investigated (Cornmell et al., 2008; Pfruender et al., 2006; Allen et al., 2014; Dennewald et al., 2012).

The following sections provide an overview of the advantages and disadvantages of using different enzymes in ILs. Furthermore, the classification of biocatalysts will be described in order to explore the wide range of enzymatic resources available to be used as versatile catalysts in organic chemistry. This chapter provides a nonexhaustive overview of biocatalysis and biotransformations in ILs, but attempts to demonstrate the advantages of combining ILs and enzymes for the development of successful transformation with a full potential for synthetic purposes.

2.1.4 Ionic Liquids and Hydrolases

Hydrolases are a class of enzymes that traditionally have catalyzed bond cleavage reactions using water as a nucleophile with no cofactor requirements. These biocatalysts have often shown great catalytic efficiency in the hydrolysis of acid derivatives in aqueous medium, but the application in synthetic reactions has been also demonstrated in organic solvents since the discovery in 1985 that these enzymes can catalyze reverse hydrolytic reactions in nonaqueous media. Thus, among other conventional reactions, aminolysis, ammonolysis, esterification, perhydrolysis, polymerization, thiolysis, and especially transesterification processes have been extensively reported in the literature (Bornscheuer and Kazlauskas, 2006). In this context, the use of neoteric solvents such as ILs or supercritical fluids for hydrolase-catalyzed reactions has been reviewed in the 2000s (Lozano, 2010; Fan and Qian, 2010; Yu et al., 2014; Hernández-Fernández et al., 2010), their use as stabilizing agents or additives being possible (Yang, 2009; Zhao, 2005; Yang and Pan, 2005; Patel et al., 2014), but also as cosolvent or sole media for selected biotransformations (Domínguez de María, 2008; van Rantwijk and Sheldon, 2007; Gorke et al., 2010; Sureshkumar and Lee, 2009; Zhao, 2012).

In this part we will review some examples of the use of hydrolases, paying close attention to those exploiting the selectivity of this class of enzymes under mild reaction conditions. Hence, the examples shown below are divided according to the type of reaction studied, mainly focusing on hydrolytic and transesterification reactions but also considering less exploited aminolysis, esterification, peptide synthesis, polymerization, or sequential chemo-enzymatic oxidative transformations. Since Sheldon and coworkers reported the first examples of CAL-B catalyzed transesterification, ammonolysis, and epoxidation reactions in ILs (Madeira Lau et al., 2000), their use has allowed the development of a vast number of both hydrolytic and nonreversible hydrolysis processes depending on the reaction conditions, allowing the production of high added-value compounds such as drug precursors or biodiesel (Lai et al., 2012a).

2.1.5 Hydrolytic Reactions

The structure of the IL is an important factor to consider when studying a hydrolytic reaction; therefore, an exhaustive analysis of the hydrolase activity in these systems is crucial. Many stability and thermal studies have been performed by using hydrolytic processes as model reactions. Some of the most recent achievements are described here for nonasymmetric transformations, including the possibility to carry out regioselective hydrolytic reactions.

CAL-B is probably the most common hydrolase for both hydrolytic and synthetic processes, and its stability in different hydrophilic ILs has been studied toward the hydrolysis of *p*-nitrophenyl laurate (Scheme 2.1) (Ventura et al., 2012). The influence of the anion and cation structure has been extensively analyzed, finding that cations with longer alkyl chains decrease the enzyme activity through the obstruction of the nonpolar active site because of the establishment of van der Waals interactions between the alkyl chains and the nonpolar domains of the lipase. On the other hand, interactions between CAL-B and the anion contributed to a dramatic loss of the activity due to

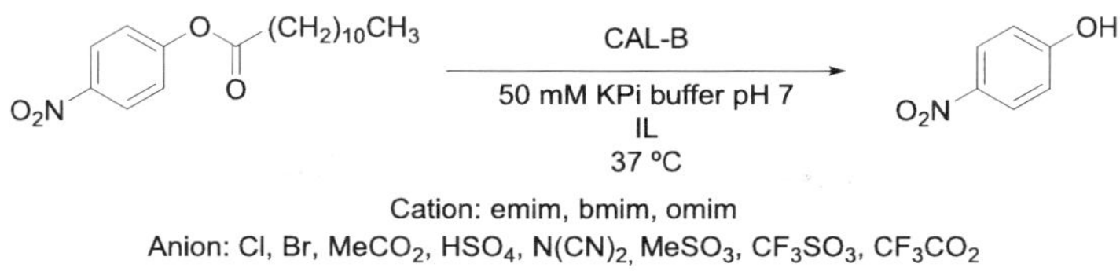

SCHEME 2.1 *Candida antarctica* lipase type B hydrolysis of *p*-nitrophenyl laurate in different buffer–ionic liquids mixtures.

dispersion forces and hydrogen bonding. It is for that reason that usually high substrate concentrations can cause a loss in the enzyme activity by decreasing its water activity.

The hydrolytic performance of *Thermomyces lanuginosus* lipase (TLL) was analyzed in the presence of a protic IL such as triethylammonium mesylate [tea][Ms] (Akanbi et al., 2012). This IL is able to change the secondary and tertiary structure of the lipase, enhancing the lipase activity for the hydrolysis of ϖ-3-fatty acids such as (5Z,8Z,11Z,14Z,17Z)-5,8,11,14,17-icosapentanoic acid, also known as eicosapentanoic acid, additionally finding higher thermal stabilities in the presence of [tea][Ms].

From the lipases toolbox, the chemical modification of porcine pancreas lipase (PPL) has been done using functional ILs, studying the activity of the enzyme in the hydrolysis of *p*-nitrophenyl palmitate at different temperatures (Scheme 2.2) (Jia et al., 2014). The chemical modification has been done by treatment of the enzyme with carbonyldiimidazole and water miscible [mmim][$MeSO_4$], improving the activity and thermostability of the enzyme compared to the native enzyme.

Among proteases, trypsin has been extensively applied in peptide synthesis. Saraiva and coworkers have explored the stability of trypsin for the hydrolysis of *N*-α-benzoyl-DL-arginine-*p*-nitroanilide in ILs combined with Tris–HCl buffer at pH 8 using batch and flow modes (Pinto et al., 2012a). An enhanced catalytic efficiency was found for the immobilized enzyme in comparison with the free enzyme. The utilization of concentrations lower than 25–30% of [bmim][BF_4] led to excellent enzyme stability values, which make this enzyme suitable for industrial processes.

Moreover, the use of ILs has been largely applied in sugar synthesis (Yang and Huang, 2012). For instance, the regioselective hydrolysis of a per-*O*-acetylated lactal has been possible through the use of an immobilized form of *Rhizomucor miehei* lipase (RML) in acetate buffer (Scheme 2.3) with 3% of acetonitrile and just a low amount of ILs (5 equivalents) (Filice et al., 2010).

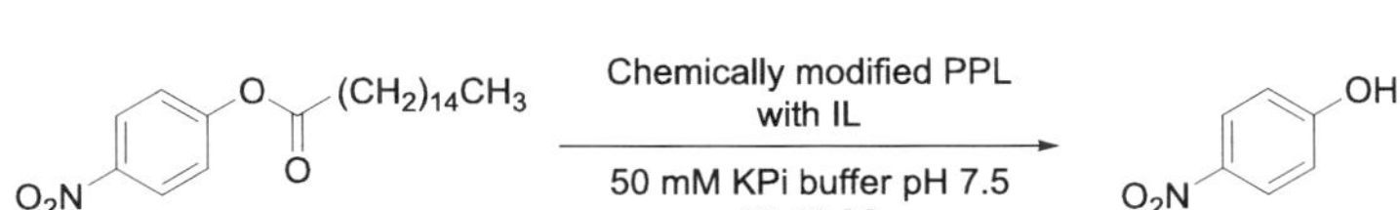

SCHEME 2.2 Hydrolysis of *p*-nitrophenyl palmitate with porcine pancreas lipase modified with functional ionic liquids.

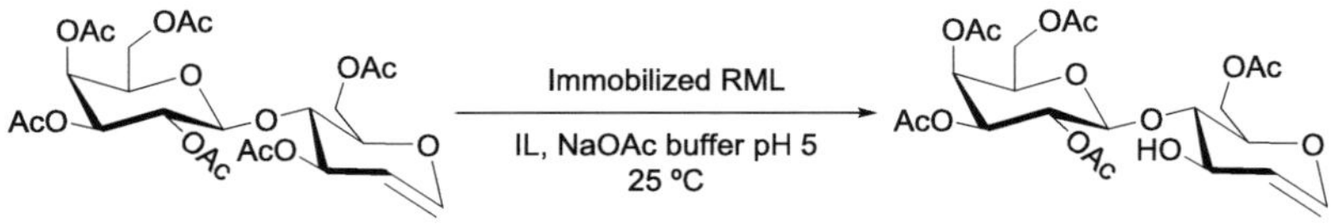

SCHEME 2.3 Regioselective hydrolysis of a per-*O*-acetylated lactal using *Rhizomucor miehei* lipase in mixtures of acetate buffer and ionic liquid.

Thus, complete conversions were achieved with different imidazolium-based ILs obtaining the hydrolysis of just one of the acetyl groups.

Alternatively the combination of ILs and lipases led to excellent results in the deacetylation of methyl 2,3,5-tri-*O*-acetyl-α,β-D-furanosides (Scheme 2.4) (Gudiño et al., 2012). Iglesias and coworkers have described a regioselective hydrolysis processes using *Candida rugosa* lipase (CRL) as the biocatalyst in a mixture of phosphate buffer at pH 7 and different ILs, namely [bmim][BF_4] and [bmim][PF_6], leading to the formation of the corresponding methyl 2,3-di-*O*-acetyl-α,β-D-furanosides after hydrolysis of the acetyl group in the 5-position.

The versatility of other hydrolases has been demonstrated in interesting hydrolytic industrial processes such as the action of β-galactosidase in dodecylsulfate–IL mixed micelles in variable compositions and concentrations for the hydrolysis of 4-methylumbelliferyl-α-D-galactopyranoside in the synthesis of glycol-oligosaccharides (Pinto et al., 2012b), or the hydrolysis of rutin into isoquerticitrin catalyzed by a hesperidinase using a mixture (10:90) of [bmim][BF_4]-glycine hydroxide buffer at pH 9 (Wang et al., 2013).

The use of hydrolases reports important benefits in terms of selectivity since not only can regioselective processes be carried out but also the development of enantioselective processes is possible through classical kinetic resolutions, dynamic kinetic resolutions (DKRs), or desymmetrization processes. For instance, the use of an IL such as [bmim][BF_4] in the hydrolysis of racemic amino acid esters can improve the initial rate and the enantioselectivity of the papain-catalyzed kinetic resolution process compared with the results obtained in the absence of cosolvents or in the presence of organic cosolvents (acetonitrile, acetone, 2-propanol, or *tert*-butanol) (Liu et al., 2005). Remarkably, better enantioselectivities were obtained in the presence of

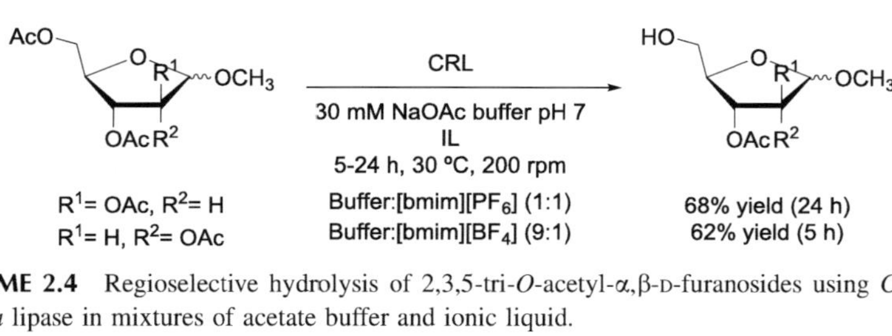

SCHEME 2.4 Regioselective hydrolysis of 2,3,5-tri-*O*-acetyl-α,β-D-furanosides using *Candida rugosa* lipase in mixtures of acetate buffer and ionic liquid.

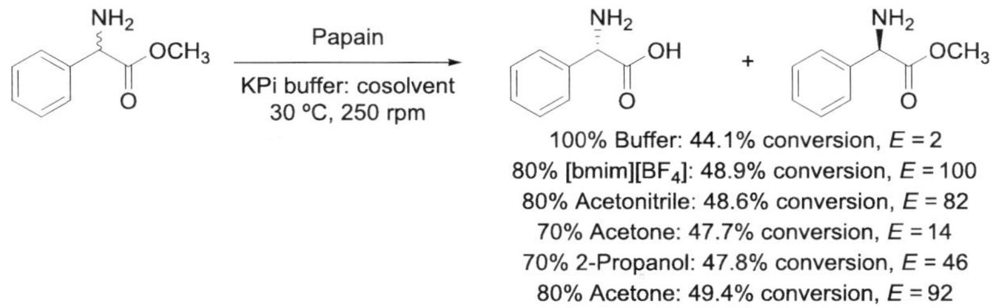

SCHEME 2.5 Papain-catalyzed kinetic resolution of D,L-phenylglycine methyl ester through a hydrolytic process.

[bmim][BF_4] for the hydrolysis of D,L-phenylglycine methyl ester (Scheme 2.5), D,L-methionine methyl ester, D,L-4-chlorophenylalanine ethyl ester, and D,L-β-phenylalanine methyl ester.

The asymmetric hydrolysis of epoxides is a highly challenging task in organic synthesis, epoxide hydrolases (EHs) being valuable enzymes for these regio- and stereoselective ring-opening reactions. This family of hydrolases usually acts with good selectivity for less hindered substrates, and their action in ILs was demonstrated for the first time by Chiappe and coworkers using soluble EHs in the hydrolysis of *trans*-β-methylstyrene oxide using [bmim][PF_6], [bmim][NTf_2] and [bmim][BF_4] as reaction media, yielding the (1*S*,2*R*)-erythro-1-phenylpropane-1,2-diol through a stereoconvergent process with moderate selectivities (Scheme 2.6) (Chiappe et al., 2004).

The Mung bean EH catalyzed the asymmetric hydrolysis of styrene oxide into (*R*)-1-phenyl-1,2-ethanediol in a hexane/buffer biphasic system containing various hydrophobic (Chen et al., 2012a) and hydrophilic ILs. Remarkably, the minimization of the nonenzymatic background reaction was observed, allowing the development of more selective processes just with a low loading of ILs (0−10%). The hydrophilic ILs with cations containing an alkanol group such as [C_2OHmim][BF_4], [C_2OHmim][OTf], [Ch][Arg], and [Ch][Pro] were found to be the most suitable for this process in terms of reaction rate and thermostability, all of them leading to similar selectivity values (Chen et al., 2012b). This Mung bean EH has been effectively immobilized as a cross-linked enzyme aggregate (CLEA) displaying a good selectivity in the hydrolysis of styrene oxide using ILs as cosolvents in biphasic systems

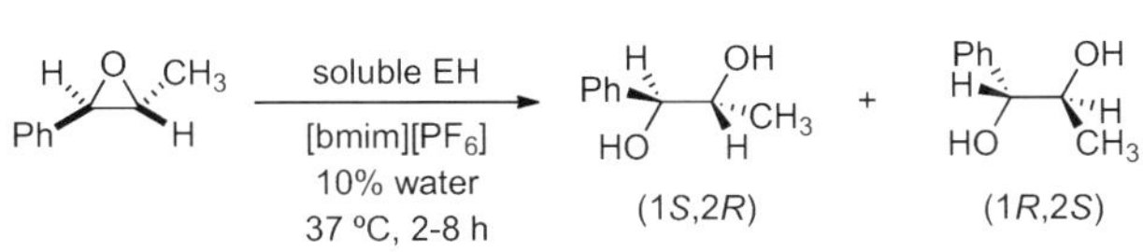

SCHEME 2.6 Epoxide hydrolase-catalyzed hydrolysis of *trans*-β-methylstyrene oxide using [bmim][PF_6].

(Yu et al., 2014). The best results were found in the system containing [bmim][PF_6] at 40°C and pH 7.5, the reaction at 500 mL scale being successfully achieved.

2.1.6 Transesterification Reactions

Here, we have summarized examples for the use of hydrolytic enzymes in transesterification reactions that can be also referred to as acylation processes, which have been performed both in a nonselective or selective manner. 3-Phenylpropionic acid esters are relevant compounds for fragrance and flavor chemistry, and the transesterification reaction of ethyl 3-phenylpropionate with *Pseudomonas cepacia* lipase (PSL) has been described using three different ILs, namely [bmim][BF_4], [bmim][PF_6], and [bmim][NTf_2] and 11 lineal alcohols (Scheme 2.7) (Vidya and Chadha, 2009). Best results were obtained with the [bmim][NTf_2] obtaining the corresponding esters in 54–96% yield after 24 h. Remarkably, the IL stabilized the lipase to a great extent since when the enzyme was incubated for 20 days in [bmim][NTf_2] before using it, a notable increase in the new ester yields was achieved (65–98%).

Similarly, the transesterification of butyl propionate was studied using nine different ILs with a variety of lipases (CAL-A, CAL-B, TLL, and RML), finding a clear dependence of the nature of the ions with respect to the enzyme activity (Scheme 2.8), however, it seems that there is no simple rule for predicting the enzyme behavior in each system (Diego et al., 2009). Nevertheless, this methodology seems to be adequate for producing alkyl ester starting from their vinyl analogs.

Carrying out regioselective processes using transesterification reactions in ILs is also possible. For instance, Gotor and coworkers described the PSL-catalyzed acylation of 1,2-dideoxy-D-ribose with 1.1 equivalents of vinyl benzoate in [bmim][PF_6], obtaining selectively the desired monoester in 5′-position with >98% conversion after 10 h at 50°C (Scheme 2.9) (Martínez-Montero et al., 2011).

Remarkably, the intrinsic chirality of enzymes has allowed the development of enantioselective processes, the acetylation of alcohols and amines being extensively reported in organic media. The use of ILs for kinetic resolution processes has been known since the 2000s, enabling the recovery of two

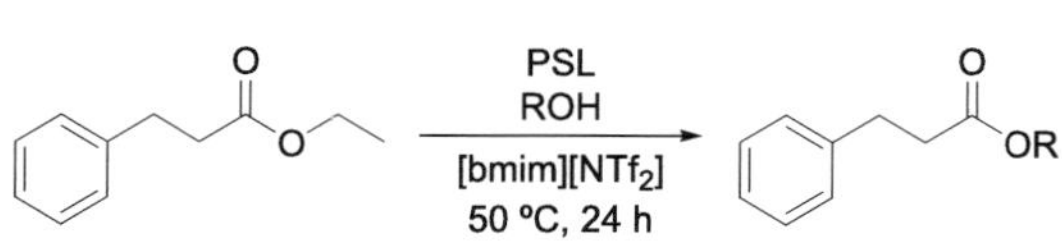

R= CH_3, $CH_3(CH_2)_2$, $(CH_3)_2CH$, $CH_3(CH_2)_3$, $CH_3(CH_2)_4$, $CH_3(CH_2)_5$, $CH_3(CH_2)_6$, $CH_3(CH_2)_7$, $CH_3(CH_2)_9$, $(CH_3)_2CH(CH_2)_2$, $C_6H_5CH_2$

SCHEME 2.7 *Pseudomonas cepacia* lipase–catalyzed transesterification of ethyl 3-phenylpropionate with linear alcohols.

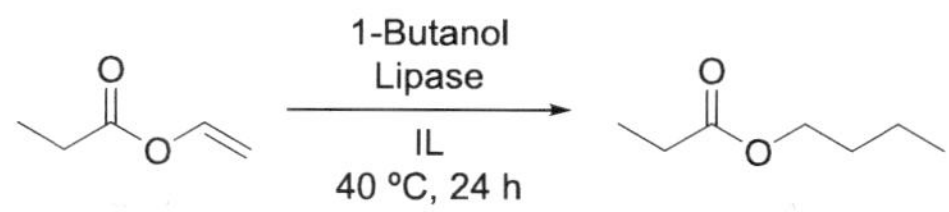

SCHEME 2.8 Lipase-catalyzed transesterification of vinyl propionate with 1-butanol for the production of butyl propionate.

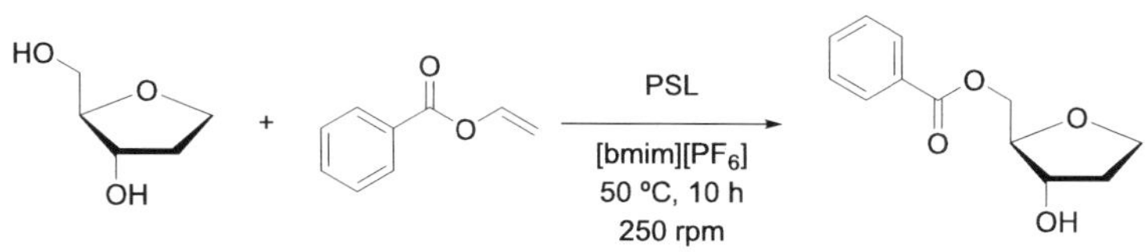

SCHEME 2.9 Regioselective acylation of 1,2-dideoxy-D-ribose using vinyl benzoate and *Pseudomonas cepacia* lipase in [bmim][PF_6].

different products in enantiopure form with opposite stereochemistry. This is the case of the kinetic resolution for the most common model substrate for this type of reactions, 1-phenylethanol. Among other enzymes, CAL-B was able to perform the kinetic resolution of this racemic alcohol in an enantioselective manner using two equivalents of vinyl acetate and different protic ILs composed by tertiary amines and carboxylic acids (Scheme 2.10) (de los Ríos et al., 2012). Both the free enzyme and the one immobilized in CLEAs led to the (*S*)-alcohol and the (*R*)-acetate in excellent enantiomeric excess. Additionally, excellent results were achieved with other hydrolases such as PSL, currently known as *Burkhloderia cepacia* lipase, which has catalyzed the enantioselective acetylation of 1-phenylethanol using the enzyme immobilized on KynolTM ACC 507-15 active carbon cloth (Hara et al., 2010) or as CLEA (Hara et al., 2009) using [emim][NTf_2] or mixtures of ILs and organic solvents, respectively. Similarly the effectiveness of this process has been demonstrated in phosphonium ILs (Abe et al., 2012). Synthesized phosphonium ILs such as 2-methoxyethyl(tri-*n*-butyl)phosphonium bis(trifluoromethanesulfonyl)imide [$MeBu_3P$][NTf_2] displayed an excellent compatibility with PSL and CAL-B in the kinetic resolution of a variety of secondary alcohols by using 1.5 equivalents of vinyl acetate at 35°C (Abe et al., 2008). Several reactions in this IL proceeded faster than the ones carried out in diisopropylether, a common organic solvent used in lipase-catalyzed acylation reactions.

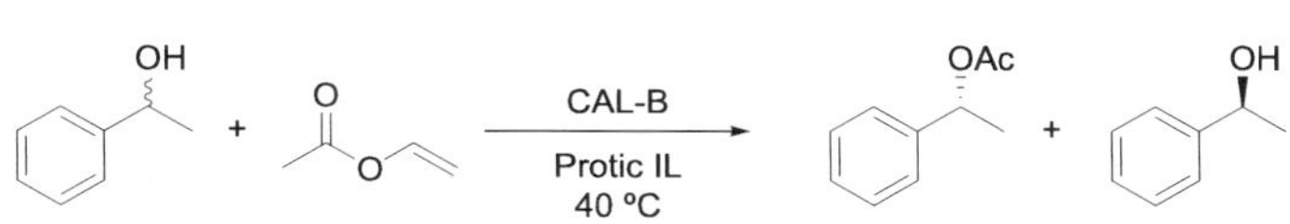

SCHEME 2.10 Kinetic resolution of 1-phenylethanol through acetylation with vinyl acetate in a variety of protic solvents.

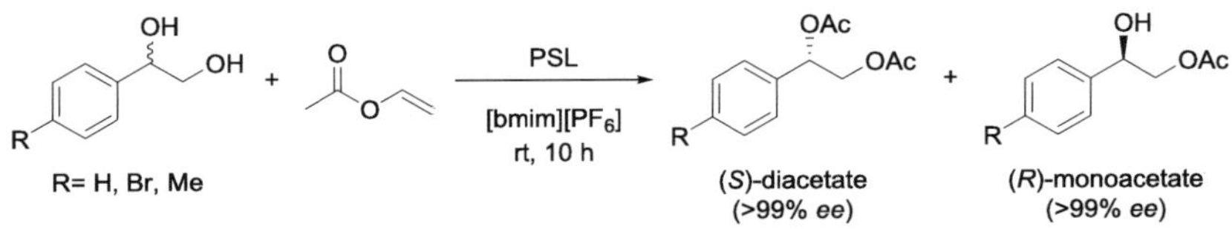

SCHEME 2.11 *Pseudomonas cepacia* lipase–catalyzed kinetic resolution of racemic aryl-1,2-ethanediol using vinyl acetate and [bmim][PF_6].

A family of racemic aryl-1,2-ethanediols employing the PSL and six equivalents of vinyl acetate in [bmim][PF_6] at room temperature was investigated (Kamal and Chouhan, 2004). The acetylation process took place regioselectively on the primary hydroxyl group to afford the corresponding monoacetates in 1–2 h (Scheme 2.11). Furthermore, diacetate products were obtained in an enantioselective manner after 10 h by the acetylation of the (*S*)-monoacetates, affording the (*R*)-monoacetates and the (*S*)-diacetates in enantiomerically pure form at the same time.

Interestingly, the kinetic resolution of racemic alcohols possessing more than one stereocenter is also possible. In this context, PPL has catalyzed the multigram kinetic resolution of (±)-*cis*-4-*O*-TBS-2-cyclopenten-1,4-diol through acetylation with five equivalents of vinyl acetate in [omim][PF_6] (Scheme 2.12) (Das et al., 2008). Remarkably, the biocatalyst can be used five times without significant loss of activity and enantioselectivity.

The development of DKR has allowed the design of synthetic strategies to overcome the limitations of traditional individual steps such as a maximum 50% yield for an enantiopure compound in classical kinetic resolutions. This is possible through the compatibility of an enzyme with a biochemical or metal racemization agent in most of the cases. In this context, the successful acylation of secondary alcohols was reported using a hydrolase and a ruthenium catalyst in [bmim][PF_6] in combination with three equivalents of acyl donor in the presence of triethylamine at room temperature (Scheme 2.13) (Kim et al., 2004). The (*R*)-acetates were obtained in 99% *ee* and 85–92% yield after 2–3 days when selecting a lipase (LSP-TN-M) and 1,1,1-trifluoroethyl acetate as acyl donor. Moreover, the (*S*)-butanoates were isolated in 82–99% *ee* and 80–91% yield using subtilisin that displayed an opposite selectivity in comparison with lipases and 1,1,1-trifluoroethyl butyrate as acyl donor after 6 days at room temperature.

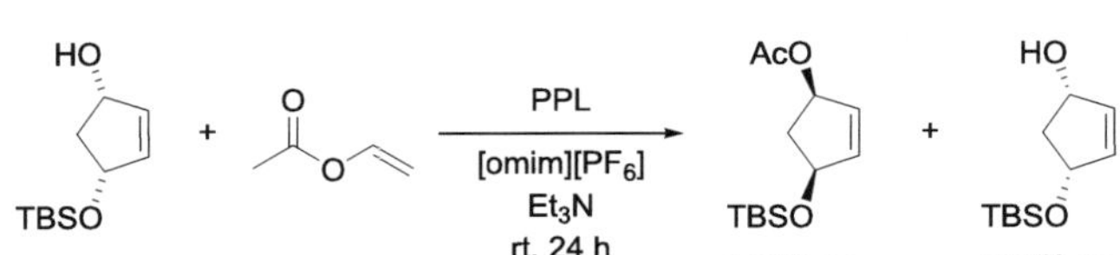

SCHEME 2.12 Enantioselective acetylation of (±)-*cis*-4-*O*-TBS-2-cyclopenten-1,4-diol using porcine pancreas lipase in [omim][PF_6].

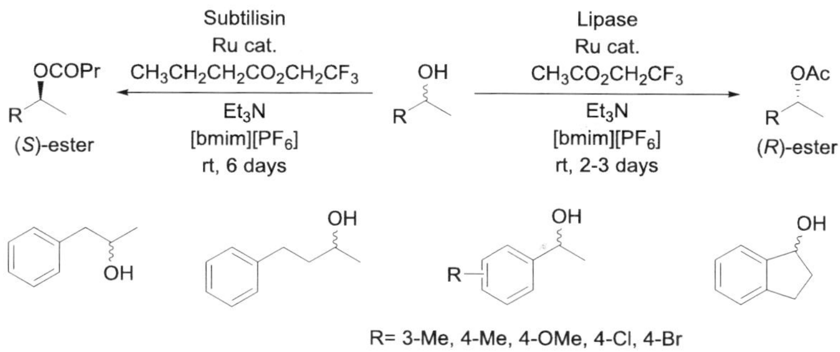

SCHEME 2.13 Dynamic kinetic resolution of secondary alcohols using a combination of hydrolases and a ruthenium catalyst in [bmim][PF_6].

2.1.7 Hydrolases and Ionic Liquids: Other Applications and Reaction Types

Apart from hydrolytic and reversed transesterification processes, hydrolases have efficiently catalyzed a myriad of reactions using ILs as reaction media, yielding high added value in excellent conversions and with good selectivity. This part will deal with the recent application of ILs as solvents in different reactions (esterification, amidation, polymerization, etc.), and in lipid and peptide synthesis.

Multiple examples of esterification reactions have been reported in the literature using a carboxylic acid and an alcohol. For instance, the synthesis in [emim][NTf_2] of the antiinflammatory caffeic acid phenyl ester has been possible through the CAL-B catalyzed esterification of caffeic acid (Ha et al., 2013). An exhaustive response surface methodology studied was performed including four variables: (1) temperature (65–75°C); (2) reaction time (36–60 h); (3) molar ratio of alcohol versus acid (20:1 to 40:1); and (4) ratio in weight of CAL-B versus caffeic acid (10:1 to 20:1). Optimal conditions for a quantitative conversion into the caffeic acid phenyl ester were found for the reaction using 27.1 equivalents of phenethylalcohol with a 17.8 excess of enzyme (w/w) with respect to caffeic acid at 73.7°C after 60 h (Scheme 2.14). The esterification of α-lipoic acid was also successfully carried out with pyrodoxime and tyrosol in the presence of CAL-B and a variety of ILs (Papadopoulou et al., 2013), finding the methyltrioctylammonium bis(trifluoromethylsulfonyl)imide [mtoa][NTf_2] to be the best solvent, whereas the use of a DES such as choline chloride:urea (ChCl:U) led to the lowest conversion values. Interestingly, the amidation reaction of α-lipoic acid with tyramine yielded the tyramine-8-*N*-lipoate, the ILs showing a similar trend in the conversion values compared with the ones obtained in the esterification processes. Lozano and coworkers have reported the versatility of hydrophobic ILs for producing flavor esters using switchable IL/solid phases as reaction/separation media in the presence of CAL-B at an ideal temperature of 50°C

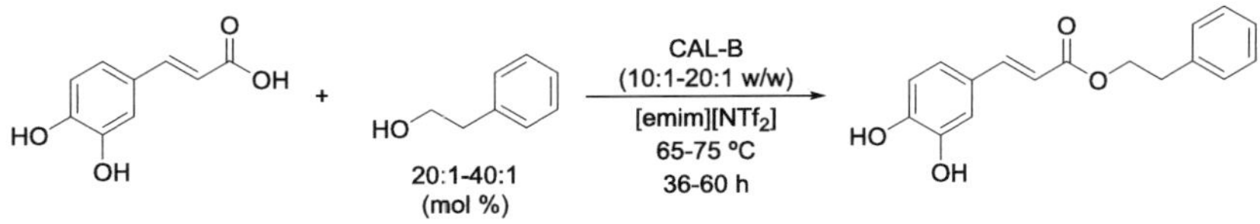

SCHEME 2.14 *Candida antarctica* lipase type B-catalyzed esterification of caffeic acid with phenethylalcohol in [emim][NTf_2].

(Lozano et al., 2012). Esterification reactions proceeded to completion using an aliphatic carboxylic acid (acetic, propionic, butyric, or valeric) and a flavor alcohol (isoamyl alcohol, nerol, citronellol, or geraniol in *N*, *N*′, *N*″, *N*‴-hexadecyltrimethylammonium bis(trifluoromethylsulfonylimide)), achieving a clean purification and recycling of the catalytic system by a cooling/centrifugation method.

Regioselective esterification processes have been described for sugar synthesis. For instance, Fischer and coworkers reported the lipase-mediated synthesis of mono 6 and 6′-*O*-linoleyl-α-D-maltose using volatile organic solvents (hexane, toluene, dimethylformamide (DMF), tetrahydrofuran, acetone, acetonitrile, dimethyl sulfoxide (DMSO), and *tert*-butanol), imidazolium-based ILs containing different anions (PF_6, NTf_2, $MeSO_3$, and $EtSO_4$), and binary systems (Fischer et al., 2013). From a set of lipases, CAL-B and PSL displayed the higher activities for the selective esterification of maltose with two equivalents of linoleic acid in the presence of molecular sieves at 65°C (Scheme 2.15); unfortunately separation of the regioisomers must be improved for industrial purposes.

The development of enantioselective esterification processes is also possible, for instance in the reaction between racemic 2-substituted-propanoic acids and 1-butanol (Scheme 2.16) (Ulbert et al., 2004). The use of [bmim][PF_6]

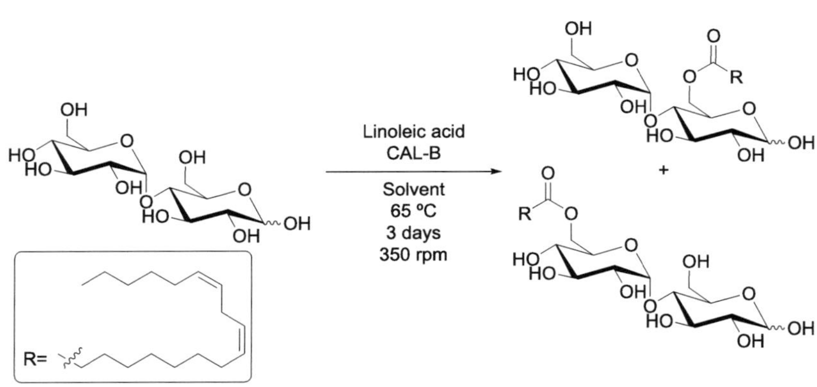

SCHEME 2.15 Lipase-catalyzed esterification of maltose with linoleic acid in different organic solvents, ionic liquids, and binary mixtures.

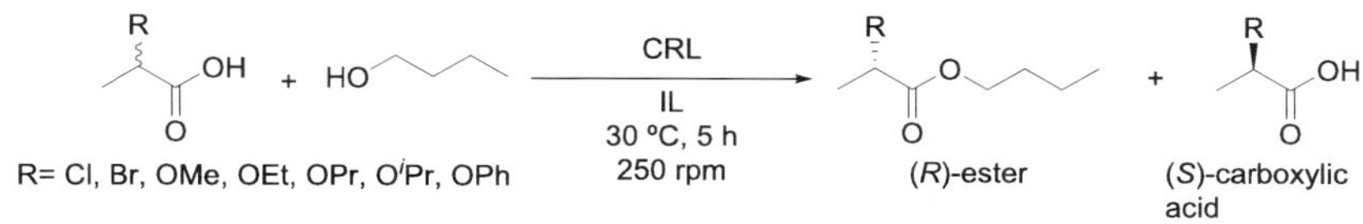

SCHEME 2.16 Kinetic resolution of 2-substituted-propanoic acids with 1-butanol using *Candida rugosa* lipase as biocatalyst in [bmim][PF_6] or [omim][PF_6].

and [omim][PF_6] is beneficial for the enzyme activity but also for the enantioselectivity of the process when compared with the use of hexane or toluene. The recycling of the enzyme was also found easier and possible for at least five cycles without significant loss of activity.

Peptide synthesis is also one of the challenging tasks for the organic chemist, where the use of enzymes has been largely expanded mainly using acylases and proteases. The synthesis of amoxicillin is highly appealing in this context (Scheme 2.17), and for this purpose immobilized Penicillin G acylase (PGA) has resulted as an excellent biocatalyst using mixtures of phosphate buffer at pH 6.5 and an IL in different proportions (25–75%) under mild reaction conditions (25°C) and short reaction times (3 h) (Pereira et al., 2012). The use of chymotrypsin, trypsin, and other proteases has allowed the ligation of cleavage-sensitive peptide and protein fragments in IL-containing solvent systems (Wehofsky et al., 2008). The application of a [mmim][Me_2PO_4]/buffer system presents real advantages in comparison with pure water and traditional organic solvents, improving the enzyme activity and suppressing competitive proteolytic side reactions.

Modification of lipids is a challenging task because of the importance of synthetic lipids in different areas such as biofuels, pharmaceutical and food industry, the use of ILs (Lue et al., 2007), and enzyme-catalyzed processes (Jala et al., 2012) helping to achieve relevant transformations. For instance, the production of biodiesel was successfully achieved from microalgae *Botrycoccus braunii*, *Chlorella vulgaris*, and *Chlorella pyrenoidosa* (Lai et al., 2012b). The conversion of microalgal oil to biodiesel was exhaustively analyzed after 48 h for *C. pyrenoidosa* using a (3:1) or (4:1) MeOH:oil molar ratio. The use of [bmim][PF_6] at 50°C led to better results (>86% yield) in comparison with the ones obtained in *tert*-butanol at 40°C (<50% yield) employing two different hydrolases (CAL-B and *Penicillium expansum* lipase).

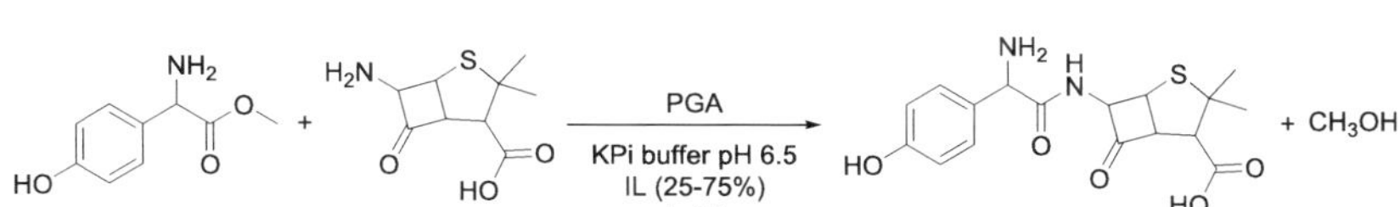

SCHEME 2.17 Synthesis of amoxicillin catalyzed by Penicillin G acylase in mixtures of buffer and ionic liquid.

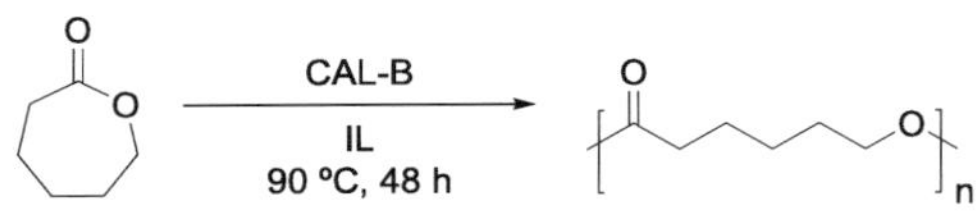

SCHEME 2.18 *Candida antarctica* lipase type B-catalyzed ring-opening polymerization of ε-caprolactone in monocationic and dicationic ionic liquids.

Enzymatic polymerization is also an attractive and environmentally friendly method for polyester synthesis in contrast to traditional chemical methods, which usually require harsh reaction conditions and metal catalysts. Monocationic and dicationic imidazolium-based ILs have been used to promote the ring-opening polymerization of ε-caprolactone using CAL-B as biocatalyst (Scheme 2.18) (Wu et al., 2013). Hydrophilic ILs containing the BF_4 anion did not lead to the desired polymers, while the use of ILs with the PF_6 led to excellent results, and the polymerization reactions were optimized in terms of IL nature, temperature, and lipase loading, obtaining polymers of different molecular weights.

Chemoenzymatic oxidation reactions have been also carried out through the combination of lipases and ILs. These are sequential processes, in which the lipase catalyzes the perhydrolysis of a carboxylic acid intermediate to form a peracid, which is responsible for the final chemical and nonselective oxidation. Remarkably, the development of alkene epoxidations and Baeyer–Villiger oxidations of cyclic ketones has been successfully achieved using a 50% H_2O_2 solution and octanoic acid as precursor of the reacting peracid in the presence of CAL-B (Kotlewska et al., 2011). Hydrogen bond donating ILs exhibited a substantial positive solvent effect on the chemical oxidation step. Nascimiento and coworkers performed an exhaustive analysis of the reaction conditions for the epoxidation of citronellol, studying the influence of the oxidizing agent (hydrogen peroxide or the less hazardous urea–hydrogen peroxide (UHP) complex), reaction time, carboxylic acid intermediate, temperature, and solvent (Scheme 2.19) (Da Silva and Nascimiento, 2012). The citronellol oxide was obtained in excellent conversions at

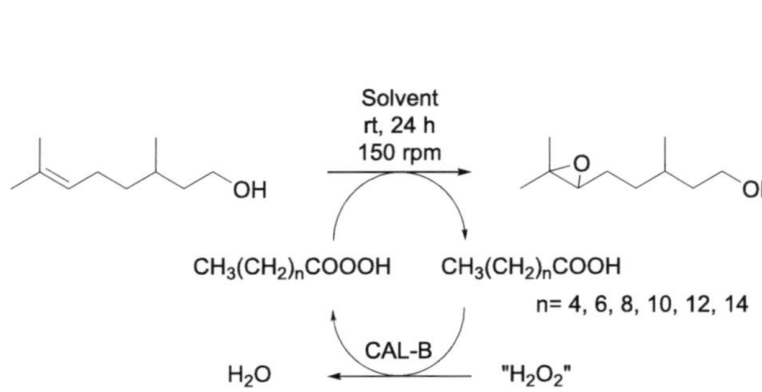

SCHEME 2.19 Chemoenzymatic epoxidation of citronellol using *Candida antarctica* lipase type B in organic solvents and ionic liquids.

20°C when using the UHP complex and octanoic acid as carboxylic acid intermediate in organic solvents such as chloroform or dichloromethane, whereas the use of binary systems composed by organic solvents and 1-butyl-3-methylimidazolium (bmim)-based ILs led to mixture of products favoring the epoxidation and the subsequent acylation of the free hydroxyl group.

Biocatalytic promiscuity is an emerging area in biocatalysis since the discovery that enzymes can catalyze nonconventional reactions (Bornscheuer and Kazlauskas, 2004; Kazlauskas, 2005; Hult and Berglund, 2007). Among others, the development of lipase-catalyzed oxidative reactions and carbon–carbon and carbon–heteroatom bond formations has been extensively studied (Busto et al., 2010; Wu et al., 2010; Humble and Berglund, 2011; Kapoor and Bupta, 2012). IL has also played an important role in these types of transformations. For instance, Sinha and coworkers reported for the first time the combination of a lipase (PPL) and a neutral IL [hmim][Br] for the oxidation of aryl alcohols into ketones using a 30% hydrogen peroxide solution or UHP at 60°C and short reaction times (8–12 h) (Sharma et al., 2009). Remarkably, the reusability of the lipase was demonstrated up to 10 times in the oxidation of 1-(4-methoxyphenyl)propan-1-ol with a little decrease in the yield (96–89%). On the other hand PPL has efficiently catalyzed asymmetric cross aldol reactions of aromatic and heteroaromatic aldehydes with different ketones in [bmim][PF_6] and low amount of water (<10%) at 37°C, obtaining similar results to those achieved using traditional organic solvents (Zhang et al., 2014).

2.1.8 Nonsolvent Applications of Ionic Liquids in Hydrolase-Catalyzed Reactions

In previous sections, the use of ILs as (co)solvents for hydrolase-catalyzed reactions has been fully explored and demonstrated. Here, biocatalytic reactions will be presented, the role of the ILs being more unconventional. Thus, the following processes are described, including the kinetic resolution of racemic ILs, methods to prepare optically active ILs through chemoenzymatic approaches, and also their use as acylating agents.

Salunkhe and coworkers developed the synthesis and kinetic resolution of an IL-anchored ibuprofen, which was enantioselectively hydrolyzed using lipases for the production of (*S*)-ibuprofen (Scheme 2.20) (Naik et al., 2007). From a set of lipases, CAL-B showed the best activity for the hydrolysis of the ester functionality using a mixture of phosphate buffer pH 7 and DMSO, yielding with moderate selectivity ($E = 35$) the (*S*)-ibuprofen in 86% *ee* and the remaining IL anchored (*R*)-ibuprofen ester that after chemical hydrolysis led to (*R*)-ibuprofen.

In addition, the synthesis of chiral imidazole and triazole-based ILs can be easily achieved by convenient resolution of the alcohol precursors by means of lipase-catalyzed acetylation processes using vinyl acetate (Scheme 2.21)

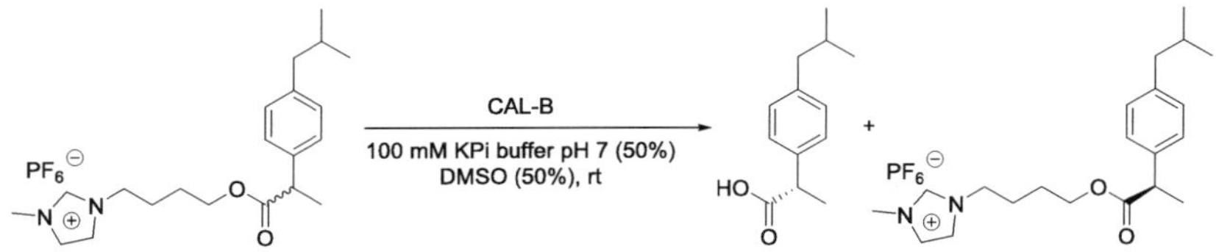

SCHEME 2.20 *Candida antarctica* lipase type B-catalyzed enantioselective hydrolysis of ionic liquid-anchored ibuprofen ester.

(Busto et al., 2007; Ríos-Lombardía et al., 2010; Borowiecki et al., 2013). Furthermore, quaternization and metathesis reactions can allow the production of wide families of chiral ILs and salts with applications as additives (Porcar et al., 2013) or organocatalysts (Ríos-Lombardía et al., 2011; Paul et al., 2012) in selected asymmetrical C—C bond formation reactions.

The use of ILs as acylating agents for the enzymatic kinetic resolution of secondary alcohols has been also described by Lourenço and coworkers (Lourenço et al., 2010; Teixeira and Lourenço, 2014). These task-specific ILs have allowed the enantioselective acylation of the corresponding alcohols in mild reaction conditions. For instance, 1-methyl-3-(11-ethoxycarbonylundecyl)imidazole hexafluorophosphate was synthesized and used as an efficient acylating agent for the kinetic resolution of 2-hydroxycyclohexanecarbonitrile with good stereopreference (Scheme 2.22) (Lourenço et al., 2010). The process was carried out in [bmim][PF_6] using equimolecular amounts of racemic alcohol and acyl donor at 35°C, recovering after the enzymatic process the enantioenriched (1*R*,2*S*)-alcohol, while its (1*S*,2*R*)-alcohol antipode was obtained after the transesterification of the resulting ester. Similarly, an IL anhydride was prepared and successfully used

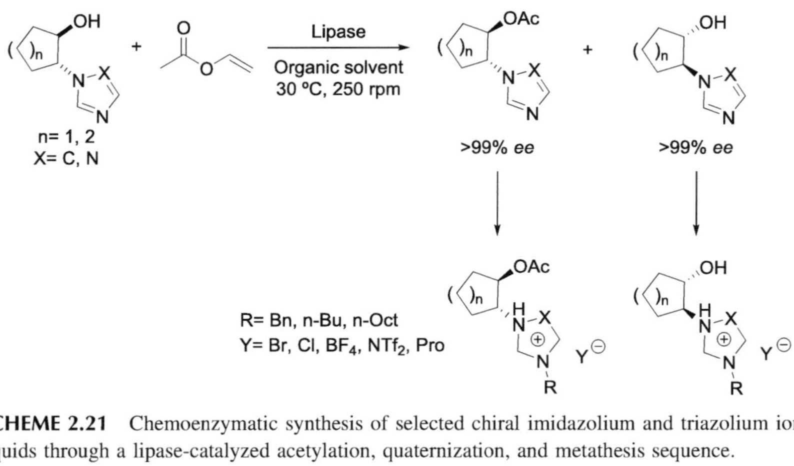

SCHEME 2.21 Chemoenzymatic synthesis of selected chiral imidazolium and triazolium ionic liquids through a lipase-catalyzed acetylation, quaternization, and metathesis sequence.

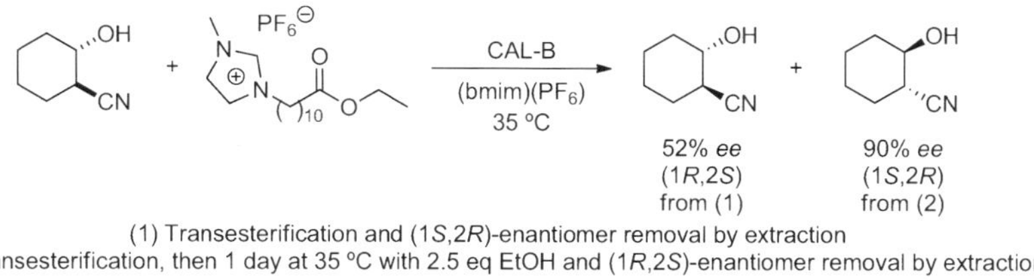

SCHEME 2.22 Use of ionic liquids as acylating agents in the kinetic resolution.

in the CAL-B-catalyzed kinetic resolution of a family of secondary alcohols such as 1-phenylethanol, 4-phenyl-2-butanol, 1-cyclohexylethanol, 2-pentanol, 2-octanol, and 6-methyl-5-hepten-2-ol with excellent selectivities ($E > 144$) (Teixeira and Lourenço, 2014).

2.2 IONIC LIQUIDS AND OXIDOREDUCTASES

Enantioenriched compounds are increasingly obtained via enzyme-catalyzed reduction-oxidation (redox) reactions. The required classes of enzymes for this synthetic challenge are oxidoreductases, which are being increasingly used in industrial processes for the pharmaceutical and fine chemical industries. This section will describe the use and application of various oxidoreductases in ILs (Pinto et al., 2008).

2.2.1 Reductions and Oxidations: Dehydrogenases

Formate dehydrogenases (FDHs, EC 1.2.1.2) catalyze the irreversible oxidation reaction of formate to carbon dioxide, whereas alcohol dehydrogenases (ADHs, EC 1.1.1.1, and 1.1.1.2) perform stereoselective carbonyl reductions and enantioselective alcohol oxidations. ADHs are part of the large family of redox enzymes that can reversibly reduce a C=O double bond (ketones, aldehydes, keto esters) to the corresponding reduced hydroxy compound. These enzymes display a certain preference for the nicotinamide cofactor NAD(P)H, and in the case of ADHs show either (*R*)- or (*S*)-stereospecificity. The group of Kragl was the first to report the use of isolated dehydrogenases with ILs as cosolvents (De Diego et al., 2004). Thus, the influence of various ILs on the activity of FDH from *Candida boidinii*, yeast ADH (YADH), and an ADH from *Candida parapsilosis* (*Cp*ADH) was evaluated. The results showed a dependence of the activity on the content of IL in the buffer solution. The ILs screened, [mmim][$MeSO_4$], [Et_3NH][$MeSO_4$], [Et_3NMe][$MeSO_4$], [$PrNH_3$][NO_3], [bmim][BF_4], [emim][$PhCO_2$], [bmim][CF_3SO_3], and [bmim][$OctSO_4$], were added in different proportions: 25%, 50%, and 75%. With FDH, 98% residual activity was achieved in 75% [mmim][$MeSO_4$] IL:buffer.

The group of Kragl was also the first to use an IL in a two-phase system to improve an ADH-catalyzed reduction reaction (Eckstein et al., 2004). Thus, the ADH from *Lactobacillus brevis* (*Lb*ADH) was employed to catalyze the conversion of 2-octanone to (*R*)-2-octanol, using 2-propanol as a cosubstrate for NADPH regeneration (Scheme 2.23).

The reaction rate was significantly higher with [bmim][$(CF_3SO_2)_2N$] than with methyl *tert*-butyl ether (MTBE) as a cosolvent (88% compared to 61% conversion after 180 min). This system gave the advantage of biphasic conditions; that is, using the organic (IL) phase as a reservoir for the hydrophobic substrate and product, without the thermodynamic limitations found with recycling systems with ADHs.

Additionally, *Lb*ADH has been used in the presence of several water-miscible ILs to reduce hardly water-soluble ketones such as 2-nonanone and 2-decanone, or employing an IL as a two-phase system to efficiently achieve the bioreduction of 2-octanone due to favorable partition coefficients of the cosubstrate (2-propanol) utilized and the coproduct (acetone) obtained. More importantly, the use of these cosolvents has led to the development of efficient systems to obtain high added-value compounds in a selective manner. The activity of *Lb*ADH in 10 different ILs was thus studied for the reduction of highly hydrophobic ketones to their corresponding enantiopure alcohol product, (*R*)-3-octanol, (*R*)-2-octanol, (*R*)-2-nonanol, and (*R*)-2-decanol. The IL improved the storage stability as well as the enzyme activity, and reduced undesired product inhibition (Kohlmann et al., 2011).

Further investigations on the effect of ILs on the catalytic characteristics of ADH were carried out on horse liver ADH (HLADH) with different methyl-imidazolium IL derivatives: [bmim] as the cation, and either [Cl], [Br], [PF_6], [BF_4], or [OTf] as anion (Scheme 2.24) (Shi et al., 2005, 2006).

HLADH clearly displayed higher activity and stability in certain ILs, [bmim][Cl], [emim][Cl], [bmim][Br], and [bmim][PF_6] when compared to buffer, and lower activity when [BF_4] and [OTf] were used as anions.

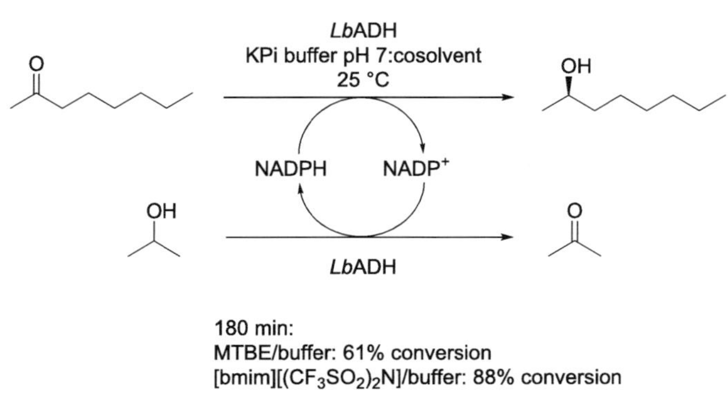

SCHEME 2.23 Alcohol dehydrogenase–catalyzed reduction of 2-octanone in ionic liquid/buffer mixtures.

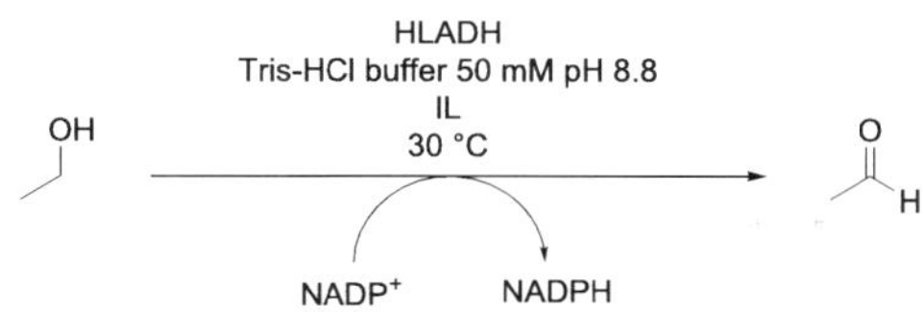

SCHEME 2.24 HLADH (alcohol dehydrogenase from horse liver)-catalyzed oxidation of ethanol to acetaldehyde measuring HLADH activity in different ionic liquids, varying their content from 0 to 0.40 g/mL.

The ILs were added in small amounts (0.025–0.05 g/mL) for increased activity, since a higher content of IL led to enzyme inactivation due to a protein structure shift and the higher viscosity, which essentially limits the mass transfer of the substrates and products. A detailed study on the enzyme kinetics of YADH was performed to evaluate how the enzymatic activity can be influenced by various ILs with different cations and anions (Weibels et al., 2012).

In another extensive study, the enzymatic reduction of ketones was performed by ADH-A from *Rhodococcus ruber*, in order to evaluate the activity and stability of this enzyme in water-miscible ILs (Scheme 2.25) (de Gonzalo et al., 2007).

The ILs were hydroxy-functionalized and water-miscible in an effort to increase the solubility of the substrates. The IL tris(2-hydroxyethyl)methylammonium methylsulfate [MTEOA][$MeSO_4$] afforded good results, even when added at 90% v/v (in Tris–HCl buffer pH 7.5), allowing substrate

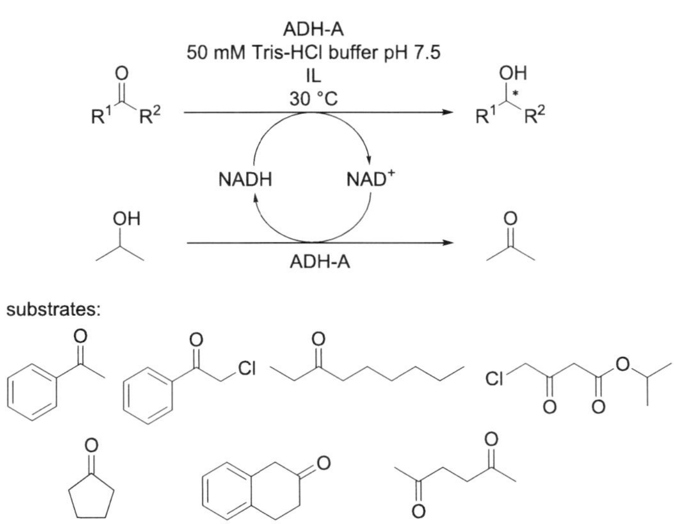

SCHEME 2.25 Alcohol dehydrogenase from *Rhodococcus ruber* (ADH-A)–catalyzed reduction of ketones in hydroxy-functionalized water-miscible ionic liquids.

concentrations from 1.2 to 1.5 mol/L. These results showed that higher hydrophobic substrates concentrations can be used with ADHs in high ILs:buffer content.

ADHs have been extensively investigated for their role in redox reactions and are now used in multiple industrial processes (Hall and Bommarius, 2011). Recently, various studies have used ILs as additives with the aim of using an IL as the sole medium in enzymatic reactions with several ADHs (Paul et al., 2014). In another study, whole cells of *Acetobacter* sp. CCTCC M209061 were used to catalyze the asymmetric reduction of ethyl acetoacetate in water-miscible ILs (Wang et al., 2013). This work was further extended to cover other prochiral ketones for their anti-Prelog ADH-catalyzed reduction in ILs (Xiao et al., 2012). Moran and coworkers have screened various microorganisms for the reduction of (*Z*)-3-halo-4-phenyl-3-buten-2-one to the corresponding halohydrins in an IL:water biphasic system. The results showed enhanced enantio- and diastereoselectivity in the [bmim][PF_6]-water biphasic system than in water alone (Zampieri et al., 2013).

The FDH from *C. boidinii* was used in a study to improve its stability and activity. The native FDH was inactivated when more than 50% v/v of the water-miscible IL [mmim][Me_2PO_4] was used (Bekhouche et al., 2010). Through a chemical modification of the lysines in the FDH, to which cations such as cholinium, hydroxyethyl-methylimidazolium, and hydroxypropyl-methylimidazolium were covalently attached, the overall performance, activity, and stability of the FDH was modestly improved.

Ene-reductases (ERs), more specifically old yellow enzymes (OYEs, EC 1.6.99.1), catalyze the asymmetric hydrogenation of activated C=C bonds, thus producing an alkane with up to two new stereocenters. The application of these enzymes in ILs had scarcely been shown before 2015. Brenna and coworkers have used OYE1 (from *Saccharomyces pastorianus*), OYE2, and OYE3 (from *Saccharomyces cerevisiae*) for the preparation of enantiopure (*R*)-arylpropanenitriles and obtained good conversions (up to >99% with OYE1) and excellent enantioselectivity when using the IL [bmim][PF_6] in a phosphate buffer at pH 7 (Scheme 2.26) (Brenna et al., 2014). The IL hydrophobic phase in this biphasic system was essentially used as a reservoir for the substrate, and improved the workup of the reaction. This work is the first example of isolated OYEs in ILs.

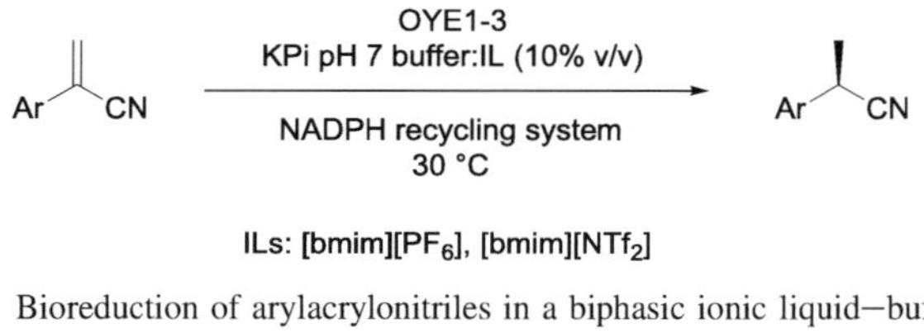

SCHEME 2.26 Bioreduction of arylacrylonitriles in a biphasic ionic liquid–buffer system.

2.2.2 Oxidations and Hydroxylations: Oxidases, Monooxygenases, and Peroxidases

The activity of oxidative enzymes, laccases (Shipovskov et al., 2008), and peroxidases (Tavares et al., 2012a) in ILs have also been studied. Laccases (EC 1.10.3.2), multicopper oxidases, are cofactor independent and catalyze one-electron oxidations using molecular oxygen as electron acceptor. Peroxidases, on the other hand, use hydrogen peroxide (H_2O_2) to catalyze the oxidation of different substituted phenols and aromatic amines. Within the first few years of discovering the use of ILs in biotransformations, laccase C from *Trametes* sp., horseradish peroxidase (HRP), and soybean peroxidases were shown to have catalytic activity in various systems containing the ILs 4-methyl-*N*-butylpyridinium [4-mbp][BF_4] or [bmim][PF_6] in various proportions with water (Hinckley et al., 2002). Peroxidases and oxidases were also investigated for the chemo- and stereoselective oxidation of sulfides (Scheme 2.27) (Okrasa et al., 2003; Chiappe et al., 2006). These substrates being hydrophobic, the addition of ILs such as [bmim][PF_6] enabled higher substrate concentration. Sanfilippo and coworkers reported activity with a chloroperoxidase (CPO) from *Caldariomyces fumago*, in up to 30% v/v [bmim][$MeSO_4$] or [mmim][$MeSO_4$] (Sanfilippo et al., 2004). In another study, HRP was used in combination with [bmim][BF_4] or [bmim][Cl] mixed in different proportions with respect to the phosphate buffer, from 5% to 25% v/v. With 20% IL, the activity of the enzyme was maintained, and with 5% and 10% v/v of [bmim][BF_4] the thermal stability of HRP was even improved (Machado and Saraiva, 2005).

In a 2013 work, a laccase previously studied in the presence of various ILs and cosolvents (Tavares et al., 2008) was immobilized on modified silica and its characteristics were studied (Tavares et al., 2013). The stability of the laccase, as well as its activity, was found to be improved with [emim][$EtSO_4$], when compared to the free enzyme. Furthermore, the group of Stephens and coworkers have developed a biphasic system for laccases with hydrophobic ILs (Rehmann et al., 2014). This biphasic approach essentially protects the laccase (in this study from *Trametes versicolor*) from inactivation by the mediator used in these reactions, such as (2,2,6,6-tetramethyl-1-piperidinyl) oxidanyl (TEMPO) and 2,2′-azino-bis-(3-ethylbenzo-thiazoline-6-sulfonic acid) (ABTS), by portioning it between the water and IL phases. Therefore, the laccase had only minimal contact with the mediator and had an extended lifetime. The most suitable IL to achieve the best results had to be matched according to the mediator used in the reaction. As a result, the laccase retained

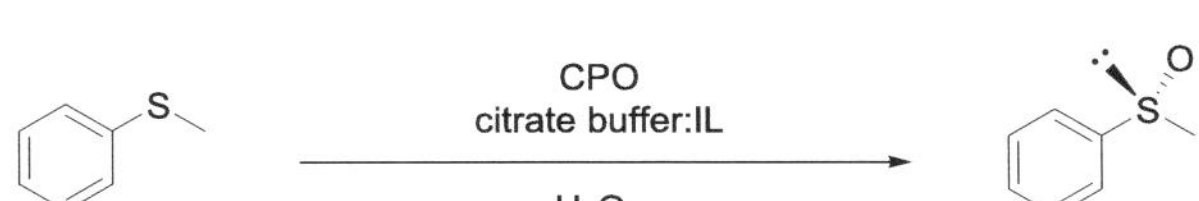

SCHEME 2.27 Chloroperoxidase-catalyzed oxidation of thioanisole.

a higher activity with [hmim][AOT] than without when using ABTS, 4-hydroxybenzyl alcohol, phenothiazine, or 2-hydroxybiphenyl as a mediator, whereas [mtoa][Sac] was best with TEMPO.

Since discovering that certain ILs can dissolve lignin and lignocellulose, they have been investigated as an efficient reaction medium for wood-based polymers. Hence, the laccase from *Melanocarpus albomyces* was used with the IL [amim][Cl] for the formation of the coniferyl alcohol dehydropolymer (DHP) (Lahtinen et al., 2013). With an increasing concentration of [amim][Cl] from 0% to 40%, the activity of the laccase decreased, whereas the insoluble fraction of the DHP formed in the presence of 40% [amim][Cl] had higher molecular density.

The initial activity and stability of a commercially available peroxidase was studied in the presence of five water miscible imidazolium ILs. When adding a moderate amount (5–10% v/v) of IL [bmim][Cl], [hmim][Br], and [omim][Cl], the initial peroxidase activities gave good results. The peroxidase was also more thermostable with [bmim][Br] (Tavares et al., 2012b).

Recent studies on ILs have also included Baeyer–Villiger monooxygenases (BVMOs) (Melgarejo-Torres et al., 2011). Gotor and coworkers focused their attention on a thermostable phenylacetone monooxygenase (PAMO) from *Thermobifida fusca*. Several ILs screened such as [bmp][PF_6], [bmim][PF_6], [hmim][PF_6], and [bmim][$MeSO_4$] were found to actually improve the enantioselectivity of the PAMO-catalyzed oxidation of substrates, namely α-acetylphenylacetonitrile and 3-methyl-4-phenylbutan-2-one affording full conversion (Scheme 2.28) (Rodriguez et al., 2010).

Tyrosinase (EC 1.14.18.1) belongs to the type 3 copper enzyme family and catalyzes the hydroxylation of phenols. Tyrosinase activity in ILs was investigated as well (Yang et al., 2008). In a later study, a tyrosinase from *Bacillus megaterium*, TyrBm, was used to determine the effect of various ILs, [bmim][BF_4], [bmim][Cl], [emim][$EtSO_4$], and [ea][NO_3] (Goldfeder et al., 2013). Thus, for the hydroxylation of L-tyrosine in 40% [bmim][BF_4] the specificity of the reaction (monophenolase versus diphenolase activity) was improved greatly.

2.2.3 Cofactor Recycling Using Ionic Liquids

The β-nicotinamide adenine dinucleotide cofactor is the central redox cofactor in all living cells. It exists in phosphorylated (NADP) and nonphosphorylated form (NAD) as well as in reduced form (NAD(P)H). The nicotinamide

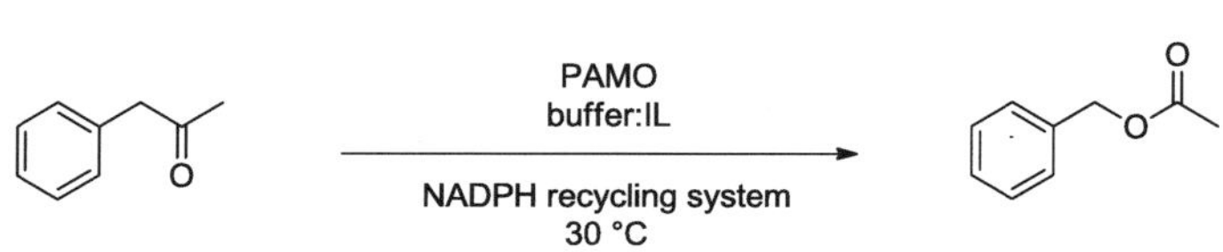

SCHEME 2.28 Phenylacetone monooxygenase-catalyzed oxidation of phenylacetone in a buffer/ionic liquid system.

cofactor provides the necessary reducing equivalents for a redox reaction, acting as electron donor or acceptor. NAD(P)H-dependent oxidoreductases such as ADHs, ERs, and BVMOs (vide supra) require this costly nicotinamide cofactor. A substantial amount of nicotinamide cofactor recycling systems have already been established and extensively reviewed (Chenault and Whitesides, 1987; Rodriguez et al., 2012; Wu et al., 2013; Kara et al., 2014; Paul et al., 2013). In principle, an ideal cofactor regeneration system should meet the following criteria: (1) stable and inexpensive enzymes, (2) high specific enzymatic activity, (3) simple and inexpensive reagents that do not interfere with the isolation of the desired product or with the stability of the enzyme, (4) high turnover number, and (5) a reachable favorable equilibrium for product formation.

The most important strategies that have been developed for the in situ regeneration of nicotinamide cofactors with ADHs are (1) using another enzyme or cosubstrate in an enzyme-coupled or substrate-coupled system, (2) engineering whole cells to overexpress both the ADH and the cofactor regenerating enzyme, (3) electrochemical, and (4) chemical.

Glucose dehydrogenase (GDH) in particular is a stable and readily available enzyme, has a good specific activity, and is relatively low cost. However, these enzymes prove to be somewhat unstable and sensitive to product inhibition. The substrate-coupled system employing 2-propanol as a cosubstrate requires a surplus to shift the equilibrium, although it is highly convenient for economic reasons.

For ADHs, the use of acetone or acetaldehyde in a huge molar excess has been demonstrated in a coupled-substrate approach, but in many cases these conditions are not compatible with the stability of the enzyme. In 2004, Walker and Bruce synthesized a range of hydrophilic ILs, derivatives of [bmim][PF_6], and screened them with the NADP-dependent morphine dehydrogenase (MDH) from *Pseudomonas putida* M10 with a cofactor recycling system using either an ADH from *Thermoanaerobium brockii* or GDH from *Cryptococcus uniguttulatus* (Walker and Bruce, 2004a). In total, six functionalized ILs were evaluated—[bmim][PF_6], [bmim][glycolate], [bdmim][PF_6], [C_3OHmim][PF_6], [C_3OHmim][glycolate], [C_3OHmim][Cl]—and were compared to three organic solvents, hexane, *t*-butanol, dimethyl carbonate (Me_2CO_3), at different water contents, varying from <100 ppm, 1−40% water. The conversion of codeine to codeinone was measured after 1, 4, and 24 h (Scheme 2.29).

The enzyme activity varied noticeably when using the various ILs, solvents, or different water contents. The use of organic solvents led to no conversion, whereas using a 40% water:IL mixture afforded similar product yield as with 100% water. The added advantage of using these ILs was the suppression of product solvolysis, and higher isolable yield. When optimizing reaction conditions, 20% of the codeinone was obtained with GDH as the coenzyme for the cofactor recycling system (Walker and Bruce, 2004b).

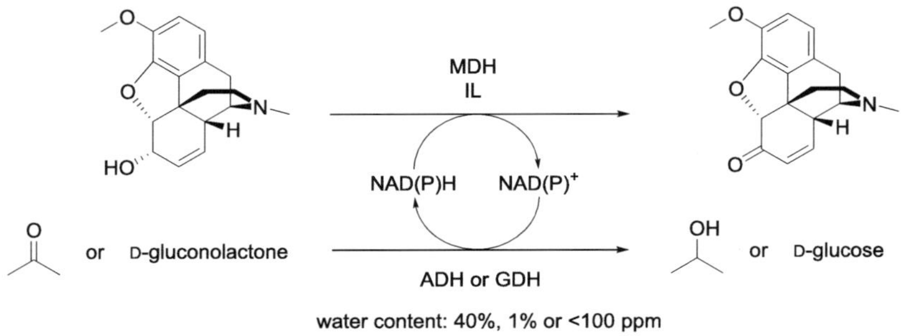

SCHEME 2.29 Oxidation of codeine to codeinone using morphine dehydrogenase in mixtures of ionic liquids and water.

Enzymatic ketone reductions with cofactor recycling were also investigated to improve reactions with ILs as cosolvents (Hussain et al., 2008). The ADH from *Rhodococcus erythropolys* (ADH*Re*) was used to catalyze the reduction of 4′-bromo-2,2,2-trifluoroacetophenone using GDH103 as a coenzyme for the NADH recycling system (Scheme 2.30). Using a hydrophobic IL, butylmethylpyrrolidinium [BMP][NTf_2], significantly improved the performance of the reaction, increasing the half-life of the enzymes to 250 h (using 10% v/v IL), increasing the initial rate of the reaction by four, and improving product isolation.

Another recycling system was designed for the oxidation and reduction of alcohols and ketones, respectively, by various ADHs. In this study, Gotor and coworkers synthesized two ILs, 1-(2-hydroxypropyl)-3-methylimidazolium chloride and 3-methyl-1-(2-oxopropyl)imidazolium chloride, to be used as a cosubstrate in an NAD(P)H recycling system (Paul et al., 2014). Thus, seven different ADHs were screened (ADH-A from *R. ruber*, *Lb*ADH from *L. brevis*, *Lk*ADH from *Lactobacillus kefir*, *Ras*ADH from *Ralstonia* sp., *Sy*ADH from *Sphingobium yanoikuyae*, *Tes*ADH from *Thermoanaerobacter ethanolicus*, and ADH-T from *Thermoanaerobium* sp.) with these ILs as cosubstrates or as additives. The results showed that these imidazolium ILs were not accepted as cosubstrates by the ADHs (Scheme 2.31), however, when used as an additive

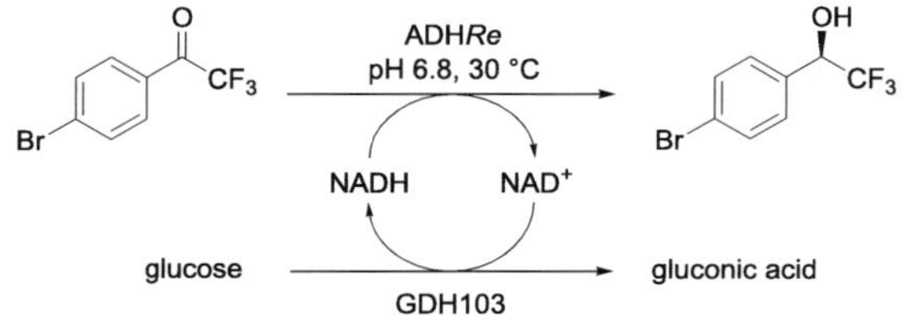

SCHEME 2.30 ADH*Re*-catalyzed asymmetric reduction of a 4′-bromo-2,2,2-trifluoroacetophenone with NADH recycling using glucose dehydrogenase (GDH) 103.

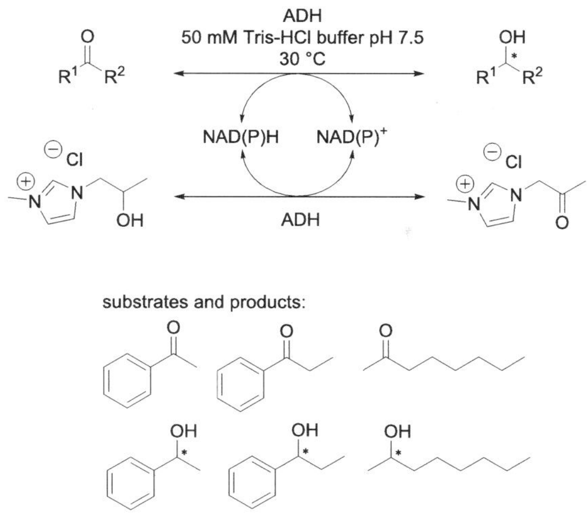

SCHEME 2.31 Alcohol dehydrogenase–catalyzed reduction and oxidation using ionic liquids as cosubstrate for NAD(P)H recycling.

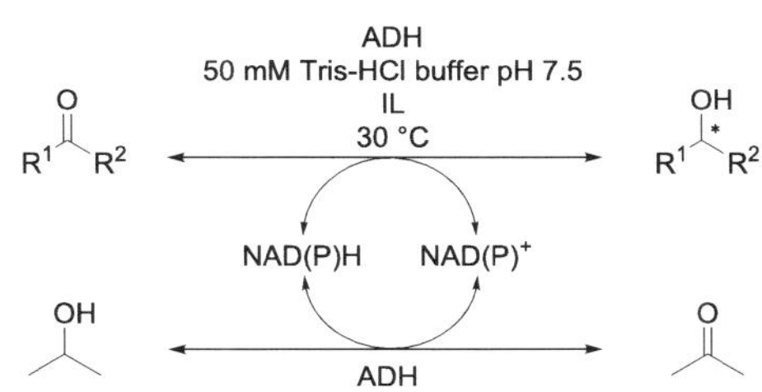

SCHEME 2.32 Alcohol dehydrogenase–catalyzed reduction and oxidation using ionic liquids as cosolvent.

from 0% to 90% v/v in buffer, in up to 20% v/v the conversions were improved significantly, also retaining the perfect enantioselectivity these enzymes display (Scheme 2.32).

2.3 IONIC LIQUIDS AND OTHER CLASSES OF ENZYMES

Oxynitrilases (hydroxynitrile lyases (HNLs), EC 4.1.2.47) catalyze the reversible reaction of addition of HCN to aldehydes and ketones to afford cyanohydrins. These enzymes are gaining in popularity with an increasing number of academic and industrial applications (Kassim and Rumbold, 2014). The first HNL-catalyzed cyanohydrin formation in ILs was achieved in a biphasic system in which the IL phase essentially prevents the undesired chemical addition of HCN to the carbonyl group from occurring while keeping the HNL activity intact (Scheme 2.33) (Gaisberger et al., 2004).

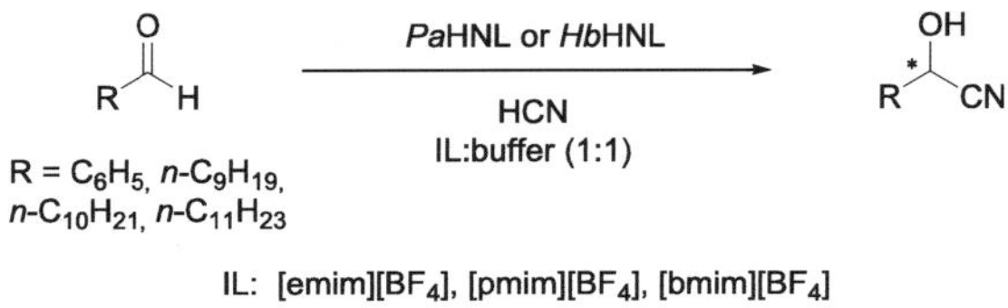

SCHEME 2.33 Synthesis of cyanohydrins by *Prunus amigdalus* hydroxynitrile lyase (*Pa*HNL) and *Hevea brasiliensis* hydroxynitrile lyase (*Hb*HNL)-catalyzed reactions in a two-phase solvent system with buffer and ionic liquid (1:1).

Using 1:1 mixtures of buffer:IL using [emim], [pmim], or [bmim] as a cation and $[BF_4]$ as an anion, enhanced the rate of the reaction compared to that with MTBE mixtures. Another study on *Pa*HNL with similar ILs showed that adding 2–6% IL in an aqueous medium increased the activity of the enzyme and gave higher enantiomeric excesses (Scheme 2.34) (Lou et al., 2005).

The enzyme *Me*HNL from *Manihot esculenta* was also screened but lost activity even with the smallest percentage of IL. Nevertheless, both *Pa*HNL and *Me*HNL displayed higher thermal stability in [bmim]$[BF_4]$ with respect to organic solvents. In 2014, the benzaldehyde lyase (BAL) was investigated in a series of DES-buffer mixtures, which are discussed in the following section (Maugeri and Domínguez de María, 2014a).

Concerning the use of isomerases, glucose isomerase (GI) catalyzes the isomerization of glucose to fructose, which is involved in the industrial production of high-fructose corn syrup used in food products. An immobilized GI from *Streptomyces murinus* (Sweetzyme®) was screened in various ILs with different IL:water ratios and was found to catalyze the isomerization of glucose into fructose and xylose to xylulose in IL:water with a lower or higher activity depending on the IL and substrate (Stahlberg et al., 2012). Xylose being the natural substrate of GI, reaction rates were faster in IL:water than with glucose, with respect to the reaction in just water. This important study contributes to exploring potential biocatalytic industrial applications.

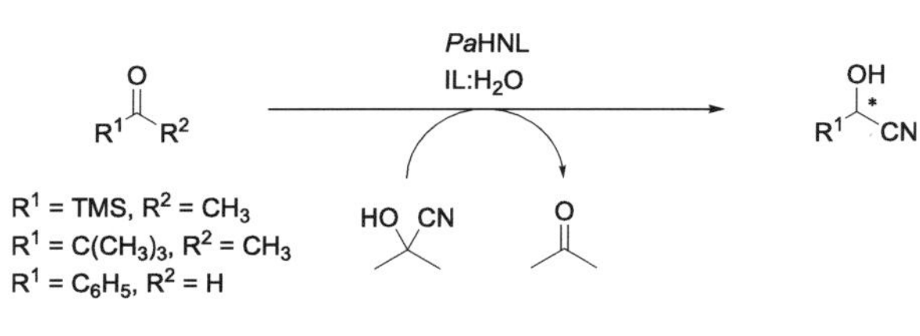

SCHEME 2.34 *Pa*HNL(*Prunus amigdalus* hydroxynitrile lyase)-catalyzed transcyanation with acetone cyanohydrin with ionic liquids.

2.4 DEEP EUTECTIC SOLVENTS IN BIOCATALYSIS

DESs are formed by mixing and complexation of a quaternary ammonium salt such as choline chloride with an adequate hydrogen bond donor (HBD) compound (eg, alcohols, amides, amines, carboxylic acids, etc.). The mixing of the ammonium salt and the HBD leads to a disruption in the crystalline structure of the quaternary ammonium salt, triggering a decrease in the melting point in comparison with the individual components, thus generating a liquid at room temperature. This occurs because of the charge delocalization through hydrogen bonding between the halide anion and the HBD. The general structures of DES are depicted in Fig. 2.2.

DESs are typically considered as biorenewable and green solvents, although studies related to their potential toxicity have appeared (Hayyan et al., 2013a,b). Their applicability in different areas has been demonstrated (Smith et al., 2014; García-Álvarez &, 2014), and among others it is noteworthy to mention their use in extraction processes (Pena-Pereira and Namieśnik, 2014; Maugeri et al., 2012; Krystof et al., 2013), production of materials (Wagle et al., 2014), and their valuable role as reaction media in synthetic transformations (Zhang et al., 2012), including also biotransformations (Domínguez de María and Maugeri, 2011).

In this section, we have summarized a representative number of biotransformations carried out in DES. Most of these studies have paid special attention to their role in hydrolase-catalyzed reactions, although the action of oxidoreductases and carboligases in these neoteric solvents also has been scarcely reported in the literature.

2.4.1 Conventional Hydrolase-Catalyzed Processes in Deep Eutectic Solvent

In 2008 Kazlauskas and coworkers reported for the first time the activity of hydrolases in DES (Gorke et al., 2008). Initially, the lipase-catalyzed transesterification of ethyl valerate with 1-butanol was studied in eight different DESs using CAL-A, CAL-B, and PSL at 60°C (Scheme 2.35). Interestingly,

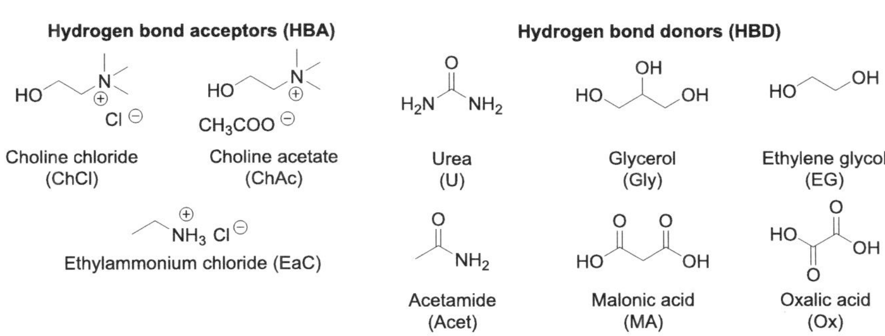

FIGURE 2.2 Representative structures of deep eutectic solvent and their corresponding abbreviations.

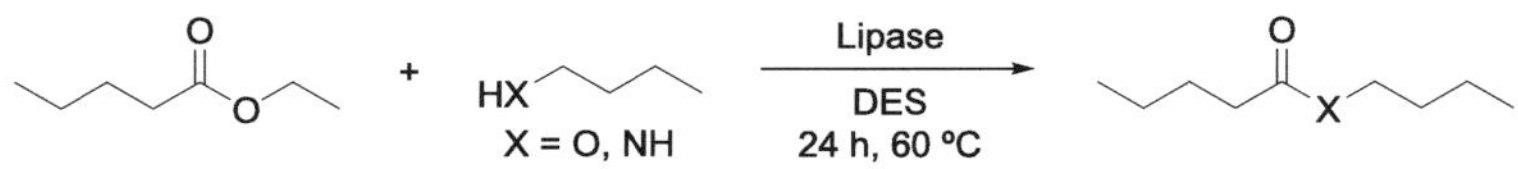

SCHEME 2.35 Lipase-catalyzed transesterification and aminolysis reactions in deep eutectic solvent of ethyl valerate with *n*-butanol (X = O) and 1-butylamine (X = NH), respectively.

CAL-B acted with similar activity in five DESs compared to the performance in a hydrophobic solvent such as toluene, whereas CAL-A and PSL showed poorer activities in most of them. This study was successfully extended to the aminolysis reaction of ethyl valerate using a little excess of 1-butylamine instead of *n*-butanol, leading to >90% conversion in ChCl:Gly, ChCl:U, and toluene. Remarkably, the specific activity of CAL-B for the aminolysis reaction in DES was higher than the one achieved in a representative number of ILs such as [bmim][BF_4] or [bmim][NTf_2].

The same authors also reported for the first time the use of DES as a cosolvent for ester and epoxide hydrolysis, increasing the hydrolase-catalyzed reaction of *p*-nitrophenyl acetate with 10% ChCl:Gly up to threefold compared to the reaction using pig liver esterase in aqueous system, or until 20 times for an EH-catalyzed reaction of styrene oxide using 25% ChCl:Gly (Gorke et al., 2008).

From there on a number of biotransformations have been developed using DES. Herein, we have organized these examples depending on the enzymatic source, hence, a variety of enzymes and synthetic reactions will be discussed. Some of these examples are limited by the necessity of using water to ensure that no enzyme deactivation is occurring. A few examples deal with the use of DES as a cosolvent reaction instead of pure solvents, although in some cases the sole addition of a minimum water content (<10%) is enough to find an excellent enzyme activity.

Probably the most common hydrolase-catalyzed reaction is the reaction between an alcohol and an ester, the alkoxy group being exchanged between both molecules. This reaction has received several names such as transesterification, alcoholysis, and acylation, and lipases and proteases are able to catalyze these transformations in organic solvents with high selectivity. Previously in this chapter, we have shown a representative number of examples using RT ILs as solvents. Similarly, DESs have also been served for the formation of alcohols and esters mainly in a nonselective fashion.

A couple of proteases have been effectively tested in the transesterification of *N*-acetyl-L-phenylalanine ethyl ester using 1-propanol using low water contents (2–4%) (Zhao et al., 2011a). The best results were achieved with immobilized enzyme preparations, which led to total selectivity of the propyl ester over the hydrolytic by-product using different choline-based HBA:Gly mixtures (1:1.5 or 1:2, Scheme 2.36). In this context, it is important to note that the structure of the DES is crucial in order to avoid other undesired background reactions. For instance, the presence of malonic acid, oxalic acid,

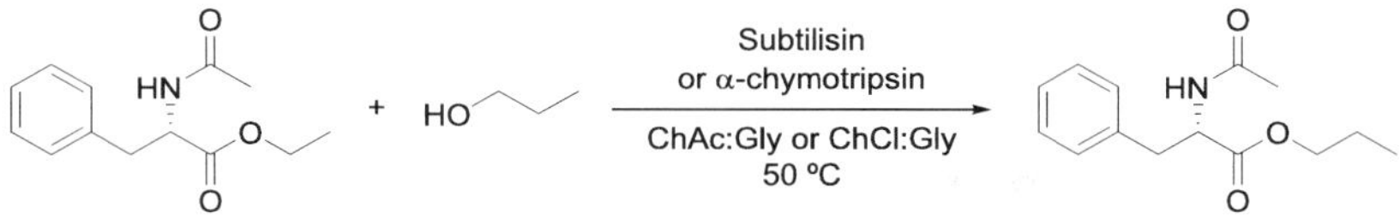

SCHEME 2.36 Protease-catalyzed transesterification of *N*-acetyl-L-phenylalanine ethyl ester with 1-propanol in pure deep eutectic solvent.

or ethylene glycol (EG) in ChCl-based DES can lead to a competition with the substrates in lipase-catalyzed alcoholysis reactions between an alcohol and an ester, leading not only to by-product formations but also to DES destruction (Durand et al., 2012).

On the other hand, Durand and coworkers observed that some organic compounds such as phenolic esters are difficult to extract when they are immersed in a DES. In these cases, the use of a protic cosolvent such as water allows the transesterification of esters with excellent conversions, as water creates hydrogen bonding more easily than the corresponding esters (Durand et al., 2013). For instance, the reaction between methyl *p*-coumarate and six equivalents of 1-octanol catalyzed by CAL-B favored the formation of the corresponding octyl ester in complete conversion after 72 h at 60°C using a (92:8) mixture of ChCl:U and water (Scheme 2.37). This process did not occur in any extension in the absence of water, while higher water contents (10–20%) led to complete conversions; the side-hydrolytic reaction is also observed in appreciable extent.

The search for environment-friendly strategies for the biodiesel production is a challenging task. Lipase-catalyzed transesterifications offer serious advantages with traditional chemical methods such as mild reaction conditions, low energy demand, and waste reduction. Zhao and coworkers have reported the successful use of CAL-B in combination with DES for the transesterification of triglycerides using methanol (Zhao et al., 2011b). Under optimal conditions, the Miglyol oil has been produced in short reaction times (1–3 h) and high conversions (82–97%) at 50°C, using ChAc:Gly (1:1.5) as solvent with 1% in water. The same research group also reported the lipase-catalyzed transesterification of soybean oil with methanol in DES with 1% of water at 50°C (Zhao et al., 2013). CAL-B was again found as the best biocatalyst in a mixture of ChCl:Gly (1:2), yielding 81% of product after 24 h,

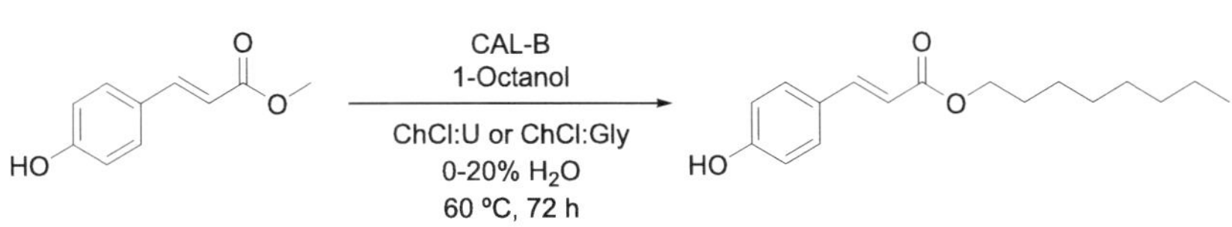

SCHEME 2.37 Transesterification of methyl *p*-coumarate with 1-octanol using *Candida antarctica* lipase type B in low water content deep eutectic solvent.

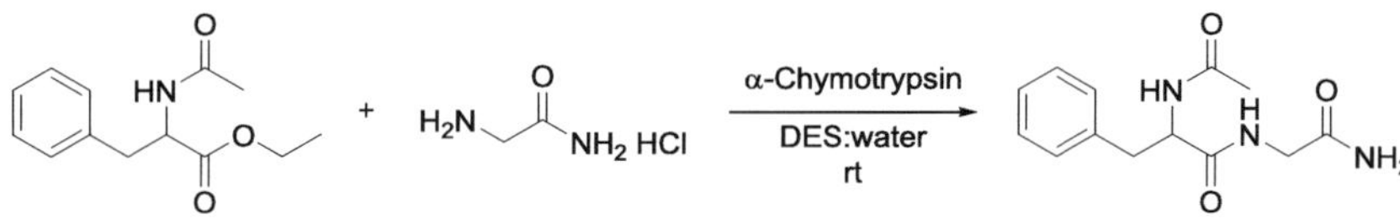

SCHEME 2.38 Protease-catalyzed peptide synthesis between *N*-acetyl-L-phenylalanine ethyl ester and glycinamide hydrochloride in deep eutectic solvent:water mixtures.

conditions that were further optimized until 88% triglyceride conversion with 0.2% of water.

Lipases and proteases can also catalyze aminolysis reactions for the formation of amides and peptides, and in these cases the ester is reacted with an amine instead of an alcohol. α-Chymotripsin has efficiently catalyzed the reaction between racemic *N*-acetyl-L-phenylalanine ethyl ester and glycinamide hydrochloride in ChCl-based DES (Scheme 2.38), obtaining the corresponding peptides with excellent conversions at low water concentrations (10−25%) (Maugeri et al., 2013). Higher contents in water (ie, 50%) led to a significant amount of the corresponding carboxylic acid resulting from the hydrolysis reaction.

One of the main advantages of the use of DES is the possibility of carrying out reactions at very high substrate concentrations. In this context, Widersten and coworkers reported the use of ChCl:EG, ChCl:Gly, and ChCl:U in the hydrolysis of 1,2-*trans*-2-methylstyrene oxide using potato EH StEH1 (Lindberg et al., 2010). The DES solutions dissolved 1.5-fold higher epoxide concentrations in comparison with the phosphate buffer. In addition, beneficial effects were also observed in the acceleration of the reaction and the regioselectivity toward the epoxide ring opening at the benzylic carbon (Scheme 2.39). More recently, the kinetic resolution of optically active tetrahydropyrazolo[1,5-α]pyrimidines has been carried via lipase-catalyzed hydrolytic reactions using phosphate buffer as hydrolytic agent and a variety of organic solvents (DMSO or dioxane) or DES (ChCl:Gly, ChCl:U or CHCl:Xylitol) as cosolvent in 10% v/v ratio (Fernández-Álvaro et al., 2014). Excellent selectivities were found using CAL-B as biocatalyst for the

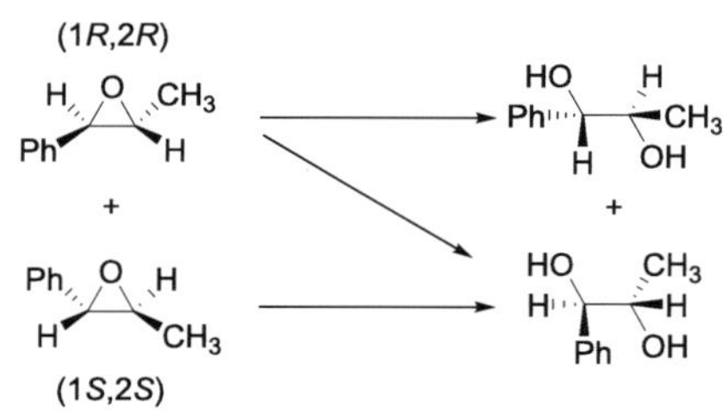

SCHEME 2.39 Stereoselective outcome of the epoxide hydrolase catalyzed-reaction of 2-methylstyrene oxide enantiomers using deep eutectic solvent as cosolvents.

hydrolysis of the ethyl ester group, offering an attractive alternative to current existing chiral HPLC preparative separations.

2.4.2 Nonconventional and Tandem Hydrolase-Catalyzed Processes in Deep Eutectic Solvent

Catalytic promiscuity is an emerging area in biocatalysis, illustrating the potential of enzymes for unexpected reactions (Hult and Berglund, 2007). Hydrolases are a class of enzymes with multiple applications in both hydrolytic and reverse hydrolytic process, however, their use in nonconventional C—C and C—heteroatom bond formation, oxidation reaction, and polymerization have attracted great attention in 2010 (Busto et al., 2010). In some cases, the lipase is responsible for this promiscuous transformation but in other cases, its action is part of a more complex mechanism via a tandem process. A couple of examples will be discussed in this area.

A series of dihydropyrimidin-2(1*H*)-one derivatives has been prepared by reacting aromatic aldehydes such as ethyl acetoacetate or pentane-2,4-dione, and urea or thiourea using *Rhizopus oryzae* lipase (ROL) as biocatalyst and ChCl:U (1:2) as solvent (Scheme 2.40) (Borse et al., 2012). This Biginelli reaction occurred with low conversion in the absence of the lipase, so the ROL accelerated the reaction to a high extent, the recyclability of the DES and lipase also being possible, obtaining a range of 75—95% yield in the model reaction after four recycling uses. The combination of the lipase with the ChCl:U (1:2) gave better results in comparison with the use of other more conventional solvents such as water, methanol, dioxane, or DMF.

Based on the excellent activity of CAL-B in DES, an elegant approach was developed for the in situ acetaldehyde production through a transesterification reaction between vinyl acetate and 2-propanol (Scheme 2.41) (Müller et al., 2014). This acetaldehyde served for the cross aldol reaction with 4-nitrobenzaldehyde using proline-derived derivatives leading to the corresponding hydroxyketone, which was subsequently extracted in ethyl acetate and chemically reduced toward the diol in 92% conversion and 95% *ee*. The recycling of the system was satisfyingly achieved for the DES and CAL-B, requiring the addition of fresh organocatalyst in order to reach good

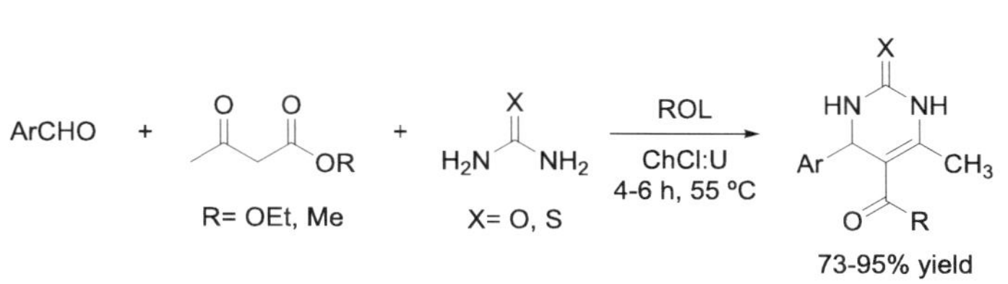

SCHEME 2.40 Biginelli reaction using the *Rhizopus oryzae* lipase and ChCl:U (1:2).

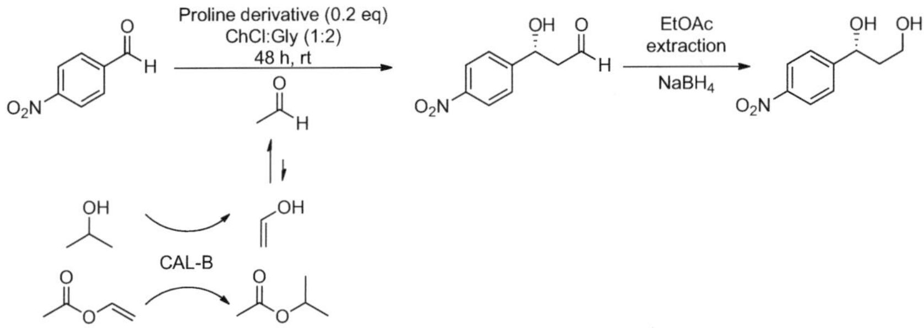

SCHEME 2.41 Tandem lipase-proline derived organocatalyst approach for cross-aldol reactions using acetaldehyde generated in situ.

activity values. This approach was extended to a family of another seven aldehydes, with the obtained conversions being highly dependent on the substrate structure.

2.4.3 Deep Eutectic Solvent in Other Biocatalytic Processes

Nowadays, the use of DES for hydrolase-catalyzed reactions is approaching a mature stage, while the use of other enzymes is still in its infancy. Remarkably, the enantioselective bioreduction of ethyl acetoacetate is possible using Baker's yeast in different mixtures of water and DES (Maugeri and Domínguez de María, 2014b). The stability at long reaction times (>250 h) has been proven, observing a complete inversion of the enantioselectivity in pure water where the (*S*)-alcohol is obtained, compared to a pure DES such as ChCl:Gly (1:2), which favors the formation of the (*R*)-antipode (Scheme 2.42). Alternatively, the racemic alcohol can be obtained at a (30:70) DES-water mixture. The results are explained by the use of a whole-cell enzyme, the DES being able to inhibit some of the (*S*)-oxidoreductases present in the enzymatic preparation.

In addition, the use of DES enables the improvement in the activity and stability of HRP based on experimental and structural studies (Wu et al., 2014). An exhaustive study related to the choice of the ammonium salt (ChCl or ChCl:Ac), the HBD compound (urea, glycerol, acetamide, and EG) at 3 M ratios (1:2, 1:1, and 2:1) was performed, finding different trends after studying

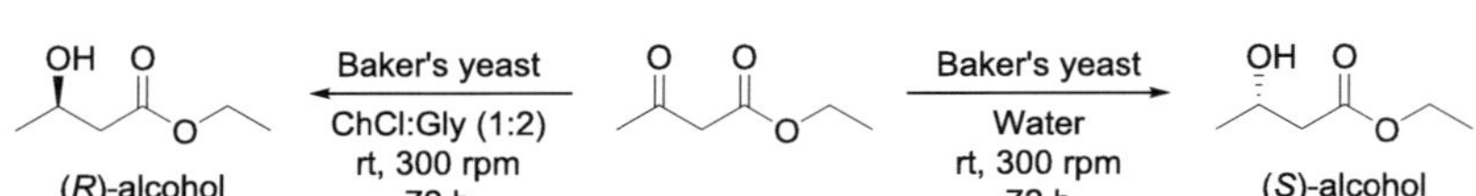

SCHEME 2.42 Enantioselective Baker's yeast-catalyzed bioreduction of ethyl acetoacetate in deep eutectic solvent or water.

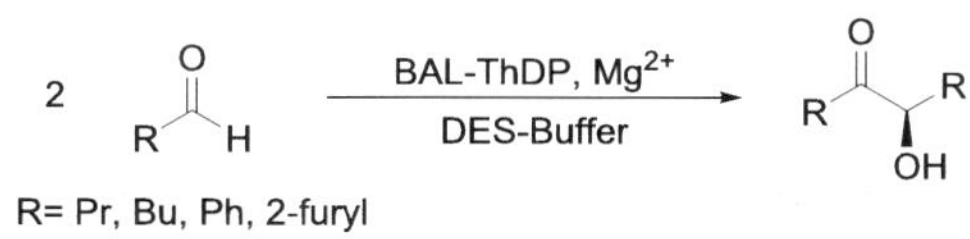

SCHEME 2.43 Benzaldehyde lyase–catalyzed enantioselective carboligation in deep eutectic solvent–aqueous mixtures.

the behavior of 24 DES. For instance: (1) ChCl-based DES promote to a high extent the peroxidase activity in comparison with ChAc-based DES; (2) for DES composed of the same quaternary ammonium salt, a higher content was more beneficial for the HRP activity (2:1 > 1:1 > 1:2); (3) DES capable of providing the enzyme with a higher α-helix content and a slightly more relaxed tertiary structure may facilitate the HRP activity.

Maugeri and Domínguez de María reported the use of a BAL for the carboligation of aldehydes in DES-buffer mixtures (Maugeri and Domínguez de María, 2014a). For this study two molecules of representative aldehydes such as butyraldehyde, valeraldehyde, benzaldehyde, and 2-furaldehyde were considered for the homocoupling reaction (Scheme 2.43), leading to high conversions (75–98%) for the formation of the enantioenriched (*R*)-products (27–99% *ee*) after 24 h at room temperature when using a 60:40 mixture of ChCl:Gly and phosphate buffer at pH 8. These results were quite similar to the ones obtained with 2-methyl-tetrahydrofuran as cosolvent. The behavior of BAL was observed to be very sensitive to the amount of DES used, noting a decrease in the enzyme activity at higher DES concentrations, although the system 70:30 ChCl:U and phosphate buffer was still a good system for biocatalytic carboligation reactions.

2.5 CONCLUSIONS AND OUTLOOK

ILs have been widely used in biotransformations and biocatalysis, from the first proof-of-concept to direct applications. Their ability to improve enzymatic processes in many different ways, in terms of activity, stability, and thermal stability, has greatly contributed to their increasing use in chemoenzymatic applications (Roosen et al., 2008). The improved activity, stability, and enantioselectivity displayed by oxidoreductases in biphasic systems, for example, have led to many analytical and synthetic applications, whereas the use of lipases in pure ILs for nonhydrolytic reactions have provided sometimes better results than those obtained in traditional organic solvents. The application of novel biocatalytic reactions is a question of time, since amino transferases have gained great attention in recent years and no examples are yet reported. We expect further interesting developments to be seen in the near future, especially using DES and other environmentally friendly bio-based ILs. The implementation in the industrial sector of biocatalytic process in ILs is still pending, but there are many advantages which can lead to a proximate successful development.

REFERENCES

Ahrenberg, M., Brinckmann, M., Schmelzer, J.W.P., Beck, M., Schmidt, C., Kessler, O., Kragl, U., Verevkin, S.P., Schick, C., 2014. Determination of volatility of ionic liquids at the nanoscale by means of ultra-fast scanning calorimetry. Phys. Chem. Chem. Phys. 16, 2971–2980.

Attri, P., Choi, E.H., 2013. Influence of reactive oxygen species on the enzyme stability and activity in the presence of ionic liquids. PLoS One 8, e75096.

Allen, C.C.R., Boudet, C.J., Hardacre, C., Migaud, M.E., 2014. Enhancement of whole cell dioxygenase biotransformations of haloarenes by toxic ionic liquids. RSC Adv. 4, 19916–19924.

Akanbi, T.O., Barrow, C.J., Byrne, N., 2012. Increased hydrolysis by *Thermomyces lanuginosus* lipase for omega-2-fatty acids in the presence of a protic ionic liquid. Catal. Sci. Technol. 2, 1839–1841.

Abe, Y., Yagi, Y., Hayase, S., Kawatsura, M., Itoh, T., 2012. Ionic liquid engineering for lipase-mediated optical resolution of secondary alcohols: design of ionic liquids applicable to ionic liquid coated-lipase catalyzed reaction. Ind. Eng. Chem. Res. 51, 9952–9958.

Abe, Y., Kude, K., Hayase, S., Kawatsura, M., Tsunashima, K., Itoh, T., 2008. Design of phosphonium ionic liquids for lipase-catalyzed transesterification. J. Mol. Catal. B Enzym. 51, 81–85.

Bräutigam, S., Dennewald, D., Schurmann, M., Lutje-Spelberg, J., Pitner, W.R., Weuster-Botz, D., 2009. Whole-cell biocatalysis: evaluation of new hydrophobic ionic liquids for efficient asymmetric reduction of prochiral ketones. Enzym. Microb. Technol. 45, 310–316.

Bornscheuer, U.T., Kazlauskas, R.J., 2006. Hydrolases in Organic Synthesis: Regio- and Stereoselective Biotransformations, second ed. Wiley-VCH.

Bornscheuer, U.T., Kazlauskas, R.J., 2004. Catalytic promiscuity in biocatalysis: using old enzymes to form new bonds and follow new pathways. Angew. Chem. Int. Ed. 43, 6032–6040.

Busto, E., Gotor-Fernández, V., Gotor, V., 2010. Hydrolases: catalytically promiscuous enzymes for non-conventional reactions in organic synthesis. Chem. Soc. Rev. 39, 4504–4523.

Busto, E., Gotor-Fernández, V., Ríos-Lombardía, N., García-Verdugo, E., Alfonso, I., García-Granda, S., Menéndez-Velázquez, A., Burguete, M.I., Luis, S.V., Gotor, V., 2007. Simple and straightforward synthesis of novel enantiopure ionic liquids via efficient enzymatic resolution of (±)-2-(1*H*-imidazol-1-yl)cyclohexanol. Tetrahedron Lett. 48, 5251–5254.

Borowiecki, P., Milner-Krawczyk, M., Plenkiewicz, J., 2013. Chemoenzymatic synthesis and biological evaluation of enantiomerically enriched 1-(β-hydroxypropyl)imidazolium- and triazolium-base ionic liquids. Belstein. J. Org. Chem. 9, 516–5225.

Bekhouche, M., Doumeche, B., Blum, L.J., 2010. Chemical modifications by ionic liquid-inspired cations improve the activity and the stability of formate dehydrogenase in [MMIm][Me_2PO_4]. J. Mol. Catal. B Enzym. 65, 73–78.

Brenna, E., Crotti, M., Gatti, F.G., Manfredi, A., Monti, A., Parmeggiani, F., Santangelo, S., Zampieri, D., 2014. Enantioselective synthesis of (*R*)-2-arylpropanenitriles catalysed by ene-reductases in aqueous media and in biphasic ionic liquid-water systems. ChemCatChem 6, 2425–2431.

Borse, B.N., Borude, V.S., Shukla, S.R., 2012. Synthesis of novel dihydropyrimidin-2(1H)-ones derivatives using lipase and their antimicrobial activity. Curr. Chem. Lett. 1, 59–68.

Chiappe, C., Rajamani, S., 2011. Structural effects on the physico-chemical and catalytic properties of acidic ionic liquids: an overview. Eur. J. Org. Chem. 5517–5539.

Cull, S.G., Holbrey, J.D., Vargas-Mora, V., Seddon, K.R., Lye, G.J., 2000. Room-temperature ionic liquids as replacements for organic solvents in multiphase bioprocess operations. Biotechnol. Bioeng. 69, 227–233.

Cantone, S., Hanefeld, U., Basso, A., 2007. Biocatalysis in non-conventional media – ionic liquids, supercritical fluids and the gas phase. Green Chem. 9, 954–971.

Cornmell, R.J., Winder, C.L., Schuler, S., Goodacre, R., Stephens, G., 2008. Using a biphasic ionic liquid/water reaction system to improve oxygenase-catalysed biotransformation with whole cells. Green Chem. 10, 685–691.

Chiappe, C., Leandri, E., Lucchesi, S., Pieraccini, D., Hammock, B.D., Morisseau, C., 2004. Biocatalysis in ionic liquids: the stereoconvergent hydrolysis of *trans*-β-methylstyrene oxide catalyzed by soluble epoxide hydrolase. J. Mol. Catal. B Enzym. 27, 243–248.

Chen, W.-J., Lou, W.-Y., Zong, M.-H., 2012a. Efficient asymmetric synthesis hydrolysis of styrene oxide catalyzed by mung bean epoxide hydrolases in ionic liquid-based biphasic systems. Bioresour. Technol. 115, 58–62.

Chen, W.-J., Lou, W.-Y., Yu, C.-Y., Wu, H., Zong, M.-H., Smith, T.J., 2012b. Use of hydrophilic ionic liquids in a two-phase system to improve mung bean epoxide hydrolases-mediated asymmetric hydrolysis of styrene oxide. J. Biotechnol. 162, 183–190.

Chiappe, C., Neri, L., Pieraccini, D., 2006. Application of hydrophilic ionic liquids as co-solvents in chloroperoxidase catalyzed oxidations. Tetrahedron Lett. 47, 5089.

Chenault, H.K., Whitesides, G.M., 1987. Regeneration of nicotinamide cofactors for use in organic-synthesis. Appl. Biochem. Biotechnol. 14, 147–197.

Da Silva, J.N.R., Nascimiento, M.G., 2012. Chemoenzymatic epoxidation of citronellol catalyzed by lipases. Process. Biochem. 47, 517–522.

Das, S., Chandrashekar, S., Yadav, J.S., Rao, A.R.V., Grée, R., 2008. An efficient process for the resolution of *cis*-4-*O*-protected-2-cyclopenten-1,4-diol using pancreatin lipase in [C_8mim][PF_6] as a reusable system. Tetrahedron Asymmetry 19, 2543–2545.

Domínguez de María, P., 2012. Ionic Liquids in Biotransformations and Organocatalysis: Solvents and Beyond. John Wiley & Sons.

Domínguez de María, P., 2008. "Nonsolvent" applications of ionic liquids in biotransformations and organocatalysis. Angew. Chem. Int. Ed. 47, 6960–6968.

Domínguez de María, P., Maugeri, Z., 2011. Ionic liquids in biotransformations: from proof-of-concept to emerging deep-eutectic-solvents. Curr. Opin. Chem. Biol. 15, 220–225.

De Diego, T., Lozano, P., Gmouh, S., Vaultier, M., Iborra, J.L., 2004. Fluorescence and CD spectroscopic analysis of the α-chymotrypsin stabilization by the ionic liquid, 1-ethyl-3-methylimidazolium bis[(trifluoromethyl)sulfonyl]amide. Biotechnol. Bioeng. 88, 916–924.

De Diego, T., Lozano, P., Gmouh, S., Vaultier, M., Iborra, J.L., 2005. Understanding structure – stability relationships of *Candida antartica* lipase B in ionic liquids. Biomacromolecules 6, 1457–1464.

Dreyer, S., Kragl, U., 2008. Ionic liquids for aqueous two-phase extraction and stabilization of enzymes. Biotechnol. Bioeng. 99, 1416–1424.

Dennewald, D., Hortsch, R., Weuster-Botz, D., 2012. Evaluation of parallel milliliter-scale stirred-tank bioreactors for the study of biphasic whole-cell biocatalysis with ionic liquids. J. Biotechnol. 157, 253–257.

Diego, T.D., Lozano, P., Abad, M.A., Stefffensky, K., Vaultier, M., Iborra, J.L., 2009. On the nature of ionic liquids and their effects on lipases that catalyze ester synthesis. J. Biotechnol. 140, 234–241.

Durand, E., Lecomte, J., Baréa, B., Piombo, G., Dubreucq, E., Villeneuve, P., 2012. Evaluation of deep eutectic solvents as new media for *Candida antarctica* B lipase catalyzed reactions. Process. Biochem. 47, 2081–2089.

Durand, E., Lecomte, J., Baréa, B., Dubreucq, E., Lortie, R., Villeneuve, P., 2013. Evaluation of deep eutectic solvent-water binary mixtures for lipase-catalyzed lyophilization of phenolic acids. Green Chem. 15, 2275–2282.

Earle, M.J., Esperanca, J.M.S.S., Gilea, M.A., Lopes, J.N.C., Rebelo, L.P.N., Magee, J.W., Seddon, K.R., Widegren, J.A., 2006. The distillation and volatility of ionic liquids. Nature 439, 831–834.

Eckstein, M., Sesing, M., Kragl, U., Adlercreutz, P., 2002. At low water activity α-chymotrypsin is more active in an ionic liquid than in non-ionic organic solvents. Biotechnol. Lett. 24, 867–872.

Eckstein, M., Villela, M., Liese, A., Kragl, U., 2004. Use of an ionic liquid in a two-phase system to improve an alcohol dehydrogenase catalysed reduction. Chem. Commun. 1084–1085.

Erbeldinger, M., Mesiano, A.J., Russell, A.J., 2000. Enzymatic catalysis of formation of Z-aspartame in ionic liquid – an alternative to enzymatic catalysis in organic solvents. Biotechnol. Prog. 16, 1129–1131.

Feher, E., Major, B., Belafi-Bako, K., Gubicza, L., 2007. On the background of enhanced stability and reusability of enzymes in ionic liquids. Biochem. Soc. Trans. 35, 1624–1627.

Fan, L.-L., Li, H.-J., Chen, Q.-H., 2014. Applications and mechanisms of ionic liquids in whole-cell biotransformation. Int. J. Mol. Sci. 15, 12196–12216.

Fan, Y., Qian, J., 2010. Lipase catalysis in ionic liquids/supercritical carbon dioxide and its applications. J. Mol. Catal. B Enzym. 66, 1–7.

Filice, M., Guisan, J.M., Palomo, J.M., 2010. Effect of ionic liquids as additives in the catalytic properties of different immobilized preparations of *Rhizomucor miehei* lipase in the hydrolysis of peracetylated lactal. Green Chem. 12, 1365–1369.

Fischer, F., Happe, M., Emery, J., Fornage, A., Schültz, R., 2013. Enzymatic synthesis of 6- and 6′-*O*-linoleyl-α-D-maltose: from solvent-free to binary ionic liquid reaction media. J. Mol. Catal. B Enzym. 90, 98–106.

Fernández-Álvaro, E., Esquivias, J., Pérez-Sánchez, M., Domíguez de María, P., Remuiñán-Blanco, M.J., 2014. Assessing biocatalysis for the synthesis of optically active tetrahydropyrazolo[1,5-α]pyrimidines (THPPs) as novel therapeutic agents. J. Mol. Catal. B Enzym. 100, 1–6.

Goldfeder, M., Fishman, A., 2014. Modulating enzyme activity using ionic liquids or surfactants. Appl. Microbiol. Biotechnol. 98, 545–554.

Gorke, J., Srienc, F., Kazlauskas, R., 2010. Toward advanced ionic liquids. Polar, enzyme-friendly solvents for biocatalysis. Biotechnol. Bioprocess Eng. 15, 40–53.

Garcia, S., Lourenco, N.M.T., Lousa, D., Sequeira, A.F., Mimoso, P., Cabral, J.M.S., Afonso, C.A.M., Barreiros, S., 2004. A comparative study of biocatalysis in non-conventional solvents: ionic liquids, supercritical fluids and organic media. Green Chem. 6, 466–470.

Gangu, S.A., Weatherley, L.R., Scurto, A.M., 2009. Whole-cell biocatalysis with ionic liquids. Curr. Org. Chem. 13, 1242–1256.

Gudiño, E.D., Iribarren, A.M., Iglesias, L.E., 2012. Regioselectivity of lipase-catalyzed deacetylation of methyl 2,3,5-tri-*O*-acetyl-α,β-D-furanosides in ionic liquid. Curr. Catal. 1, 19–23.

de Gonzalo, G., Lavandera, I., Durchschein, K., Wurm, D., Faber, K., Kroutil, W., 2007. Asymmetric biocatalytic reduction of ketones using hydroxy-functionalised water-miscible ionic liquids as solvents. Tetrahedron Asymmetry 18, 2541–2546.

Goldfeder, M., Egozy, M., Ben-Yosef, V.S., Adir, N., Fishman, A., 2013. Changes in tyrosinase specificity by ionic liquids and sodium dodecyl sulfate. Appl. Microbiol. Biotechnol. 97, 1953–1961.

Gaisberger, R.P., Fechter, M.H., Griengl, H., 2004. The first hydroxynitrile lyase catalysed cyanohydrin formation in ionic liquids. Tetrahedron Asymmetry 15, 2959–2963.

García-Álvarez, J., 2014. Deep eutectic solvents and their applications as new green and biorenewable reaction media. In: Wypych, G. (Ed.), Handbook of Solvents, Use, Health, and Environment, second ed., vol. 2. ChemTec Publishing, Toronto.

Gorke, J.T., Srienc, F., Kazlauskas, R.J., 2008. Hydrolase-catalyzed biotransformations in deep eutectic solvents. Chem. Commun. 1235–1237.

Hallett, J.P., Welton, T., 2011. Room-temperature ionic liquids: solvents for synthesis and catalysis. 2. Chem. Rev. 111, 3508–3576.

Husum, T.L., Jorgensen, C.T., Christensen, M.W., Kirk, O., 2001. Enzyme catalysed synthesis in ambient temperature ionic liquids. Biocatal. Biotransform. 19, 331–338.

Harjani, J.R., Naik, P.U., Nara, S.J., Salunkhe, M.M., 2007. Enzyme mediated reactions in ionic liquids. Curr. Org. Synth. 4, 354–369.

Hernández-Fernández, F.J., de los Ríos, A.P., Lozano-Blanco, L.J., Godínez, C., 2010. Biocatalytic ester synthesis in ionic liquids media. J. Chem. Technol. Biotechnol. 85, 1423–1435.

Hara, P., Mikkola, J.-P., Murzin, D.Y., Kanerva, L.T., 2010. Supported ionic liquids in *Burkholderia cepacia* lipase-catalyzed asymmetric acylation. J. Mol. Catal. B Enzym. 67, 129–134.

Hara, P., Hanefeld, U., Kanerva, L.T., 2009. Immobilised *Burkholderia cepacia* lipase in dry organic solvents and ionic liquids: a comparison. Green Chem. 11, 250–256.

Ha, S.H., Anh, T.V., Koo, Y.-M., 2013. Optimization of lipase-catalyzed synthesis of caffeic acid phenethyl ester in ionic liquids by response surface methodology. Bioprocess Biosyst. Eng. 36, 799–807.

Hult, K., Berglund, P., 2007. Enzyme promiscuity: mechanism and applications. Trends Biotechnol. 25, 231–238.

Humble, M.S., Berglund, P., 2011. Biocatalytic promiscuity. Eur. J. Org. Chem. 3391–3401.

Hall, M., Bommarius, A.S., 2011. Enantioenriched compounds via enzyme-catalyzed redox reactions. Chem. Rev. 111, 4088–4110.

Hinckley, G., Mozhaev, V.V., Budde, C., Khmelnitsky, Y.L., 2002. Oxidative enzymes possess catalytic activity in systems with ionic liquids. Biotechnol. Lett. 24, 2083–2087.

Hussain, W., Pollard, D.J., Truppo, M., Lye, G.J., 2008. Enzymatic ketone reductions with co-factor recycling: improved reactions with ionic liquid co-solvents. J. Mol. Catal. B Enzym. 55, 19–29.

Hayyan, M., Hashim, M.A., Hayyan, A., Al-Saadi, M.A., AlNashef, I.M., Mirghani, M.E.S., Saheed, O.K., 2013a. Are deep eutectic solvents benign or toxics? Chemosphere 90, 2193–2195.

Hayyan, M., Hashim, M.A., Hayyan, A., Al-Saadi, M.A., AlNashef, I.M., Mirghani, M.E.S., 2013b. Assessment of cytotoxicity and toxicity for phosphonium-based deep eutectic solvents. Chemosphere 93, 455–459.

Jia, R., Hu, Y., Huang, H., 2014. Improving the catalytic performance of porcine pancreatic lipase in the presence of [MMim][$MeSO_4$] with the modification of functional ionic liquids. Process. Biochem. 49, 668–672.

Jala, R.C.R., Hu, P., Yang, T., Jiang, Y., Zheng, Y., Xu, X., 2012. Lipases as biocatalysts for the synthesis of structured lipids. In: Sandoval, G. (Ed.), Lipases and Phospholipases. Methods and Protocols, Methods in Molecular Biology, vol. 861. Humana Press, New York.

Kosmulski, M., Gustafsson, J., Rosenholm, J.B., 2004. Thermal stability of low temperature ionic liquids revisited. Thermochim. Acta 412, 47–53.

Kragl, U., Eckstein, M., Kaftzik, N., 2002. Enzyme catalysis in ionic liquids. Curr. Opin. Biotechnol. 13, 565–571.

Klibanov, A.M., 2001. Improving enzymes by using them in organic solvents. Nature 409, 241–246.

Kaftzik, N., Wasserscheid, P., Kragl, U., 2002. Use of ionic liquids to increase the yield and enzyme stability in the β-galactosidase catalysed synthesis of *n*-acetyllactosamine. Org. Process. Res. Dev. 6, 553–557.

Kim, K.W., Song, B., Choi, M.Y., Kim, M.J., 2001. Biocatalysis in ionic liquids: markedly enhanced enantioselectivity of lipase. Org. Lett. 3, 1507–1509.

Kragl, U., Kaftzik, N., Schofer, S.H., Eckstein, M., Wasserscheid, P., Hilgers, C., 2001. Enzyme catalysis in the presence of ionic liquids. Chim. Oggi 19, 22–24.

Klembt, S., Dreyer, S., Eckstein, M., Kragl, U., 2008. In: Ionic Liquids in Synthesis. Wiley-VCH Verlag GmbH & Co. KGaA, pp. 641–661.

Kamal, A., Chouhan, G., 2004. Chemoenzymatic synthesis of enantiomerically pure 1,2-diols employing immobilized lipase in the ionic liquid [bmim][PF_6]. Tetrahedron Lett. 45, 8801–8805.

Kim, M.-J., Kim, H.M., Kim, D., Ahn, Y., Park, J., 2004. Dynamic kinetic resolution of secondary alcohols by enzyme-metal combinations in ionic liquid. Green Chem. 6, 471–474.

Kotlewska, A.J., van Rantwijk, F., Sheldon, R.A., Arends, I.W.C.E., 2011. Epoxidation and Baeyer-Villiger oxidation using hydrogen peroxide and a lipase dissolved in ionic liquids. Green Chem. 13, 2154–2160.

Kazlauskas, R.J., 2005. Enhancing catalytic promiscuity for biocatalysis. Curr. Opin. Chem. Biol. 9, 195–201.

Kapoor, M., Bupta, M.N., 2012. Lipase promiscuity and its biochemical applications. Process. Biochem. 47, 555–569.

Kohlmann, C., Robertz, N., Leuchs, S., Dogan, Z., Lutz, S., Bitzer, K., Na'amnieh, S., Greiner, L., 2011. Ionic liquid facilitates biocatalytic conversion of hardly water soluble ketones. J. Mol. Catal. B Enzym. 68, 147–153.

Kara, S., Schrittwieser, J.H., Hollmann, F., Ansorge-Schumacher, M.B., 2014. Recent trends and novel concepts in cofactor-dependent biotransformations. Appl. Microbiol. Biotechnol. 98, 1517–1529.

Kassim, M.A., Rumbold, K., 2014. HCN production and hydroxynitrile lyase: a natural activity in plants and a renewed biotechnological interest. Biotechnol. Lett. 36, 223–228.

Krystof, M., Pérez-Sánchez, M., Domínguez de María, P., 2013. Lipase-catalyzed (Trans)esterification of 5-Hydroxy-methylfurfural and separation from HMF esters using deep-eutectic solvents. ChemSusChem 6, 630–634.

Ludwig, R., Kragl, U., 2007. Do we understand the volatility of ionic liquids? Angew. Chem. Int. Ed. 46, 6582–6584.

Lou, W.Y., Zong, M.H., Wu, H., Xu, R., 2004. Recent progress on biocatalysis and biotransformations in ionic liquids. Chin. J. Chem. Eng. 12, 543–549.

Lou, W.Y., Xu, R., Zong, M.H., 2005. Hydroxynitrile lyase catalysis in ionic liquid-containing systems. Biotechnol. Lett. 27, 1387–1390.

Lozano, P., de Diego, T., Guegan, J.P., Vaultier, M., Iborra, J.L., 2001a. Stabilization of α-chymotrypsin by ionic liquids in transesterification reactions. Biotechnol. Bioeng. 75, 563–569.

Lozano, P., De Diego, T., Carrie, D., Vaultier, M., Iborra, J.L., 2001b. Over-stabilization of *Candida antarctica* lipase B by ionic liquids in ester synthesis. Biotechnol. Lett. 23, 1529–1533.

Lozano, P., De Diego, T., Gmouh, S., Vaultier, M., Iborra, J.L., 2005. Dynamic structure-function relationships in enzyme stabilization by ionic liquids. Biocatal. Biotransform. 23, 169–176.

Lai, J.Q., Li, Z., Lu, Y.H., Yang, Z., 2011. Specific ion effects of ionic liquids on enzyme activity and stability. Green Chem. 13, 1860–1868.

Liu, Q.B., Janssen, M.H.A., van Rantwijk, F., Sheldon, R.A., 2005. Room-temperature ionic liquids that dissolve carbohydrates in high concentrations. Green Chem. 7, 39–42.

Lozano, P., 2010. Enzymes in neoteric solvents: from one-phase to multiphase systems. Green Chem. 12, 555–569.

Lai, J.-Q., Hu, Z.-L., Sheldon, R.A., Yang, Z., 2012a. Catalytic performance of cross-linked enzyme aggregates of *Penicillium expansum* lipase and their use as catalyst for biodiesel production. Process. Biochem. 47, 2058–2063.

Liu, Y.-Y., Lou, W.-Y., Zong, M.-H., Xu, R., Hing, X., Wu, H., 2005. Increased enantioselectivity in the enzymatic hydrolysis of amino acid esters in the ionic liquid 1-Butyl-3-methyl-imidazolium tetrafluoroborate. Biocatal. Biotransf. 23, 89–95.

Lozano, P., Bernal, J.M., Navarro, A., 2012. A clean enzymatic process for producing flavour esters by direct esterification in switchable ionic liquid/solid phases. Green Chem. 14, 3026–3033.

Lue, B.M., Guo, Z., Xu, X., 2007. Lipid processing in ionic liquids. Lipid Technol. 19, 204–207.

Lai, J.-Q., Hu, Z.-L., Wang, P.-W., Yang, Z., 2012b. Enzymatic production of biodiesel in ionic liquid [bmim][PF_6]. Fuel 95, 329–333.

Lourenço, N.M.T., Monteiro, C.M., Afonso, C.A.M., 2010. Ionic acylating agents for the enzymatic resolution of *sec*-alcohols in ionic liquids. Eur. J. Org. Chem. 6938–6943.

Lahtinen, M., Viikari, L., Karhunen, P., Asikkala, J., Kruus, K., Kilpelainen, I., 2013. On the reactivity of the *Melanocarpus albomyces* laccase and formation of coniferyl alcohol dehydropolymer (DHP) in the presence of ionic liquid 1-allyl-3-methylimidazolium chloride. J. Mol. Catal. B Enzym. 85-86, 169–177.

Lindberg, D., Revenga, M.F., Widersten, M., 2010. Deep eutectic solvents (DES) are viable cosolvents for enzyme-catalyzed epoxide hydrolysis. J. Biotechnol. 147, 169–171.

Madeira Lau, R., van Rantwijk, F., Seddon, K.R., Sheldon, R.A., 2000. Lipase-catalyzed reactions in ionic liquids. Org. Lett. 2, 4189–4191.

Moon, Y.H., Lee, S.M., Ha, S.H., Koo, Y.M., 2006. Enzyme-catalyzed reactions in ionic liquids. Korean J. Chem. Eng. 23, 247–263.

Miyako, E., Maruyama, T., Kamiya, N., Goto, M., 2003. Enzyme-facilitated enantioselective transport of (*S*)-ibuprofen through a supported liquid membrane based on ionic liquids. Chem. Commun. 2926–2927.

Moniruzzaman, M., Nakashima, K., Kamiya, N., Goto, M., 2010. Recent advances of enzymatic reactions in ionic liquids. Biochem. Eng. J. 48, 295–314.

Martínez-Montero, S., Fernández, S., Sanghvi, Y.S., Gotor, V., Ferrero, M., 2011. An expedient biocatalytic procedure for abasic site precursors useful in oligonucleotide synthesis. Org. Biomol. Chem. 9, 5960–5966.

Machado, M.F., Saraiva, J.M., 2005. Thermal stability and activity regain of horseradish peroxidase in aqueous mixtures of imidazolium-based ionic liquids. Biotechnol. Lett. 27, 1233–1239.

Melgarejo-Torres, R., Torres-Martinez, D., Gutierrez-Rojas, M., de Jesus, A.G., Lye, G.J., Huerta-Ochoa, S., 2011. Regime analysis of a Baeyer-Villiger bioconversion in a three-phase (air-water-ionic liquid) stirred tank bioreactor. Biochem. Eng. J. 58-59, 87–95.

Maugeri, Z., Domínguez de María, P., 2014a. Benzaldehyde lyase (BAL)-catalyzed enantioselective C–C bond formation in deep-eutectic-solvents-buffer mixtures. J. Mol. Catal. B Enzym. 107, 120–123.

Maugeri, Z., Leitner, W., Domínguez de María, P., 2012. Practical separation of alcohol-ester mixtures using deep-eutectic solvents. Tetrahedron Lett. 53, 6968–6971.

Maugeri, Z., Leitner, W., Domínguez de María, P., 2013. Chymotrypsin-catalyzed peptide synthesis in deep eutectic solvents. Eur. J. Org. Chem. 4223–4228.

Müller, C.R., Meiners, I., Domínguez de María, P., 2014. Highly enantioselective tandem enzyme-organocatalyst crossed aldol reactions with acetaldehyde in deep-eutectic-solvents. RSC Adv. 4, 46097–46101.

Maugeri, Z., Domínguez de María, P., 2014b. Whole-cells biocatalysis in deep-eutectic-solvents/aqueous mixtures. ChemCatChem 6, 1535–1537.

Nakashima, K., Kamiya, N., Koda, D., Maruyama, T., Goto, M., 2009. Enzyme encapsulation in microparticles composed of polymerized ionic liquids for highly active and reusable biocatalysts. Org. Biomol. Chem. 7, 2353–2358.

Naik, P.U., Nara, S.J., Harjani, J.R., Salunkhe, M.M., 2007. Ionic liquid anchored substrate for enzyme catalysed kinetic resolution. J. Mol. Catal. B Enzym. 44, 93–98.

Oppermann, S., Stein, F., Kragl, U., 2011. Ionic liquids for two-phase systems and their application for purification, extraction and biocatalysis. Appl. Microbiol. Biotechnol. 89, 493–499.

Okrasa, K., Guibé-Jampel, E., Therisod, M., 2003. Ionic liquids as a new reaction medium for oxidase peroxidase-catalyzed sulfoxidation. Tetrahedron Asymmetry 14, 2487–2490.

Passos, H., Freire, M.G., Coutinho, J.A.P., 2014. Ionic liquid solutions as extractive solvents for value-added compounds from biomass. Green Chem. 16, 4786–4815.

Park, S., Kazlauskas, R.J., 2003. Biocatalysis in ionic liquids – advantages beyond green technology. Curr. Opin. Biotechnol. 14, 432–437.

Pfruender, H., Jones, R., Weuster-Botz, D., 2006. Water immiscible ionic liquids as solvents for whole cell biocatalysis. J. Biotechnol. 124, 182–190.

Patel, R., Kumari, M., Khan, A.B., 2014. Recent advances in the application of ionic liquids in protein stability and activity: a review. Appl. Biochem. Biotechnol. 172, 3701–3720.

Pinto, P.C.A.G., Costa, S.P.F., Costa, A.D.F., Passos, M.L., Lima, J.L.F.C., Saraiva, M.L.M.F.S., 2012a. Trypsin activity in imidazolium based ionic liquids: evaluation of free and immobilized enzyme. J. Mol. Liquids 171, 16–22.

Pinto, P.C.A.G., Costa, S.P.F., Lima, J.L.F.C., Saraiva, M.L.M.F.S., 2012b. β-Galactosidase activity in mixed micelles of imidazolium ionic liquids and sodium dodecylsulfate: a sequential injection kinetic study. Talanta 96, 26–33.

Papadopoulou, A.A., Katsoura, M.H., Chatzikonstantinou, A., Kyriakou, E., Polydera, A.C., Tzakos, A.G., Stamatis, H., 2013. Enzymatic hybridization of α-lipoic acid with bioactive compounds in ionic solvents. Bioresour. Technol. 136, 41–48.

Pereira, S.C., Bussamara, R., Marin, G., Giordano, R.L.C., Dupont, J., Giordano, R.C., 2012. Enzymatic synthesis of amoxicillin by penicillin g acylase in the presence of ionic liquids. Green Chem. 14, 3146–3156.

Porcar, R., Ríos-Lombardía, N., Busto, E., Gotor-Fernández, V., Gotor, V., García-Verdugo, E., Burguete, M.I., Luis, S.V., 2013. Chemoenzymatic synthesis of optically active 2-(2′- or 4′-substituted-1*H*-imidazol-1-yl)cycloalkanols: chiral additives for (L)-proline. Catal. Sci. Technol. 3, 2596–2601.

Paul, C.E., Gotor-Fernández, V., Lavandera, I., Montejo-Bernardo, J., García-Granda, S., Gotor, V., 2012. Chemoenzymatic preparation of optically active 3-(1*H*-imidazol-1-yl) cyclohexanol-based ionic liquids: application in organocatalysis and toxicity studies. RSC Adv. 2, 6455–6463.

Pinto, P.C.A.G., Saraiva, M.L.M.F.S., Lima, J.L.F.C., 2008. Oxidoreductase behavior in ionic liquids: a review. Anal. Sci. 24, 1231–1238.

Paul, C.E., Lavandera, I., Gotor-Fernández, V., Gotor, V., 2014. Imidazolium-based ionic liquids as non-conventional media for alcohol dehydrogenase-catalysed reactions. Top. Catal. 57, 332–338.

Paul, C.E., Lavandera, I., Gotor-Fernández, V., Kroutil, W., Gotor, V., 2013. *Escherichia coli*/ADH-A: an all-inclusive catalyst for the selective biooxidation and deracemisation of secondary alcohols. ChemCatChem 5, 3875–3882.

Pena-Pereira, F., Namieśnik, J., 2014. Ionic liquids and deep eutectic solvents mixtures: sustainable solvents for extraction processes. ChemSusChem 7, 1784–1800.

van Rantwijk, F., Madeira Lau, R., Sheldon, R.A., 2003. Biocatalytic transformations in ionic liquids. Trends Biotechnol. 21, 131–138.

van Rantwijk, F., Sheldon, R.A., 2007. Biocatalysis in ionic liquids. Chem. Rev. 107, 2757–2785.

Ranke, J., Stolte, S., Stormann, R., Arning, J., Jastorff, B., 2007. Design of sustainable chemical products – the example of ionic liquids. Chem. Rev. 107, 2183–2206.

Reetz, M.T., Wiesenhofer, W., Francio, G., Leitner, W., 2002. Biocatalysis in ionic liquids: batchwise and continuous flow processes using supercritical carbon dioxide as the mobile phase. Chem. Commun. 992–993.

de los Ríos, A.P., van Rantwijk, F., Sheldon, R.A., 2012. Effective resolution of 1-phenyl ethanol by *Candida antarctica* lipase B catalysed acylation with vinyl acetate in protic ionic liquids (PILs). Green Chem. 14, 1584–1588.

Ríos-Lombardía, N., Busto, E., Gotor-Fernández, V., Gotor, V., Porcar, R., García-Verdugo, E., Luis, S.V., Alfonso, I., García-Granda, S., Menéndez-Velázquez, A., 2010. From salts to ionic liquids by systematic structural modifications: a rational approach towards the efficient modular synthesis of enantiopure imidazolium salts. Chem. Eur. J. 16, 836–847.

Ríos-Lombardía, N., Porcar, R., Busto, E., Alfonso, I., Montejo-Bernardo, J., García-Granda, S., Gotor, V., Luis, S.V., García-Verdugo, E., Gotor-Fernández, V., 2011. Enantiopure triazolium salts: chemoenzymatic synthesis and applications in organocatalysis. ChemCatChem 3, 1921–1928.

Rehmann, L., Ivanova, E., Gunaratne, H.Q.N., Seddon, K.R., Stephens, G., 2014. Enhanced laccase stability through mediator partitioning into hydrophobic ionic liquids. Green Chem. 16, 1462–1469.

Rodriguez, C., de Gonzalo, G., Fraaije, M.W., Gotor, V., 2010. Ionic liquid for enhancing the enantioselectivity of isolated BVMO-catalysed oxidations. Green Chem. 12, 2255–2260.

Rodriguez, C., Lavandera, I., Gotor, V., 2012. Recent advances in cofactor regeneration systems applied to biocatalyzed oxidative processes. Curr. Org. Chem. 16, 2525–2541.

Roosen, C., Muller, P., Greiner, L., 2008. Ionic liquids in biotechnology: applications and perspectives for biotransformations. Appl. Microbiol. Biotechnol. 81, 607–614.

Sheldon, R.A., 2001. Catalytic reactions in ionic liquids. Chem. Commun. 2399–2407.

Sheldon, R.A., Madeira Lau, R., Sorgedrager, M.J., van Rantwijk, F., Seddon, K.R., 2002. Biocatalysis in ionic liquids. Green Chem. 4, 147–151.

Sate, D., Janssen, M.H.A., Stephens, G., Sheldon, R.A., Seddon, K.R., Lu, J.R., 2007. Enzyme aggregation in ionic liquids studied by dynamic light scattering and small angle neutron scattering. Green Chem. 9, 859–867.

Schofer, S.H., Kaftzik, N., Wasserscheid, P., Kragl, U., 2001. Enzyme catalysis in ionic liquids: lipase catalysed kinetic resolution of 1-phenylethanol with improved enantioselectivity. Chem. Commun. 425–426.

Sheldon, R.A., 2010. Cross-linked enzyme aggregates as industrial biocatalysts. Org. Process. Res. Dev. 15, 213–223.

Sureshkumar, M., Lee, C.-K., 2009. Biocatalytic reactions in hydrophobic ionic liquids. J. Mol. Catal. B: Enzym. 60, 1–12.

Sharma, U.K., Sharma, N., Kumar, R., Kumar, R., Sinha, A.K., 2009. Biocatalytic promiscuity of lipase in chemoselective oxidation of aryl alcohols/acetates: a unique synergism of CAL-B and [hmim]Br for the metal-free H_2O_2 activation. Org. Lett. 11, 4846–4848.

Shi, X.A., Zong, M.H., Meng, C., Guo, Y.H., 2005. Catalytic characteristics of horse liver alcohol dehydrogenase in a medium containing ionic liquid [bmim]Cl. Chin. J. Catal. 26, 982–986.

Shi, X.A., Zong, M.H., Lou, W.Y., 2006. Effect of ionic liquids on catalytic characteristics of horse liver alcohol dehydrogenase. Chin. J. Chem. 24, 1643–1647.

Shipovskov, S., Gunaratne, H.Q.N., Seddon, K.R., Stephens, G., 2008. Catalytic activity of laccases in aqueous solutions of ionic liquids. Green Chem. 10, 806–810.

Sanfilippo, C., D'Antona, N., Nicolisi, G., 2004. Chloroperoxidase from *Caldariomyces fumago* is active in the presence of an ionic liquid as co-solvent. Biotechnol. Lett. 26, 1815–1819.

Stahlberg, T., Woodley, J., Riisager, A., 2012. Enzymatic isomerization of glucose and xylose in ionic liquids. Catal. Sci. Technol. 2, 291–295.

Smith, E.L., Abbot, A.P., Ryder, K.S., 2014. Deep eutectic solvents (DESs) and their applications. Chem. Rev. 114, 11060–11082.

Teixeira, R., Lourenço, N.M.T., 2014. Enzymatic kinetic resolution of *sec*-alcohols using an ionic liquid anhydride as acylating agent. Tetrahedron Asymmetry 25, 944–948.

Tavares, A.P.M., Rodriguez, O., Macedo, E.A., 2012a. Peroxidase biocatalysis in water-soluble ionic liquids: activity, kinetic and thermal stability. Biocatal. Biotransform. 30, 417–425.

Tavares, A.P.M., Rodriguez, O., Macedo, E.A., 2008. Ionic liquids as alternative co-solvents for laccase: study of enzyme activity and stability. Biotechnol. Bioengin. 101, 201–207.

Tavares, A.P.M., Rodriguez, O., Fernandez-Fernandez, M., Dominguez, A., Moldes, D., Sanroman, M.A., Macedo, E.A., 2013. Immobilization of laccase on modified silica: stabilization, thermal inactivation and kinetic behaviour in 1-ethyl-3-methylimidazolium ethylsulfate ionic liquid. Bioresour. Technol. 131, 405–412.

Tavares, A.P.M., Pinho, B., Rodriguez, O., Macedo, E.A., 2012b. Biocatalysis in ionic liquid: degradation of phenol by laccase. Proced. Engin. 42, 226–230.

Ulbert, O., Fráter, T., Bélafi-Bakó, K., Gubicza, L., 2004. Enhanced enantioselectivity of *Candida rugosa* lipase in ionic liquids as compared to organic solvents. J. Mol. Catal. B Enzym. 31, 39–45.

Ventura, S.P.M., Santos, L.D.F., Saraiva, J.A., Coutinho, J.A.:P., 2012. Concentration effect of hydrophilic ionic liquids on the enzymatic activity of *Candida antarctica* lipase B. World J. Microbiol. Biotechnol. 28, 2303–2310.

Vidya, P., Chadha, A., 2009. The role of different anions in ionic liquids on *Pseudomonas cepacia* lipase catalyzed transesterification and hydrolysis. J. Mol. Catal. B: Enzym. 57, 145–148.

Wasserscheid, P., Welton, T., 2008. Ionic liquids in synthesis. Wiley Online Lib. 1.

Wood, N., Ferguson, J.L., Gunaratne, H.Q.N., Seddon, K.R., Goodacre, R., Stephens, G.M., 2011. Screening ionic liquids for use in biotransformations with whole microbial cells. Green Chem. 13, 1843–1851.

Wang, J., Sun, G.-X., Yu, L., Wu, F.-A., Guo, X.-J., 2013. Enhancement of the selective enzymatic biotransformation of rutin to isoquercitrin using an ionic liquid as cosolvent. Bioresour. Technol. 128, 156–163.

Wehofsky, N., Wespe, C., Cerovsky, V., Pech, A., Hoess, E., Rudolph, R., Bordusa, F., 2008. Ionic liquids and proteases: a clean alliance for semisynthesis. ChemBioChem 9, 1493–1499.

Wu, C., Zhang, Z., He, F., Zhuo, R., 2013. Enzymatic synthesis of poly(ε-caprolactone) in monocationic and dicationic ionic liquids. Biotechnol. Lett. 35, 879–885.

Wu, Q., Liu, B.-K., Lin, X.-F., 2010. Enzymatic promiscuity for organic synthesis and cascade process. Curr. Org. Chem. 14, 1966–1988.

Weibels, S., Syguda, A., Herrmann, C., Weingartner, H., 2012. Steering the enzymatic activity of proteins by ionic liquids. A case study of the enzyme kinetics of yeast alcohol dehydrogenase. Phys. Chem. Chem. Phys. 14, 4635–4639.

Wang, X.T., Yue, D.M., Zong, M.H., Lou, W.Y., 2013. Use of ionic liquid to significantly improve asymmetric reduction of ethyl acetoacetate catalyzed by *acetobacter* sp. CCTCC M209061 cells. Ind. Eng. Chem. Res. 52, 12550–12558.

Wu, H., Tian, C.Y., Song, X.K., Liu, C., Yang, D., Jiang, Z.Y., 2013. Methods for the regeneration of nicotinamide coenzymes. Green Chem. 15, 1773–1789.

Walker, A.J., Bruce, N.C., 2004a. Cofactor-dependent enzyme catalysis in functionalized ionic solvents. Chem. Commun. 2570–2571.

Walker, A.J., Bruce, N.C., 2004b. Combined biological and chemical catalysis in the preparation of oxycodone. Tetrahedron 60, 561–568.

Wagle, D.V., Zhao, H., Baker, G.A., 2014. Deep eutectic solvents: sustainable media for nanoscale and functional materials. Acc. Chem. Res. 47, 2299–2308.

Wu, B.-P., Wen, Q., Xu, H., Yang, Z., 2014. Insights into the impact of deep eutectic solvents on horseradish peroxidase: activity, stability and structure. J. Mol. Catal. B: Enzym. 101, 101–107.

Xiao, Z.-J., Du, P.-X., Lou, W.-Y., Wu, H., Zong, M.-H., 2012. Using water-miscible ionic liquids to improve the biocatalytic anti-prelog asymmetric reduction of prochiral ketones with whole cells of *acetobacter* sp. CCTCC M209061. Chem. Engin. Sci. 84, 695–705.

Yang, Z., 2009. Hofmeister effects: an explanation for the impact of ionic liquids on biocatalysis. J. Biotechnol. 144, 12–22.

Yang, Z., Yue, Y.J., Huang, W.C., Zhuang, X.M., Chen, Z.T., Xing, M., 2009. Importance of the ionic nature of ionic liquids in affecting enzyme performance. J. Biochem. 145, 355–364.

Yang, Z., Pan, W.B., 2005. Ionic liquids: green solvents for nonaqueous biocatalysis. Enzym. Microb. Technol. 37, 19–28.

Yu, X., Sun, Y., Xue, L., Huang, X., Qu, Y., 2014. Strategies for improving the catalytic performance of an enzyme in ionic liquids. Top. Catal. 57, 923–934.

Yang, Z., Huang, Z.-L., 2012. Enzymatic synthesis of sugar fatty acid esters in ionic liquids. Catal. Sci. Technol. 2, 1767–1775.

Yu, C.-Y., Wei, P., Li, X.-F., Zong, M.-H., Lou, W.-Y., 2014. Using ionic liquid in a biphasic system to improve asymmetric hydrolysis of styrene oxide catalyzed by cross-linked enzyme aggregates (CLEAs) of mung bean epoxide hydrolases. Ind. Eng. Chem. Res. 53, 7923–7930.

Yang, Z., Yue, Y.-J., Xing, M., 2008. Tyrosinase activity in ionic liquids. Biotechnol. Lett. 30, 153–158.

Zaks, A., Klibanov, A.M., 1988. Enzymatic catalysis in nonaqueous solvents. J. Biol. Chem. 263, 3194–3201.

Zhao, H., 2005. Effect of ions and other compatible solutes on enzyme activity, and its implication for biocatalysis using ionic liquids. J. Mol. Catal. B Enzym. 37, 16–25.

Zhang, Q.H., Vigier, K.D., Royer, S., Jerome, F., 2012. Deep eutectic solvents: syntheses, properties and applications. Chem. Soc. Rev. 41, 7108–7146.

Zhao, H., Olubajo, O., Song, Z.Y., Sims, A.L., Person, T.E., Lawal, R.A., Holley, L.A., 2006a. Effect of kosmotropicity of ionic liquids on the enzyme stability in aqueous solutions. Bioorg. Chem. 34, 15–25.

Zhao, H., Song, Z.Y., 2007. Nuclear magnetic relaxation of water in ionic-liquid solutions: determining the kosmotropicity of ionic liquids and its relationship with the enzyme enantioselectivity. J. Chem. Technol. Biotechnol. 82, 304–312.

Zhao, H., Campbell, S., Solomon, J., Song, Z.Y., Olubajo, O., 2006b. Improving the enzyme catalytic efficiency using ionic liquids with kosmotropic anions. Chin. J. Chem. 24, 580–584.

Zhao, H., Baker, G.A., Song, Z.Y., Olubajo, O., Crittle, T., Peters, D., 2008. Designing enzyme-compatible ionic liquids that can dissolve carbohydrates. Green Chem. 10, 696–705.

Zhao, H., 2012. Ionic liquids as (co-)solvents for hydrolytic enzymes. In: Domínguez de María, P. (Ed.), Ionic Liquids in Biotransformations and Organicatalysis. John Willey & Sons, pp. 151–227.

Zhang, Y., Wang, N., Xie, Z.-B., Zhou, L.-H., Yu, X.-Q., 2014. Ionic liquids as a recyclable and efficient medium for lipase-catalyzed asymmetric cross aldol reaction. J. Mol. Catal. B Enzym. 110, 100–110.

Zampieri, D.S., de Paula, B.R.S., Zampieri, L.A., Vale, J.A., Rodrigues, J.A.R., Moran, P.J.S., 2013. Enhancements of enantio and diastereoselectivities in reduction of (*Z*)-3-halo-4-phenyl-3-buten-2-one mediated by microorganisms in ionic liquid/water biphasic system. J. Mol. Catal. B Enzym. 85-86, 61–64.

Zhao, H., Baker, G.A., Holmes, S., 2011a. Protease activation in glycerol-based deep eutectic solvents. J. Mol. Catal. B Enzym. 72, 163–167.

Zhao, H., Baker, G.A., Holmes, S., 2011b. New eutectic ionic liquids for lipase activation and enzymatic preparation of biodiesel. Org. Biomol. Chem. 9, 1908–1916.

Zhao, H., Zhang, C., Crittle, T.D., 2013. Choline-based deep eutectic solvents for enzymatic preparation of biodiesel from soybean oil. J. Mol. Catal. B Enzym. 85-86, 243–247.

Chapter 3

Lipase Production and Purification from Fermentation Broth Using Ionic Liquids

Sónia P.M. Ventura, João A.P. Coutinho
CICECO, University of Aveiro, Department of Chemistry, Aveiro, Portugal

3.1 INTRODUCTION

The downstream processing of biotechnological products consists of different unit operations, which depend on the product characteristics, the required level of purification, and whether the product is intracellular (formed inside the cell) or extracellular (secreted into the surrounding medium). If biological cells are not involved in the production stage, the reaction medium can be directly concentrated and the target product purified; otherwise, a first step of cellular lysis may be required to release the target product, followed by a cell, or cell debris, separation.

The first step thus involves the separation of cells from the fermentation broth, which is normally carried by common techniques such as centrifugation and filtration (Krieger et al., 1999). The various stages of the downstream processing include the concentration, extraction, purification, and polishing of the compound of interest. The main difference between the product concentration and purification is that the latter involves the separation of unwanted components, which closely resemble the product in its physical and chemical form. Because the number of techniques required for the complete purification is often high and complex, this step is the most expensive in a (bio)product production. The final task is the (bio)product polishing, that is, the preparation of the purified product in a high purity level or in other words, the complete isolation of the target product from the main components/contaminants present, in a stable form, portable and convenient to use. A scheme of a bioprocess is summarized in Fig. 3.1.

In 2014, nearly 6500 enzymes were known, of which more than 200 were commercialized. The outstanding increase in the enzymes sale, allied to the improved understanding of their production biochemistry, fermentation process design, and recovery technologies, has promoted an increase in the number of

Ionic Liquids in Lipid Processing and Analysis. http://dx.doi.org/10.1016/B978-1-63067-047-4.00003-9

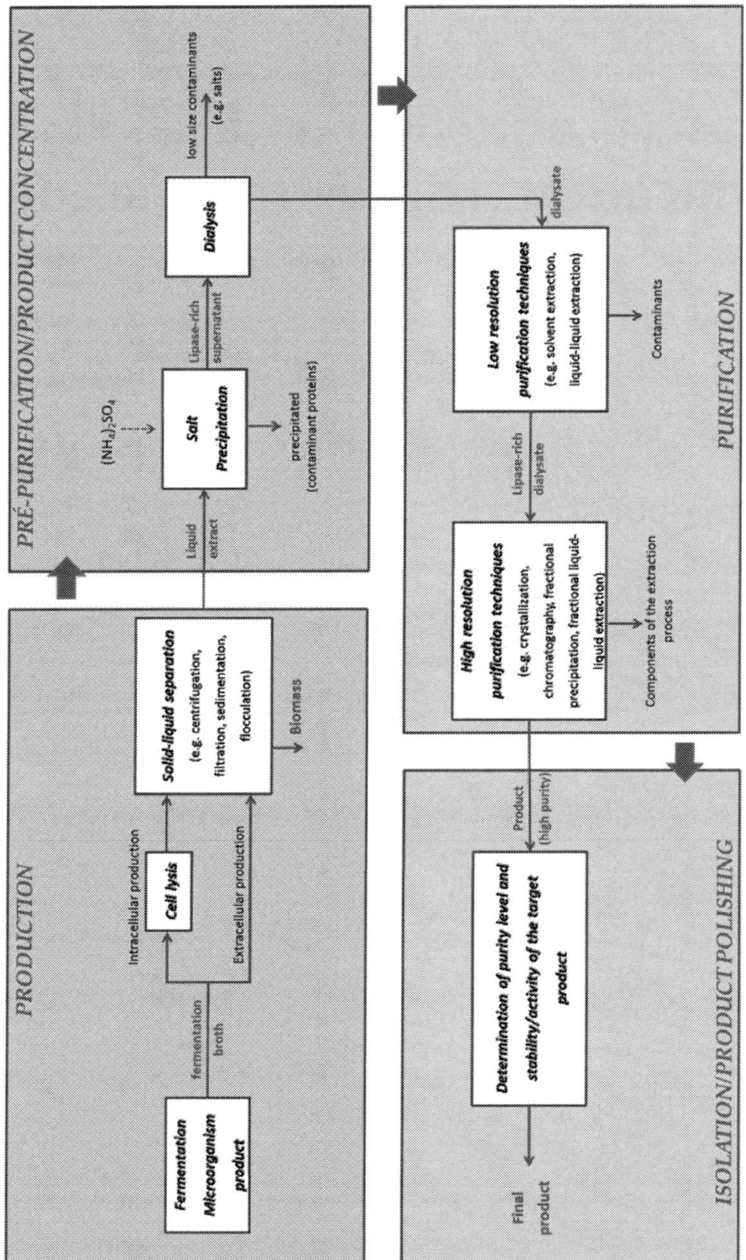

FIGURE 3.1 Scheme representing the production and purification steps of a general target microbial product.

enzymes that can be affordably produced. Lipases (triacylglycerol acylhydrolases, E.C. 3.1.1.3), which hydrolyze ester linkages of glycerides at the water—oil interface, are one of the most important and well-known classes of enzymes already in literature (data collected from the *ISI Web of Knowledge* indicates that more than 122,000 articles were published with *lipase* as one of the topics, and from this set 54 are reviews focused in the *lipase + production + purification* (Isi Web)). Considering the wide number of lipase applications (Sharma et al., 2001), namely in leather and cosmetics processing, food, animal feed, pulp and paper processing, textile industry, biotechnological and chemical fields, and their ability to catalyze a plethora of reactions in aqueous and nonaqueous media (esterification, alcoholysis, and acidolysis (Sharma et al., 2001; Otero et al., 2005; Byun et al., 2007)), the focus on these compounds has been increasing in the last years, enabling them to emerge as a key class of enzymes. This accrued interest is also related with their chemo-, regio- and enantio-specific behavior. Meanwhile, the downstream processing of lipases is equally difficult and expensive, since this step is crucial to obtain the enzymes with a high purity level while maintaining their enzymatic activity and stability behavior. The most conventional extraction technologies are still being extensively used for protein purification; however, these methodologies are characterized by a significant number of drawbacks, namely difficulties in scale-up, high production costs, and lack of suitable biocompatible solvents. These issues raise concerns not only from scientists working in the field, but also from the industries responsible for the lipase production and the companies that actually use (or could use) lipases in their processes, due to the consequent cost of this raw material. Therefore, cost-effective methods that can continuously separate, concentrate, and purify proteins are of great commercial interest. Because lipases are of microbial origin (Thakur, 2012), their production is done by means of microbial fermentation processes (Saxena et al., 2003). Most commercial applications do not require homogeneous lipase preparations, however a certain degree of purity can enable their efficient and successful usage. Despite the versatility of the lipase production and purification conditions, these steps were not studied and optimized until recently.

The traditional methods to purify macromolecules involve, as mentioned above, several stages, normally following the ammonium sulfate precipitation, dialysis, ionic and affinity chromatography and/or electrophoresis. The search for inexpensive production systems capable of producing large quantities of lipases has resulted in the development of new technology platforms using different steps of purification, based on diverse extraction methodologies. The main developments were carried taking into account some heuristic rules that will dictate the success of the extraction/purification technology applied, namely:

1. The mass transfer and equilibrium should be fast and reached by relatively low energy inputs.
2. The extraction process should be quick and highly selective.

3. The separation process should be preferentially carried out at room temperature to minimize the energy costs of the purification step.
4. Novel extraction processes should be more economical than those currently applied.
5. The purification processes should not affect the chemical structure, activity, and stability of lipases.
6. The scale-up process should be possible and predictable from the lab to the pilot and industrial scales.
7. The continuous mode should be possible to apply, aiming at promoting the processing of higher amounts of the fermentation broth and the continuous purification, causing a consequent decrease in the overall costs associated.
8. The extraction processes should be designed as much as possible to be implemented and (easily) incorporated along with the production step, aiming not only at the continuous purification but also simultaneously promoting the continuous production and purification tasks (extractive fermentation).

To better understand the idea behind this work, it is necessary to briefly distinguish the different techniques already applied in the extraction/purification of microbial lipases, mainly considering their division in technologies of low and high resolution. In this context, a brief description and analysis of the principal techniques (chromatographic processes) will be presented with the objective of highlighting the importance of seeking new alternative purification platforms for the extraction/purification of lipases.

The most common purification processes used at the time of this writing are the chromatographic and liquid–liquid extraction techniques, commonly represented by polymeric or alcoholic aqueous biphasic systems (ABS), aqueous micellar two-phase systems (AMTPS), or aqueous two-phase flotation (ATPF) systems. Some reviews (Saxena et al., 2003; Antonian, 1988; Aires-Barros et al., 1994; Palekar et al., 2000) have already described the results achieved with these various technologies and it is not our intention to extensively discuss these processes here. This work will focus on the use of ABS based in ionic liquids (ILs) on lipase purification presenting the use of IL-based ABS, their advantages, disadvantages, and the main results reported in literature.

Most lipases produced by microorganisms are extracellular, implying that, independently of the microorganism used, the step of cellular lysis is avoided and only a first step of cell separation is required, which is included in the pretreatment/prepurification of the fermentation broth (Fig. 3.1; Cramer and Holstein, 2011). This step is normally carried out by low resolution processes, such as a centrifugation or filtration, to eliminate the biomass. Following the removal of the cells, a precipitation is applied in the liquid extract, as described in Fig. 3.1, by saturation with salts (commonly ammonium sulfate) or some organic solvents (normally ethanol or acetone), aiming at the elimination of a significant fraction of contaminants, in particular proteins (Cramer and Holstein, 2011). Some authors propose the use of ultrafiltration to promote the concentration of the fermentation broth, followed by the precipitation (Palekar et al., 2000). The lipase-rich supernatant obtained is then introduced in a dialysis system to eliminate small contaminants, namely cell debris or salts used in the fermentation process and/or in the precipitation of proteins. As final output of the pretreatment task, a dialysate is obtained for posterior use in the purification step, richer in lipase, and free of the most common and highly concentrated contaminants, which will then be used in the purification process. The extent of the separation/purification processes varies with the order and resolution of each purification step. The most common techniques used in the purification are the chromatographic processes, membranes and immunopurification, and the focus of this work, ABS, AMTPS, and ATPF. Various works (including reviews (Saxena et al., 2003; Palekar et al., 2000; Gupta et al., 2004)) have already demonstrated the capabilities of those different purification processes, although the number of works dealing with the use of membranes and techniques of immunopurification (Saxena et al., 2003) is very limited.

Summing up, this chapter focuses on the description and critical analysis of the most recommended techniques already employed in the lipase production and mainly in its purification. Associated or not with the utilization of the vulgar technologies, the common liquid–liquid extraction techniques normally applied in lipase purification and their suitability are assessed and properly discussed. This chapter then discusses improvements in the performance of the liquid–liquid technologies by including ILs as separation and extraction agents. Finally, a critical analysis based on the main conclusions will be presented and future perspectives and challenges will be addressed.

3.2 COMMON METHODS OF LIPASE EXTRACTION

3.2.1 Common Separation Techniques

In most lipase purification plants, chromatography is used to achieve the level of purity required, which is entirely dependent on the final application. Hence, the use and combination of different chromatographic processes may be considered (Krieger et al., 1999). Ion exchange chromatography, gel filtration, and affinity chromatography are three of the best known and most applied chromatographic techniques (Table 3.1). The ion exchange practice (Krieger et al., 1999; Veeraragavan et al., 1990) is extensively used, due to its high applicability.

Other strong ion exchanger groups and Q-Sepharose (Palekar et al., 2000) were also reported. The ion exchange is most commonly used due to its high capacity for loaded protein (Pimentel et al., 1994). This technique is normally

TABLE 3.1 Chromatography Techniques Commonly Applied in the Protein Purification According to Their Main Properties

Properties	Technique
Size	Gel filtration (size exclusion) chromatography
Charge	Ion exchange chromatography
Hydrophobicity	Hydrophobic interaction chromatography
	Reversed phase chromatography
Ligand specifity	Affinity chromatography

characterized by the great degree of lipase purification, due to the establishment of strong electrostatic interactions between the enzyme and the gel (Krieger et al., 1999). In the work reported by Krieger et al. (1999), the lipase was recovered in the form of a high molecular aggregate after gel filtration chromatography, which is the second chromatographic technique most investigated in terms of purification, sometimes being used more than once per process. Schmidt-Dannert et al. (1994, 1996) started with the analysis of a lipase from the thermophilic *Bacillus thermocatenulatus* (DMS 730) by the application of some chromatographic techniques, namely the ion exchange chromatography on Q-sepharose. The authors concluded that by applying one step of hexane extraction, methanol precipitation, and ion exchange chromatography, the lipase was 67-fold more concentrated (Schmidt-Dannert et al., 1994). Some other works (Veeraragavan et al., 1990; Schmidt-Dannert et al., 1996; Borkar et al., 2009; Kumar et al., 2012; Ghanem et al., 2000; Imamura and Kitaura, 2000; Litthauer et al., 2002; Snellman et al., 2002; Abdou, 2003; Taipa et al., 1992) appeared with similar approaches using chromatographic techniques; the results were considered by the authors a success (purification factors using the ammonium sulfate precipitation and different chromatography schemes between 10- and 3028-fold). Gel filtration and affinity chromatography are also important processes used in more than 60% and 27% of the purification apparatus, respectively (Saxena et al., 2003; Palekar et al., 2000; Gupta et al., 2004). The number of works dealing with these techniques is significant, with the chromatographic techniques being used as purification platforms, principally for biopharmaceuticals (some of them also enzymes) where the level of purity required is normally extremely high and the techniques are well described, and for which the knowledge generated allows the manipulation of the process even at large scales.

However, for the lipase, a different scenario is found; the usual procedures are of deficient performance, mainly because they promote a decrease in the lipase activity (Gupta et al., 2004), they are difficult to manipulate, principally due to the high number of chromatographic units connected in the same process, which results in time-consuming processes with low final yields. The affinity chromatography can be applied in an early stage of the purification process, however the materials are expensive and thus, gel filtration or ion exchange chromatography are preferred due to the lower costs associated with their consumables. Moreover, although gel filtration represents lower purification capacities for loaded proteins, it can be applied in both cases, as an initial step of concentration or in the final contribution for the product polishing.

3.2.2 Liquid–Liquid Extraction

The liquid–liquid extraction technologies described here are represented by three particular types of ABS, particularly those based in alcohols, polymers (systems normally called by ABS), copolymers (systems normally referred to as aqueous two-phase flotation or ATPF), or surfactants (used in the study of aqueous micellar two-phase systems—AMTPS). ABS are normally described as systems formed as the result of the incompatibility between aqueous solutions of two polymers, or a polymer and a salt of high ionic strength. The works on the subject suggest that when the two polymers (most often one of them being polyethylene glycol (PEG) (Saxena et al., 2003; Molino et al., 2013; Mazzola et al., 2008; Hamel and Hunter, 1990; Raja et al., 2011)) are mixed, large aggregates are formed, their affinity for the water molecules is changed, and thus, the polymers tend to separate into two different aqueous phases due to steric exclusion. The same exclusion behavior can be achieved when one polymer is mixed with a high concentration of salt since the salt will interact with the water molecules present, promoting in the same way the phase separation (Diamond and Hsu, 1992). With promising results found for these techniques when applied in the extraction/purification of proteins, these systems were widely investigated and explored, considering not only the system components, but also the main conditions applied in the purification processes, namely temperature, pH, and concentration of the main components used in the preparation of the extraction system (Zaslavsky et al., 1986). The ABS are thermodynamically described by the phase diagrams, presenting the binodal curves and the respective tie-lines (Mazzola et al., 2008).

The binodal curve (red line in Fig. 3.2) represents the borderline between the monophasic and biphasic regions, which means that from these data it is possible to determine the entire region with extraction potential (biphasic). The tie-lines (black lines in Fig. 3.2) allow the description of the compositions of the phases (top and bottom) in equilibrium. The phase diagrams can be used for two different purposes/interpretations. The thermodynamic approach is the first, and the second, no less important, is the interpretation from the point of view of the partition/purification of molecules. The first idea that should be

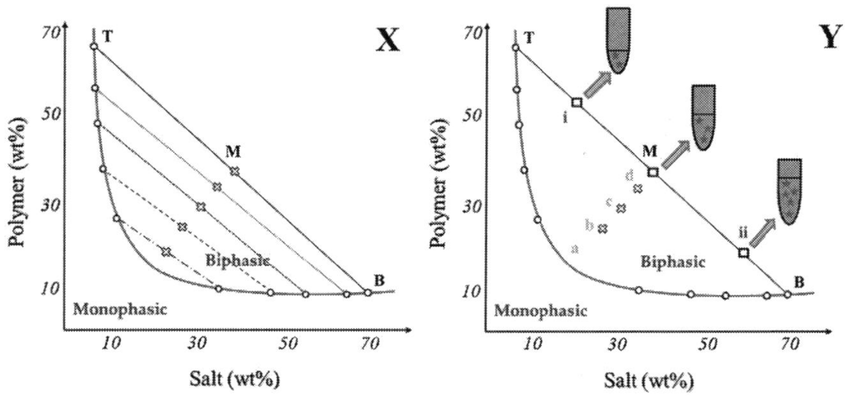

FIGURE 3.2 Illustration of a general phase diagram in mass fraction percentage (wt%) composed of a generic polymer and a salt. In the system, the binodal curve (*red line* (dark gray in print versions)) and different tie-lines (*black lines*) are described for a fixed temperature. (X) In the phase diagram ***T*** represents the top phase composition, ***B*** the bottom phase composition, and ***M*** the overall system composition. (Y) The small letters ***i*** and ***ii*** represent extraction points (biphasic point) in the same tie-line of ***M*** mixture point, ***a***, ***b***, ***c***, and ***d*** also represent extraction points, but now corresponding to different tie-lines.

clarified is that each point in the biphasic region is different and is part of an extensive number of potential points of extraction with different properties and characteristics, which leads to a large range of conditions to be applied in the extraction of a target molecule. Specifically, two distinct scenarios can occur when discussing a phase diagram. First of all, if different mixture points are considered randomly in the biphasic region (points ***a***, ***b***, ***c***, ***d***, ***M***), different extraction systems will be formed, as well as different tie-lines, which means that distinct ABS are formed with distinct phase compositions. From the point of view of the extraction process, different extractions are promoted, with different partition and purification phenomena of the solute occurring, since the interactions in both phases and between all system components are different. On the other hand (second scenario), if the same tie-line is considered (Fig. 3.2Y) and different points are chosen (points ***M***, ***i***, and ***ii***), distinct extraction points are obviously formed and characterized by the same top and bottom phases in terms of compositions. In this second scenario, the extraction of one solute is also promoted, but the interaction between all components and the solute will be the same.

In both scenarios some consideration should be taken into account: first, these rules are applied if the same solute is being considered or if the same conditions of temperature and pH are promoted; otherwise, different interactions and (entropic or hindrance) effects can occur (Willauer et al., 2002). Second, the same components can be used, meaning the same pairs of polymer/polymer, polymer/salt, alcohol/salt, IL/salt, or IL/polymer; if not, different phase diagrams will be investigated and consequently, a different response will be achieved in terms of partition/purification process. Third, it is mandatory to take into account the complexity of the sample feed where the solute of interest is present (in particular when the extraction or purification represents the capture of the solute from the fermentation broth), since the balance of interactions between the target molecule to be purified, the main contaminants, and each one of the components of the extraction system may be different, consequently imposing new rules in the choice of the ABS type and the extraction point in the biphasic region. Summing up, different conditions, components, and solutes may give a distinct response in terms of partition, concentration, and purification, which is normally pointed out as one of the most important reasons for the use of ABS as part of downstream processing or as downstream processes (Molino et al., 2013). Moreover, ABS are generally pointed out as processes capable of providing a mild environment (meaning higher amounts of water), they are nonvolatile, and they require the use of (sometimes) cheap components, allowing an easy scalability. However, they continue to be underestimated as relevant separation techniques by the academic and industrial communities, mainly due to the system complexity (at least three components to form the system), and poor understanding of the partition mechanisms (several interactions controlling the partition phenomenon, few techniques providing a deep and fundamental analysis of those interactions, allowing a complete and controlled manipulation of this extraction platform), leading to a labor-intensive, time-consuming, and high-cost process development (Rosa et al., 2010).

The discrepancies between opinions of different authors concerning the use and applicability of ABS are apparent. Notwithstanding, as shown in Fig. 3.3, the number of studies using these systems, and thus their potential applications, is increasing, which means that the confidence of the scientific community in their potential and applicability keeps growing, and this explains why the number of relevant publications keeps increasing each year. Fig. 3.3 gives evidences that the number of articles dealing with ABS in different fields is significant (1338 reported in the *ISI Web of Knowledge* by July, 2014). These works report different analysis, from the fundamental (based principally in the intermolecular interactions) point of view, passing through the thermodynamic analysis, and culminating in the study and design of distinct applications. The domain of application is one of the key fields, because the importance of the fundamental point of view and the thermodynamic design appears to understand how to manipulate only the systems having as background the final application. The idea of describing a new methodology is of crucial importance, the capacity of the user to properly control the technology is of extreme importance but only taking into consideration the success of the application. In the field of ABS the set of applications is summarized in three principal domains: as product concentration, extraction, and purification methodologies (Molino et al., 2013; Freire et al., 2012; Goja et al., 2013).

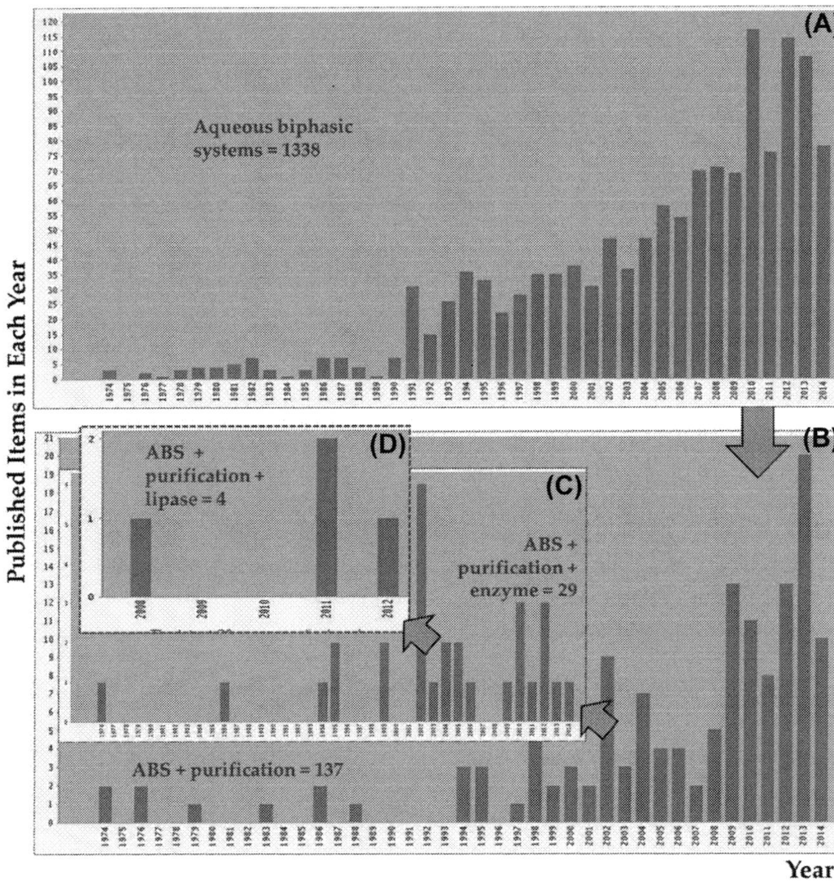

FIGURE 3.3 Brief search in the *ISI Web of Knowledge* on the domains: (A) Aqueous biphasic systems (1338 works); (B) Aqueous biphasic systems + purification (137 works); (C) Aqueous biphasic systems + purification + enzyme (29 works); (D) Aqueous biphasic systems + purification + lipase (4 works). The graphics exposed show the evolution of the ABS field since 1974 and demonstrate the development considering the technique and its application in different fields (1338 works), including the extraction of enzymes (29 works) and, in particular, lipases (4 works).

Considering the purification, a vast range of products have been studied, namely virus, biopharmaceuticals, genetic material (eg, nucleic acids), pollutants (eg, endocrine disruptors), added-value compounds from terrestrial and marine raw materials (eg, fucoxanthin and R-phycoerythrin from brown and red macroalgae), therapeutic proteins, and finally, microbial enzymes (Molino et al., 2013; Freire et al., 2012; Goja et al., 2013). Currently those are the main focus on the use of ABS basically because they are describing compounds whose purification is still very complicated and expensive, imposing the demand for other technologies with high efficiency and capable of maintaining the activity and chemical structure of the products purified. Most reports in the field describe the use of polymeric systems, where the impact of distinct polymers, their molecular weights, polymer + polymer and polymer + salt combinations, temperatures, and pH values were evaluated (Aires-Barros et al., 1994; Gupta et al., 2004; Molino et al., 2013; Raja et al., 2011; Rosa et al., 2010, 2011).

The partitioning of enzymes in ABS depends on (1) (intrinsic conditions) the physicochemical properties of the solute (eg, protein hydrophobicity/hydrophilicity, charge, and size), and (2) (extrinsic conditions) the ABS type, components, and concentrations of the components applied in the extraction. A plethora of ABS will be shown and discussed in this work. Their selection was based on systems already applied in terms of lipase purification (Zhang and Liu, 2010; Ooi et al., 2009a; Bassani et al., 2010; Zhou et al., 2013; Khayati and Alizadeh, 2013; Li et al., 2010; Ooi et al., 2011a). Thus, ABS with alcohols (Table 3.2), polymers (Table 3.3), surfactants (Table 3.4), and copolymers (Table 3.5) will be reviewed here.

3.2.2.1 Alcohol/Salt-Based ABS

Ooi et al. (2009b) have applied for the first time ABS based in alcohols and salt systems to recover a lipase derived from *Burkholderia pseudomallei*. In this work, nine ABS comprised of ethanol, 2-propanol and 1-propanol; and ammonium sulfate, potassium phosphate, and sodium citrate, were evaluated for their effectiveness/success in recovering the lipase. The lipase from *B. pseudomallei* was successfully purified using a 2-propanol/potassium phosphate ABS in a single-step procedure. A purification factor of 13.5 and a yield of 99% were achieved, without compromising the enzyme activity.

TABLE 3.2 Extraction and Purification of Lipase Using Alcohol-Based ABS

Lipase derived from *Burkholderia pseudomallei* (Ooi et al., 2009b)	
Phase former agents	1-propanol, 2-propanol, ethanol + potassium phosphate, ammonium sulfate, sodium sulfate, NaCl (additive)
Conditions studied	Concentration of the phase former agents Addition of various concentrations of NaCl (additive)
Advantages	Easy recovery of the alcohol, lower costs, enzyme activity is maintained
Disadvantages	Lower purification factors
Maximum extraction and purification performance	PF = 13.5, yield = 99%

TABLE 3.3 Extraction and Purification of Lipase Using Polymer-Based ABS

Intracellular lipase from *Trichosporon laibacchii* (Zhang and Liu, 2010)	
Phase former agents	PEG, K_2HPO_4, NaCl (additive)
Conditions studied	PEG molecular weight, different PEG 4000/K_2HPO_4 combinations, pH
Advantages	Combination of extraction and enzyme immobilization approaches Ease of scale-up Possible recycling of the phase components
Disadvantages	20% of enzyme activity loss
Maximum extraction and purification performance	$K = 7.61$; PF = 5.84
Lipase from *B. pseudomallei* (Ooi et al., 2009a)	
Phase former agents	PEG, K_2HPO_4, NaCl (additive)
Conditions studied	PEG molecular weight, phase composition, tie-line length, volumetric ratio, feed of crude fermentation broth, pH
Advantages	Single-step approach
Disadvantages	
Maximum extraction and purification performance	PF = 12.42; yield = 93%
Lipase from *B. pseudomallei* (Ooi et al., 2011a)	
Phase former agents	PEG, dextran T500, ammonium sulfate, sodium phosphate, magnesium sulfate, potassium phosphate, sodium citrate
Conditions studied	PEG molecular weight, PEG concentration
Advantages	Coupling the upstream and the downstream by applying ABS; extractive fermentation approach was capable of maintaining the viability of the cells in 7 cycles of the repetitive ATPS fermentation Effective approach in terms of process costs and processing time
Disadvantages	Only dextran T500 is suitable to perform the extraction
Maximum extraction and purification performance	Yield = 92.1% (single step)
***C. rugosa* (Lip1) (Bassani et al., 2010)**	
Phase former agents	PEG, potassium phosphate salt, NaCl (additive)
Conditions studied	PEG molecular weight, PEG concentration, temperature, NaCl concentration, volumetric ratio, feed of crude fermentation broth

TABLE 3.3 Extraction and Purification of Lipase Using Polymer-Based ABS—cont'd

Advantages	The polymer is not promoting any changes in the secondary and tertiary structures or in the enzyme activity
Disadvantages	Lower purification factors
Maximum extraction and purification performance	Enzyme recovery = 78.3%, PF = 2.3
Porcine pancreatic lipase (PPL) (Zhou et al., 2013)	
Phase former agents	PEG, potassium phosphate dibasic, potassium phosphate buffer
Conditions studied	PEG molecular weight, PEG and salt concentration, temperature
Advantages	
Disadvantages	
Maximum extraction and purification performance	$K = 12.7$, extraction efficiency = 94.7%, PF = 4
Lipase from *R. aglutinis* (Khayati and Alizadeh, 2013)	
Phase former agents	PEG, potassium oxalate, potassium sodium tartrate
Conditions studied	PEG molecular weight, salt, feed of crude fermentation broth
Advantages	
Disadvantages	
Maximum extraction and purification performance	PF = 13.9, yield = 71.2%
Porcine pancreatic lipase (Li et al., 2010)	
Phase former agents	P_{NNC}, P_{ADB}
Conditions studied	P_{NNC} concentration, pH, neutral salts (additives) and their concentrations, pure and crude lipase
Advantages	The recycling of the phase former agents is performed by pH adjustments
Disadvantages	
Maximum extraction and purification performance	PF = 18, recovery of lipase = 83.7%

TABLE 3.4 Extraction and Purification of Lipase Using AMTPS

Lipases A and B from *Chromobacterium viscosum* (Vicente et al., 1990)	
Phase former agents	AOT-based reverse micelles, benzene as the organic phase
Conditions studied	Single step extraction of lipase A, back extraction of lipase B, number of steps
Details of the process	Two lipases were removed
Maximum extraction and purification performance	Lipase A: PF = 4.3 and enzyme recovery = 91% Lipase B: PF = 3.7 and enzyme recovery = 76%
Yeast-lipase (Yu et al., 2003)	
Phase former agents	AOT-based reverse micelles, isooctane as the organic phase
Conditions studied	pH, AOT concentration, salt concentration (ionic strength), phase–volume ratio, stirring time
Details of the process	In this work the back extraction of the lipase was studied keeping into consideration the following conditions: Ethanol (cosolvent) concentration, pH, stirring time, and pH
Maximum extraction and purification performance	Enzyme extraction = 100% Back-extraction: Yield = 68%
Lipase from *Aspergillus niger* (Nandini and Rastogi, 2010)	
Phase former agents	CTAB, isooctane, potassium phosphate buffer
Conditions studied	Forward and backward extraction were studied, taking into account the salt and surfactant concentrations and salt concentration and pH, respectively
Details of the process	Both forward and backward extractions were studied and optimized
Maximum extraction and purification performance	Lipase activity recovery ≥ 78%, PF ≥ 4.0
Lipase from the *Burkholderia* sp. ST8 strain (Ooi et al., 2011b)	
Phase former agents	Triton X-114, Pluronic L31, Pluronic L61, Pluronic L81, Pluronic L121
Conditions studied	Pluronic L81 concentration, addition of additives (salts), back extraction of lipase
Details of the process	The cloud-point temperatures were determined for the surfactant and copolymers
Maximum extraction and purification performance	PF = 7.2 single step of purification, polishing step yield = 89%, K between 0.34 and 4.5

TABLE 3.5 Extraction and Purification of Lipase Using ATPF Systems	
Lipase from *Burkholderia cepacia* strain ST8 (Show et al., 2011, 2012, 2013)	
Phase former agents	EOPO copolymer
Conditions studied	Polymer molar mass, ammonium sulfate concentration, pH, the amount of the loaded crude feedstock, initial volume of EOPO phase situated in the top of the aqueous solution, concentration of EOPO, initial volume of aqueous phase, nitrogen flow rate, flotation time
Details of the process	The recycling of the components was also studied
Maximum extraction and purification performance	Extraction efficiency = 99%

The authors believe that the use of alcohol/salt-based ABS proved to be effective for the purification of solvent-tolerant lipase, not only because of the significant purification parameters obtained without compromising the enzyme activity and structure, but also because the recovery of the alcohol is easy, potentially decreasing the costs associated with the entire process of production and purification. Unfortunately, no more studies were developed and of course, the effect of other process conditions remain unknown, namely, pH values and temperatures of the extraction system, the isolation of the lipase, and the scale-up of the process, making possible the study and implementation of the commercial recovery process.

3.2.2.2 Polymer/Salt-Based ABS

Zhang and Liu (2010) reported the partial purification of an intracellular lipase from *Trichosporon laibachii* by applying different polymeric-based ABS. In this work, the impact of different conditions, in particular the PEG molecular weight, the system's phase compositions, and the pH were evaluated on the lipase partition. The best results obtained in this work reported a partition coefficient of 7.61, with an activity recovery of 80.4%, and a purification factor of 5.84 (PEG 4000 (12 wt%) + K_2HPO_4 (13 wt%), at pH 7 and with 2.0 wt% of NaCl). In this work, a new approach was reported consisting of the combination of ABS with the enzyme immobilization. The main results suggest that this new integrated methodology can be an advantageous and useful technique for the purification of lipases (the in situ immobilization of the lipase in the PEG phase resulted in a highest immobilized lipase activity of 1114.6 U/g). The authors (Zhang and Liu, 2010) claim that this process can be considered cost-effective as well as time-saving, because it is fast and simple, involving a small number of steps (the immobilization step is directly performed in the top PEG-rich phase to achieve an in situ immobilization).

Zhang and Liu (2010) ended their analysis calling the attention to the relative easiness of the scale-up process and possible recycling of the phase components, evidencing that the whole process can be designed to be cost-effective.

In the same line of work, Ooi et al. (2009a) have determined the effect of different ABS based in distinct PEG polymers in terms of their purification capacity. The main focus of this work was the direct purification of a lipase produced by *B. pseudomallei*. The migration behavior of the lipase was observed taking into consideration several parameters such as the phase composition, the tie-line length, the volumetric ratio (V_r), the feed of crude fermentation broth, the system pH and, finally, the addition of neutral salts. The optimum conditions for the direct purification of lipase from the fermentation broth were optimized as follows: ABS based in PEG 6000 + potassium phosphate system at pH 7; a TLL of 42.2 wt%, allowing the fixation of the compositions of the phases in terms of all components and, consequently, of the interactions; a V_r of 2.70, where the same components' compositions in both top and bottom phases and the reconcentration of the lipase in the polymeric phase are achieved; 1 wt% of NaCl used as additive (acts in the polarity of the phases, allowing the differential migration of the lipase); for 20 wt% of crude load. Based on this system, the maximum lipase purification factor achieved was 12.42-fold, with a yield of 93%. The same group developed (Ooi et al., 2011a) an extractive fermentation process for the same lipase, using polymeric-based ABS. This process consisted in the simultaneous development of the cell cultivation and the downstream processing of the extracellular lipase derived from *B. pseudomallei*, in two distinct partitioning systems. The best results found for this integrated process were achieved for the ABS composed of 9.6 wt% of PEG 8000 and 1.0 wt% of dextran T500. The process allowed the biomass accumulation in the bottom phase whereas the lipase was preferentially concentrated into the top phase. A yield of 92.1% was achieved in a single step. The potential of coupling the upstream fermentation with the downstream processing by applying ABS is very attractive, since the purification of the lipase in repeated cycles of extractive fermentation can be achieved, which is ideally effective in terms of process costs (recycling of the cells/biomass for subsequent fermentations) and processing time (the simultaneous purification during batch fermentation was promoted).

Bassani et al. (2010) have determined the interaction between a lipase from *Candida rugosa* (Lip1) and PEG polymers of different molecular weights, through the utilization of fluorescence and circular dichroism. These studies were then applied to the analysis of the enzyme partition mechanism by applying distinct ABS based in PEG (2000, 4000, 8000, and 10,000) and the potassium phosphate salt (at pH 7). The main results presented in this work (Bassani et al., 2010) describe a decrease of the partition coefficients with the PEG molecular weight showing that the enzyme migration is driven by the excluded volume effect and not by the unfavorable Lip1−PEG interactions (notice the preferential migration of the lipase for the PEG-rich phase). The presence of the polymer was not responsible for any significant changes in the secondary and tertiary lipase chemical structures or in its biological activity. Lip1 was preferentially concentrated in the PEG-rich layer being the PEG 2000, the system in which the best enzyme recovery (78.3%) and the highest purification factor (2.3) were achieved (Bassani et al., 2010).

Zhou et al. (2013) have also studied the use of polymeric-based ABS for the purification of porcine pancreatic lipase (PPL) from crude PPL using PEG and the inorganic salt potassium phosphate dibasic or potassium phosphate buffer (pH 7). Behind the phase diagrams study at room temperature, the authors have evaluated the preferential partition of PPL for the PEG-rich phase, and have concluded that the preference of the lipase for the more hydrophobic phase was independent of the molecular weight of the PEG, when PEG 1000 and PEG 1500 were applied. However, for systems based in PEG 4000, the lipase was more concentrated in the phosphate-rich phase, indicating that important interactions between the lipase and the salt and/or water were created. The authors (Zhou et al., 2013) concluded that at low PEG molecular weight, the PEG-lipase interactions controlled the partition phenomenon, but at high PEG molecular weights, the exclusion phenomenon promoted by the polymer is more important than the polymer—lipase interaction, since the lipase was excluded from the polymer-rich phase. The authors mentioned that the enzyme was efficiently purified in PEG 1500/potassium phosphate (17/13 wt%) systems at pH 7 and 4°C, with an enzyme partition coefficient of 12.7, an extraction efficiency of 94.7%, and a purification factor around 4.

In the same year Khayati and Alizadeh (2013) studied the application of ABS, again based in PEG and various salts (potassium oxalate and potassium sodium tartrate), at pH 6.6 and 24 °C, and directly applied them to the cell-free fermentation broth containing the lipase of interest produced by *Rhodotorul aglutinis*. The most appropriate system to perform the purification of the lipase from the fermentation broth (12.5%, v/w) was composed by PEG 4000 (17.5 wt%) + potassium oxalate (12.5 wt%), being the enzyme partitioned for the PEG (top)-rich phase with a maximum purification factor of 13.9 and lipase yield of 71.2%. The authors concluded that the partition coefficient of the lipase (K_e) decreased with the PEG molecular weight increase. Having these results in mind, it is possible to conclude that probably the fermentation contaminants are somehow forcing the lipase salting-out from the phase where the main contaminant proteins are presented, which means the increase of the purification factors, when these are compared with the purification factors obtained in previous works.

One key problem of the ABS is that the phase-forming polymers could not be efficiently recycled (Li et al., 2010). The lack of efficient recycling strategies imposes high costs to the process and increased environmental pollution. In this context, Li et al. (2010) studied the use of ABS composed by the pH-sensitive copolymer, the P_{ADB}, and the light-sensitive copolymer, P_{NNC}. The top phase is rich in P_{NNC} and the bottom phase is enriched in P_{ADB}. In this work, the impact of various process parameters, such as the concentration of the phase-forming copolymers, the system pH, and the different types and concentrations of salts on the migration profile of the lipase were evaluated. Considering the main results of the direct lipase purification from the crude material, it was found that the lipase was purified from crude material with 83.7% of recovery and a purification factor of approximately 18-fold (Li et al., 2010). The authors concluded that the lipase partition coefficient could be effectively controlled by using some inorganic salts at certain concentrations. Because those systems were designed by using a light-sensitive copolymer (P_{NNC}) and a pH-sensitive copolymer (P_{ADB}), the recycle of P_{ADB} was carried out by adjusting the pH of the system to its isoelectric point ($pI = 4.1$). On the other hand, the light-sensitive P_{NNC} was eliminated by laser irradiation at 488 nm (Li et al., 2010). In conclusion the authors proposed novel ABS with potential application in the biotechnological industry in particular, as a downstream technology, with reduced costs and limiting any environmental impact.

3.2.2.3 Surfactant-Based ABS: Aqueous Micellar Two-Phase Systems

The liquid–liquid extraction process by applying reverse micelles consists of two fundamental steps: (1) a forward extraction in which the lipase is transferred from an aqueous solution into a reverse micellar organic phase and (2) a back extraction process, where the protein is released from the reverse micelles and transferred into an aqueous phase to be recovered (Taipa et al., 1992; Lee and Chong, 2011). The general process is described by the entrapment of the lipase units in the water pools inside the micelle structure, thus avoiding the direct contact of the macromolecule with the organic solvent, and avoiding one of the most important sources of nonfavorable interactions with the enzyme responsible for its denaturation. In this context, some authors truly believe that this technology can be extensively applied as a downstream process (Molino et al., 2013; Basheer and Thenmozhi, 2010). In 1990, Vicente et al. studied the selective separation and purification of a lipolytic preparation from *Chromobacterium viscosum* by applying AOT-based reverse micelles with benzene as the organic phase. During the production via fermentation, two lipases were produced; lipase A was purified from the original crude enzyme preparation by 4.3-fold with a recovery of 91% and lipase B by 3.7-fold with a recovery of 76%. Later, Yu et al. (2003) used sodium bis(2-ethylhexyl) sulfosuccinate (AOT) reverse micelles in isooctane for the separation of yeast-lipase by a two-step procedure. Different conditions were used in this study, namely the pH and ionic strength of the aqueous phase, the surfactant concentration, the phase volume ratio, the temperature, the stirring time, and finally the cosolvent concentration used in the back-extraction of the lipase. The complete extraction (100%) of the lipase was achieved in the forward extraction using 250 mmol/

L of AOT (surfactant), 0.05 mol/L of KCl in the initial aqueous phase, a stirring time of 10 min, and room temperature. The back extraction was also optimized, resulting in 68% of back extraction yield and 45% of activity recovery yield for the lipase, obtained for 0.5 mol/L of KCl in the second aqueous phase, 3% (v/v) of ethanol, a stirring time ranging from 15 to 30 min at room temperature, and pH 8.0 (Yu et al., 2003).

Later in 2010, Nandini and Rastogi investigated AMTPS based in the cationic surfactant cetyltrimethylammonium bromide (CTAB). They studied the effect of various process parameters on both forward (salt and surfactant concentrations) and backward (salt concentration and pH) extraction processes of a lipase from its crude extract. Both the yield and purification parameters were optimized using a response surface methodology. The results showed that the optimum processing conditions were salt concentration of 0.16 M, surfactant concentration of 0.20 M, and pH 9 for forward extraction, and salt concentration of 0.80 M and pH of 7.23 for the backward extraction, leading to a value of lipase activity recovery ($\geq$78%) and a purification factor $\geq$ 4.0 (Nandini and Rastogi, 2010).

Ooi et al. (2011b) have studied the efficiency of a temperature-induced AMTPS composed of a single nonionic surfactant (Triton X-114) and Pluronics (triblock copolymers) on the partitioning and recovery of a lipase from the *Burkholderia* sp. ST8 strain. In this work, the clouding phenomenon (cloud-point temperature) and the performance of the lipase partitioning in these AMTPS were the principal parameters investigated. Pluronic L81 showed the best results in terms of the lipase partition efficiency to the micelle-rich phase. Based on AMTPS composed of 24 wt% of Pluronic L81 and 0.5 wt% of potassium chloride (KCl), a purification factor of 7.2 was achieved. Meanwhile, the lipase was consecutively extracted from the micelle-rich phase to a new aqueous solution aiming to eliminate the presence of the surfactant from the lipase bulk, as part of the polishing approach, by applying a potassium thiocyanate (KSCN) solution. The yield of the polishing step obtained was around 89% and the partition coefficients of 0.34 and 4.50 were achieved for lipase and surfactant, respectively. The separation of the surfactant from the lipase was achieved ($K_{surfactant} = 4.50$) and its back extraction was accomplished in 80% (Ooi et al., 2011b).

3.2.2.4 Aqueous Two-Phase Flotation Systems

ATPF is a purification methodology, which combines the use of ABS and solvent sublation (Bi et al., 2010a,b). This technique is based on the use of surface-active compounds with a hydrophilic group (hydroxyl or glucosan) and a hydrophobic group (phenyl or alkyl) in water, which are adsorbed to the surface of nitrogen bubbles of an ascending gaseous stream. The bubbles are then dissolved in the polymer placed on the top of the aqueous solution as depicted in Fig. 3.4 (scheme adapted from (Bi et al., 2010a; Show et al., 2011)). The main advantages of the technique are related with the high

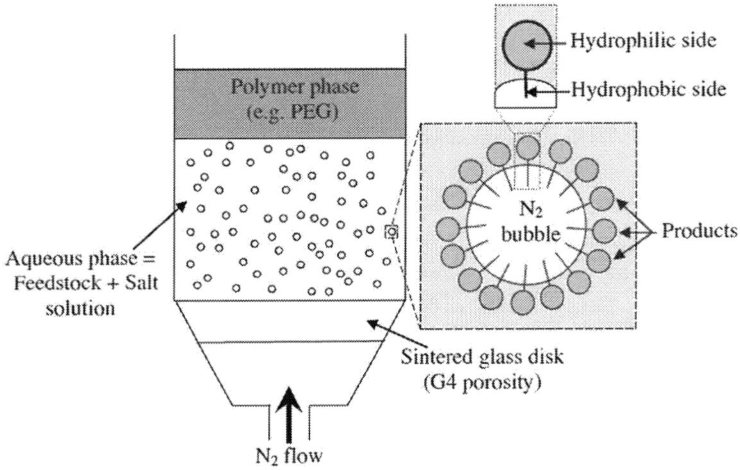

FIGURE 3.4 Illustrative scheme of the ATPF technique representing the use of a nonrecyclable polymer and a nonspecific salt solution. *Adaptation from Show, P.L., Tan, C.P., Anuar, M.S., Ariff, A., Yusof, Y.A., Chen, S.K., Ling, T.C., 2011. Direct recovery of lipase derived from* Burkholderia cepacia *in recycling aqueous two-phase flotation. Sep. Purif. Technol. 80, 577–584.*

concentration coefficient, soft separation, the low dosage of organic solvents required, high separation efficiency, simple operation, and, according to some experts, the low environmental impact (Bi et al., 2009).

Show et al. (2011, 2012, 2013) used this technique having as the main purpose the recovery and purification of lipases. In the first approach, the authors have used ATPF systems composed of a thermo-sensitive ethylene oxide-propylene oxide (EOPO) copolymer and one ammonium sulfate salt for the recovery of lipase from *Burkholderia cepacia* strain ST8 directly from the fermentation broth. Different strategies were considered in terms of the conditions studied, namely the variation of the polymer molar mass, the concentration of ammonium sulfate, the pH of the system, the amount of the loaded crude feedstock, the initial volume of EOPO phase situated in the top of the aqueous solution, the concentration of EOPO, the initial volume of aqueous phase, the nitrogen flow rate, and finally, the flotation time. According to the authors, the lipase was successfully purified from the fermentation broth and then easily separated by using the EOPO copolymer in the design of the ATPF systems. Considering the optimal ATPF' conditions, the authors were capable of efficiently separating and purifying the lipase between 13% and 99%, respectively (Show et al., 2011, 2012, 2013).

These results showed that there was no relevant variation of the lipase specific activities between the products recovered by the application of ATPF systems with fresh and recycled chemicals. Another important approach developed in this work was the recycling of the ATPF phase-forming components, polymer and salt, considered effective in terms of costs associated, processing time (reduction of the operation time), and environmental impact (the phase components are introduced again in the formation of new ATPF systems) (Show et al., 2011, 2012, 2013).

3.3 LIPASE EXTRACTION BY IL-BASED ABS

In the mid-2010s, novel ABS based in ILs were proposed by Rogers and coworkers (Gutowski et al., 2003). Commonly described as room temperature liquid salts with uncommon properties (Earle et al., 2006; Esperança et al., 2010; Seddon, 2003), ILs have attracted attention from both academia and industry in downstream processes and analytical techniques. Their crescent interest relies on their unique properties, such as the negligible vapor pressure and lack of flammability, as well as their high chemical and thermal stabilities, low melting points, and a large liquidus temperature range (Freire et al., 2012; Brennecke and Maginn, 2001; Aparicio et al., 2010). However their major advantage is a unique capability to solvate a huge variety of solutes, of a wide range of polarities, and to have a significant impact upon their solubility in water. ILs are thus capable of solubilizing several classes of compounds, from polar to nonpolar, organic to inorganic, biocompounds to metals, which means that they can cover the whole hydrophilicity—hydrophobicity (polarity) range. By fine-tuning their characteristics and properties it is possible to develop novel and more effective ABS for a given separation, which is a crucial issue for their use as part of the downstream processing of biocompounds.

It has been shown that the application of ILs as phase forming agents in ABS boosts the extractive performance and the selectivity parameters of a wide range of compounds (Freire et al., 2012). Since the first report on the use of ILs to form ABS by Rogers and coworkers (Gutowski et al., 2003), a large number of works have been dealing with the study of IL-based ABS and prompted the publication of a major critical review (Freire et al., 2012) of the field. This review addresses three major points concerning the IL-based ABS: (1) the interpretation of the main interactions contributing to the formation of ABS; (2) the study of a large plethora of ABS with different components, considering in particular the effect of the IL structure and various types of *salting-out* agents on the phase equilibria of ABS; the most important (3) applications of IL-based ABS, in which the extraction of biomolecules and other added-value compounds are included. Despite the number of articles published on the subject, the authors suggest that there is still room for improvement that should be accomplished regarding not only the study of new IL-ABS, but also their potential applications, principally in terms of exploring and understanding their extractive potential. While most studies are focused on the variation of the structural features of the IL and the type of salt employed (Freire et al., 2012), these could focus on cheaper and more environmentally friendly compounds using either quaternary ammonium and cholinium-based ILs (e Silva et al., 2014; Pereira et al., 2013a; Shahriari et al., 2013) and citrate-based or other organic salts of a renewable nature (e Silva et al., 2014; Passos et al., 2012). It should be emphasized, however, that the IL-based ABS known in 2015 are well beyond the IL-salt combinations, already being reported the use of amino acids (Domínguez-Pérez et al., 2010; Zhang et al., 2007), carbohydrates (Freire et al., 2011; Wu et al., 2008a,b,c; Chen et al., 2010), polymers (Pereira et al., 2013b,c; Tomé et al., 2014), and even surfactants (Bhatt et al., 2013; Gong and Zhu, 2014; Vicente et al., 2014).

Despite the number of publications on IL-based ABS, the works regarding the use of IL-based ABS in the extraction and purification of lipases are few. Deive et al. (2011) proposed the use of IL-based ABS to extract the *Thermomyces lanuginosus* lipase (*Tl*L). The enzyme activity was monitored for the systems studied, based in imidazolium ILs with distinct alkyl chain lengths and combined with the chloride, alkylsulfate, alkylsulfonate, and acetate anions. Several operating conditions influencing the lipase activity and the ABS partition were studied, namely the temperature, the pH, the deactivation kinetics, and the water content. The kinetic of the *Tl*L deactivation was investigated and ATR-FTIR studies were carried, aiming at the identification of the *Tl*L structure when exposed to the selected ILs. The main results reported allowed the identification of the optimal conditions, which were identified as the use of ABS based in the 1-ethyl-3-methylimidazolium ethylsulfate ($[C_2MIM][EtSO_4]$) combined with the potassium carbonate (K_2CO_3)—Fig. 3.5—due to the capacity to maintain the enzyme native

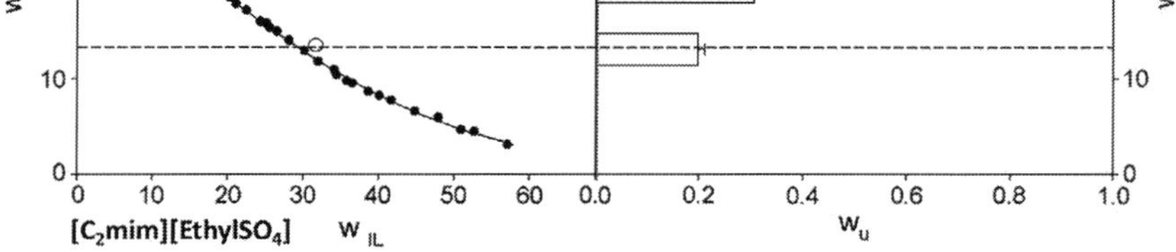

FIGURE 3.5 Summary of the results by Deive et al. (2011), taking into account the high enzyme activity achieved. (A) Phase diagram of the ABS composed of $[C_2mim][EthylSO_4] + K_2CO_3 +$ water. The *black dots* are the experimental data and the *solid line* the fitting of the binodal curve. The *white and striped dots* represent the compositions of the biphasic mixtures used in the extractions. (B) In the extraction experiments, the *white bars* represent the enzyme recovered in the IL-rich phase and the *hatched bar* represents the enzyme recovered at 25% salt content and 27% of IL content. *Adapted from Deive, F.J., Rodriguez, A., Pereiro, A.B., Araujo, J.M.M., Longo, M.A., Coelho, M.A.Z., Lopes, J.N.C., Esperanca, J.M.S.S., Rebelo, L.P.N., Marrucho, I.M., 2011. Ionic liquid-based aqueous biphasic system for lipase extraction. Green Chem. 13, 390–396.*

structure (the same structure found for the enzyme in water), which was confirmed by the ATR-FTIR data and the deactivation kinetics analysis. The authors concluded from the lipase extraction efficiencies found (99% of activity) that the use of IL-based ABS constitutes a powerful and promising separation alternative to conventional methods (Deive et al., 2011).

In the same year, Ventura et al. (2011, 2012a) developed an integrated work of optimization (Ventura et al., 2011) and application to the extraction and purification of a lipase from the fermentation broth (Ventura et al., 2012a). In a first step, the authors studied the *Candida antarctica* lipase B (CaLB) partitioning in several ABS composed of distinct IL chemical structures and three phosphate inorganic salts, aiming at the identification of the best IL for the enzyme purification (Ventura et al., 2011). For that purpose, different families of ILs were tested, namely pyridinium, piperidinium, pyrrolidinium, and imidazolium. Included in the imidazolium structures, different anions, namely triflate [CF_3SO_3], dicyanamide [$N(CN)_2$], methanesulfonate [CH_3SO_3], and chloride (Cl), and several alkyl chain lengths, from C_2 to C_8, were investigated. For each system studied, the enzyme partitioning between the two aqueous phases was measured and the purification factor and the enzyme recovery parameters were determined. The authors identified that the maximum lipase purification and recovery were obtained for an octyl side chain associated to the imidazolium cation, the [$N(CN)_2$] anion and ILs belonging to the aromatic pyridinium family. However, the additive characteristics were not observed for the extraction parameter, since the tailored [C_8pyr][$N(CN)_2$] was tested, and the extraction results were not efficient (purification factor $= 0.998 \pm 0.002$). In this context, the best results were achieved with [C_8mim]Cl with a purification factor of 2.6 ± 0.1 and an enzyme recovery of $(95.9 \pm 0.2)\%$ at the salt-rich phase.

Following this optimization study and considering the ABS with the best results in terms of purification factor and extraction efficiency (Fig. 3.5), a new study was conducted, aiming at the production and purification of an extracellular lipolytic enzyme produced by *Bacillus* sp. ITP-001 (Ventura et al., 2012a). The direct contact of the bacteria with the ILs was avoided and the separation and purification steps were performed using the fermentation broth after the end of the production stage. This work was the first report of a comprehensive study where the integration of IL-based ABS in the process of production and purification of a lipase was shown. The first step on the prepurification stage was a salt precipitation with ammonium sulfate $(NH_4)_2SO_4$, followed by dialysis. During the salt precipitation using $(NH_4)_2SO_4$, a significant amount of contaminant proteins was removed and the lipase was concentrated in the supernatant. Following the salt precipitation, the supernatant was passed through a dialysis system aimed at removing the low molecular weight compounds, including the inorganic salts used in the salt precipitation and fermentation processes. After this step, the results showed a small decrease in the enzymatic activity, justified by losses of enzyme during the dialysis process. The purification step was then performed by applying four distinct ABS based in four hydrophilic ILs, namely 1-butyl-3-methylimidazolium chloride [C_4mim]Cl, 1-butyl-3-methylimidazolium chloride [C_4mim]Cl, 1-butyl-3-methylimidazolium dicyanamide [C_4mim][$N(CN)_2$], 1-methyl-3-octylimidazolium chloride [C_8mim]Cl, and 1-butyl-3-methylpyridinium chloride [C_4mpyr]Cl, conjugated with the phosphate buffer solution composed of K_2HPO_4 and KH_2PO_4 at pH 7. In this work, a buffer was used to keep the pH during the entire process of purification, since the pH can have a big influence in both the enzyme activity and its partition (Ventura et al., 2011; Barbosa et al., 2011; Perumalsamy et al., 2007).

Finally, the ability of IL-based ABS in the purification of the lipase produced by submerged fermentation was evaluated and compared against some conventional PEG-based ABS. The results suggest that the enzyme purification was mainly controlled by the alkyl chain length, followed by the cation core and the anion moiety. Both high purification factors and enzyme recovery efficiencies at the salt-rich phase were obtained for all systems ($90.6 \pm 0.1 < R_B^E < 96.14 \pm 0.08$)%, in accordance with the previous optimization report (Ventura et al., 2011). The maximum purification and recovery parameters were obtained for the [C_8mim]Cl-based ABS (Fig. 3.6). The purification of the *Bacillus* sp. ITP-001 lipase was also investigated, using ABS based in polyethylene glycol (PEG 8000) and potassium phosphate (Barbosa et al., 2011). Despite some differences in the operation conditions of the two systems (distinct inorganic salts used, different temperature and pH), the results obtained when IL-based ABS are investigated are in general superior, with purification factors ranging from 37 to 51, against purification factors below 30 for the PEG-based ABS.

As highlighted in these works (Ventura et al., 2012a; Barbosa et al., 2011), another advantage of the IL-based ABS when compared with PEG-based systems is the low viscosity of the aqueous phases on these systems. The dynamic viscosity of the top and bottom phases was measured at room temperature for several IL-based and polymer-based ABS. The results suggest that the viscosity of the salt-rich (bottom) phase for IL-ABS (6.11−19.70 mPa s) is low and similar to the polymer-based ABS (6.96−8.27 mPa s). Meanwhile, the viscosity of the IL-rich phases (4.96−8.91 mPa s) is comparable with those of the salt-rich phases (6.11−19.70 mPa s), while the viscosity of the PEG-rich phase is normally higher, and in this particular case is larger by an order of magnitude (26.67−134.57 mPa s) (Ventura et al., 2012a). This is a relevant advantage, since the lower viscosities make the fluid transport and the mass transfer between both phases easier, helping in the solute partition. These results, associated with the performance of the IL-based ABS, appear as excellent advantages of this purification technology.

One of the disadvantages often cited about the use of IL-based ABS is the high cost of some ILs. To overcome this problem we are currently studying novel approaches using ABS based in ILs, in which lower amounts of ILs are

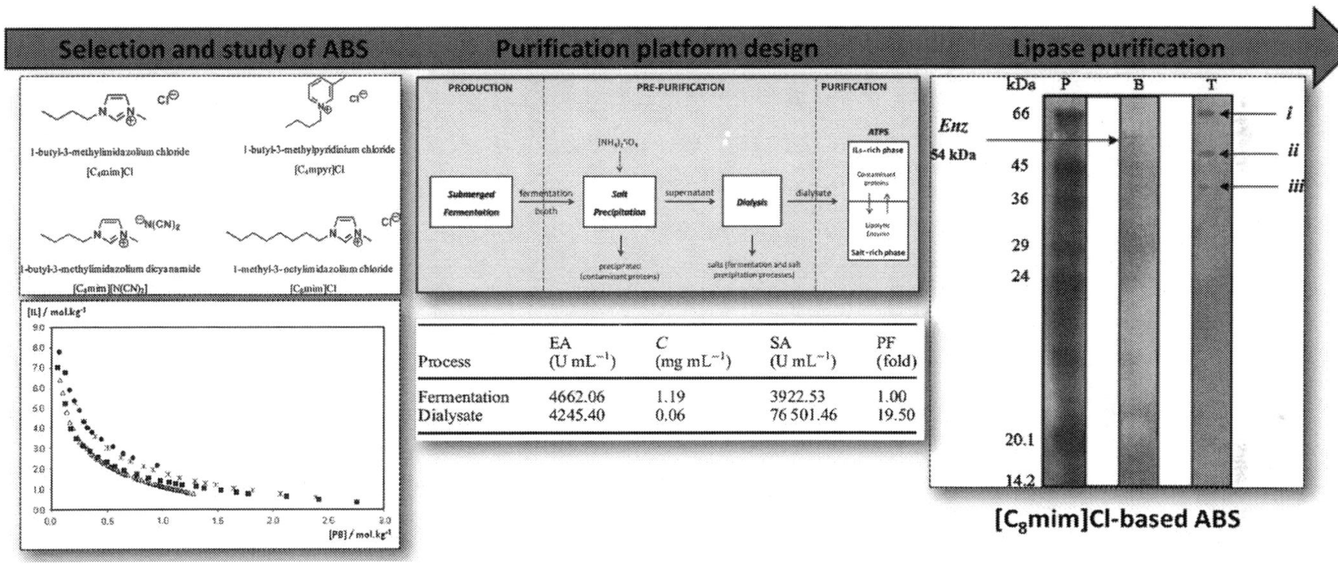

Process	EA (U mL⁻¹)	C (mg mL⁻¹)	SA (U mL⁻¹)	PF (fold)
Fermentation	4662.06	1.19	3922.53	1.00
Dialysate	4245.40	0.06	76 501.46	19.50

FIGURE 3.6 Main route selected for the study of the production and purification of a lipolytic enzyme produced by *Bacillus* sp. ITP-001 via submerged fermentation applying IL-based ABS (Ventura et al., 2012).

required. In this context, ILs are applied as additives or adjuvants in the formation of ABS (Pereira et al., 2010; de Souza et al., 2014a). In a recent work (de Souza et al., 2014b) imidazolium-based ILs as adjuvants (at a concentration of 5 wt%) in ABS composed of PEG polymers (1500, 4000, 6000, and 8000 g/mol) and potassium phosphate buffer (pH 7) were applied in the purification of the *Bacillus* sp. ITP-001 lipase. After a preliminary optimization carried out with commercial CaLB (de Souza et al., 2014b), the best results were transposed to the purification of the lipolytic enzyme from *Bacillus* sp. ITP-001. The results indicate that it was possible to purify the commercial lipase, despite its high purity, 5.2 times and the produced lipase 254 times, using [C_6mim]Cl as adjuvant. Besides the modification proposed by this work (de Souza et al., 2014b) concerning the amount of IL used in the preparation of the extractive system, another change was possible. This new study considered the analysis of the importance of the prepurification route of salt precipitation and dialysis. The ABS with [C_6mim]Cl as adjuvant (system with best purification capacity) was thus tested in terms of extraction efficiency and purification factor for both routes (de Souza et al., 2014b). ***Route ii*** is described by the direct use of the fermentation broth in the preparation of the ABS, as depicted in Fig. 3.7. This figure shows the two approaches investigated. The numerical results reported in Fig. 3.7 indicate that in both cases a significant purification of the lipase is achieved, since the purification factors are very high. When the two routes are compared, it is possible to evaluate the effect of the prepurification step, since the purification factors can be significantly improved when the salt precipitation and the dialysis are excluded from the purification process (de Souza et al., 2014b) ($PF_{\boldsymbol{Route\ i}} = 103.5 \pm 1.2$-fold and $PF_{\boldsymbol{Route\ ii}} = 245.9 \pm 9.5$-fold). Moreover, a comparison between the various works investigating the purification of this lipolytic enzyme from *Bacillus* sp. ITP-001 (Ventura et al., 2011; de Souza et al., 2014b) shows that the highest purification factors were achieved by the application of ABS using ILs as adjuvants ($PF_{\boldsymbol{Route\ i}} = 245.9 \pm 9.5$-fold and $PF_{\boldsymbol{Route\ ii}} = 103.5 \pm 1.2$-fold) (de Souza et al., 2014b) when compared with the IL-ABS ($PF = 51 \pm 2$-fold) (Ventura et al., 2012a) and polymer-based ABS ($PF = 30$-fold).

3.4 MAIN CONCLUSIONS

In this work, some of the recent developments in downstream processing with an emphasis on the purification of lipases are addressed. Bacterial lipases are mostly extracellular bioproducts and their production by fermentation is a strategic factor for their future commercialization on a large scale. However, their production is affected by various conditions, namely nutritional and physicochemical factors, such as temperature, pH, nitrogen and carbon sources, presence of lipids, inorganic salts, agitation, and dissolved oxygen concentration (Aires-Barros et al., 1994). Lipases are important for many

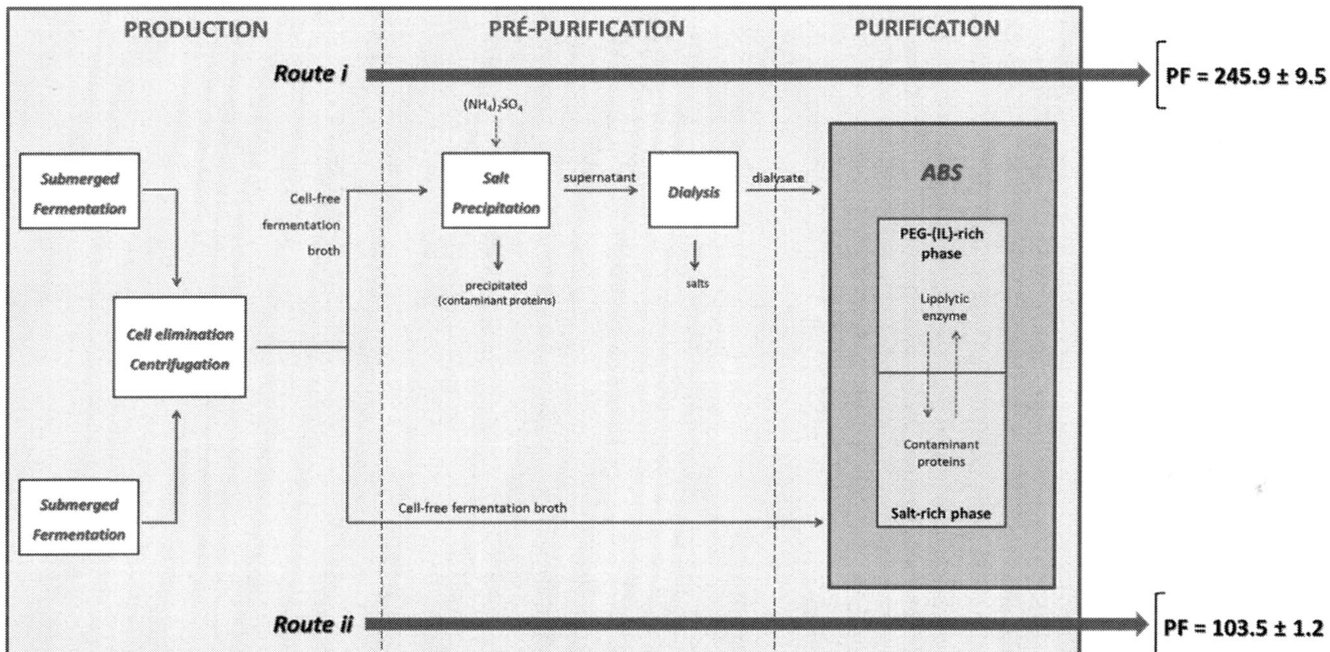

FIGURE 3.7 Schematic representation of the two routes considered in the most recent works for the production and purification of the lipolytic enzyme produced by *Bacillus* sp. ITP-001 via submerged fermentation: ***Route i*** represents the classic approach, characterized by a prepurification step, including precipitation with $(NH_4)_2$ SO_4 followed by dialysis; ***Route ii*** represents the novel approach without prepurification.

applications and their purification is, consequently, a field of utmost importance. Despite the fact that some of the commercial applications do not require highly pure homogeneous lipase preparations, an adequate degree of purity is essential for their industrial application, because it enables their most efficient and successful use in industries such as fine chemicals, pharmaceuticals, and cosmetics. The industrial strategies for downstream processing, in particular for the purification scheme employed, should be inexpensive or low cost, fast, with high yields easy scale-up. Until 2015, the principal technologies considered are the chromatographic techniques associated with the use of salt precipitations. However, liquid–liquid extraction, in particular the use of aqueous biphasic systems, aqueous two-micellar systems, and aqueous two-phase flotation systems, may lead to improved downstream processes. The results of the purifications achieved when those techniques are considered can be very diverse, as shown in Table 3.6. The main results discussed here also suggest that the chromatographic techniques, depending on their position on the process and the complexity of the fermentation broth, can be advantageous, in particular when these techniques are used in the final steps of purification (product polishing). The results reported in Table 3.6 suggest that the best results regarding the purification of lipases were obtained by using

TABLE 3.6 Purification Factors Obtained by the Various Works Reviewed, Taking into Account the Distinct Liquid–Liquid Purification Techniques Employed

Purification Technique	Purification Factor Range	References
Chromatography	10–3028	Saxena et al. (2003), Palekar et al. (2000), Gupta et al. (2004), Veeraragavan et al. (1990), Schmidt-Dannert et al. (1996), Schmidt-Dannert et al. (1994), Borkar et al. (2009), Kumar et al. (2012), Ghanem et al. (2000), Imamura and Kitaura (2000), Litthauer et al. (2002), Snellman et al. (2002), Abdou (2003), and Taipa et al. (1992)
Common ABS		
Polymer-ABS	4–30	Zhang and Liu (2010), Ooi et al. (2009a), Bassani et al. (2010), Zhou et al. (2013), Khayati and Alizadeh (2013), and Li et al. (2010)
Alcohol-ABS	13.5	Ooi et al. (2009b)
Surfactant-ABS	3.7–7.2	Vicente et al. (1990), Yu et al. (2003), Nandini and Rastogi (2010), and Ooi et al. (2011b)
IL-ABS	37–246	Deive et al. (2011), Ventura et al. (2012a), and Ventura et al. (2011)

chromatographic techniques. However, it should be noticed that these values were the result of a complex scheme of purification obtained by the combination of several steps of purification based on salt precipitation and various chromatographic techniques.

Concerning the polymeric-ABS, the main conclusions are that the partitioning phenomenon is normally influenced by the type and molecular weight of the polymers, the pH, the addition of salts, or by temperature modifications, among others (Molino et al., 2013). The advantages of ABS are the reduction on the volume treated, their high extraction capacity, the low time required to promote the partition of the solute, and also the fact that these technologies are considered relatively straightforward to scale up. Meanwhile, from all technologies employed, the use of IL-based ABS are one of the most promising approaches shown, which is justified by the high purification factors achieved, though these should be taken with care given the limited number of studies and systems investigated.

Different approaches were proposed regarding the use of IL-based ABS, namely the simple ABS composed of ILs + salt + water, or ABS using ILs as additives composed of polymer + salt + water + IL. The results analyzed suggest that the most promising techniques are based on the use of ABS where the ILs are applied as adjuvants (higher purification factors). However, the number of works contemplating the application of IL-based ABS are still limited and more studies are required to create a more complete picture, to allow the development of a deeper understanding on the mechanisms of interaction between the main components of the ternary/quaternary systems and the partition of the solvents, and the optimization of the best IL systems to be applied. For example, the study of more hydrophobic ILs as adjuvants should be carried by applying non-aromatic acyclic families such as quaternary ammonium, phosphonium, and cholinium. Moreover, the operational conditions should be investigated, namely the pH, temperature, or even other polymeric-ABS. Besides these studies, the design of new schemes (conjugation of various purification techniques defined for a specific lipase with a specific application) should be addressed.

The purification schemes should have potential for being applied in a continuous process, with extraction capacity and high selectivity for the desired product. As aforementioned, several purification approaches were already proposed in literature (Saxena et al., 2003; Antonian, 1988; Aires-Barros et al., 1994; Gupta et al., 2004). From those, the number of works investigating the use of conventional chromatographic techniques and common aqueous biphasic systems, based principally in polymers and simple salts, is significant, despite the lower purification factors achieved. This means that these works need to be planned taking into consideration the process design, because it is notorious that the use of isolated techniques was not very successful until now. A deeper understanding is required for the success of the lipase purification schemes or for the success of purification processes in general, which we believe must be based in an appropriate case-by-case design and development of the purification platform. Most works analyzed here provide clear evidence that the extent of purification varies with the number and the order of the purification steps, and also that these aspects have been evaluated through different purification protocols pursued by several researchers. The main constraints about the purification schemes being proposed is that they often rely on purification strategies with low yields, long time periods, and low selectivity.

3.5 CRITICAL ANALYSIS AND FUTURE CHALLENGES

The main objective of this work was the description and analysis of novel methodologies based on ABS used to purify lipases after their production by microorganisms. Despite the significant efforts made by various researchers, the number of works reported is limited but the results are promising. In this sense, the selection or design of the best purification scheme for a specific lipase is yet extremely difficult. As a result, more studies are required and should be performed to fulfill the experimental gaps and the lack of knowledge. As discussed, the chromatographic techniques can be advantageous not only due to its most popular characteristics, namely the high yields of extraction and the reduced time spent in the extraction operations, but also because of the possibility to conjugate several techniques performing different schemes of purification, allowing the optimization of the purification capacities (Krieger et al., 1999). However, the best results are not satisfactory and, in this context, new alternatives of purification or even new schemes are demanded, principally to replace those with negative effects on the lipase activity and stability. Besides the decrease in the enzyme performance, some of these technologies are not easy to manipulate, due to the high number of purification units connected in the same purification process, which results in the development of more time-consuming processes with low final yields. Alternative and more amenable techniques are being developed, mainly considering the application of liquid—liquid extraction technologies.

The use of ABS (a widely used technology) was validated and widely applied in the extraction, concentration, and purification of many solutes, including lipases. Associated with these technologies, several properties are often highlighted as advantages of ABS, namely the low interfacial tension, lower process time and energy, and the higher amounts of water present (diminishing the of enzyme inactivation) and their capacity to manipulate various solutes (depending of the main interactions acting in the partition phenomenon). Moreover, ABS are advantageous downstream processes because they are appropriate for a continuous operation regime, they have scale-up potential, it is easy to incorporate them in general purification platforms, and finally, the phase forming components are nontoxic and highly biodegradable. However, the selectivity, one of the most important problems associated with the purification strategies, remained a major challenge. In this sense, IL-based ABS with a great variety of different cations, anions, and alkyl chains to be conjugated, and different salts, carbohydrates, and amino acids, to be used as salting-out species, appeared and have been investigated (Freire et al., 2012). One of the major advantages of IL-based ABS is the large range of polarities achieved when different anion/cation/alkyl chain combinations are conjugated (Freire et al., 2012). Playing with the polarity of both aqueous phases, the principal interactions can be manipulated and, consequently, the selectivity enhanced. Other aspects were also improved by the combination of ILs and ABS, namely the lower viscosities of the phases (Freire et al., 2012). However, for certain cases, the lipase activity and stability can be negatively affected due to the presence of some particular ILs and also, when the main contaminants present in the fermentation broth are very similar in terms of chemical structures, the extraction/purification selectivity remains an issue.

An advantageous procedure, in our opinion, is the development and use of new ABS based in ILs with reduced capacity to interact with proteins and lipases, and ILs with surfactant nature, capable of auto-aggregating in micellar systems. The main idea behind these systems is the same described for regular AMTPS, the formation of micellar systems able to separate into two aqueous phases, promoting the migration of the lipase between the two phases, depending on the main interactions in control, but this time, by adding ILs with long alkyl chains (with tensioactive nature). The first efforts in this direction were recently carried where different long alkyl chain-based ILs were applied as cosurfactants in the formation of AMTPS based in the nonionic surfactant Triton X-114 and the McIlvaine buffer (Vicente et al., 2014). This work looks at ILs as a new class of tunable cosurfactants, three distinct families being studied here, namely imidazolium, phosphonium, and quaternary ammonium. These families were conjugated with different lengths of alkyl chain and various anions aimed at studying the binodal curves of the novel IL-AMTPS. In this work, the impact of the IL absence/presence, the concentration and the structural features of distinct ILs on the binodal curves construction, was studied. The main results obtained regarding the binodal curves studied provide evidence that the presence of ILs has an important effect on the T_{cloud}, since the binodal curves position seems to be highly dependent on the ILs hydrophobic/hydrophilic character. Aiming at evaluating their applicability as extraction systems, studies considering the partition of two model (bio)molecules, namely the protein Cytochrome c and the dye Rhodamine 6G, were performed. The results reported in this work put in evidence of the potential that small amounts of IL can have in the selective purification of different molecules, which is in our opinion opening new doors in the proper development of successful downstream technologies.

Besides the usual conditions that can be changed in all classes of ABS described in this work, namely temperature, pH, concentration of phase former components, type of polymer, salt, IL, and different cation/anion/alkyl chain combinations, some important issues still need to be addressed regarding the final objective of the downstream processing, the industrial application. The use of ILs needs to be improved, since some of the most common ILs used to prepare ABS can deactivate lipases (Ventura et al., 2012b; de los Rios et al., 2007; Kaar et al., 2003; Klähn et al., 2011). Another question normally raised by industry is the high price of some ILs, which means that at lab scale it is possible to apply the ILs as solvents, but when the scale is increased, the costs associated may not be compatible with industrial use, even if the target product is an added-value compound. The scale-up is one big issue and four major questions need to be answered before the purification process design: (1) how much lipase is required and what is the demand in terms of purity level; (2) what is the source of the lipase; (3) what is the scheme optimized at lab scale; and (4) what equipment is available at the industrial facilities. The design is clearly very important, and we believe that the concept of extractive fermentation may be advantageous regarding the purification of lipases.

As well-known, the fermentation processes are constantly hampered by a variety of problems, side reactions, production of intermediates or contaminants, originated from the accumulation of products in the bioreactor. Thus, the integration of fermentation and a primary product separation step will have a positive impact on the productivity/yield of the fermentation processes (Ooi et al., 2011a), and in the overall costs, since the product recovery costs and effluent treatment costs will be reduced, as a result of the use of a more concentrated feedstock. Meanwhile, several conditions need optimization, such as the selection of a suitable solvent, taking into account the biocompatibility between solvents and microorganisms; the development of appropriate models to the mass transfer; the heat transfer; the partitioning of different solutes and contaminants; the best conditions of temperature, agitation, and pH; and also the incorporation of models and parameters into a process simulator aiming at the optimization of the whole process. With all these requirements satisfied, it will be possible in the future to design more efficient integrated platforms of production and purification of lipases (capable to guarantee their required level of purify). The process modeling may have a crucial role in the purification of lipases, not only in the implementation of extractive fermentation processes, but also in the study and implementation of a continuous purification regime. Since ABS and AMTPS are liquid–liquid separation processes, they are excellent candidates for microfluidic separation techniques, a field that is in quick expansion since 2000.

Miniaturization, as a process intensification methodology (which means increased mass and heat transfer), allows for the binding of the high surface-to-volume ratio, aiming at developing efforts to achieve higher yields over shorter periods of time, higher product purity, and better process control along with the reduction of cost and equipment associated with downstream processing (Woodcock et al., 2008; Keoschkerjan et al., 2004) and the possibility for parallel operations (Berg et al., 2010; Jahnisch et al., 2004;

Teh et al., 2008). Other economic benefits, improvement of intrinsic safety, and reduction of the environmental impact can be achieved, as well as the benefits of moving from batch to continuous mode. These microfluidic devices have important advantages such as savings in time, space, materials, costs, and a higher control of the operator over the purification system. Due to their flexibility as manufacturing processes, they can be potentially designed for the required chemistry in contrast to the conventional technologies (Dietrich, 2009). Despite the reduced number of studies using this technology, the attempts reported in literature were a huge success in terms of selectivity and yield (Meagher et al., 2008; Pohar et al., 2012). Recently, microfluidic separation devices and ABS based in ILs were already tested and applied in purification schemes (Novak et al., 2012). This microscale equipment seems also to be a promising device for parallel processing (SooHoo and Walker, 2009; Znidarsic-Plazl and Plazl, 2007), which could be applied with success in the large-scale purification of lipases from the fermentation broth.

Studies contemplating the recycling of the phase components (polymers, ILs, and salts) are being developed (Zhang and Liu, 2010; Freire et al., 2012). In the particular case of ILs, three main approaches are being considered, namely their recycling, regeneration, or reuse (Freire et al., 2012; Claudio et al., 2013, 2014; Alvarez-Guerra et al., 2014). The results suggest that it is possible to successfully recycle and reuse various ILs applied in the extraction of several biomolecules, including proteins and even when these ILs are applied as solvents in solid–liquid extraction technologies (Claudio et al., 2013). Meanwhile, more studies are required, contemplating the reuse, recycling, and regeneration of these solvents and other phase separation agents (polymers, salts and surfactants), mainly considering the particular case of processes used to purify lipases. In this context, other studies are mandatory in our opinion, namely the investigation on the recovery of the lipase from the presence of the ABS and AMTPS separation agents, which will be an important task to be developed regarding the lipase polishing. The treatment of effluents originated from the lipase polishing should also be considered for future studies, because this is not only important in terms of technological scheme but also regarding the total costs of these purification processes. Indeed, not only the technological background is important but also the economic point of view; engineers and researchers should take into account the necessity to properly develop the economic analysis in parallel with the technological description. In this sense, the number of sequential operations necessary to achieve the desired purity of a lipase should be taken into account, since this is one of the main conditions with relevant contribution to the overall cost of the downstream scheme. Furthermore, the capital investment and the amount of consumables needed for each step as well as the individual time required for each operation and the amount and expertise of the human resources are key factors promoting the economic profile of the downstream processes.

Summing up, it will be possible in the future to project the production and purification of specific lipases for a certain application with total control of the technological, economic, and quality conditions required for the success of each industrial process, under the concept of Quality by Design, defined as "a systematic approach to development that begins with predefined objectives and emphasizes product and process understanding and process control, based on sound science and quality risk management," adapted from (Rathore and Winkle, 2009). The main aim will be to design a quality product and establish a robust manufacturing process that can consistently deliver the intended performance of the product.

REFERENCES

Abdou, A.M., 2003. Purification and partial characterization of psychrotrophic *Serratia marcescens* lipase. J. Dairy Sci. 86, 127–132.

Aires-Barros, M.R., Taipa, M.A., Cabral, J.M.S., 1994. Isolation and purification of lipases. In: Woolley, P., Petersen, S.B. (Eds.), Lipases: Their Structure, Biochemistry and Application. Cambridge Univ. Press, Cambridge, pp. 243–270.

Alvarez-Guerra, E., Ventura, S.P.M., Coutinho, J.A.P., Irabien, A., 2014. Ionic liquid-based three phase partitioning (ILTPP) systems: ionic liquid recovery and recycling. Fluid Phase Equilib. 371, 67–74.

Antonian, E., 1988. Recent advances in the purification, characterization and structure determination of lipases. Lipids 23, 1101–1106.

Aparicio, S., Atilhan, M., Karadas, F., 2010. Thermophysical properties of pure ionic liquids: review of present situation. Ind. Eng. Chem. Res. 49, 9580–9595.

Barbosa, J.M.P., Souza, R.L., Fricks, A.T., Zanin, G.M., Soares, C.M.F., Lima, Á.S., 2011. Purification of lipase produced by a new source of *Bacillus* in submerged fermentation using an aqueous two-phase system. J. Chromatogr. B 879, 3853–3858.

Basheer, S.A., Thenmozhi, M., 2010. Reverse micellar separation of lipases: a critical review. Int. J. Chem. Sci. 8, S57–S67.

Bassani, G., Fuciños, P., Picó, G., Farruggia, B., 2010. *Candida rugosa* lipase Lip1–polyethyleneglycol interaction and the relation with its partition in aqueous two-phase systems. Colloids Surf. B: Biointerfaces 75, 532–537.

Berg, A., Craigheadb, H.G., Yang, P., 2010. From microfluidic application to nanofluidic phenomena issue. Chem. Soc. Rev. 39, 899–900.

Bhatt, D., Maheria, K., Parikh, J., 2013. Studies on surfactant-ionic liquid interaction on clouding behaviour and evaluation of thermodynamic parameters. J. Surfact. Deterg. 16, 547–557.

Bi, P.-Y., Li, D.-Q., Dong, H.-R., 2009. A novel technique for the separation and concentration of penicillin G from fermentation broth: aqueous two-phase flotation. Sep. Purif. Technol. 69, 205–209.

Bi, P.-Y., Dong, H.-R., Dong, J., 2010a. The recent progress of solvent sublation. J. Chromatogr. A 1217, 2716–2725.

Bi, P.-Y., Dong, H.-R., Yuan, Y.-C., 2010b. Application of aqueous two-phase flotation in the separation and concentration of puerarin from *Puerariae* extract. Sep. Purif. Technol. 75, 402–406.

Borkar, P.S., Bodade, R.G., Rao, S.R., Khobragade, C.N., 2009. Purification and characterization of extracellular lipase from a new strain: *Pseudomonas aeruginosa* SRT 9. Braz. J. Microbiol. 40, 358–366 [online].

Brennecke, J.F., Maginn, E.J., 2001. Ionic liquids: innovative fluids for chemical processing. AIChE J. 47, 2384–2389.

Byun, H.G., Eom, T.K., Jung, W.K., Kim, S.K., 2007. Lipase-catalyzed hydrolysis of fish oil in an optimum emulsion system. Biotechnol. Bioprocess Eng. 12, 484–490.

Chen, Y., Meng, Y., Zhang, S., Zhang, Y., Liu, X., Yang, J., 2010. Liquid–liquid equilibria of aqueous biphasic systems composed of 1-butyl-3-methyl imidazolium tetrafluoroborate + sucrose/maltose + water. J. Chem. Eng. Data 55, 3612–3616.

Claudio, A.F.M., Ferreira, A.M., Freire, M.G., Coutinho, J.A.P., 2013. Enhanced extraction of caffeine from guarana seeds using aqueous solutions of ionic liquids. Green Chem. 15, 2002–2010.

Claudio, A.F.M., Marques, C.F.C., Boal-Palheiros, I., Freire, M.G., Coutinho, J.A.P., 2014. Development of back-extraction and recyclability routes for ionic-liquid-based aqueous two-phase systems. Green Chem. 16, 259–268.

Cramer, S.M., Holstein, M.A., 2011. Downstream bioprocessing: recent advances and future promise. Curr. Opin. Chem. Eng. 1, 27–37.

de los Rios, A.P., Hernandez-Fernandez, F.J., Martinez, F.A., Rubio, M., Villora, G., 2007. The effect of ionic liquid media on activity, selectivity and stability of *Candida antarctica* lipase B in transesterification reactions. Biocatal. Biotransform. 25, 151–156.

de Souza, R.L., Campos, V.C., Ventura, S.P.M., Soares, C.M.F., Coutinho, J.A.P., Lima, Á.S., 2014a. Effect of ionic liquids as adjuvants on PEG-based ABS formation and the extraction of two probe dyes. Fluid Phase Equilib. 375, 30–36.

de Souza, R.L., Campos, V.C., Ventura, S.P.M., Soares, C.M.F., Coutinho, J.A.P., Lima, Á.S., 2014b. Purification of Lipase Using Ionic Liquids as Adjuvants in Aqueous Two-Phase Systems.

Deive, F.J., Rodriguez, A., Pereiro, A.B., Araujo, J.M.M., Longo, M.A., Coelho, M.A.Z., Lopes, J.N.C., Esperanca, J.M.S.S., Rebelo, L.P.N., Marrucho, I.M., 2011. Ionic liquid-based aqueous biphasic system for lipase extraction. Green Chem. 13, 390–396.

Diamond, A.D., Hsu, J.T., 1992. Aqueous two-phase systems for biomolecule separation. In: Tsao, G.T. (Ed.), Bioseparation. Springer, Berlin, Heidelberg, pp. 89–135.

Dietrich, T.R., 2009. Microchemical Engineering in Practice, first ed. Wiley & Sons, NJ, pp. 1–29.

Domínguez-Pérez, M., Tomé, L.I.N., Freire, M.G., Marrucho, I.M., Cabeza, O., Coutinho, J.A.P., 2010. (Extraction of biomolecules using) aqueous biphasic systems formed by ionic liquids and aminoacids. Sep. Purif. Technol. 72, 85–91.

e Silva, F.A., Sintra, T., Ventura, S.P.M., Coutinho, J.A.P., 2014. Recovery of paracetamol from pharmaceutical wastes. Sep. Purif. Technol. 122, 315–322.

Earle, M.J., Esperança, J.M.S.S., Gilea, M.A., Canongia Lopes, J.N., Rebelo, L.P.N., Magee, J.W., Seddon, K.R., Widegren, J.A., 2006. The distillation and volatility of ionic liquids. Nature 439, 831–834.

Esperança, J.M.S.S., Canongia Lopes, J.N., Tariq, M., Santos, L.M.N.B.F., Magee, J.W., Rebelo, L.P.N., 2010. Volatility of aprotic ionic liquids – a review. J. Chem. Eng. Data 55, 3–12.

Freire, M.G., Louros, C.L.S., Rebelo, L.P.N., Coutinho, J.A.P., 2011. Aqueous biphasic systems composed of a water-stable ionic liquid + carbohydrates and their applications. Green Chem. 13, 1536–1545.

Freire, M.G., Claudio, A.F.M., Araujo, J.M.M., Coutinho, J.A.P., Marrucho, I.M., Lopes, J.N.C., Rebelo, L.P.N., 2012. Aqueous biphasic systems: a boost brought about by using ionic liquids. Chem. Soc. Rev. 41, 4966–4995.

Ghanem, E., Al-Sayed, H., Saleh, K., 2000. An alkalophilic thermostable lipase produced by a new isolate of *Bacillus alcalophilus*. World J. Microbiol. Biotechnol. 16, 459–464.

Goja, A.M., Yang, H., Cui, M., Li, C., 2013. Aqueous two-phase extraction advances for bioseparation. J. Bioprocess. Biotech. 4. http://dx.doi.org/10.4172/2155-9821.1000140.

Gong, A., Zhu, X., 2014. Surfactant/ionic liquid aqueous two-phase system extraction coupled with spectrofluorimetry for the determination of dutasteride in pharmaceutical formulation and biological samples. Fluid Phase Equilib. 374, 70–78.

Gupta, R., Gupta, N., Rathi, P., 2004. Bacterial lipases: an overview of production, purification and biochemical properties. Appl. Microbiol. Biotechnol. 64, 763–781.

Gutowski, K.E., Broker, G.A., Willauer, H.D., Huddleston, J.G., Swatloski, R.P., Holbrey, J.D., Rogers, R.D., 2003. Controlling the aqueous miscibility of ionic liquids: aqueous biphasic systems of water-miscible ionic liquids and water-structuring salts for recycle, metathesis, and separations. J. Am. Chem. Soc. 125, 6632–6633.

Hamel, J.-F.P., Hunter, J.B., 1990. Modeling and applications of downstream processing: a survey of innovative strategies. In: Downstream Processing and Bioseparation, ACS Symposium Series. American Chemical Society, Washington, DC. http://dx.doi.org/10.1021/bk-1990-0419.ch1001.

Imamura, S., Kitaura, S., 2000. Purification and characterization of a monoacylglycerol lipase from the moderately thermophilic *Bacillus* sp. H-257. J. Biochem. 127, 419–425.

ISI Web of Knowledge accessed at July 13th, including the words 'lipase' or 'lipase + production + purification'.

Jahnisch, K., Hessel, V., Löwe, H., Baerns, M., 2004. Chemistry in microstructured reactors. Angew. Chem. Int. Ed. 43, 406–446.

Kaar, J.L., Jesionowski, A.M., Berberich, J.A., Moulton, R., Russell, A.J., 2003. Impact of ionic liquid physical properties on lipase activity and stability. J. Am. Chem. Soc. 125, 4125–4131.

Keoschkerjan, R., Richter, M., Boskovic, D., Schnürer, F., Löbbecke, S., 2004. Novel multifunctional microreaction unit for chemical engineering. Chem. Eng. J. 101, 469–475.

Khayati, G., Alizadeh, S., 2013. Extraction of lipase from *Rhodotorula glutinis* fermentation culture by aqueous two-phase partitioning. Fluid Phase Equilib. 353, 132–134.

Klähn, M., Lim, G.S., Wu, P., 2011. How ion properties determine the stability of a lipase enzyme in ionic liquids: a molecular dynamics study. Phys. Chem. Chem. Phys. 13, 18647–18660.

Krieger, N., Taipa, M.A., Melo, E.H.M., Lima-Filho, J.L., Aires-Barros, M.R., Cabral, J.M.S., 1999. Purification of a *Penicillium citrinum* lipase by chromatographic processes. Bioprocess Eng. 20, 59–65.

Kumar, R., Sharma, A., Kumar, A., Singh, D., 2012. Lipase from *Bacillus pumilus* RK31: production, purification and some properties. World Appl. Sci. J. 16, 940–948.

Lee, K.L., Chong, C.C., 2011. The Use of Reverse Micelles in Downstream Processing of Biotechnological Products. http://arxiv.org/ftp/arxiv/papers/1109/1109.1641.pdf.

Li, X., Wan, J., Cao, X., 2010. Preliminary application of light-pH sensitive recycling aqueous two-phase systems to purification of lipase. Process Biochem. 45, 598–601.

Litthauer, D., Ginster, A., van Eeden Skein, E., 2002. *Pseudomonas luteola* lipase: a new member of the 320-residue Pseudomonas lipase family. Enzyme Microb. Technol. 30, 209–215.

Mazzola, P.G., Lopes, A.M., Hasmann, F.A., Jozala, A.F., Penna, T.C.V., Magalhaes, P.O., Rangel-Yagui, C.O., Pessoa Jr., A., 2008. Liquid–liquid extraction of biomolecules: an overview and update of the main techniques. J. Chem. Technol. Biotechnol. 83, 143–157.

Meagher, R.J., Light, Y.K., Singh, A.K., 2008. Rapid, continuous purification of proteins in a microfluidic device using genetically engineered partition tags. Lab Chip 8, 527–532.

Molino, J.V.D., Viana Marques, D. de A., Júnior, A.P., Mazzola, P.G., Gatti, M.S.V., 2013. Different types of aqueous two-phase systems for biomolecule and bioparticle extraction and purification. Biotechnol. Prog. 29, 1343–1353.

Nandini, K.E., Rastogi, N., 2010. Separation and purification of lipase using reverse micellar extraction: optimization of conditions by response surface methodology. Biotechnol. Bioprocess Eng. 15, 349–358.

Novak, U., Pohar, A., Plazl, I., Žnidaršič-Plazl, P., 2012. Ionic liquid-based aqueous two-phase extraction within a microchannel system. Sep. Purif. Technol. 97, 172–178.

Ooi, C.W., Tey, B.T., Hii, S.L., Ariff, A., Wu, H.S., Lan, J.C.W., Juang, R.S., Kamal, S., Ling, T., 2009a. Direct purification of *Burkholderia pseudomallei* lipase from fermentation broth using aqueous two-phase systems. Biotechnol. Bioprocess Eng. 14, 811–818.

Ooi, C.W., Tey, B.T., Hii, S.L., Kamal, S.M.M., Lan, J.C.W., Ariff, A., Ling, T.C., 2009b. Purification of lipase derived from *Burkholderia pseudomallei* with alcohol/salt-based aqueous two-phase systems. Process Biochem. 44, 1083–1087.

Ooi, C.W., Hii, S.L., Kamal, S.M.M., Ariff, A., Ling, T.C., 2011a. Extractive fermentation using aqueous two-phase systems for integrated production and purification of extracellular lipase derived from *Burkholderia pseudomallei*. Process Biochem. 46, 68–73.

Ooi, C.W., Tan, C.P., Hii, S.L., Ariff, A., Ibrahim, S., Ling, T.C., 2011b. Primary recovery of lipase derived from *Burkholderia* sp. ST8 with aqueous micellar two-phase system. Process Biochem. 46, 1847–1852.

Otero, C., Perez, M.F., Gil, J.P., 2005. Effects of interactions with micellar interfaces on the activity and structure of different lipolytic isoenzymes from *Candida rugosa*. Enzyme Microb. Technol. 32, 14–26.

Palekar, A.A., Vasudevan, P.T., Yan, S., 2000. Purification of lipase: a review. Biocatal. Biotransform. 18, 177–200.

Passos, H., Ferreira, A.R., Cláudio, A.F.M., Coutinho, J.A.P., Freire, M.G., 2012. Characterization of aqueous biphasic systems composed of ionic liquids and a citrate-based biodegradable salt. Biochem. Eng. J. 67, 68–76.

Pereira, J.F.B., Lima, A.S., Freire, M.G., Coutinho, J.A.P., 2010. Ionic liquids as adjuvants for the tailored extraction of biomolecules in aqueous biphasic systems. Green Chem. 12, 1661–1669.

Pereira, J.F.B., Vicente, F., Santos-Ebinuma, V.C., Araújo, J.M., Pessoa, A., Freire, M.G., Coutinho, J.A.P., 2013a. Extraction of tetracycline from fermentation broth using aqueous two-phase systems composed of polyethylene glycol and cholinium-based salts. Process Biochem. 48, 716–722.

Pereira, J.F.B., Ventura, S.P.M., e Silva, F.A., Shahriari, S., Freire, M.G., Coutinho, J.A.P., 2013b. Aqueous biphasic systems composed of ionic liquids and polymers: a platform for the purification of biomolecules. Sep. Purif. Technol. 113, 83–89.

Pereira, J.F.B., Rebelo, L.P.N., Rogers, R.D., Coutinho, J.A.P., Freire, M.G., 2013c. Combining ionic liquids and polyethylene glycols to boost the hydrophobic-hydrophilic range of aqueous biphasic systems. Phys. Chem. Chem. Phys. 15, 19580–19583.

Perumalsamy, M., Bathmalakshmi, A., Murugesan, T., 2007. Experiment and correlation of liquid–liquid equilibria of an aqueous salt polymer system containing PEG6000 + sodium citrate. J. Chem. Eng. Data 52, 1186–1188.

Pimentel, M.C.B., Krieger, N., Coelho, L.C.C.B., Fontana, J.D., Melo, E.H.M., Ledingham, W.M., Lima-Filho, J.L., 1994. Lipase from a Brazilian strain of *Penicillium citrinum*. Appl. Biochem. Biotechnol. Adv. 49, 59–69.

Pohar, A., Žnidaršič-Plazl, P., Plazl, I., 2012. Integrated system of a microbioreactor and a miniaturized continuous separator for enzyme catalyzed reactions. Chem. Eng. J. 189–190, 376–382.

Raja, S., Murty, V.R., Thivaharan, V., Rajasekar, V., Ramesh, V., 2011. Aqueous two phase systems for the recovery of biomolecules – a review. Sci. Technol. 1, 7–16.

Rathore, A.S., Winkle, H., 2009. Quality by design for biopharmaceuticals. Nat. Biotech. 27, 26–34.

Rosa, P.A.J., Ferreira, I.F., Azevedo, A.M., Aires-Barros, M.R., 2010. Aqueous two-phase systems: a viable platform in the manufacturing of biopharmaceuticals. J. Chromatogr. A 1217, 2296–2305.

Rosa, P.A.J., Azevedo, A.M., Sommerfeld, S., Bäcker, W., Aires-Barros, M.R., 2011. Aqueous two-phase extraction as a platform in the biomanufacturing industry: economical and environmental sustainability. Biotechnol. Adv. 29, 559–567.

Saxena, R.K., Sheoran, A., Giri, B., Davidson, W.S., 2003. Purification strategies for microbial lipases. J. Microbiol. Methods 52, 1–18.

Schmidt-Dannert, C., Sztajer, H., Stöcklein, W., Menge, U., Schmid, R.D., 1994. Screening, purification and properties of a thermophilic lipase from *Bacillus thermocatenulatus*. Biochim. Biophys. Acta 1214, 43–53.

Schmidt-Dannert, C., Rúa, M.L., Atomi, H., Schmid, R.D., 1996. Thermoalkalophilic lipase of *Bacillus thermocatenulatus*. I. Molecular cloning, nucleotide sequence, purification and some properties. Biochim. Biophys. Acta 1301, 105–114.

Seddon, K.R., 2003. Ionic liquids: a taste of the future. Nat. Mater. 2, 363–365.

Shahriari, S., Tome, L.C., Araujo, J.M.M., Rebelo, L.P.N., Coutinho, J.A.P., Marrucho, I.M., Freire, M.G., 2013. Aqueous biphasic systems: a benign route using cholinium-based ionic liquids. RSC Adv. 3, 1835–1843.

Sharma, R., Chisti, Y., Banerjee, U.C., 2001. Production, purification, characterization, and applications of lipases. Biotechnol. Adv. 19, 627–662.

Show, P.L., Tan, C.P., Anuar, M.S., Ariff, A., Yusof, Y.A., Chen, S.K., Ling, T.C., 2011. Direct recovery of lipase derived from *Burkholderia cepacia* in recycling aqueous two-phase flotation. Sep. Purif. Technol. 80, 577–584.

Show, P.L., Tan, C.P., Anuar, M.S., Ariff, A., Yusof, Y.A., Chen, S.K., Ling, T.C., 2012. Primary recovery of lipase derived from *Burkholderia cenocepacia* strain ST8 and recycling of phase components in an aqueous two-phase system. Biochem. Eng. J. 60, 74–80.

Show, P.L., Ooi, C.W., Anuar, M.S., Ariff, A., Yusof, Y.A., Chen, S.K., Annuar, M.S.M., Ling, T.C., 2013. Recovery of lipase derived from *Burkholderia cenocepacia* ST8 using sustainable aqueous two-phase flotation composed of recycling hydrophilic organic solvent and inorganic salt. Sep. Purif. Technol. 110, 112–118.

Snellman, E.A., Sullivan, E.R., Colwell, R.R., 2002. Purification and properties of the extracellular lipase, Lip A, of *Acinetobacter* sp. RAG-1. Eur. J. Biochem. 269, 5771–5779.

SooHoo, J.R., Walker, G.M., 2009. Microfluidic aqueous two phase system for leukocyte concentration from whole blood. Biomed. Microdevices 11, 323–329.

Taipa, M.A., Aires-Barros, M.R., Cabral, J.M.S., 1992. Purification of lipases. J. Biotechnol. 26, 111–142.

Teh, S.Y., Lin, R., Hung, L.H., Lee, A.P., 2008. Droplet microfluidics. Lab Chip 8, 198–220.

Thakur, S., 2012. Lipases, its sources, properties and applications: a review. Int. J. Sci. Eng. Res. 3, 1–29.

Tomé, L., Pereira, J.F.B., Rogers, R.D., Freire, M.G., Gomes, J.R.B., Coutinho, J.A.P., 2014. "Washing-out" ionic liquid from polyethylene glycol to form aqueous biphasic systems. Phys. Chem. Chem. Phys. 16, 2271–2274.

Veeraragavan, K., Colpitts, T., Gibbs, B.F., 1990. Purification and characterization of two distinct lipases from *Geotrichum candidum*. Biochim. Biophys. Acta 1044, 26–33.

Ventura, S.P.M., Sousa, S.G., Freire, M.G., Serafim, L.S., Lima, Á.S., Coutinho, J.A.P., 2011. Design of ionic liquids for lipase purification. J. Chromatogr. B 879, 2679–2687.

Ventura, S.P.M., de Barros, R.L.F., de Pinho Barbosa, J.M., Soares, C.M.F., Lima, A.S., Coutinho, J.A.P., 2012. Production and purification of an extracellular lipolytic enzyme using ionic liquid-based aqueous two-phase systems. Green Chem. 14, 734–740.

Ventura, S.M., Santos, L.F., Saraiva, J., Coutinho, J.P., 2012. Concentration effect of hydrophilic ionic liquids on the enzymatic activity of *Candida antarctica* lipase B. World J. Microbiol. Biotechnol. 28, 2303–2310.

Vicente, M.L.C., Aires-Barros, M.R., Cabral, J.M.S., 1990. Purification of *Chromobacterium viscosum* lipases using reversed micelles. Biotechnol. Tech. 4, 137–142.

Vicente, F.A., Malpiedi, L.P., e Silva, F.A., Pessoa Jr., A., Coutinho, J.A.P., Ventura, S.P.M., 2014. Design of novel aqueous micellar two-phase systems using ionic liquids as co-surfactants for the selective extraction of (bio)molecules. Sep. Purif. Technol. 135, 259–267 (accepted).

Willauer, H.D., Huddleston, J.G., Rogers, R.D., 2002. Solute partitioning in aqueous biphasic systems composed of polyethylene glycol and salt: the partitioning of small neutral organic species. Ind. Eng. Chem. Res. 41, 1892–1904.

Woodcock, L.L., Wiles, C., Greenway, G.M., Watts, P., Wells, A., Eyley, S., 2008. Enzymatic synthesis of a series of alkyl esters using novozyme 435 in a packed-bed, miniaturized, continuous flow reactor. Biocatal. Biotransform. 26, 501–507.

Wu, B., Zhang, Y., Wang, H., 2008a. Phase behavior for ternary systems composed of ionic liquid + saccharides + water. J. Phys. Chem. B 112, 6426–6429.

Wu, B., Zhang, Y., Wang, H., Yang, L., 2008b. Temperature dependence of phase behavior for ternary systems composed of ionic liquid + sucrose + water. J. Phys. Chem. B 112, 13163–13165.

Wu, B., Zhang, Y.M., Wang, H.P., 2008c. Aqueous biphasic systems of hydrophilic ionic liquids + sucrose for separation. J. Chem. Eng. Data 53, 983–985.

Yu, Y.-C., Chu, Y., Ji, J.-Y., 2003. Study of the factors affecting the forward and back extraction of yeast-lipase and its activity by reverse micelles. J. Colloid Interface Sci. 267, 60–64.

Zaslavsky, B.Y., Bagirov, T.O., Borovskaya, A.A., Gasanova, G.Z., Gulaeva, N.D., Levin, V.Y., Masimov, A.A., Mahmudov, A.U., Mestechkina, N.M., Miheeva, L.M., Osipov, N.N., Rogozhin, S.V., 1986. Aqueous biphasic systems formed by nonionic polymers I. Effects of inorganic salts on phase separation. Colloid Polym. Sci. 264, 1066–1071.

Zhang, Y-y., Liu, J-h., 2010. Purification and in situ immobilization of lipase from of a mutant of *Trichosporon laibachii* using aqueous two-phase systems. J. Chromatogr. B 878, 909–912.

Zhang, J., Zhang, Y., Chen, Y., Zhang, S., 2007. Mutual coexistence curve measurement of aqueous biphasic systems composed of [bmim][BF4] and glycine, l-serine, and l-proline, respectively. J. Chem. Eng. Data 52, 2488–2490.

Zhou, Y.-J., Hu, C.-L., Wang, N., Zhang, W.-W., Yu, X.-Q., 2013. Purification of porcine pancreatic lipase by aqueous two-phase systems of polyethylene glycol and potassium phosphate. J. Chromatogr. B 926, 77–82.

Znidarsic-Plazl, P., Plazl, I., 2007. Steroid extraction in a microchannel system-mathematical modelling and experiments. Lab Chip 7, 883–889.

Chapter 4

Lipase Activation and Stability Enhancement in Ionic Liquids

Hua Zhao
Savannah State University, Department of Chemistry and Forensic Science, Savannah, GA, United States

4.1 INTRODUCTION

Ionic liquids (ILs) have adjustable physical properties (such as polarity, hydrophobicity, and hydrogen-bond basicity) through a rational design and combination of cations and anions. These tunable properties are very important for biocatalytic systems; as a result, many enzymatic reactions have been examined in different types of IL media (Kragl et al., 2002; Park and Kazlauskas, 2003; van Rantwijk and Sheldon, 2007; Moniruzzaman et al., 2010). In particular, a number of hydrolase-catalyzed reactions have been evaluated in ILs, which has been reviewed in depth by several groups (van Rantwijk and Sheldon, 2007; Moniruzzaman et al., 2010; Sureshkumara and Lee, 2009). This chapter focuses on the lipase activation and stabilization following two major themes: (1) the effect of physical properties of ILs on lipase activity and stability (Section 4.2), and (2) methods to improve lipase activity and stability in ILs (Section 4.3). In addition, this chapter also briefly discusses the use of an IL analog, deep eutectic solvents (DES), in lipase-catalyzed reactions (Section 4.4).

4.2 EFFECT OF PHYSICAL PROPERTIES OF ILs ON LIPASE ACTIVITY AND STABILITY

The activity of an enzyme in ILs can be influenced by common factors including water activity, pH, excipients, and impurities (Yang and Pan, 2005). More importantly, unique physical properties of ILs have been found playing crucial roles on the lipase's activity, stability, and specificity. Some of these key solvent properties are discussed below with examples.

Ionic Liquids in Lipid Processing and Analysis. http://dx.doi.org/10.1016/B978-1-63067-047-4.00004-0

4.2.1 IL Polarity (also see: van Rantwijk et al., 2003)

There are several common parameters for quantifying the solvent polarity, such as dielectric constants (ε_r), Hildebrandt solubility (δ), dipole moments (μ), and the solvatochromic polarity scales (such as E_T^N and Kamlet–Taft scales). The E_T^N scale is a normalized polarity scale, which sets tetramethylsilane as 0.0 and water as 1.0 (Reichardt, 1994). The E_T^N values of some ILs are summarized in Table 4.1 from literature (Reichardt, 2005). Based on the solvatochromic scale, most ILs are marked as moderately polar being close to lower alcohols (Carmichael and Seddon, 2000; Aki et al., 2001) or formamide.

A practical enzymology often suggests that less polar, hydrophobic organic solvents lead to a higher enzyme activity and stability, especially at high temperatures. However, Narayan and Klibanov (1993) also indicated that the solvent polarity and water-miscibility may not have a direct correlation with the enzymatic activities of three lipases and one protease in organic solvents. A few groups have found that the lipase activities in ILs can be related to the IL polarity. Park and Kazlauskas (2001) observed a higher lipase (from *Pseudomonas cepacia*) activity in ILs with higher polarity in the acetylation of racemic 1-phenylethanol with vinyl acetate; for example, the initial reaction rate in less polar [BMIM][PF_6] was three times slower than that in more polar [EMIM][BF_4]. Mutschler et al. (2009) conducted the esterification of methyl-α-D-glucopyranoside with fatty acids catalyzed by immobilized lipase B from *Candida antarctica* (Novozym® 435), and found that the ester conversion increased with the IL polarity when ILs were employed as liquid film-coating (under solvent-free conditions), but decreased with the IL polarity when ILs were used as solvents. However, the correlation between IL polarity and enzyme activity has not been clearly established for other enzymatic reactions performed in ILs (Yang and Pan, 2005; Kaar et al., 2003; Schöfer et al., 2001; Zhao et al., 2006a). Relying on the E_T^N polarity scale in Table 4.1, some enzyme-denaturing ILs (such as [EMIM][dca], [BMIM]Cl, and [BMIM][OAc]) have about the same polarities as those enzyme-compatible ILs based on Tf_2N^- and PF_6^-.

4.2.2 Hydrogen-Bond (H-Bond) Basicity and Nucleophilicity of Anions

Although hydrogen-bond basicity and nucleophilicity are two different concepts, they are often relevant to each other. The stronger base is usually the stronger nucleophile in aprotic solvents for molecules containing the same nucleophilic atoms of the same charge. Based on the solvatochromic data, several groups have concluded the order of anions' basicity (nucleophilicity) in a decreasing order as follows:

Basicity series #1: (Muldoon et al., 2001) $OTf^-(CF_3SO_3^-) > Tf_2N^- > PF_6^-$

TABLE 4.1 Polarity of Ionic Liquids as Evaluated by the E_T^N Scale

Solvent	E_T^N (25°C)	References
Formamide	0.775	Reichardt (1994)
Methanol	0.762	Reichardt (1994)
Ethanol	0.654	Reichardt (1994)
1-Propanol	0.617	Reichardt (1994)
[EMIM][Tf_2N]	0.676	Selected value by Reichardt (2005)
[BMIM][Tf_2N]	0.642	Muldoon et al. (2001)
[HMIM][Tf_2N]	0.654	Selected value by Reichardt (2005)
[BMIM][PF_6]	0.667	Muldoon et al. (2001)
[EMIM][BF_4]	0.710	Park and Kazlauskas (2001)
[BMIM][BF_4]	0.673	Muldoon et al. (2001)
[BMIM][OTf]	0.667	Selected value by Reichardt (2005)
[BMIM][NO_3]	0.651	Selected value by Reichardt (2005)
[BMIM][OAc]	0.571	Selected value by Reichardt (2005)
[BMIM][CF_3COO]	0.630	Selected value by Reichardt (2005)
[EMIM][dca]	0.648	Selected value by Reichardt (2005)
[BMIM]Cl	0.614	Selected value by Reichardt (2005)

Basicity series #2: (Linert et al., 1993) $Cl^- > Br^- > SCN^- > OAc^- > I^- > NO_3^- > OTf^- > ClO_4^- > BF_4^-$
Basicity series #3: (Camard et al., 2005) $Cl^- > Br^- > OAc^- > OTf^- > ClO_4^- > BF_4^-$
Basicity series #4: (Oehlke et al., 2006) $Cl^- > Br^- > CH_3OSO_3^- > SCN^- > BF_4^- \sim OTf^- > PF_6^-$

Using the above series and some discussions in literature (MacFarlane et al., 2006; Pagni, 2002), we summarized the basicity of selected anions as shown in Fig. 4.1. These anions can be divided into three categories (basic, neutral, and acidic), and some of them are arranged in the order of basicity. Basic anions include halides, acetate, dicyanamide (dca^-), lactate, and methyl sulfate; these anions are good hydrogen-bond donors and tend to form hydrogen-bonds with proteins leading to enzyme denaturation and/or inactivation. Neutral anions include those forming hydrophobic ILs (Tf_2N^- and PF_6^-) and others forming hydrophilic ILs (BF_4^-, OTf^-, SCN^-, NO_3^-, and $CH_3SO_3^-$). These anions have weak abilities in forming hydrogen bonds.

Basic anions	Neutral anions	Acidic anions
←——→	←——→	←——→
$Cl^- > Br^- > OAc^- >$	$OTf^- > Tf_2N^- > BF_4^- > PF_6^- >$	$H_2PO_4^-$, HSO_4^- (amphoteric)
dca⁻, lactate⁻, $MeSO_4^-$	SCN^-, NO_3^-, $CH_3SO_3^-$	

← **Increasing Nucleophilicity**

FIGURE 4.1 Comparison of hydrogen-bond basicity of selected anions in ionic liquids.

Acidic anions (eg, amphoteric $H_2PO_4^-$ and HSO_4^-) are not common anions of ILs used in biocatalysis.

Bernson and Lindgren (1994) found that lithium salts LiX could be dissolved in poly(propylene glycol) (MW = 3000) with hydroxy end-groups, and observed that the shifts of −OH stretching band depend on the strength of H-bonds formed between the −OH group and the anion, as well as the coordination of cations with the −OH group, by using IR spectroscopy. The strength of anion coordination is thus dependent on the hydrogen-bond basicity of the anion, and can be deduced from the IR band shifts as follows (in an increasing order):

$$PF_6^- < BF_4^- < ClO_4^- < OTf^- < I^- < Br^- < Cl^-$$

This basicity series is generally in line with the basicity order from solvatochromic measurements (Fig. 4.2). Based on the IR and ESI-MS data, Dupont (2004) suggested an increasing order of the strength of H-bond basicity:

$$BPh_4^- < PF_6^- < BF_4^- < CF_3COO^-$$

Meanwhile, Henderson (2006) summarized the *ionic association strength* of LiX salts in a variety of aprotic solvents including glymes (see a short review in the Supporting Information of Ref Henderson, 2006). The approximate ionic association strength in aprotic solvents is illustrated in an increasing order (Henderson, 2006, 2007):

$$\text{beti}^-, Tf_2N^- < PF_6^- < ClO_4^-, I^- < SCN^- < BF_4^- < CF_3SO_3^- < Br^- < NO_3^- < CF_3COO^- < Cl^-$$

This order denotes the strength of an anion in interacting with the solvated cations via ionic attraction, or could be implied to represent the strength of interactions between the anions and charged regions of macromolecules (such as proteins/enzymes). Although the exact mechanism of this ion−protein interaction is not well established, this ionic association strength series resembles the anion's hydrogen-bond basicity order in Fig. 4.1.

To demonstrate the effect of nucleophilicity and basicity of anions on the lipase activity and stability, we discuss a number of enzymatic reactions carried out in ILs. The first set of examples emphasizes the impact of the anion's nucleophilicity. Free *Candida rugosa* lipase was found only active in hydrophobic [BMIM][PF_6], but inactive in all hydrophilic ILs based on NO_3^-, OAc^-, and CF_3COO^- during the transesterification of methylmethacrylate with 2-ethyl-1-hexanol (Kaar et al., 2003). The latter three anions are more nucleophilic than PF_6^-, and thus could interact with the enzyme causing the protein conformation changes. In this reaction, the solvent hydrophobicity is another important factor in influencing the enzyme activity (Section 4.2.6). Hernández-Fernández et al. (2009) observed that the stability of CALB (lipase B from *C. antarctica*) in ILs follows a decreasing order of [HMIM][PF_6] > [HMIM][Tf_2N] > [HMIM][BF_4], and [BMIM][PF_6] > [BMIM][dca], while the stability of Penicillin G acylase follows a similar order of [BMIM][Tf_2N] > [BMIM][PF_6] > [BMIM][BF_4]. It was explained that the decreasing stability was generally in agreement with the increasing order of nucleophilicity in Fig. 4.1 ($PF_6^- < BF_4^- < Tf_2N^- < dca^-$), where the more nucleophilic anions tend to interact with the positively charged sites in enzymes and to modify the enzyme's conformation. However, they also suggested that the enzyme's stability is consistent with the hydrophobicity of ILs: both enzymes were more stable in hydrophobic ILs than in hydrophilic ones (Section 4.2.6). Lee et al. (2008) determined the initial transesterification rates of three lipases (Novozym® 435, *Rhizomucor miehei* lipase, and *C. rugosa* lipase) in different ILs under the same water activity (a_w), and suggested the anion effect on the initial rates follows a decreasing order of $Tf_2N^- > PF_6^- > OTf^- > SbF_6^- \sim BF_4^-$. The likely reason is that OTf^- and BF_4^- are more nucleophilic than PF_6^-. Another contributing factor could be the IL hydrophobicity (Section 4.2.6) since the lipases appeared to be more active in hydrophobic ILs than in hydrophilic ones. Nevertheless, another study reported some contradictory results. Irimescu and Kato (2004) performed the CALB-catalyzed enantioselective acylation of 1-phenylethylamine with 4-pentenoic acid, and observed that the reaction rates depended on the type of IL anions (reaction rates in a decreasing order of $OTf^- > BF_4^- > PF_6^-$, same cations). This example implies a higher anion nucleophilicity actually leading to a higher enzymatic activity. In another acylation reaction of 2-phenyl-1-propylamine with 4-pentenoic acid, however, Irimescu and Kato (2004) reported that PF_6^- based ILs afforded fastest reaction rates, followed by OTf^- and BF_4^- based ILs. The puzzling results could be due to the fact that the enzymatic reaction is influenced by multiple factors of ILs such as nucleophilicity, hydrophobicity, viscosity, and impurity.

The second set of examples examined the effect of anion's hydrogen-bond basicity. Lee et al. (2006) noticed a drastic drop of the lipase activity in [OMIM][Tf_2N] with an increasing addition of [OMIM]Cl. Through using the multiple salvation interactions, Anderson et al. (2002) concluded that [BMIM]Cl exhibits the largest hydrogen-bond basicity among ILs considered and consequently could dissolve complex polar molecules including cyclodextrins and antibiotics (Armstrong et al., 1999). Lou et al. (2005)

observed no Novozym® 435 activity toward the ammonolysis of (*R*, *S*)-*p*-hydroxyphenylglycine methyl ester in [BMIM]Br and [BMIM][NO_3], suggesting these two ILs are lipase-denaturing. Lau et al. (2004) found a low CALB activity in [BMIM][lactate], which was caused by the hydrogen-bond formation between lactate anions and peptide chains leading to the secondary structure changes of the protein. Dicyanamide (dca^-)-based ILs including [BMIM][dca] could dissolve carbohydrates (Forsyth et al., 2002; Liu et al., 2005); however, [BMIM][dca] is an enzyme-inactivating IL (Toral et al., 2007; Zhao et al., 2008, 2009a) likely because of the high hydrogen-bond basicity of the anion. Our group (Zhao et al., 2009b) also observed that both free and immobilized CALB in [EMIM][OTf] and [BMIM][dca] were less active than in hydrophobic ILs. Bermejo et al. (2008) indicated that free CALB lost 35% of its initial activity upon dissolution in [HOPMIm][NO_3], although it maintained 80% of the remaining activity after 3 months of incubation in this IL. The loss of CALB activity in [HOPMIm][NO_3] is mainly induced by the denaturing effect of NO_3^-. The Xu group (Lue et al., 2010) carried out the acylation of flavonoids (esculin and rutin) with long-chain fatty acids in various ILs catalyzed by Novozym 435, and found the lipase was active in ILs carrying anions of low hydrogen-bond basicity (Tf_2N^-, PF_6^-, and BF_4^-) but inactive in those containing anions of high hydrogen-bond basicity (such as Cl^-, CF_3COO^-, $OctSO_4^-$, and $CF_3SO_3^-$).

4.2.3 IL Network

It has been reported that ILs could form hydrogen-bonded polymeric supramolecules, so-called organized nano-structures, with polar and nonpolar regions in solid, liquid, and solution states, or even in the gas phase (Dupont, 2004; Schröder et al., 2000). Dupont (2004) indicated that the aqueous solution of free enzymes could be surrounded by the IL network, which helps to

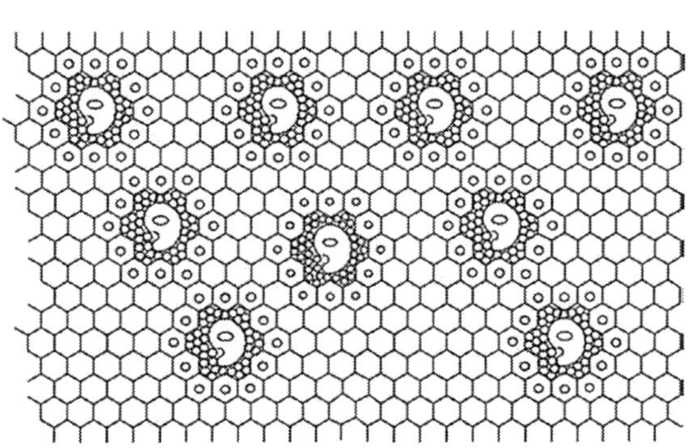

FIGURE 4.2 Enzymes with a small amount of water are firmly trapped in the network of ionic liquids. *Reproduced by permission, from Fehér, E., Major, B., Bélafi-Bakó, K., Gubicza, L., 2007. On the background of enhanced stability and reusability of enzymes in ionic liquids. Biochem. Soc. Trans. 35, 1624–1627, © the Biochemical Society.*

retain the native structure of proteins by preserving the essential water of proteins and the preferential solvophobic interactions. For imidazolium-based ILs, each cation coordinates with at least three anions, whereas each anion coordinates with three cations, resulting in a polymeric network. The inclusion of other molecules and macromolecules into this polymeric network leads to the formation of polar and nonpolar regions (Dupont, 2004). When the enzyme-in-water droplets are dissolved (or dispersed) into the IL network (in polar regions), the enzyme's active conformation can be preserved by the network (see Fig. 4.2) (Fehér et al., 2007). The inclusion of enzyme molecules in such highly ordered supramolecular structures of ILs prevents the protein from thermal unfolding (Lozano et al., 2005). However, since enzymes are not soluble in most common ILs, enzyme molecules (in particular, immobilized enzymes) are practically suspended in reaction media with low or little water; as a result, the IL network theory is not always suitable for explaining the enzyme activity and stability. The impact of individual anions on enzyme inactivation also cannot be explained by the IL network.

4.2.4 Ion Kosmotropicity

In diluted aqueous solutions, hydrophilic ILs become dissociated at least partially and solvated ions, and these ions may interact with the enzyme directly. The ion specificity was proposed by Franz Hofmeister in his observation of ions showing different abilities in precipitating the proteins (globulins from blood serum and hen's egg) (Hofmeister, 1888; Kunz et al., 2004). The sequence of these ions in salting out proteins is known as the Hofmeister series (Fig. 4.3). Several theories have been actively discussed to explain the Hofmeister effect, which include salt-in and salt-out interactions (Baldwin, 1996; Arakawa and Timasheff, 1984), water-structure changes (low/high

← Protein stabilization Protein destabilization →

Anions (*kosmotropic*) PO_4^{3-} > SO_4^{2-} > $EtSO_4^-$ > OAc^- > $MeSO_4^-$ > Cl^- > Br^- > I^- > BF_4^- > PF_6^- (*chaotropic*)

B-coefficients: 0.495→0.206→0.265→0.246→0.188→-0.005→-0.033→-0.073→-0.093→-0.21

Cations: (*chaotropic*) $(CH_3)_4N^+$ > K^+ > Na^+ > Li^+ > Ca^{2+} > Mg^{2+}> Al^{3+} (*kosmotropic*)

B-coefficients: 0.123→0.009→0.085→0.146→0.284→0.385→0.744

FIGURE 4.3 The Hofmeister series as an order of the ion effect on protein stability (Collins and Washabaugh, 1985; von Hippel and Schleich, 1969) (viscosity *B*-coefficients in dm^3/mol at 25°C are taken from the Marcus collection (Jenkins and Marcus, 1995) except those of $EtSO_4^-$ and $MeSO_4^-$ were from Tamaki et al. (1973); the positions of $EtSO_4^-$ and $MeSO_4^-$ are based on the consideration of *B*-coefficients, NMR *B*′-coefficients (Zhao and Song, 2007), and enzyme stability studies (Zhao et al., 2006b; Constantinescu et al., 2007; Fujita et al., 2007; Yang et al., 2009)).

density water) and protein preferential hydration (Collins, 2004; Bauduin et al., 2004; Collins and Washabaugh, 1985; Frank and Wen, 1957; Franks, 1973; von Hippel and Wong, 1965; Wiggins, 1997), hydrophobic interactions (Wiggins, 1997; Tanford, 1978; Némethy, 1967), excluded volume (Minton, 1981; Davis-Searles et al., 2001; Schellman, 2003), preferential interactions (Robinson and Jencks, 1965; Courtenay et al., 2001; Myers et al., 1995), electrostatic interactions (Honig and Nicholls, 1995; Perutz, 1978), and so on. However, there is no single theory that can fully interpret the Hofmeister effect due to the complex nature of ion–protein interactions.

Despite the debates over these theories, the protein stability is often associated with the hydration behavior of ions (Baldwin, 1996; Boström et al., 2004). Strongly hydrated ions (such as Mg^{2+}, Ca^{2+}, Li^{+}, CH_3COO^{-}, SO_4^{2-}, and HPO_4^{2-}) have strong interactions with water molecules and increase the structuring of water, resulting in a lower fluidity (or a higher viscosity) of the solution than that of pure water. Therefore, these ions are named structure-makers or kosmotropes (see Fig. 4.3). On the other hand, some other ions are weakly hydrated in aqueous solutions, such as SCN^{-}, I^{-}, NO_3^{-}, BF_4^{-}, Cs^{+}, $(NH_2)_3C^{+}$ (guanidinium), and $(CH_3)_4N^{+}$ (tetramethylammonium). They have weak interactions with water molecules and reduce the structuring of water, resulting in a higher fluidity of the solution. Therefore, this effect is known as negative hydration (Samoilov, 1953, 1957), and these ions are often known as structure-breakers or chaotropes (see Fig. 4.3).

The capacity of an ion in strengthening the water structure, known as kosmotropicity (vs chaotropicity), therefore, is directly associated with the degree of ion hydration. As discussed in our review (Zhao, 2006a), the ion kosmotropicity can be measured by different thermodynamic parameters including viscosity B-coefficients, structural entropies, structural volumes, structural heat capacities, NMR B'-coefficients, ion mobility, and so forth. These parameters provide valuable information of the interactions involved in the ion hydration from different aspects, and possibly reveal the mechanism behind some phenomena and properties. Jones-Dole viscosity B-coefficients are the most frequently used and widely available parameters for comparing the ion kosmotropicity. The B-coefficients can be computed from the Jones-Dole empirical equation (Eq. [1]) of the relative viscosities of electrolyte solutions as functions of their concentrations (Jones and Dole, 1929):

$$\eta/\eta_0 = 1 + Ac^{1/2} + Bc + Dc^2 \ldots \qquad [1]$$

where η is the viscosity of the solution and η_0 is the viscosity of the solvent (both of them have the same unit, for example Pa s), while c is the molar concentration (mol/cm^3). The A-coefficients (also known as the Falkenhagen coefficient (Ali et al., 2005)), representing the solute–solute or electrostatic interactions, can be calculated theoretically. However, A-values are usually small and negligible for nonelectrolytes (Jenkins and Marcus, 1995);

therefore, they are often omitted in calculations. The *B*-coefficients represent the solute–solvent interactions (short-range dispersion forces), while *D*-coefficients indicate the solute–solute interactions as well as the solute–solvent interactions (Belibagli and Ayranci, 1990). For most salts at low concentrations [(<0.5 M) (Jenkins and Marcus, 1995) or (<0.1 M for binary strong electrolytes) (Collins, 1997)], the *D* or higher coefficients can be neglected although they are needed at higher concentrations (Jenkins and Marcus, 1995). Positive *B*-values typically indicate ions as kosmotropes since strongly hydrated ions exhibit a larger change in viscosity with concentration, while negative *B*-coefficients imply chaotropes for weakly hydrated ions (Jenkins and Marcus, 1995). However, hydrophobic solutes tend to have unusually large *B*-coefficients due to the so-called hydrophobic hydration (Zhao, 2006a). For example, tetramethylammonium cation (Me_4N^+) has a positive *B*-value as high as 0.123 (Jenkins and Marcus, 1995), but this ion is considered as a structure-breaker (Frank and Wen, 1957; Marcus, 1994; Kay et al., 1966; Frank and Evans, 1945; Nightingale, 1962; Frank, 1963). Some researchers (Jenkins and Marcus, 1995; Sarma and Ahluwalia, 1973; Tsangaris and Martin, 1965; Devine and Lowe, 1971) recommend the use of first derivatives of *B*-values over temperature because the sign of dB/dT might be more indicative for measuring the structure-making or breaking ability than the sign or quantity of *B*-coefficients. The negative sign of dB/dT means structure-making (kosmotropic) while the positive sign suggests structure-breaking (chaotropic).

In aqueous solutions of inorganic salts, many studies (see our review (Zhao, 2005)) have concluded that the ion effect on the enzyme activity followed the ion kosmotropicity (Hofmeister series): *kosmotropic anions and chaotropic cations stabilize the enzyme, while chaotropic anions and kosmotropic cations destabilize it.* A series of studies (Zhao et al., 2006a,b,c,d,e, Zhao and Song, 2007; Zhao, 2006a,b) in our laboratory have demonstrated that the same principle is loosely applicable to the enzyme activity in IL aqueous solutions. In our first study, the activities of Amano protease P6 (from *Aspergillus melleus*) in 0.7 M IL aqueous solutions were affected by anions in a decreasing order of CH_3COO^-, $CF_3COO^- > Cl^-$, $Br^- > OTs^- > BF_4^-$ (which is in agreement with the decreasing order of anion's kosmotropicity), and affected by cations in a decreasing order of $EMIM^+$, $BuPy^+ > BMIM^+ > EtPy^+$ (Zhao et al., 2006a). In a second study (Zhao et al., 2006b) our group carried out the kinetic hydrolysis of enantiomeric phenylalanine methyl ester catalyzed by *Bacillus licheniformis* protease in aqueous solutions of several hydrophilic ILs (0.5 M). The protease enantioselectivity was in a decreasing order in the presence of these anions: $PO_4^{3-} >$ citrate^{3-}, CH_3COO^-, $EtSO_4^-$, $CF_3COO^- > Br^- >$ OTs^-, BF_4^- (decreasing kosmotropicity), and in the presence of these cations: $EMIM^+ > BMIM^+ > HMIM^+$ (decreasing chaotropicity). The overall IL kosmotropicity can be measured by the δ value (difference in viscosity *B*-coefficients of anion and cation). In general, a high enzyme enantioselectivity was

← ***Increasing kosmotropicity***

Anions Citrate^{3-} > CF_3COO^- > $EtSO_4^-$ > OAc^- > Cl^- > BF_4^-
B'-coefficients: 0.65 → 0.21 → 0.17 → 0.13 → -0.017 → -0.17
Cations Mg^{2+} > Na^+ > K^+ and
$HMIM^+$, $BuPy^+$ > $BMIM^+$ > $EMIM^+$ > Me_4N^+, $EtPy^+$
B'-coefficients: 0.60 → 0.050 → -0.017 and
0.40, 0.40 → 0.33 → 0.29 → 0.18, 0.11

FIGURE 4.4 The ion kosmotropicity based on NMR *B'*-coefficients (Zhao and Song, 2007).

observed in the solution of IL with a high δ value. After measuring the NMR *B'*-coefficients of a number of ions (see Fig. 4.4, which is consistent with Fig. 4.3), our group (Zhao and Song, 2007) further found a linear correlation between the enzyme enantioselectivity in aqueous solution and the δ' parameter (difference in NMR *B'*-coefficients of anion and cation) of ILs, suggesting that high enzyme enantiomeric ratios (*E*) could be achieved in solutions of ILs with high δ' values. Other groups (Husum et al., 2001; Kaftzik et al., 2002; Lang et al., 2006) reported low or no activities of β-glycosidase in aqueous solutions of [BMIM][BF_4], which may be explained by the chaotropic nature of BF_4^- in solutions (Lang et al., 2006). (**Note**: In neat ILs based on BF_4^-, the chaotropic property of anion may not influence the enzyme activity; therefore, many studies observed certain enzyme activities in BF_4^- based ILs.) Our group (Zhao et al., 2006c) also conducted the enzymatic hydrolysis of DL-phenylalanine methyl ester in aqueous solutions of ILs containing anions of chiral- or ω-amino acids, and reported higher ees and yields in ILs based on D-amino acids rather than in those derived from L-isomers. The likely explanation is that amino acid anions are more kosmotropic than zwitterionic amino acids (Zhao, 2006b), and D-amino acids are more kosmotropic than L-isomers (Wiggins, 2003). The use of ILs with kosmotropic anions (OAc^- and CF_3COO^-) in activating hydrolases in aqueous solutions was further demonstrated in two of our studies (Zhao et al., 2006d,e).

Recently, Fujita et al. (2005, 2006, 2007) evaluated the stability of cytochrome *c* in ILs containing 20% (wt) water and its relevance to the kosmotropicity of components ions; the cation's effect on the protein stability followed a decreasing order of choline$^+$ > $BMPyrr^+$ > $BMIM^+$, which is also a decreasing order of cation chaotropicity; the anion's effect on the protein stability followed a decreasing order of $H_2PO_4^-$ > $Bu_2PO_4^-$ > OAc^- > lactate$^-$ > $MeSO_4^-$, which is the decreasing order of anion kosmotropicity (*B*-coefficients at 25°C: $H_2PO_4^- = 0.340\ dm^3/mol$) (Jenkins and Marcus, 1995), $OAc^- = 0.246$ (Jenkins and Marcus, 1995), $MeSO_4^- = 0.188$ (Tamaki et al., 1973); lactate might be considered a kosmotropic anion (Tanaka and Oka, 2005).

Constantinescu et al. (2007, 2010) suggested that the thermal stability of ribonuclease A (RNase A) in aqueous solution of ILs (typically 0–2 M) followed the Hofmeister series. In their study, differential scanning calorimetry

(DSC) was employed to measure the effect of ILs on the thermal denaturation of RNase A near 60°C. In terms of decreasing protein stability, the cation series are

$$K^+ > Na^+ \sim Me_4N^+ > Li^+ > Et_4N^+ \sim EMIM^+ > BMPyrr^+ > BMIM^+ \sim Pr_4N^+ > HMIM^+ \sim Bu_4N^+ \text{and}$$

$$K^+ > Na^+ \sim Me_4N^+ > Cholinium^+ > EMIM^+ \sim Guanidinium^+ > BMIM^+$$

The anion series follows :

$$SO_4^{2-} > HPO_4^{2-} > Cl^- > EtSO_4^- > BF_4^- \sim Br^- > MeSO_4^- > OTf^- > SCN^- \sim dca^- > Tf_2N^-$$

The cation series suggests the higher the cation hydrophobicity (of organic cations), the higher the cation kosmotropicity, and the lower the protein stability in general. The anion series offers the opposite: the higher the anion kosmotropicity, the higher the protein stability in general (with slight differences in the position of neighboring ions from our earlier discussion).

Choline acetate is an IL consisting of a kosmotropic anion and a chaotropic cation. The Huang group (Xue et al., 2013) found that at low concentrations (up to 5 mM), this IL could improve the hydrolytic activity of *C. rugosa* lipase in AOT/water/isooctane reverse micelles (Fig. 4.5), and causes no lipase conformational changes as evidenced by fluorescence spectra. Infrared spectra suggest stronger hydrogen bonds between choline acetate and water than those between water molecules; as a result, the addition of a low content of choline acetate improves the nucleophilicity of water, accelerating the attack of water molecules on the acyl enzyme intermediate and increasing the lipase's catalytic efficiency.

An excellent review by Yang (2009) systematically discussed the possible mechanisms of Hofmeister effects of ILs on the enzyme activity and stability. The above experimental studies have shown that the kosmotropic effect of ILs on enzymes may be applicable to diluted aqueous solutions of ILs (Zhao et al., 2006a,b; Zhao, 2005) as well as some concentrated ILs (such as 20 wt % water (Fujita et al., 2007)). However, it is not quite clear if such an effect exists in neat ILs or ILs with a trace amount of water, and how the IL hydrophobicity may influence the kosmotropicity. For example, PF_6^- is a chaotropic anion (Zhao, 2006a), and denatures enzymes when dissolved in aqueous solutions as Na^+ or K^+ salt (more denaturing than BF_4^- and $MeSO_4^-$ for mushroom tyrosinase (Yang et al., 2009)). However, PF_6^- based ILs (such as [BMIM][PF_6]) are hydrophobic, and thus the solubility and degree of dissociation of ILs in water are limited. Meanwhile, it is also known that PF_6^- based ILs containing low water contents are typically enzyme stabilizing (van Rantwijk and Sheldon, 2007). Therefore, the Hofmeister effect may not be suitable for explaining the enzyme's behaviors

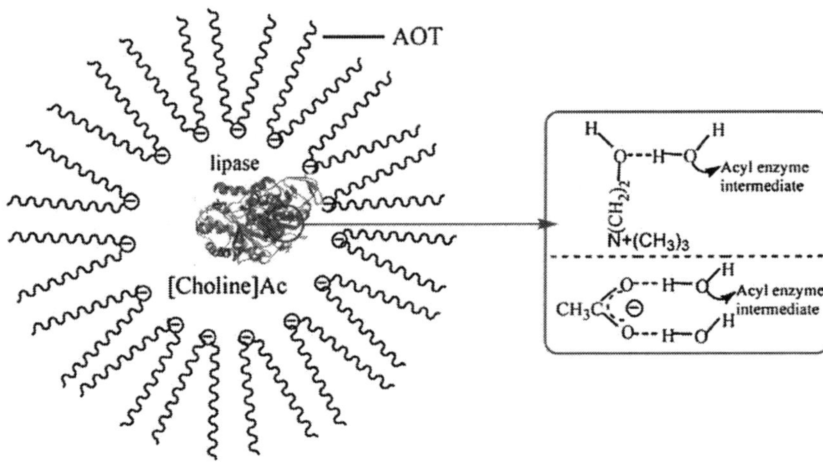

FIGURE 4.5 Illustration of choline acetate influencing the nucleophilicity of water molecules near the lipase in AOT reverse micelles. *Adapted from Xue, L., Zhao, Y., Yu, L., Sun, Y., Yan, K., Li, Y., Huang, X., Qu, Y., 2013. Choline acetate enhanced the catalytic performance of* Candida rogusa *lipase in AOT reverse micelles. Colloids Surf. B 105, 81–86, reproduced by permission of Elsevier Limited.*

in these hydrophobic ILs or their mixtures with water. Without sufficient water to hydrate them, kosmotropic or borderline anions (such as acetate, lactate, and chloride) of ILs bearing high hydrogen-bond basicities tend to interact strongly with enzymes, causing their inactivation (Section 4.2.2). Accordingly, the enzyme stabilization/activation kosmotropic anions (such as OAc^- and Cl^-) in diluted aqueous solutions become enzyme-inactivating agent in ILs containing none or low contents of water. For example, several papers (Zhao et al., 2006a,b,d; Turner et al., 2003) have mentioned the enzyme activation at low concentrations of chloride-based ILs in water, but inactivation at high concentrations.

4.2.5 IL Viscosity

ILs are typically more viscous than conventional organic solvents (see a collection of viscosity data of ILs in (Wasserscheid and Welton, 2008)); furthermore, many enzymatic reactions carried out in ILs are heterogeneous systems because of the low solubility of enzymes in many neat ILs. For these reasons, the internal and external mass-transfer limitations must be considered (Kragl et al., 2002). Eckstein et al. (2002) observed a higher enantioselectivity of lipase in [BMIM][Tf_2N] at low water activities ($a_w < 0.53$) than in methyl *tert*-butyl ether (MTBE); this is due to two likely reasons: (1) the higher viscosity of the IL (52 mPa s) than that of MTBE (0.34 mPa s) could cause a higher mass transfer limitation and a slower reaction rate; (2) the lower solubility of substrates in the IL than in MTBE might result in a lower activation energy in the IL. van Rantwijk and Sheldon (2007) suggested that the high viscosity of ILs reduces the conformation changes of proteins, enabling enzymes to retain their native structures and activity. Our group (Zhao et al., 2009a) examined the CALB-catalyzed transesterification of ethyl butyrate and 1-butanol in more than 20 ILs, and indicated that the IL viscosity could influence the reaction rates in some cases, but may not always be the controlling factor of enzyme activity.

4.2.6 IL Hydrophobicity

Hydrophobicity may be considered as a subset of polarity. In practice, it is important to differentiate hydrophobicity from polarity because the former is often referred to the miscibility with water (van Rantwijk et al., 2003). The hydrophobicity of ILs may be quantified by the log P scale, a concept derived from the partition coefficient of ILs between 1-octanol and water. The partition coefficient (K_{OW} or P) is a ratio of concentrations of un-ionized compound between the two phases. The log P is defined as the partition coefficient at the unlimited dilution concentration of solute,

$$\log P = \lim_{c \to 0} K_{OW} = \lim_{c \to 0} \frac{C^{o}}{C^{w}} \qquad [2]$$

where C^{o} is the IL concentration in the octanol phase and C^{w} is the IL concentration in the aqueous phase. Practically, very low concentrations of ILs are used in the experiment instead of extrapolating the IL concentration to zero (Eq. [2]). However, since ILs dissociate into ions in water and current K_{OW} values were reported as the ratio of concentrations of both undissociated and dissociated ILs in two phases, most log P values of ILs (Table 4.2) should be strictly called log D, where D is the distribution coefficient, the ratio of the total concentrations of all forms of IL (ionized and un-ionized) between two phases. Alternatively, the intrinsic partition coefficients of ILs should be calculated from the apparent partition coefficients (D) (Lee and Lee, 2009).

The log P values (or log K_{OW} at low concentrations) of ILs in Table 4.2 are informative for comparing the hydrophobicity of ILs with conventional organic solvents. Overall, ILs are very hydrophilic in nature based on the negative log P values (or log K_{OW}) of most ILs (including some water-immiscible Tf_2N^- and PF_6^- ILs). The discrepancy between different data of the same IL might be due to different initial concentrations of ILs (as high concentrations leading to higher K_{OW} values (Lee and Lee, 2009; Ropel et al., 2005)) and different experimental techniques.

A number of studies have evaluated the effect of IL hydrophobicity on the lipase stabilization. The Russell group (Kaar et al., 2003) determined the log P values (<-2.0) for several ILs, and indicated that these ILs are very hydrophilic in nature based on the Laane's scale (Hilhorst et al., 1984;

TABLE 4.2 log *P* (or log K_{OW} at Low Concentrations[a]) Values of Ionic Liquids at 25°C

	Solvent	log *P* (or log K_{OW})	References
1	Dichloromethane	1.25	Selected value by Sangster (1989)
2	THF	0.46	Selected value by Sangster (1989)
3	*t*-Butanol	0.35	Selected value by Sangster (1989)
4	Acetone	−0.24	Selected value by Sangster (1989)
5	Acetonitrile	−0.34	Selected value by Sangster (1989)
6	[EMIM][Tf_2N]	−1.18 log K_{OW} (−1.05 to −0.96) (0.28–2.8 mM)	Lee et al. (2007a) Calculated from Ropel et al. (2005)
7	[BMIM][Tf_2N]	0.11 log K_{OW} (−0.96 to −0.21) (0.15–2.2 mM) 0.33 −1.74	Zhao et al. (2009a) Calculated from Ropel et al. (2005) Lee et al. (2005) Lee and Lee (2009)
8	[HMIM][Tf_2N]	0.64 log K_{OW} (0.15–0.22) (0.32 –0.38 mM) 0.65	Zhao et al. (2009a) Calculated from Ropel et al. (2005) Lee et al. (2005)
9	[OMIM][Tf_2N]	0.79 log K_{OW} (0.80–1.05) (0.099–0.21 mM)	Lee et al. (2007a) Calculated from Ropel et al. (2005)
10	[EMMIM][Tf_2N]	log K_{OW} (−1.15 to −0.92) (0.32–2.9 mM)	Calculated from Ropel et al. (2005)
11	[PMMIM][Tf_2N]	log K_{OW} (−0.92 to −0.62) (1.4–2.8 mM)	Calculated from Ropel et al. (2005)
12	[HMMIM][Tf_2N]	log K_{OW} (0.13–0.25) (0.36 –0.49 mM)	Calculated from Ropel et al. (2005)
13	[BMIM][PF_6]	−1.66 −2.39 −2.38 −2.06 −2.35	Calculated from Ropel et al. (2005) Kaar et al. (2003) and Lee et al. (2005) Lee and Lee (2009) and Ulbert et al. (2004) Lee et al. (2007a) Lou and Zong (2006)

TABLE 4.2 log *P* (or log K_{OW} at Low Concentrations[a]) Values of Ionic Liquids at 25°C—cont'd

	Solvent	log *P* (or log K_{OW})	References
14	[HMIM][PF_6]	−1.86	Lee et al. (2005)
15	[OMIM][PF_6]	−0.35 −1.33	Lee et al. (2007a) Lee et al. (2005)
16	[OMIM][PF_6]	−2.19	Ulbert et al. (2004)
17	[BMIM]Cl	−2.40	Calculated from Ropel et al. (2005)
18	[BMIM]Br	−2.48	Calculated from Ropel et al. (2005)
19	[EMIM][OAc]	−2.53	Zhao et al. (2009a)
20	[BMIM][OAc]	−2.77	Kaar et al. (2003)
21	[EMIM][CF_3COO]	−2.75	Zhao et al. (2009a)
22	[HMIM][CF_3COO]	−2.30	Zhao et al. (2009a)
23	[BMIM][NO_3]	−2.90 −2.42	Kaar et al. (2003) Calculated from Ropel et al. (2005)
24	[BMIM][dca]	−2.32	Zhao et al. (2009a)
25	[EMIM][BF_4]	−2.57	Zhao et al. (2009a)
26	[BMIM][BF_4]	−2.51 −2.44 −2.52	Zhao et al. (2009a) Ulbert et al. (2004) and Lou and Zong (2006) Calculated from Ropel et al. (2005)
27	[OMIM][BF_4]	−1.34 −1.14	Zhao et al. (2009a) Lee et al. (2007a)
28	[EtPy][CF_3COO]	−2.57	Zhao et al. (2009a)
29	[EtPy][Tf_2N]	−0.90	Zhao et al. (2009a)
30	[BuPy][Tf_2N]	−0.26	Zhao et al. (2009a)
31	[Choline][Tf_2N]	log K_{OW} = −0.57 (calculated value)	Chapeaux et al. (2007)[b]

[a]*log K_{OW} values calculated from Ropel et al. (2005) were converted from initial values of K_{OW} measured at room temperature (24 ± 2°C), and the concentration range given for each log K_{OW} was the IL concentration range in the aqueous phase.*
[b]*This reference also provides K_{OW} values for a number of pyridinium and imidazolium ILs based on Tf_2N^- and $B(CN)_4^-$.*

Laane et al., 1985, 1987); this group also found that free lipase (*C. rugosa*) was only active in hydrophobic [BMIM][PF_6] (log $P = -2.39$), but inactive in other hydrophilic ILs including [BMIM][CH_3COO] (log $P = -2.77$), [BMIM][NO_3] (log $P = -2.90$), and [BMIM][CF_3COO] (Kaar et al., 2003). Another study by Nara et al. (2002) reported higher transesterification activities of lipases in [BMIM][PF_6] than in [BMIM][BF_4]. The Goto group (Nakashima et al., 2006a) also observed higher activities of poly(ethylene glycol) (PEG)-modified lipase in more hydrophobic ILs (ie, [EMIM][Tf_2N]). Lou and Zong (2006) examined the lipase-catalyzed enantioselective acylation of (*R,S*)-1-trimethylsilylethanol with vinyl acetate in several ILs, and concluded that the activity, enantioselectivity, and thermostability of Novozym® 435 increased with the IL hydrophobicity ([BMIM][PF_6] > [OMIM][BF_4] > [C_7MIM][BF_4] > [HMIM][BF_4] > [C_5MIM][BF_4] > [BMIM][BF_4]). A 2007 study (de los Ríos et al., 2007) on the CALB-catalyzed alcoholysis of vinyl butyrate and 1-butanol revealed that the lipase activities were generally much lower in water-miscible ILs (such as BF_4^-, dca$^-$, NO_3^-, OAc$^-$, etc.) than in water-immiscible ones (PF_6^- and Tf_2N^-), and the enzyme's activities increased with the cation's hydrophobicity ($EMIM^+ < BMIM^+ < HMIM^+ < OMIM^+$). Ha et al. (2008) also noticed that Novozym® 435 was less active and stable in hydrophilic ILs (BF_4^- and OTf$^-$) than in other hydrophobic ILs (Tf_2N^- and PF_6^-). Lee et al. (2008) observed that Novozym® 435 was more thermally stable in hydrophobic ILs than in hydrophilic ILs following the order of [BMIM][Tf_2N] > [BMIM][PF_6] > [BMIM][OTf] > [BMIM][BF_4] > [BMIM][SbF_6]. Shen et al. (2008) obtained a high enantioselectivity (80% ee_p) of Amano lipase PS in hydrophobic [OMIM][PF_6], but poor enantioselectivities (<5% ee_p) in hydrophilic [HMIM][BF_4] and [HMIM]Cl (Section 4.2.8.1) during the kinetic resolution of racemic cyanohydrins. Hernández-Fernández et al. (2009) suggested that both free CALB and penicillin G acylase (PGA) were more stable in hydrophobic ILs than in hydrophilic ones; for CALB, the stability was in a decreasing order of [HMIM][PF_6] > [HMIM][Tf_2N] > [HMIM][BF_4], and [BMIM][PF_6] > [BMIM][dca], as well as [OMIM][PF_6] > [HMIM][PF_6] > [BMIM][PF_6]; for PGA, the stability was in a decreasing order of [BMIM][Tf_2N] > [BMIM][PF_6] > [BMIM][BF_4]. On the other hand, the hydrophobic cations showed a different effect on the PGA stability: [EMIM][Tf_2N] > [BMIM][Tf_2N], and [BMIM][PF_6] > [OMIM][PF_6]. The effect of nucleophilicity of these anions has been discussed in Section 4.2.2. These studies indicated that the high hydrophobicity (large log *P*) of ILs could be beneficial to the enzyme stabilization.

Our group (Zhao et al., 2009a) systematically evaluated the Novozym® 435-catalyzed transesterification in over 20 ILs, and found that the lipase activity increased with the log *P* value of ILs to a maximum, and then declined with a further increase in log *P* (a bell shape). Similarly, Lou et al. (2005)

found the initial rates of Novozym® 435-catalyzed ammonolysis of (*R*,*S*)-*p*-hydroxyphenylglycine methyl ester increased with the hydrophobicity of BF_4^- based ILs to a maximum, and then decreased with a further increase in the IL hydrophobicity. Kim and Koo (2012) studied the CALB-catalyzed transesterification of vinyl acetate and *n*-butanol in OTf^- anion–containing ILs, and observed the initial rate increased with the IL alkyl chain length (from $EMIM^+$ to $BMIM^+$), and then decreased with the alkyl chain length ($HMIM^+$ and $OMIM^+$). They explained that both the viscosity and hydrophobicity of ILs could be important factors for this enzyme activity trend. These results implied that the enzyme is active in hydrophobic solvents (with high log *P*). However, a higher log *P* of the solvent also leads to a more thermodynamic ground-state stabilization of substrates (Ryu and Dordick, 1992), which decreases the conversion of substrates. This could be one of the explanations of the decreasing reaction rate in very hydrophobic ILs.

This section intends to offer some mechanistic discussions on why a higher hydrophobicity of ILs leads to a higher enzyme activity (up to the highest peak of activity). It is known that polar organic solvents strip off the essential water from enzyme molecules, resulting in their inactivation (Laane et al., 1987; Zaks and Klibanov, 1988). This might be a key factor for many of the above reactions. However, relatively high enzyme activities could still be achieved in [BMIM][Tf_2N] when substrates, IL, and CALB were all exclusively dried (Zhao et al., 2009a) (CALB is active in organic solvents containing little or no water (van Rantwijk et al., 2006; Réjasse et al., 2004, 2006)). Thus, the solvent stripping capacity is unlikely the only mechanism of Novozym® 435 inactivation by hydrophilic ILs. An alternative explanation may be understood from the correlation between log *P* and the solvent's hydrogen-bond basicity. A general regression equation has been established as the following (Abraham et al., 1994a):

$$\log P = c + rR + s\pi^* + a\alpha + b\beta + vV_x \qquad [3]$$

where *R* is the excess molar refraction obtained from refractive index measurements, V_x is the McGowan's characteristic volume or simply the intrinsic volume of the solute, α reflects the H-bond acidity (H-bond donating ability, property of IL cation), β reflects the H-bond basicity (H-bond accepting ability, property of IL anion), and π^* reflects the dipolarity/polarizability. The correlations of 613 organic solutes yielded the coefficient values as $c = 0.088$, $r = 0.562$, $s = -1.054$, $a = 0.034$, $b = -3.460$, and $v = 3.814$ (V_x in unit of $cm^3\ mol^{-1}/100$) (Abraham et al., 1994a,b). The near-zero *a*-coefficient indicates that the hydrogen-bond acidity does not contribute much to the partition. The *s*- and *b*-coefficients are negative since water is more dipolar and is a stronger hydrogen-bond acid than wet octanol (Abraham et al., 1999). The large and positive *v*-coefficient suggests that larger solute molecules are more hydrophobic and tend to partition into the octanol layer. Other correlations using different solutes reached similar conclusions (Abraham, 1993;

Kamlet et al., 1988; El Tayar et al., 1991). Although (Eq. [3]) was correlated from non-IL organic solutes, those coefficients generated from 613 organic solutes offer some insight into solute–solvent interactions. Based on (Eq. [3]), the higher the hydrogen-bond basicity of an IL anion, the lower the log *P* value, and thus the lower the enzyme activity due to the hydrogen-bonding interactions between the anion and enzyme. This elucidates the increasing trend of the enzyme activity with the log *P* value up to a certain value. Therefore, the IL hydrophobicity is closely related to the hydrogen-bond basicity of IL anions.

Then what is the reason of the lipase activity decreasing with a further increase in log *P*? Since *v*-coefficient is quite positive in (Eq. [3]), an increase in cation's size results in the increase of log *P* value. As mentioned earlier, the stabilization of substrates could be one reason. However, the possible hydrophobic interactions between large IL molecules and the enzyme could be another factor. For example, the Atkin group (Mann et al., 2009) investigated the stability and activity of hen's egg white lysozyme in aqueous solutions of four protic ILs (25–75 wt%); the protein denaturing–renaturing circular dichroism (CD) experiments and the activity measurements of lysozyme indicated that the highest catalytic activity and most complete refolding was achieved in solutions of $[(EtOH)NH_3][HCOO]$, followed by $[PrNH_3][HCOO]$, and then $[EtNH_3][HCOO]$ and $[(MeOEt)NH_3][HCOO]$. It is believed that the protein-IL interactions include the electrostatic interaction of IL cations with negatively charged residues in the protein, hydrogen-bonds between amine protons and the protein, as well as the hydrophobic interactions between alkyl chains in ILs and hydrophobic regions of the protein. Since electrostatic interactions between $[(EtOH)NH_3]^+$ and lysozyme is about the same as for $[EtNH_3]^+$, the hydroxyl group in $[(EtOH)NH_3]^+$ probably reduces the strength of hydrophobic interactions with the protein. Another possibility is that the hydroxyl group interacts with the anion formate via hydrogen bonds, reducing the interaction of formate with the protein. The IL viscosity-induced mass transport was not a limiting factor in the study because $[(EtOH)NH_3][HCOO]$ is several times more viscous than the other three ILs. In summary, the hydrophobicity factor of ILs is a combination effect of anion's hydrogen-bond basicity and cation's hydrophobic effect.

Since hydrophobicity is not the sole factor in manipulating the lipase activity, confusing results of some biocatalytic reactions may be difficult to explain. De Diego et al. (2009) carried out the transesterification of vinyl propionate and 1-butanol catalyzed by free and immobilized lipases from *C. antarctica* (CALA and CALB), *Thermomyces lanuginosus* (TLL), and *Rhizomucor miehei* (RML). Most of the enzyme preparations (except free CALA) exhibited higher activities in more hydrophobic $[OMIM][PF_6]$ than in $[BMIM][PF_6]$, but lower activities in other more hydrophobic-based ILs ($[OMIM][BF_4] < [HMIM][BF_4] < [BMIM][BF_4]$, and $[BDMIM][PF_6] < [BDMIM][BF_4]$). Another study by Irimescu and Kato (2004) on the lipase-catalyzed acylation of primary amines suggested lower reaction rates in

ILs with longer alkyl chains in cations, and the water miscibility of ILs was not a key factor in controlling the reaction rate. Some studies also obtained relatively high enzyme activities in hydrophilic ILs (such as [BMIM][BF_4], [EMIM][BF_4], [BMIM][OTf], and [MMIM][$MeSO_4$]) (Park and Kazlauskas, 2001; Lau et al., 2000, 2004; Itoh et al., 2003; Kaftzik et al., 2003; Lozano et al., 2003a). Therefore, multiple factors must be used to interpret those complex enzymatic systems.

4.2.7 Enzyme Dissolution

Hydrophobic ILs (such as those carrying anions of PF_6^- and Tf_2N^-) cannot dissolve a significant amount of enzymes; the aqueous enzyme in hydrophobic ILs (PF_6^- and Tf_2N^-) may be considered as the enzyme inclusion by IL network (Section 4.2.3) rather than the dissolution in the media (Lozano et al., 2001a). Meanwhile, hydrophilic ILs (such as those based on NO_3^-, lactate, $EtSO_4^-$, and CH_3COO^-) may dissolve some enzymes (There are also some exceptions. For example, BF_4^- based ILs are hydrophilic but do not dissolve the enzyme (Lau et al., 2004).); however, most of them tend to strongly interact with the protein (such as through hydrogen bonds), leading to enzyme inactivation (Lau et al., 2004; Toral et al., 2007; Zhao et al., 2008, 2009a; Turner et al., 2003; de los Ríos et al., 2007). For example, cellulase was dissolved but inactivated in concentrated solutions of [BMIM]Cl (Turner et al., 2003). Erbeldinger et al. (2000) noticed that thermolysin was inactive when fully dissolved in [BMIM][PF_6] containing 5% (v/v) water (enzyme solubility up to 3.2 mg/mL), while the suspended thermolysin remained active in the IL. At the time of this writing, there are only a few neat (or concentrated) ILs that are reported to dissolve considerable amounts of proteins/enzymes but do not completely denature/inactivate them. For example, choline dihydrogen phosphate (m.p. 119°C) containing 20% (wt) water is able to dissolve and stabilize cytochrome *c* (cyt *c*) (Fujita et al., 2005, 2007); triethylmethylammonium methyl sulfate ([Et_3MeN][$MeSO_4$]) was reported capable of dissolving >1.2 mg/mL CALB and retaining its catalytic capability (Lau et al., 2004; van Rantwijk et al., 2006). Our group prepared a series of ether-functionalized ILs that are able to dissolve enzymes (>5 mg/mL CALB at 50°C) and a variety of substrates, but do not considerably inactivate the lipase (more discussion in Section 4.3.7) (Zhao et al., 2008, 2009b). Therefore, the enzyme dissolution in ILs may not indicate the enzyme inactivation. Kotlewska et al. (2011) found that two nitrate-containing ILs (namely 1-(3-hydroxypropyl)-3-methyl imidazolium nitrate and [BMIM][NO_3]) are able to dissolve CALB while maintaining the CALB activity for in situ conversion of octanoic acid to peracid during the chemo-enzymatic epoxidation of cycloalkenes and Baeyer–Villiger oxidation of (cyclic) ketones; in addition, the hydrogen-bond donating ILs considerably improve the oxidation of substrate by peracid, leading to high product yields.

4.2.8 Other Factors Influencing Lipase Activity and Stability

Since many ILs are synthesized from their halide precursors, the residual halides in final IL products are often expected. These halide impurities have a strong impact on the enzyme's performance in ILs as discussed in Section 4.2.8.1; therefore, it is important to know the halide content in ionic media before using them in enzymatic reactions. Meanwhile, water content is critical to many enzymatic reactions in nonaqueous solvents (Halling, 1992; Bell et al., 1995); therefore, it is necessary to control the water content (or thermodynamic water activity) for biocatalysis in ILs (Section 4.2.8.2). Especially, this is crucial for synthetic reactions with hydrolysis side reactions. Another reason to control the water content is that ILs often absorb small amounts of water as another impurity (the hygroscopic nature of ILs is due to the preference of ion pair association (Spickermann et al., 2008)).

4.2.8.1 Halide Impurities in ILs

The purity of ILs can influence their physical properties, and further affect the reactions performed in ILs (Seddon et al., 2000). For example, Lee et al. (2006) observed that the activity of Novozym® 435 in [OMIM][Tf_2N] decreased linearly with the chloride content, and 1% (wt) increase in [OMIM]Cl (~1540 ppm Cl^-) led to 5% decrease in enzyme activity. However, another lipase (from *R. miehei*) could endure much less Cl^-; its activity in [OMIM][Tf_2N] with 2% [OMIM]Cl was only about 2% of the activity in neat [OMIM][Tf_2N]. In another study, Lee et al. (2008) noticed that the activity of *C. rugosa* lipase in the [OMIM][Tf_2N] containing 1% (w/w) [OMIM][Cl] was only about half of that in neat [OMIM][Tf_2N] (with <30 ppm Cl^-). Both groups of Kazlauskas (Park and Kazlauskas, 2001) and Víllora (Hernández-Fernández et al., 2007) observed higher lipase activities when the impurities were removed from ILs through washing them with aqueous solutions of $NaHCO_3$ or Na_2CO_3. Shen et al. (2008) suggested that during the kinetic resolution of racemic cyanohydrins, Amano lipase PS showed a high enantioselectivity (80% ee_p) in [OMIM][PF_6], but a poor enantioselectivity (<5% ee_p) in [HMIM]Cl. Our group (Zhao et al., 2009a) examined the transesterification between ethyl butyrate and 1-butanol catalyzed by Novozym® 435, and found that when 3000 ppm Cl^- or Br^- anions were contained in [BMIM][Tf_2N], the initial rates were about 60% of the activity in [BMIM][Tf_2N] with little halide (260 ppm Br^-). These data confirm that the degree of halide inhibition depends on specific substrates and enzymes. In our above model reaction, when the halide contents of ILs fell below 1000 ppm, there was no direct correlation between the lipase activity and the halide concentration (Zhao et al., 2009a). However, it is critical to determine the halide contents of ILs before using them in enzymatic reactions. Halides and other impurities should be removed from ILs as much as possible.

4.2.8.2 Water Activity

Many enzymatic reactions performed in ILs contain a fixed amount of water to afford the essential water desired by the enzyme. For instance, typically 1−2% (v/v) water was added into α-chymotrypsin or free CALB-catalyzed transesterification reactions in Tf_2N^-, PF_6^-, and BF_4^- based ILs (Laszlo and Compton, 2001; Lozano et al., 2001b, 2003b). However, the water content does not reflect the actual amount of water available for enzyme molecules because some water molecules may be attracted by the solvent molecules (such as through hydrogen bonds). The actual water freely available to the enzyme is often measured by a dimensionless property known as the thermodynamic water activity (a_w), whose definition is given in (Eq. [4]):

$$a_w = \frac{p}{p_0} \quad [4]$$

where p is the vapor pressure of water in the substance and p_0 is the vapor pressure of pure water at the same temperature. The water activity a_w is a function of temperature.

There are two common routes to maintain the water activity (a_w): (1) use of salt hydrate pairs, and (2) preequilibrium over saturated salt solutions.

Use of salt hydrate pairs. In this route, salt hydrate pairs (such as $NaOAc \cdot 3H_2O$ and NaOAc) are added directly into the nonaqueous solvents. This method has been well established for biocatalysis in conventional organic solvents (Halling, 1992; Bell et al., 1995; Chowdary and Prapulla, 2002). This approach can accurately control water activity in real time during the reaction, even if the reaction is generating water as the by-product (as long as a trace of each salt hydrate is present). Berberich et al. (2003) manipulated the water activity ($a_w = 0.17 - 0.78$) in [BMIM][PF_6] through different salt hydrate pairs, and indicated the initial rate of immobilized CALB-catalyzed transesterification reactions decreasing with the increase in water activity. However, one should be careful before adding salt hydrates into ILs because some ionic solvents (especially hydrophilic ones) could dissolve salts (Wasserscheid and Welton, 2008). The dissolved salt hydrates may influence the enzyme activity through binding to the enzyme surface (Berberich et al., 2003). Furthermore, Barahona et al. (2006) examined the effect of water activity on the esterification of geraniol with acetic acid in [BMIM][PF_6] catalyzed by Novozym® 435, and found a maximum initial reaction rate at $a_w = 0.6$. Yang et al. (2010) studied the *Penicillium expansum* lipase-catalyzed hydrolysis of *p*-nitrophenyl palmitate in [BMIM][PF_6] in the presence of salt hydrates, and noticed two important roles of salt hydrates in controlling the lipase activity: (1) governing the water activity, and (2) acting as Hofmeister salts (Section 4.2.4) since salt hydrates could be dissolved in microaqueous environments surrounding the enzyme molecules.

Preequilibrium over saturated salt solutions. In this method, substrates and enzymes in separate containers are placed in a large closed vessel containing saturated salt solution at a certain temperature (typically room temperature to prevent the thermal inactivation of enzyme). The equilibrium through the vapor phase can be achieved from 24 h to a few days. Some common salts and their water activities at saturation are LiCl (0.12), $MgCl_2$ (0.33), $Mg(NO_3)_2$ (0.54), NaCl (0.75), and KNO_3 (0.96) (Chowdary and Prapulla, 2002). The disadvantage of this method is that it only fixes the initial water activity (at room temperature) for substrates and enzymes, and the actual water activity during the reaction may change. This is due to two reasons: (1) the reaction temperature may not be room temperature, and (2) the reaction equilibrium has been shifted and water may be generated as the by-product. However, this method has frequently been used by biocatalysis in ILs because of its simplicity. Eckstein et al. (2002) observed a higher enantioselectivity of *Pseudomonas* sp. lipase in [BMIM][Tf_2N] at low water activities ($a_w < 0.53$) than in MTBE. Garcia et al. (2004) preequilibrated the reaction mixture components to a fixed a_w through saturated salt solutions (KOAc, $a_w = 0.22$; NaCl, $a_w = 0.75$) at 25°C for about 3 days; they found the transesterification activity of immobilized cutinase in [BMIM][PF_6] increased with an increase in a_w; however, the transesterification and hydrolysis activities of Novozym® 435 decreased with the increase in a_w. Noël et al. (2004) controlled the substrate mixture and aqueous enzyme solution in [BMIM][Tf_2N] to fixed water activities (a_w) over saturated salt solutions in closed containers (10 mL total capacity) at 25°C for 4 days; in the kinetic resolution of *rac*-2-pentanol catalyzed by free CALB, the synthetic activity increased with an increase in water activity although the selectivity decreased with an increase in water activity. Lee et al. (2008) evaluated the effect of a_w on lipase activity through the transesterification of benzyl alcohol with vinyl acetate in [OMIM][Tf_2N], and suggested the optimal initial rates of Novozym® 435 at $a_w = 0.2$, *R. miehei* lipase at $a_w = 0.5$, and *C. rugosa* lipase at $a_w = 0.4$.

4.2.9 Mechanisms of IL-Lipase Interactions

To explain the effect of these IL properties on enzyme stabilization, Klähn et al. (2011a) examined the CALB solvation in eight different ILs (cations based on imidazolium and guanidinium, and anions of NO_3^-, BF_4^-, and PF_6^-) through molecular dynamics (MD) simulation. They suggested that the predominate interaction between lipase and ILs is determined by the Coulomb interactions with anions, and then by weaker van der Waals interactions with cations. Coulomb interactions are affected by the size of the ions and the anion surface charge. The IL cations may diffuse into the active site of CALB. The same group (Klähn et al., 2011b) further pointed out that IL anions have dominant impact on the lipase stability, with a decreasing CALB stability in the order of $PF_6^- > BF_4^- >> NO_3^-$; this is consistent with their hydrogen-bond basicity series (Section 4.2.2). On the other hand, short methyl groups, other nonpolar groups, and imidazolium-based cations are more CALB stabilizing than long decyl side chains, polar methoxy groups, and guanidinium-based cations, respectively. They suggested two distinct mechanisms: (1) the protein surface can be destabilized by strong Coulomb interactions of the enzyme with denaturing anions and with polar cation groups (ie, destabilization of fold protein state), and (2) the protein core can be destabilized by long alkyl chains or direct exposure of protein core to the IL (ie, destabilization of unfolded protein state). Burney and Pfaendtner (2013) reported MD simulations of *C. rugosa* lipase in [BMIM][PF_6] and [BMIM][NO_3], and suggested the major interaction is the attraction between positively charged protein residues on the enzyme surface and IL anions. Kim et al. (2014) examined the interactions between CALB and ILs using both experimental (transesterification activity) and MD simulation tools, and found that the lipase activity decreases in the order of [BMIM][OTf] $>$ *t*-butanol $>$ [BMIM]Cl, which is consistent with the conformational changes of active site cavity. [BMIM]Cl induced a narrow cavity entrance and a low CALB activity. In addition, they pointed out that different lipase activities and structural changes are also caused by the electrostatic energy between ILs and CALB.

Since the enzymatic activity in ILs is controlled by multiple factors, there is no simple method that can stabilize or activate enzymes in all enzymatic systems. However, several methods have been implemented to improve the enzyme's stability and to increase its activity and enantioselectivity. Some representative methods are discussed in the following section.

4.3 METHODS TO IMPROVE LIPASE ACTIVITY AND STABILITY

Methods to improve enzyme activity and stability in ILs may be roughly divided into two groups: modification of enzymes and modification of solvent environments. The first group includes the enzyme immobilization (on solid support, sol−gel, or CLEA), physical or covalent attachment to PEG, chemical modifications of lipases, rinsing with *n*-propanol methods (PREP and EPRP), three-phase partitioning, enzyme/protein-coated micro-crystals, and lyophilization with cyclodextrins, and others. The second group comprises water-in-IL microemulsions, IL coating, additives in ILs, and new enzyme-compatible ionic media, among others. Methods in the first group allow the enzyme to be more tolerant to those denaturing factors of ILs, while methods in the second group reduce the denaturing properties of some ILs (such as reducing the anion's nucleophilicity and hydrogen-bonding basicity). The following sections discuss some major advances in these methods with representative examples.

4.3.1 Enzyme Immobilization

Enzyme immobilization is a common method for enzyme stabilization and activation in organic solvents and ILs. These immobilization methods can be further categorized into three types: binding to a solid carrier, sol—gel encapsulation, and protein cross-linking (a carrier-free technique) (Cao et al., 2003; Sheldon et al., 2005, 2006).

4.3.1.1 Immobilization on a Solid Carrier

The immobilization to a solid support is a routine method for improving the enzyme stability in conventional organic solvents and ILs. The immobilization techniques (both physical adsorption and covalent attachment) are usually straightforward to undertake using regular laboratory procedures, and some immobilized enzymes are commercially available (such as Novozym® 435). Two reviews (van Rantwijk and Sheldon, 2007; Moniruzzaman et al., 2010) have discussed many of these examples; therefore, this report highlights only some new carriers for enzyme immobilization, one of which is carbon nanotubes (Karajanagi et al., 2004). The high surface area and unique nanoscopic dimensions of carbon nanotubes allow a high enzyme loading and low mass-transfer resistance. For example, the noncovalent binding of proteinase K onto single-walled carbon nanotubes (SWNTs) led to a higher enzyme activity and higher thermal stability than its free form in ILs; the enzyme-SWNT conjugates were well dispersed in ILs (Eker et al., 2007). Likewise, *C. rugosa* lipase adsorbed on multiwalled carbon nanotubes showed a higher transesterification activity and enantioselectivity than the pH-tuned free form in [BMIM][PF_6] (Shah et al., 2007). More recently, another new enzyme support known as magnetic nanoparticles supported ILs was constructed by a covalent bonding of ILs-silane on magnetic silica nanoparticles; *C. rugosa* lipase immobilized on this support exhibited a higher catalytic activity and thermal stability (Jiang et al., 2009). Liu et al. (2007) coated a film of multiwall carbon nanotubes-[BMIM][PF_6] gel on a hydrophobic graphite electrode surface, and then immobilized laccase on the film modified electrode; they found this approach improved the thermal stability of laccase and achieved a high sensitivity in the detection of oxygen. In addition, Shakeri and Kawakami (2010) observed that *Rhizopus oryzae* lipase immobilized on octadecyl functionalized mesocellular foam exhibited 40—50 times higher specific activities than the free lipase in [OMIM][PF_6] and [BMIM][PF_6] during the transesterification reaction of (*S*)-glycidol and vinyl *n*-butyrate to produce (*S*)-glycidyl *n*-butyrate.

Jiang et al. (2012) used 1-butyraldehyde-3-(carbonic acid)-imidazolium salts (Cl^-, BF_4^-, and PF_6^-) as linkers to covalently immobilize *C. rugosa* lipase onto Fe_3O_4 nanoparticles; the new lipase preparation preferentially distributes at the oil—water interface, resulting in a higher activity for ester hydrolysis than the native enzyme. Interestingly, they found that a more hydrophobic IL linker leads to a lower lipase activity. The Huang group (Zou et al., 2013) immobilized porcine pancreas lipase (PPL) onto mesoporous silica SBA-15 with 98% immobilization efficiency when using 1-methyl-3-(3-(trimethoxysilyl)propyl) imidazolium lysinate as a linker, where the lysinate anion is covalently attached to lipase via glutaraldehyde; the thermal and operational stability of immobilized PPL is greatly improved. Lozano et al. (2012a) immobilized CALB on new nanostructured supports based on a polystyrene divinylbenzene (PS-DVB) porous matrix appending 1-decyl-2-methyimidazolium cations via covalent linkages (Fig. 4.6). Using the immobilized CALB, they carried out the continuous methanolysis of triolein and obtained biodiesel yields (up to 95%) and operational stability (85% yield after 45 cycles of 8–4 h) in supercritical CO_2 (45°C, 18 MPa). They also indicated that to minimize the lipase poisoning by glycerol, it is essential to use *t*-butanol as an inert cosolvent in the $scCO_2$ phase at the same concentration as triolein.

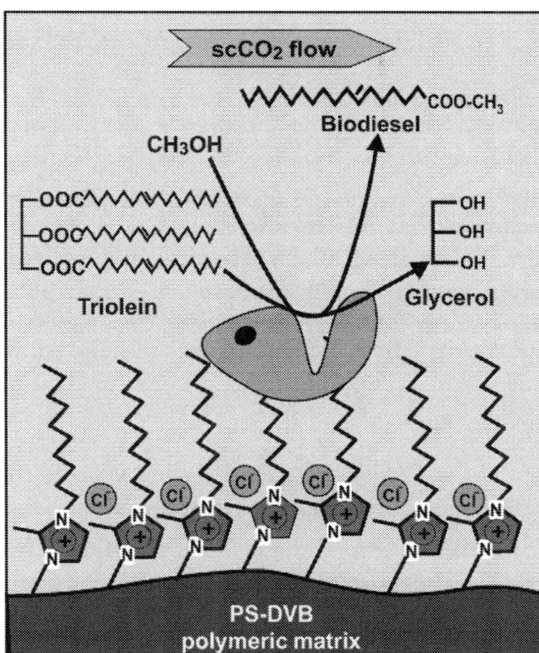

FIGURE 4.6 Nanostructured polystyrene divinylbenzene grafting with IL for CALB immobilization and the continuous methanolysis of triolein in $scCO_2$. *Adapted from Lozano, P., García-Verdugo, E., Bernal, J.M., Izquierdo, D.F., Burguete, M.I., Sánchez-Gómez, G., Luis, S.V., 2012a. Immobilised lipase on structured supports containing covalently attached ionic liquids for the continuous synthesis of biodiesel in $scCO_2$. ChemSusChem 5, 790–798, reproduced by permission of John Wiley and Sons.*

4.3.1.2 Sol–Gel Encapsulation

The encapsulation is a technique for entrapping biomolecules in a polymer matrix via noncovalent interactions between the network and the biomolecule. Sol–gel technology using ILs as additives has been actively investigated for the encapsulation of enzymes. Sol–gel encapsulation is a straightforward method (see protocols in (Reetz, 2006; Campas and Marty, 2006)), and can be used for entrapping many biological molecules including proteins, enzymes, and antibodies (Brinker and Scherer, 1990). This technique involves the acid- or base-catalyzed hydrolysis of tetraalkoxysilanes, such as tetraethyl orthosilicate or tetramethyl orthosilicate, in aqueous solutions; the subsequent cross-linking condensation produces a SiO_2 matrix to encapsulate the biomolecules (Ellerby et al., 1992; Baker, 2005). High structural rigidity of sol–gel matrixes ensures the integrity of encapsulated enzyme molecules and prevents their leaching; the mesoporous structures and high pore volume of sol–gel polymer allow the free diffusion of small substrate molecules and their effective interactions with the enzyme (Gill, 2001). However, the gel shrinkage and pore-collapse have been major issues of this method; also, there are other issues such as the alcohol release during the hydrolysis of silicon alkoxide (Avnir et al., 2006). To solve these problems, various additives (ie, sugars, amino acids, and *N*-methylglycine) have been added to minimize the gel shrinkage and control the protein hydration, and to further improve the activity and stability of enzymes (Avnir et al., 2006).

Recently, there is a rising interest in using ILs as additives (Vioux et al., 2010) for the protein/enzyme sol–gel immobilization. The main reason is that room-temperature ILs are nonvolatile, thermally stable, and are tunable to be enzyme-compatible. Earlier studies (Zhou et al., 2004; Shi et al., 2005) focused on the preparation of mesoporous silica through a high dispersion of ILs (such as [BMIM][BF_4]) in sol–gel, with the potential applications of inclusion of metal catalysts. More recently, silica xerogels were synthesized with various morphologies through the sol–gel method in the presence of ether-functionalized ILs (such as 1-triethylene glycol monomethyl ether-3-methylimidazolium methanesulfonate); these ILs act as both morphology controller and acid precatalyst (Migliorini et al., 2008; Donato et al., 2009). Another study by the Deng group (Zhang et al., 2009) further suggested that anions of ILs had a strong influence on the pore structures of silica-gel materials. Several groups have clearly shown the advantages of enzyme encapsulation in IL sol–gel hybrid carriers. The Koo group (Lee et al., 2007a,b) studied a number of ILs and their mixtures as additives for the sol–gel immobilization of *C. rugosa* lipase, and obtained higher hydrolytic and esterification activities for sol–gel encapsulated lipase with ILs than that without ILs. It is noted that the solvents for the above model reaction were either aqueous buffer or organic solvent (*n*-hexane). The Kanerva group (Hara et al., 2009) immobilized lipase PS from *Burkholderia cepacia* in a sol–gel, and achieved a higher enzyme stability in [EMIM][BF_4], [EMIM][Tf_2N], and [BMIM][PF_6] than the enzyme preparation using the cross-linked enzyme aggregates (CLEAs) method (Section 4.3.1.3). Zarcula et al. (2010) evaluated different ILs as additives for the sol−gel immobilization of lipase from *Pseudomonas fluorescens*, and found that more hydrophobic ILs lead to higher enzyme activities. Ursoiu et al. (2012) entrapped CALB in sol−gel using [OMIM][BF_4] as an additive; the immobilization resulted in high lipase stabilities in several organic solvents and a high operational stability for 15 reuse cycles during the kinetic resolution of secondary alcohols via enzymatic acylation. de Souza et al. (2013) used monoethanolamine-based protic ILs as additives in the sol−gel preparation of lipase from *B. cepacia*, and found the IL additives increased the surface area and pore size of sol−gel and induced higher enzyme activities.

Interestingly, Pavlidis et al. (2010) constructed water-in-IL microemulsion-based organogels by formulating water-in-IL microemulsions and biocompatible (hydroxypropyl)methyl cellulose (HPMC); the resulting new organogels were used as solid supports for immobilizing lipase B from *C. antarctica* and lipase from *Chromobacterium viscosum*. They observed up to 4.4-fold higher esterification activity for the new lipase preparation when compared to water-in-oil microemulsion-based organogels.

4.3.1.3 Cross-Linked Enzyme Aggregates

The cross-linked enzymes (CLEs) are the results of an early version of carrier-free immobilization, which is achieved by reacting glutaraldehyde with NH_2 groups on the protein surface (Habeeb, 1967; Jansen and Olson, 1969). However, intermolecular cross-linking of these biomolecules typically leads to a low enzyme activity, poor reproducibility, and low mechanical stability (Cao et al., 2003; Sheldon et al., 2007). Meanwhile, the cross-linking of a crystalline enzyme via glutaraldehyde was reported (Quiocho and Richards, 1964), which was further developed to a commercial immobilization technology called cross-linked enzyme crystals (CLECs or CLCs) (St. Clair and Navia, 1992; Margolin and Navia, 2001). This technique offers many advantages such as enhanced thermal, mechanical, and pH stability; designable particle sizes; and the ease of recycling (Cao et al., 2003; Sheldon et al., 2007). However, the preparation of CLECs requires a laborious purification and crystallization of enzymes. A more recent advance in cross-linking enzymes has been led by the Sheldon group (Cao et al., 2003; Sheldon et al., 2005, 2006, 2007; Sheldon, 2007), a new technique known as the cross-linked enzyme aggregates (CLEAs®). This method involves the addition of salts, organic solvents, or nonionic polymers to precipitate the enzyme as physical aggregates from aqueous solutions; the enzyme aggregates are then cross-linked by glutaraldehyde. The use of CLEAs has many advantages including the simplicity of preparation and recycling, improved enzyme activity, high stability and selectivity in organic solvents, and the possibility of immobilizing more than one enzyme (Sheldon et al., 2006; Cao et al., 2000, 2001; Lopez-Serrano et al., 2002; Schoevaart et al., 2004; Mateo et al., 2004; Yu et al., 2006).

For these reasons, a high enzyme stability is expected for CLEAs in ILs. Toral et al. (2007) immobilized CALB with two methods: CLEA method without a solid support (CALB-CL), and adsorbed/cross-linked on a polypropylene carrier (CALB-PP). They conducted several reactions catalyzed by these two enzyme preparations in denaturing ILs (such as [BMIM][dca], [BMIM][NO_3], [BMIM][OAc], and [BMIM][lactate]). In the transesterification of ethyl butyrate and 1-butanol, CALB-PP showed significantly higher catalytic activities in denaturing ILs than CALB-CL and free CALB. In addition, CALB-PP also exhibited high enantioselectivities in two resolution reactions (resolution of 1-phenylethanol in [BMIM][NO_3], and resolution of 1-phenylethylamine in [BMIM][NO_3] and [BMIM][dca]); the cross-linked lipase was not active in some ILs (such as [BMIM][OAc] and [BMIM][lactate]). Shah and Gupta (2007) found a higher enzymatic activity of *B. cepacia* lipase in [BMIM][PF_6] after the CLEA preparation than the pH-tuned lyophilized free enzyme. The Yang group (Lai et al., 2012) prepared CLEAs of lipase from *P. expansum* directly from fermentation broth, and observed an improved activity of this enzyme preparation in [BMIM][PF_6] (at up to 90°C). Furthermore, they achieved 85.7% conversion of microalgal oil to biodiesel catalyzed by CLEAs in [BMIM][PF_6]. de los Ríos et al. (2012) reported that both free and CLEA forms of CALB exhibited high activities and enantioselectivities in protic ILs based on alkylammonium cations and carboxylate anions for the kinetic resolution of 1-phenylethanol by acylation with vinyl acetate; in particular, the CLEA form showed higher initial rates than the free lipase in most of these ILs.

4.3.2 PEG-Modification

The modification of enzymes with PEG through either physical complexation or covalent interaction is a common method for stabilizing enzymes in denaturing environments. PEG molecules have both hydrophilic and hydrophobic properties (amphiphilic); therefore, modified enzymes become soluble in some organic solvents (such as benzene, toluene, and chlorinated hydrocarbons) (Inada et al., 1986) and ILs (Nakashima et al., 2006b), and show an increased stability in these solvents. The PEG-modification may be considered as a special type of enzyme immobilization methods. The most common PEG-modification is accomplished via the physical adsorption due to its simple procedures, mild conditions, and the unchanged protein structures. The Goto group (Maruyama et al., 2002, 2004) employed PEG with different molecular weights (average $M_n = 4000-35{,}000$) as the enzyme-coating amphiphile for the preparation of PEG-lipase complexes. They examined the PEG-lipases in several alcoholysis reactions in hydrophobic ILs (PF_6^- and

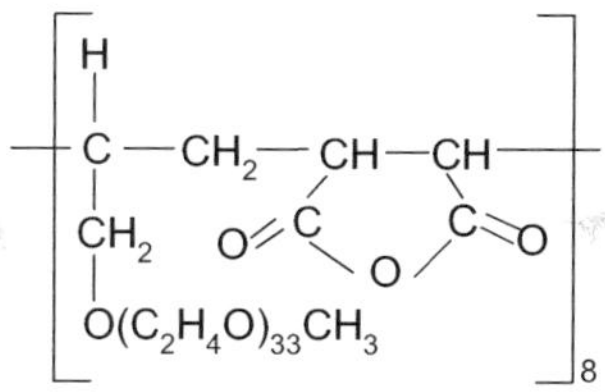

SCHEME 4.1 Structure of comb-shaped poly(ethylene glycol) PM_{13}.

Tf_2N^-), and reported higher enzyme activities (up to 14-fold) of the PEG-lipase complex than that of free form (Maruyama et al., 2002, 2004), as well as comparable or higher enantioselectivities of the enzyme complex than that of free lipase (Maruyama et al., 2004).

The second route for modifying enzymes with PEG is through the covalent attachment. For example, PEG *p*-nitrophenyl carbonate is routinely used to link PEG units with amino residues of proteins to form stable carbamates (Veronese et al., 1985; Woodward and Kaufman, 1996). Alternatively, Kaar et al. (2003) used PEG-monoisocyanate to link PEG with lysine residues of the protein to form PEG-lipase complex; however, this complex showed no improvement in the lipase activity in [BMIM][PF_6], [BMIM][NO_3], and [MMEP][OAc] when comparing with the free form. The Goto group adopted an unusual comb-shaped PEG, PM_{13} (SUNBRIGHT AM-1510K from NOF Co., Ltd., Japan) for covalently conjugating PEG with *C. rugosa* lipase (Nakashima et al., 2006a). As shown in Scheme 4.1, PM_{13} is a copolymer of PEG derivatives with maleic anhydride (molecular weight ~15,000); the acid anhydrides react with amino groups of the enzyme (Kodera et al., 1992). The PM_{13}-lipase complex showed a higher activity and stability in benzene than its free form (Hiroto et al., 1992). PM_{13}-lipase is soluble in Tf_2N^- based ILs, and exhibited a high catalytic activity and storage stability in these ionic media (Nakashima et al., 2006a).

On the other hand, there are also some disadvantages associated with the PEG-modification. For example, the preparation could be cumbersome in some cases, and the enzyme catalytic properties may vary from batch to batch during the immobilization.

4.3.3 Chemical Modifications of Lipases

ILs can be covalently attached to lipases. Jia et al. (2013a) activated five carboxylic acid-containing imidazolium ILs by carbonyldiimidazole, and then linked them to CALB, respectively, through lysine coupling (47–73% amino group modified). The IL-modified CALBs exhibited higher hydrolytic activities, higher thermal stability, and higher stability in aqueous solutions of dimethylformamide (DMF) or methanol. Following this strategy, this group (Jia et al., 2013b) also found IL-modified PPL displayed a higher hydrolytic activity (such as fourfold with [choline][H_2PO_4] modification) and an improved thermostability and enantioselectivity (12-fold thermostability improvement at 60°C and sevenfold enantioselectivity increase with [HOOC-BMIM]Cl modification). The same group (Hu et al., 2014) adopted the same technique to modify *C. rugosa* lipase (CRL), and observed that IL-modified CRLs (especially [choline][H_2PO_4]-modified CRL) showed a better thermal stability, catalytic activity in organic solvents, and higher tolerance against temperature and pH changes. The ultraviolet, CD, and fluorescence spectra indicated various degrees of enzyme conformational changes due to the chemical modification. This group (Jia et al., 2014) also chemically modified PPL by various ILs, and observed higher hydrolytic activities and thermal stabilities of the modified lipase in aqueous [MMIM] [$MeSO_4$].

Enzyme charges can be modified by chemical methods to improve their stabilities. Nordwald and Kaar (2013) modified α-chymotrypsin, *C. rugosa* lipase, and *Carica papaya* lipase by several methods: (1) acetylation, changing amines with uncharged acetyl groups; (2) succinylation, changing amines with negatively charged acid groups; (3) cationization, modifying acids to positively charged amines; and (4) neutralization, converting acids to neutral ethyl groups. They found that reducing the ratio of positive-to-negative surface charges of enzymes by chemical modifications (such as acetylation or succinylation) could improve the enzyme's stability in aqueous ILs (eg, 30–55% [BMIM]Cl and [EMIM][$EtSO_4$]) by repelling the anions from the enzyme surface. This finding affords a rational engineering of enzymes to improve their tolerance to IL solutions.

4.3.4 Enzyme Precipitated and Rinsed with *n*-Propanol

The Propanol Rinsed Enzyme Preparation (PREP) method is achieved through repeatedly rinsing the silica-immobilized enzyme with dry *n*-propanol to stabilize the enzymes (Partridge et al., 1998). The principle of this method is that *n*-propanol removes water from protein, which minimizes the protein denaturation and keeps the majority of enzyme molecules in an active conformation; only a small amount of water may be required during the reaction to activate the catalysis (Partridge et al., 1998). The PREPs of subtilisin Carlsberg (Partridge et al., 1998), α-chymotrypsin (Partridge et al., 1998), and papain (Theppakorn et al., 2004) have showed very high enzymatic activities in organic media (such as acetonitrile and *t*-butanol). A further development of PREP becomes a preparation method known as EPRP (Roy and Gupta, 2004; Solanki and Gupta, 2008). The EPRP technique is a combination of several methods: enzyme precipitation by alcohols, rinsing the precipitate with *n*-propanol, and the use of salt tuning during coprecipitation. Shah and Gupta (2007) conducted the enzymatic resolution of 1-phenylethanol in [BMIM] [PF_6] catalyzed by various lipase preparations; they found that the EPRP of

C. rugosa lipase (CRL) became more active than pH-tuned lyophilized CRL and CLEA preparation, and was more enantioselective than pH-tuned lyophilized CRL and PREP preparation. However, the EPRP of *B. cepacia* lipase lost a substantial portion of its hydrolytic activity, but two alternative formulations (ie, enzyme precipitated and rinsed with acetone (EPRA) and acetone-rinsed enzyme preparation (AREP)) resulted in high enzyme activities and enantioselectivities (Shah and Gupta, 2007).

4.3.5 Water-in-IL Microemulsions

The aforementioned methods belong to the modification of enzymes to improve their adaptability in ionic environments. The following several methods, however, aim to change the ionic media to improve their enzyme compatibility. The surfactant formation of micelles in ILs has been actively studied and reviewed (Qiu and Texter, 2008; Zech et al., 2010; Safavi et al., 2010). Recently, the Goto group (Moniruzzaman et al., 2008) reported the use of water-in-IL microemulsions as a new solvent for solubilizing various enzymes and proteins (such as lipase PS, CALB, α-chymotrypsin, horseradish peroxidase, and enhanced green fluorescent protein). The new medium was prepared by dissolving anionic surfactant sodium bis(2-ethyl-1-hexyl)sulfosuccinate (AOT) in hydrophobic [OMIM][Tf_2N] containing 10% (v/v) 1-hexanol (as a cosurfactant), followed by the addition of aqueous buffer to prepare a microemulsion. The lipase PS exhibited a higher hydrolytic activity in water-in-IL microemulsions than in water-saturated IL or in water-in-isooctane microemulsions (Moniruzzaman et al., 2008). This group further optimized the oxidation of pyrogallol in water-in-IL microemulsions catalyzed by horseradish peroxidase, and suggested that the reaction in IL microemulsions was much more effective than that in a conventional AOT/water/isooctane microemulsion (Moniruzzaman et al., 2009).

Later, the Stamatis group (Pavlidis et al., 2009) prepared the water-in-IL microemulsions by using nonionic surfactants (Tween 20 or Triton X-100) in [BMIM][PF_6]. The catalytic properties of lipases from *C. rugosa*, *C. viscosum*, and *Thermomyces lanuginosa* in these new microemulsions were evaluated via the esterification of natural fatty acids with various aliphatic alcohols, and the hydrolysis of *p*-nitrophenyl butyrate. The operational stability of these lipases in water-in-IL microemulsions was shown to be much higher than that in other microheterogeneous media. FT-IR and CD spectroscopy further confirmed that the lipase in water-in-IL microemulsions usually retains its native conformation or adapt a more rigid structure compared to the incubation in other microheterogeneous media.

The Coutinho group (Ventura et al., 2012) observed a decreased CALB hydrolytic activity in the presence of up to 0.033 M [C_{10}MIM]Cl in aqueous solutions, and a continuous increase of CALB activity (up to sixfold) in 0.033−0.090 M [C_{10}MIM]Cl. The so-called superactivity was explained that the long-chain IL self-aggregates in 0.015 M [C_{10}MIM]Cl in the presence of substrate (*p*-nitrophenyl laurate), and forms micelles and microemulsions. The microemulsion media could increase the interfacial area and thus the contact between the substrate and the CALB active site. In addition, some enzymes, including lipases, tend to gather at the interfacial region of the aggregates and become more active (Stamatis et al., 1999).

Recently, the Pandey group (Behera et al., 2009) reported a visual method for examining the formation of water-in-IL microemulsions: the Co (II) salt ($CoCl_2 \cdot H_2O$) was added into the microemulsions formed by using [BMIM][PF_6] as the oil phase and nonionic TX-100 as the surfactant; the Co (II) salt was selected as the probe due to different colors of the hexa-coordinated and tetra-coordinated complexes of the cation, depending on the solvating environment; the color change can be detected by UV–visible absorbance spectroscopy.

4.3.6 Coating Enzymes with ILs

Lee and Kim (2002) coated [PPMIM][PF_6] (PPMIM = 1-(3′-phenylpropyl)-3-methylimidazolium) onto lipase from *P. cepacia*, and obtained a high enantioselectivity and activity from this enzyme preparation. The Itoh group (Itoh et al., 2004, 2006) synthesized an imidazolium IL carrying anions of cetyl-PEG10-sulfate (Scheme 4.2), and coated it onto lipases to stabilize the enzymes in organic solvents (such as diisopropyl ether). The IL-coated lipases showed a high enantioselectivity and high reaction rates in several resolution reactions (up to 500- to 1000-fold acceleration for some substrates). The IL-coated lipase PS has become commercially available (Abe et al., 2008). *Mucor javanicus* lipase coated with various ILs was found more active and more stable than the untreated lipase in aqueous solution, and the activation factors were 1.81, 1.66, 1.56, and 1.60 for coating by [BMIM][PF_6], [EMIM][Tf_2N], [BMIM][BF_4], and [EMIM][BF_4], respectively (Dang et al., 2007). Lozano et al. (2007a) performed the enzymatic synthesis of citronellyl esters mediated by Novozym® 435, and observed the synthetic activity could be improved up to twofold by using the IL coating onto lipase; they also indicated the increase in activity following the hydrophobicity of ILs ([OMIM][PF_6] > [HMIM][PF_6] > [BMIM][PF_6]). Mutschler et al. (2009) coated various ILs onto Novozym® 435 beads, leading to higher conversions of lipase-mediated esterification of methyl-α-D-glucopyranoside with fatty acids than the uncoated lipase. Hara et al. (2010) immobilized lipase PS from *B. cepacia* on

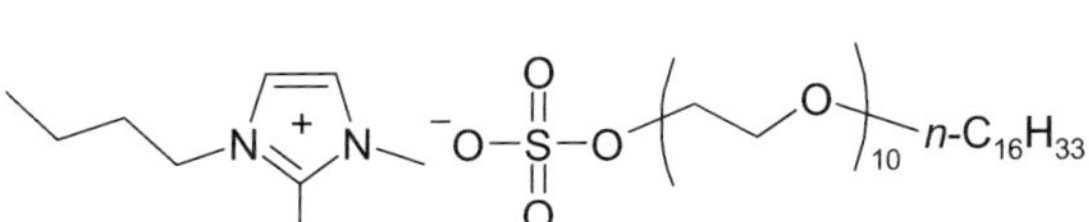

SCHEME 4.2 Structure of 1-butyl-2,3-dimethylimidazolium cetyl-PEG10-sulfate.

Kynol™ ACC 507-15 active carbon cloth in the presence of [EMIM][Tf_2N], and indicated that the use of IL support maintained the lipase activity and enantioselectivity during the acylation of 1-phenylethanol with vinyl acetate in toluene or hexane. Abdul Rahman et al. (2012) coated *C. rugosa* lipase with two amino acid-based ILs, respectively (tetraethylammonium L-histidinate and tetraethylammonium L-asparaginate), and found the coated lipase exhibited higher activities than the untreated enzyme in the esterification of oleyl alcohol with various fatty acids in hexane. Dong et al. (2012) found that Novozym 435 coated with [BMIM][PF_6] showed a high activity in the ring-opening polymerization (ROP) of 1,4-dioxan-2-one (PDO) to prepare poly(1,4-dioxan-2-one) (PPDO) (M_w up to 182,100 g/mol).

Liu et al. (2013) pretreated *C. rugosa* lipase by incubating the enzyme in ILs, organic solvents, or sub-/supercritical CO_2 (followed by rinsing with water in the case of ILs), and suggested the esterification activity of the lipase was generally lower after pretreatment by more hydrophilic and nucleophilic ILs and in organic solvents with lower log P values, but was 1.2- and 1.3-fold higher when treated by sub- and supercritical CO_2 (under 6 and 15 MPa, respectively, at 40°C for 20 min) than untreated lipase.

Monhemi et al. (2012) performed all-atom molecular dynamic simulation of mixtures of CALB, [EMIM][Tf_2N], and CO_2 molecules at 313 K and 100 bar, and reported several important results: (1) a supramolecular-like structure was formed by CALB and [EMIM][Tf_2N] in supercritical CO_2; (2) IL molecules form a coating layer on the lipase surface to prevent the enzyme from denaturing in supercritical CO_2; (3) a more native and stable conformation of lipase is found in a supercritical CO_2/IL system than in supercritical CO_2; and (4) the α-helix and β-sheet contents of lipase remain nearly unchanged in the supercritical CO_2/[EMIM][Tf_2N] system, but are reduced to some extent in supercritical CO_2.

4.3.7 Designing Lipase-Compatible ILs

As discussed earlier, the structures of ILs significantly influence the IL physical properties that are crucial to IL-enzyme interactions and enzyme stabilization. These properties include the polarity, hydrogen bond basicity, anion nucleophilicity, IL network, kosmotropicity, viscosity, and hydrophobicity. Therefore, it becomes critical to tailor the structures of ILs for particular biocatalytic applications based on our current understanding of IL structure and enzyme activity relationship. Several studies have designed the IL structures to dictate the compatibility of enzymes. The Xu group (Guo and Xu, 2005, 2006; Guo et al., 2006; Chen et al., 2008; Kahveci et al., 2009; Guo et al., 2009) selectively examined a number of commercial tetraammonium-based ILs as reaction media for the enzymatic glycerolysis. As shown in Scheme 4.3, each of these ILs is an ionic mixture containing multiple alkyloxy groups, which have both hydrophilic and hydrophobic

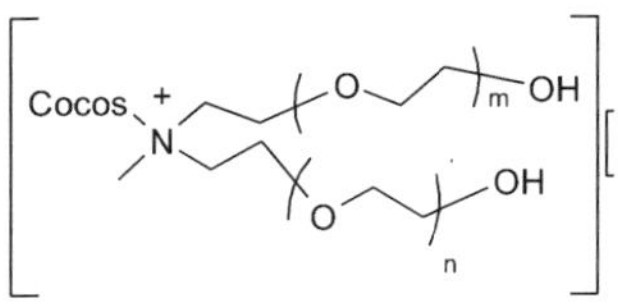

(A) Ammoeng 100
Cocos, C_{14} alkyl group; m + n =4-14

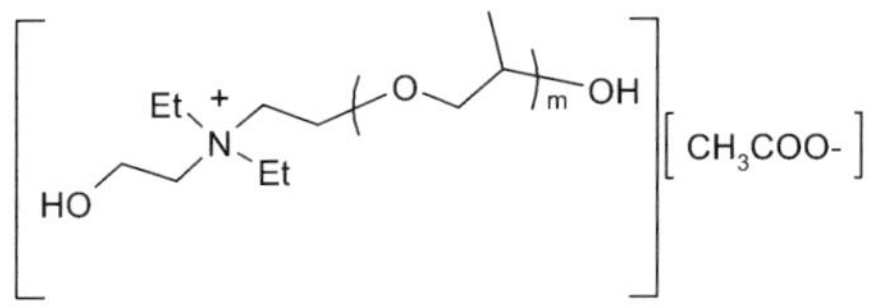

(B) Ammoeng 111
m = 50-60

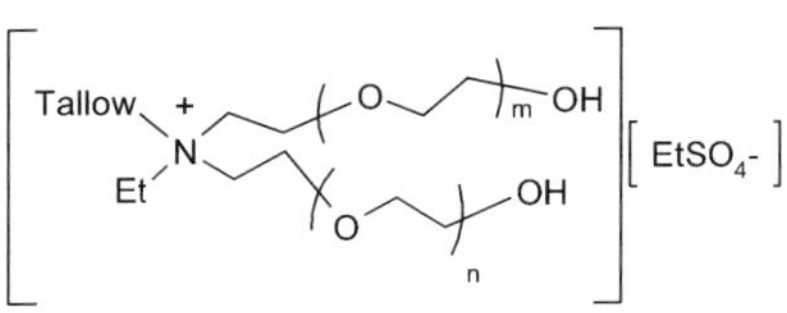

(C) Ammoeng 102
Tallow, C_{18} acyl group; m + n =14-25

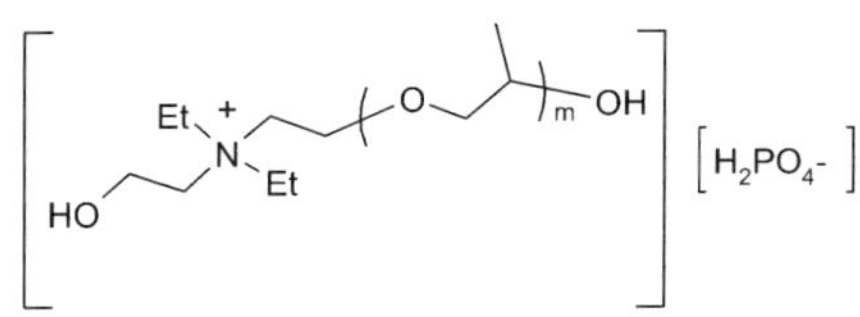

(D) Ammoeng 112
m = 50-60

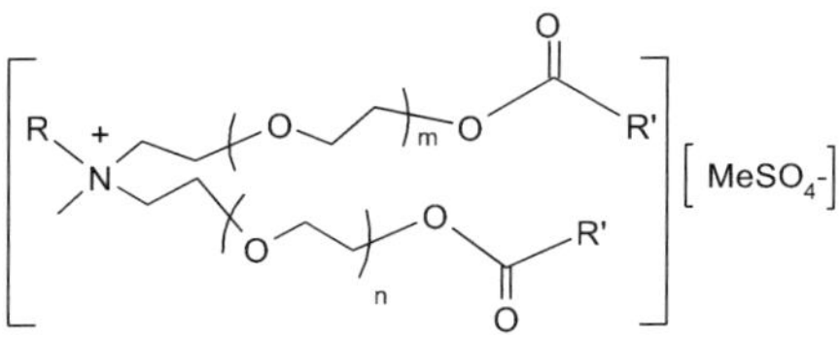

(E) Ammoeng 120
R, R', C_{18} acyl group; m, n, unavailable

SCHEME 4.3 Structures of tetraammonium-based ILs (Ammoeng® series (A) 100, (B) 111, (C) 102, (D) 112, and (E) 120).

properties like PEG. Particularly, Ammoeng 100 (also known as [CPMA][$MeSO_4$] (From the name of cocosalkyl pentaethoxy methylammonium methyl sulfate.)) and 102 could dissolve triglycerides and have been observed to be lipase-compatible in the glycerolysis reaction (Guo et al., 2006; Guo and Xu, 2006); trioctylmethylammonium bis(trifluoromethylsulfonyl)imide ([TOMA][Tf_2N]) and its mixture with Ammoeng 102 have also been evaluated as suitable solvents for the enzymatic glycerolysis (Kahveci et al., 2009, 2010; Guo et al., 2009). This group (Devi et al., 2011) further reported that Ammoeng 102 could partially mix with rapeseed oil, forming an oil/IL biphasic reaction system; such a solvent system has been shown highly effective for the Novozym 435-catalyzed preparation of biodiesel with >98% biodiesel yield and almost 100% oil conversion. De Diego et al. (2009) reported higher transesterification activities of both free and immobilized CALB in [CPMA][$MeSO_4$] than in several PF_6^- and BF_4^- based ILs; however, the other two lipases from *T. lanuginosus* (TLL) and *R. miehei* (RML) were less active in [CPMA][$MeSO_4$] than in PF_6^- and BF_4^- based ILs. Xu and coworkers (Guo et al., 2006; Chen et al., 2008) utilized the Conductor-like Screening Model for Real Solvents to derive various parameters (such as misfit, hydrogen-bonding, and van der Waals interaction energy) to understand the multiple interactions in ILs; the model also yielded some guidance for designing the structures of cations and anions (Guo et al., 2007).

The Itoh group (Abe et al., 2008) prepared an alkoxy-containing hydrophobic IL, 2-methoxyethyl(tri-*n*-butyl)phosphonium bis(trifluoromethanesulfonyl)amide ([$MEBu_3P$][Tf_2N]), and found a faster reaction rate (lipase PS-catalyzed transesterification of secondary alcohols) in this IL than in diisopropyl ether. This group (Abe et al., 2010) further synthesized a new alkoxy-functionalized IL, 2-methoxyethoxymethyl(tri-*n*-butyl)phosphonium bis(trifluoromethanesulfonyl)amide ([$P_{444}MEM$][Tf_2N]), and used it as the solvent for enantioselective acetylation of 1-phenylethanol catalyzed by alkyl-PEG sulfate IL-coated lipase PS. A rapid reaction rate and a high enantioselectivity were observed, indicating the high lipase compatibility of the alkoxy-functionalized IL. This group (Abe et al., 2012) also designed ether-functionalized phosphonium and ammonium ILs (such as [$MeOCH_2CH_2OCH_2$-Bu_3P][Tf_2N] and [$MeOCH_2CH_2OCH_2$-Et_2MeN][Tf_2N]) that led to high activities and enantioselectivities (E values >200) of an IL-coated lipase in the kinetic resolution of secondary alcohols via transesterification reactions.

Lourenço et al. (2007) observed little Novozym® 435 activity in denaturing [BMIM][dca], but a high activity and enantioselectivity in [aliq][dca] (aliq = trioctylmethylammonium—from Aliquat 336®). A likely explanation is that the denaturing anion's molar concentration in [aliq][dca] is much lower than that in [BMIM][dca] due to a higher molar mass of the former IL (Zhao et al., 2008, 2009b; Zhao and Song, 2010).

Our group prepared a series of alkyloxyalkyl-containing ILs based on acetate (Scheme 4.4) (Zhao et al., 2008). It was found these ILs could be designed to dissolve many substrates (such as cellulose, sugars, glucose, ascorbic acid, amino acids, betulinic acid, fatty acid, and triglycerides) that are not typically soluble in common organic solvents (Zhao et al., 2008, 2009b,c, 2010a,b). Additionally, the IL structures can be optimized to be compatible with CALB, and become benign solvents for preparing the derivatives of many substrates using enzymatic reactions (Zhao et al., 2008, 2009b, 2010a). Thus, it is possible to manipulate the structures of ILs to make them more compatible with enzymes, and to be able to dissolve a variety of compounds. The Yuan

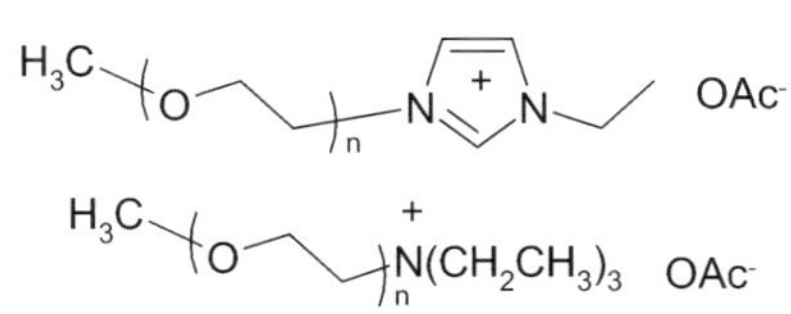

SCHEME 4.4 Imidazolium- and ammonium-based ILs consisting of alkyloxyalkyl-substituted cation and acetate anion (abbreviated as [$Me(OEt)_n$-Et-Im][OAc] and [$Me(OEt)_n$-Et_3N][OAc] respectively, $n = 2, 3, \ldots$).

group (Zhou et al., 2011) evaluated several ether-functionalized ILs for the lipase-catalyzed kinetic resolution of secondary alcohols, and achieved 99% enantioselectivity and 50% conversion of *rac*-1-phenylethanol when the reaction was catalyzed by Novozym 435 in 1-(3-ethoxypropyl)-2,3-dimethylimidazolium bis(trifluoromethylsulfonyl)imide.

Lozano et al. (2010) found that a very hydrophobic IL [C_{18}MIM][Tf_2N] (m.p. 53°C) could dissolve triolein at 60°C and promote the Novozym 435-catalyzed transesterification reaction, achieving up to 96% biodiesel yield in 6 h. Upon cooling, the IL solidifies and can be easily separated from biodiesel. This group (Lozano et al., 2012b) further investigated the use of very hydrophobic ILs as switchable liquid/solid reaction/separation media for the enzymatic synthesis of flavor esters. For example, *N*,*N′*,*N″*,*N‴*-hexadecyltrimethyl-ammonium bis(trifluoromethylsulfonyl)imide ([C_{16}tma] [Tf_2N]) is a solid at room temperature, but a liquid at 50°C when mixed with the reaction mixtures. Therefore, it is an ideal solvent for Novozym 435-catalyzed esterification reaction between an aliphatic carboxylic acid (acetic, propionic, butyric, or valeric) and a flavor alcohol (isoamyl alcohol, nerol, citronellol, or geraniol). After the reaction, through the repeated centrifugation method on the solid IL/flavor ester mixtures at controlled temperatures, the solid IL phase and the liquid flavor ester phase can be easily separated. The product yield could be improved up to 100% under optimized reaction conditions (enzyme amount, dehydrating molecular sieves, etc.) at 50°C. The lipase/IL system did not lose noticeable activity even after seven consecutive operation cycles. The same group (Lozano et al., 2013) further prepared *N*-octadecyl-*N′*,*N″*,*N‴*-trimethylammonium bis(trifluoromethylsulfonyl)imide ([C_{18}tma][Tf_2N])) as switchable IL/solid phases with temperature for Novozym 435-catalyzed methanolysis of triolein, resulting in up to 100% yield in 8 h at 60°C and the CALB half-life time up to 1370 days at 60°C. A straightforward separation of reaction products can be achieved by a cooling/centrifugation protocol.

Ou et al. (2011) added a small amount of [BMIM][H_2PO_4] as an IL buffer system into the CALB-catalyzed transesterification of ethyl butyrate and *n*-butanol in hydroxyl-functionalized ILs (ie, 1-(1-hydroxyethyl)-3-methyl-imidazolium tetrafluoroborate and 1-(1-hydroxyethyl)-3-methyl-imidazolium nitrate). They found that the IL buffer induced a more compact conformation of CALB and an improved lipase activity and stability.

An alternate approach is to integrate the IL structure with the substrates to improve the enzyme compatibility. Naik et al. (2007) attached ibuprofen with an imidazolium IL through an ester bond, followed by the hydrolysis of anchored ibuprofen by lipases in a 50/50 (v/v) mixture of organic solvent (or IL) and 0.1 M phosphate buffer (Scheme 4.5). The reaction in DMSO/buffer catalyzed by CALB produced (*S*)-ibuprofen with 86% ee and 87% isolated yield. The same reaction in [BMIM][PF_6]/buffer catalyzed by CALB

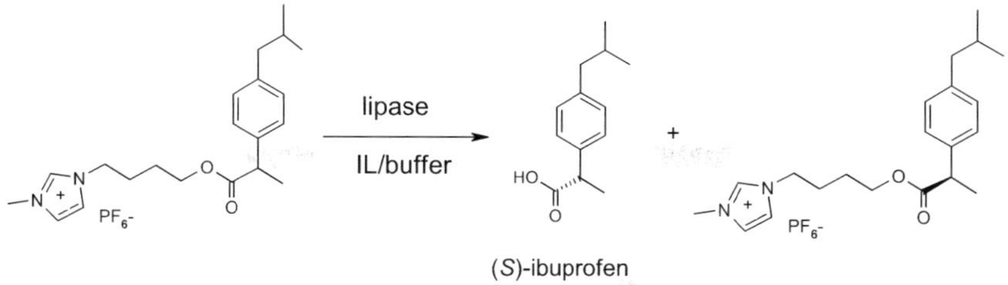

SCHEME 4.5 Lipase-catalyzed kinetic resolution of IL-anchored ibuprofen ester.

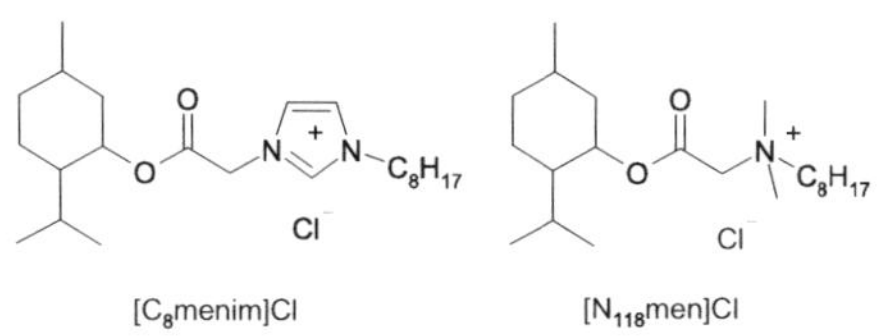

SCHEME 4.6 Functionalized ILs containing menthyl ester groups for fragrance delivery.

led to 80% ee and 80% isolated yield. Blesic et al. (2013) joined menthol with amphiphilic ILs carrying a long alkyl chain via a carboxylic ester linkage (Scheme 4.6); the enzymatic hydrolysis of the ester bond by CALB leads to a controlled fragrance delivery of menthol. This group observed a much higher enzymatic activity with [C_8menim]Cl than with [N_{118}men]Cl.

In summary, the enzyme-compatible ILs have some common structure features: (1) they usually have relatively large molecular structures so that hydrogen-bond basicity and nucleophilicity of anions are minimized, and (2) they often contain multiple ether and/or hydroxyl groups so that the solvent properties (such as IL viscosity, hydrogen-bond basicity and water affinity) can be optimized for mild IL-enzyme interactions.

4.4 DEEP EUTECTIC SOLVENTS FOR LIPASE ACTIVATION

The Abbott group (Abbott et al., 2003, 2004a, 2006) has suggested that a mixture of solid organic salt and a complexing agent could form a liquid at temperatures below 100°C, so-called deep eutectic solvents (DES). The reason is that the complexing agent (typically a hydrogen-bond donor) interacts with the anion and increases its effective size, which then decreases the anion interaction with the cation and thus decreases the freezing point (T_f) of the mixture. A typical example is the mixture of choline chloride (m.p. = 302°C, 2-hydroxyethyl-trimethylammonium chloride, Scheme 4.7) and urea (m.p. = 133°C) at a 1:2 molar ratio yielding a room-temperature liquid (T_f = 12°C) (Abbott et al., 2003). Although it is ambivalent on whether these DES should be formally classified as ILs because they contain

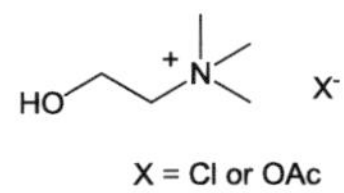

SCHEME 4.7 Structures of cholinium salts (choline chloride and choline acetate).

a significant portion of molecular component, DES definitely share a number of attractive solvent features with regular ILs. The major advantage of DES is that inexpensive and nontoxic compounds can be used and the properties of the liquid can be finely adjusted with different combinations of organic salts and complexing agents. Since many inexpensive quaternary ammonium salts are available and there is a wide choice of amides, amines, carboxylic acids, alcohols, and metal salts that can be used as complexing agents (Abbott et al., 2001, 2004a,b, 2006; Wang et al., 2011), the possibility of producing new and inexpensive DES is enormous. Particularly, choline chloride, also known as vitamin B_4, is produced on the scale of Mtonne (million metric tons) per year as an additive for chicken feed and many other applications. This ammonium salt is not only cheap and easy to make, but also nontoxic and biodegradable.

Choline chloride is also an essential micronutrient and human nutrient (Blusztajn, 1998), and cholinium alkanoates (including acetate) are environmentally benign and biodegradable (Petkovic et al., 2010). Therefore, DES based on cholinium can be biodegradable and inexpensive. In addition, DES can dissolve many metal salts, aromatic acids, amino acids, glucose, citric acid, benzoic acid, and glycerol (Abbott et al., 2003, 2004a, 2005, 2007a; Hou et al., 2008). One major application of choline-based eutectic solvents is electrodeposition and electropolishing of metals (Abbott et al., 2005, 2007b; Haerens et al., 2009), which has been summarized for several types of eutectics (Abbott and McKenzie, 2006; Abbott et al., 2008; Endres et al., 2008). In addition, these new solvents are being actively exploited as benign solvents for a number of chemical and enzymatic reactions (Tang et al., 2012).

Gorke et al. (2008) indicated that several hydrolases (CALB, CALA, and epoxide hydrolase) showed high activities in DES based on choline chloride or ethylammonium chloride (hydrogen-bond donors include acetamide, ethylene glycol, glycerol, urea, and malonic acid). Although some DES contain potentially reactive alcohols (such as ethylene glycol or glycerol), ethylene glycol was ninefold less reactive than 1-butanol, and glycerol was >600-fold less reactive than 1-butanol in CALB-catalyzed transesterifications of butyl valerate. When used as cosolvents, DES were capable of improving the reaction rate and/or conversion of hydrolases (esterases and epoxide hydrolase). The polarity study suggested these DES are more polar than common imidazolium-based ILs. The hydrogen-bond network in DES is speculated responsible for reducing the chemical potential of the components of DES and making them less reactive. Our group (Zhao et al., 2011) synthesized biocompatible DES (such as choline chloride/glycerol at 1:2 and choline acetate/glycerol at 1:1.5) that were able to maintain high CALB activity and stability, advocating their promising application in the enzymatic preparation of biodiesel. In addition, those eutectic solvents derived from cholinium salts and glycerol have lower viscosities (~80 mPa s at 50°C) than that of choline chloride/urea 1:2 (120 mPa s at 50°C). We observed the high reaction rates for Novozym 435-catalyzed transesterification of Miglyol oil 812 with methanol in choline acetate/glycerol (1:1.5 molar ratio), achieving a 97% triglyceride conversion in 3 h under optimum conditions (Zhao et al., 2011). Furthermore, choline chloride/glycerol (1:2 molar ratio) was evaluated as a cosolvent in Novozym 435-catalyzed transesterification of soybean oil with methanol to achieve up to 88% triglyceride conversions in 24 h (Zhao et al., 2013).

4.5 PROSPECTS

Another interesting development in the field is the combining use of ILs and supercritical fluids as benign media for biocatalysis (Monhemi et al., 2012; Cantone et al., 2007; Lozano et al., 2007b); this is obviously beyond the scope of this chapter. At present, there is still a lack of in-depth mechanistic understanding of IL-enzyme interactions at the molecular level. In the near future, with this information in hand, we will be able to rationally design the ionic solvents for enzyme stabilization and engineer the enzyme preparations for better compatibility with ILs. The cousin of ILs, DES, may become an inexpensive and alternative media to ILs for biocatalysis, and thus deserves further study.

ACKNOWLEDGMENTS

The author acknowledges the supports by the Henry Dreyfus Teacher-Scholar Award (2012), NIH MBRS-RISE grant (1R25GM096956), NIH NIBIB contract award (HHSN268201200011C), and the National Natural Science Foundation of China (21328601).

REFERENCES

Abbott, A.P., McKenzie, K.J., 2006. Application of ionic liquids to the electrodeposition of metals. Phys. Chem. Chem. Phys. 8, 4265–4279.

Abbott, A.P., Capper, G., Davies, D.L., Munro, H.L., Rasheed, R.K., Tambyrajah, V., 2001. Preparation of novel, moisture-stable, Lewis-acidic ionic liquids containing quaternary ammonium salts with functional side chains. Chem. Commun. 2010–2011.

Abbott, A.P., Capper, G., Davies, D.L., Rasheed, R.K., Tambyrajah, V., 2003. Novel solvent properties of choline chloride/urea mixtures. Chem. Commun. 70–71.

Abbott, A.P., Boothby, D., Capper, G., Davies, D.L., Rasheed, R.K., 2004a. Deep eutectic solvents formed between choline chloride and carboxylic acids: versatile alternatives to ionic liquids. J. Am. Chem. Soc. 126, 9142–9147.

Abbott, A.P., Capper, G., Davies, D.L., Rasheed, R., 2004b. Ionic liquids based upon metal halide/substituted quaternary ammonium salt mixtures. Inorg. Chem. 43, 3447–3452.

Abbott, A.P., Capper, G., Swain, B.G., Wheeler, D.A., 2005. Electropolishing of stainless steel in an ionic liquid. Trans. Inst. Metal Finish. 83, 51–53.

Abbott, A.P., Capper, G., Gray, S., 2006. Design of improved deep eutectic solvents using hole theory. ChemPhysChem 7, 803–806.

Abbott, A.P., Cullis, P.M., Gibson, M.J., Harris, R.C., Raven, E., 2007a. Extraction of glycerol from biodiesel into a eutectic based ionic liquid. Green Chem. 9, 868–872.

Abbott, A.P., Capper, G., McKenzie, K.J., Ryder, K.S., 2007b. Electrodeposition of zinc–tin alloys from deep eutectic solvents based on choline chloride. J. Electroanal. Chem. 599, 288–294.

Abbott, A.P., Ryder, K.S., König, U., 2008. Electrofinishing of metals using eutectic based ionic liquids. Trans. Inst. Metal Finish. 86, 196–204.

Abdul Rahman, M.B., Jumbri, K., Hanafiah, N.A.M.A., Abdulmalek, E., Tejo, B.A., Basri, M., Salleh, A.B., 2012. Enzymatic esterification of fatty acid esters by tetraethylammonium amino acid ionic liquids-coated *Candida rugosa* lipase. J. Mol. Catal. B: Enzym. 79, 61–65.

Abe, Y., Kude, K., Hayase, S., Kawatsura, M., Tsunashima, K., Itoh, T., 2008. Design of phosphonium ionic liquids for lipase-catalyzed transesterification. J. Mol. Catal. B: Enzym. 51, 81–85.

Abe, Y., Yoshiyama, K., Yagi, Y., Hayase, S., Kawatsura, M., Itoh, T., 2010. A rational design of phosphonium salt type ionic liquids for ionic liquid coated-lipase catalyzed reaction. Green Chem. 12, 1976–1980.

Abe, Y., Yagi, Y., Hayase, S., Kawatsura, M., Itoh, T., 2012. Ionic liquid engineering for lipase-mediated optical resolution of secondary alcohols: design of ionic liquids applicable to ionic liquid coated-lipase catalyzed reaction. Ind. Eng. Chem. Res. 51, 9952–9958.

Abraham, M.H., Chadha, H.S., Whiting, G.S., Mitchell, R.C., 1994a. Hydrogen bonding. 32. An analysis of water-octanol and water-alkane partitioning and the Δlog p parameter of seiler. J. Pharm. Sci. 83, 1085–1100.

Abraham, M.H., Chadha, H.S., Dixon, J.P., Leo, A.J., 1994b. Hydrogen bonding. 39. The partition of solutes between water and various alcohols. J. Phys. Org. Chem. 7, 712–716.

Abraham, M.H., Chadha, H.S., Martins, F., Mitchell, R.C., Bradbury, M.W., Gratton, J.A., 1999. Hydrogen bonding part 46: a review of the correlation and prediction of transport properties by an LFER method: physicochemical properties, brain penetration and skin permeability. Pestic. Sci. 55, 78–88.

Abraham, M.H., 1993. Scales of solute hydrogen-bonding: their construction and application to physicochemical and biochemical processes. Chem. Soc. Rev. 22, 73–83.

Aki, S.N.V.K., Brennecke, J.F., Samanta, A., 2001. How polar are room-temperature ionic liquids? Chem. Commun. 413–414.

Ali, A., Hyder, S., Akhtar, Y., 2005. Viscometric studies of α-amino acid in aqueous NaCl and $MgCl_2$ at 303 K. Indian J. Phys. 79, 157–160.

Anderson, J.L., Ding, J., Welton, T., Armstrong, D.W., 2002. Characterizing ionic liquids on the basis of multiple solvation interactions. J. Am. Chem. Soc. 124, 14247–14254.

Arakawa, T., Timasheff, S.N., 1984. Mechanism of protein salting in and salting out by divalent cation salts: balance between hydration and salt binding. Biochemistry 23, 5912–5923.

Armstrong, D.W., He, L., Liu, Y.S., 1999. Examination of ionic liquids and their interaction with molecules, when used as stationary phases in gas chromatography. Anal. Chem. 71, 3873–3876.

Avnir, D., Coradin, T., Lev, O., Livage, J., 2006. Recent bio-applications of sol–gel materials. J. Mater. Chem. 16, 1013–1030.

Baker, G.A., 2005. Sweet biofriendly silicates. Aust. J. Chem. 58, 721.

Baldwin, R.L., 1996. How Hofmeister ion interactions affect protein stability. Biophys. J. 71, 2056–2063.

Barahona, D., Pfromm, P.H., Rezac, M.E., 2006. Effect of water activity on the lipase catalyzed esterification of geraniol in ionic liquid [bmim]PF_6. Biotechnol. Bioeng. 93, 318–324.
Bauduin, P., Renoncourt, A., Touraud, D., Kunz, W., Ninham, B.W., 2004. Hofmeister effect on enzymatic catalysis and colloidal structures. Curr. Opin. Colloid Interface Sci. 9, 43–47.
Behera, K., Malek, N.I., Pandey, S., 2009. Visual evidence for formation of water-in-ionic liquid microemulsions. ChemPhysChem 10, 3204–3208.
Belibagli, K.B., Ayranci, E., 1990. Viscosities and apparent molar volumes of some amino acids in water and in 6 M guanidine hydrochloride at 25 °C. J. Solution Chem. 19, 867–882.
Bell, G., Halling, P.J., Moore, B.D., Partridge, J., Rees, D.G., 1995. Biocatalyst behaviour in low-water systems. Trends Biotechnol. 13, 468–473.
Berberich, J.A., Kaar, J.L., Russell, A.J., 2003. Use of salt hydrate pairs to control water activity for enzyme catalysis in ionic liquids. Biotechnol. Prog. 19, 1029–1032.
Bermejo, M.D., Kotlewska, A.J., Florusse, L.J., Cocero, M.J., van Rantwijk, F., Peters, C.J., 2008. Influence of the enzyme concentration on the phase behaviour for developing a homogeneous enzymatic reaction in ionic liquid–CO_2 media. Green Chem. 10, 1049–1054.
Bernson, A., Lindgren, J., 1994. Coordination of OH end-groups in the polymer electrolyte system LiX-PPG for X = PF_6^-, BF_4^-, ClO_4^-, $CF_3SO_3^-$, I^-, Br^- and Cl^-. Polymer 35, 4848–4851.
Blesic, M., Gunaratne, H.Q.N., Nockemann, P., McCarron, P., Seddon, K.R., 2013. Controlled fragrance delivery in functionalised ionic liquid-enzyme systems. RSC Adv. 3, 329–333.
Blusztajn, J.K., 1998. Choline, a vital amine. Science 281, 794–795.
Boström, M., Williams, D.R.M., Ninham, B.W., 2004. Why the properties of proteins in salt solutions follow a Hofmeister series. Curr. Opin. Colloid Interface Sci. 9, 48–52.
Brinker, C.J., Scherer, G.W., 1990. Sol-Gel Science: The Physics and Chemistry of Sol-Gel Processing. Academic Press, Boston.
Burney, P.R., Pfaendtner, J., 2013. Structural and dynamic features of *Candida rugosa* lipase 1 in water, octane, toluene, and ionic liquids BMIM-PF_6 and BMIM-NO_3. J. Phys. Chem. B 117, 2662–2670.
Camard, A., Ihara, Y., Murata, F., Mereiter, K., Fukuda, Y., Linert, W., 2005. The use of solvatochromic mixed copper(II) chelates with *N, N,N′,N′*-tetramethylethylenediamine and tropolonato or hinokitiolato ligands, [Cu(trop/hino)(tmen)]$B(C_6H_5)_4$, as indicator for Lewis basicity of solvents and low-coordinating anions. Inorg. Chim. Acta 358, 409–414.
Campas, M., Marty, J.-L., 2006. Encapsulation of enzymes using polymers and sol-gel techniques. In: Guisan, J.M. (Ed.), Immobilization of Enzymes and Cells, second ed. Humana Press, Totowa, NJ, pp. 77–85.
Cantone, S., Hanefeld, U., Basso, A., 2007. Biocatalysis in non-conventional media—ionic liquids, supercritical fluids and the gas phase. Green Chem. 9, 954–971.
Cao, L., van Rantwijk, F., Sheldon, R.A., 2000. Cross-linked enzyme aggregates: a simple and effective method for the immobilization of penicillin acylase. Org. Lett. 2, 1361–1364.
Cao, L., van Langen, L.M., van Rantwijk, F., Sheldon, R.A., 2001. Cross-linked aggregates of penicillin acylase: robust catalysts for the synthesis of beta lactam antibiotics. J. Mol. Catal. B: Enzym. 11, 665–670.
Cao, L., van Langen, L., Sheldon, R.A., 2003. Immobilised enzymes: carrier-bound or carrier-free? Curr. Opin. Biotechnol. 14, 387–394.
Carmichael, A.J., Seddon, K.R., 2000. Polarity study of some 1-alkyl-3-methylimidazolium ambient-temperature ionic liquids with the solvatochromic dye, Nile Red. J. Phys. Org. Chem. 13, 591–595.
Chapeaux, A., Simoni, L.D., Stadtherr, M.A., Brennecke, J.F., 2007. Liquid phase behavior of ionic liquids with water and 1-octanol and modeling of 1-octanol/water partition coefficients. J. Chem. Eng. Data 52, 2462–2467.

Chen, B., Guo, Z., Tan, T., Xu, X., 2008. Structures of ionic liquids dictate the conversion and selectivity of enzymatic glycerolysis: theoretical characterization by COSMO-RS. Biotechnol. Bioeng. 99, 18–29.

Chowdary, G.V., Prapulla, S.G., 2002. The influence of water activity on the lipase catalyzed synthesis of butyl butyrate by transesterification. Process Biochem. 38, 393–397.

Collins, K.D., Washabaugh, M.W., 1985. The Hofmeister effect and the behaviour of water at interfaces. Quart. Rev. Biophys. 18, 323–422.

Collins, K.D., 1997. Charge density-dependent strength of hydration and biological structure. Biophys. Chem. 72, 65–76.

Collins, K.D., 2004. Ions from the Hofmeister series and osmolytes: effects on proteins in solution and in the crystallization process. Methods 34, 300–311.

Constantinescu, D., Weingärtner, H., Herrmann, C., 2007. Protein denaturation by ionic liquids and the Hofmeister series: a case study of aqueous solutions of ribonuclease A. Angew. Chem. Int. Ed. 46, 8887–8889.

Constatinescu, D., Herrmann, C., Weingärtner, H., 2010. Patterns of protein unfolding and protein aggregation in ionic liquids. Phys. Chem. Chem. Phys. 12, 1756–1763.

Courtenay, E.S., Capp, M.W., Record, M.T.J., 2001. Thermodynamics of interactions of urea and guanidinium salts with protein surface: relationship between solute effects on protein processes and changes in water-accessible surface area. Protein Sci. 10, 2485–2497.

Dang, D.T., Ha, S.H., Lee, S.-M., Chang, W.-J., Koo, Y.-M., 2007. Enhanced activity and stability of ionic liquid-pretreated lipase. J. Mol. Catal. B: Enzym. 45, 118–121.

Davis-Searles, P.R., Saunders, A.J., Erie, D.A., Winzor, D.J., Pielak, G.J., 2001. Interpreting the effects of small uncharged solutes on protein-folding equilibria. Annu. Rev. Biophys. Biomol. Struct. 30, 271–306.

De Diego, T., Lozano, P., Abad, M.A., Steffensky, K., Vaultier, M., Iborra, J.L., 2009. On the nature of ionic liquids and their effects on lipases that catalyze ester synthesis. J. Biotechnol. 140, 234–241.

Devi, B.L.A.P., Guo, Z., Xu, X., 2011. Characterization of ionic liquid-based biocatalytic two-phase reaction system for production of biodiesel. AIChE J. 57, 1628–1637.

Devine, W., Lowe, B.M., 1971. Viscosity *B*-coefficients at 15 and 25 °C for glycine, β-alanine, 4-amino-n-butyric acid, and 6-amino-n-hexanoic acid in aqueous solution. J. Chem. Soc. A: Inorg. Phys. Theor. 2113–2116.

Donato, R.K., Migliorini, M.V., Benvegnu, M.A., Stracke, M.P., Gelesky, M.A., Pavan, F.A., Schrekker, C.M.L., Benvenutti, E.V., Dupont, J., Schrekker, H.S., 2009. Synthesis of silica xerogels with highly distinct morphologies in the presence of imidazolium ionic liquids. J. Sol-Gel Sci. Technol. 49, 71–77.

Dong, F.-X., Zhang, L., Tong, X.-Z., Chen, H.-B., Wang, X.-L., Wang, Y.-Z., 2012. Ionic liquid coated lipase: Green synthesis of high molecular weight poly(1,4-dioxan-2-one). J. Mol. Catal. B: Enzym. 77, 46–52.

Dupont, J., 2004. On the solid, liquid and solution structural organization of imidazolium ionic liquids. J. Braz. Chem. Soc. 15, 341–350.

Eckstein, M., Wasserscheid, P., Kragl, U., 2002. Enhanced enantioselectivity of lipase from *Pseudomonas* sp. at high temperatures and fixed water activity in the ionic liquid, 1-butyl-3-methylimidazolium bis[(trifluoromethyl)sulfonyl]amid. Biotechnol. Lett. 24, 763–767.

Eker, B., Asuri, P., Murugesan, S., Linhardt, R.J., Dordick, J.S., 2007. Enzyme-carbon nanotube conjugates in room-temperature ionic liquids. Appl. Biochem. Biotechnol. 143, 153–163.

El Tayar, N., Tsai, R.-S., Testa, B., Carrupt, P.-A., Leo, A., 1991. Partitioning of solutes in different solvent systems: the contribution of hydrogen-bonding capacity and polarity. J. Pharm. Sci. 80, 590–598.

Ellerby, L.M., Nishida, C.R., Nishida, F., Yamanaka, S.A., Dunn, B., Valentine, J.S., Zink, J.I., 1992. Encapsulation of proteins in transparent porous silicate glasses prepared by the sol-gel method. Science 255, 1113–1115.

Endres, F., MacFarlane, D.R., Abbott, A.P. (Eds.), 2008. Electrodeposition from Ionic Liquids. Wiley-VCH Verlag GmbH & Co. KGaA, Weinheim.

Erbeldinger, M., Mesiano, A.J., Russell, A.J., 2000. Enzymatic catalysis of formation of Z-aspartame in ionic liquids - an alternate to enzymatic catalysis in organic solvents. Biotechnol. Prog. 16, 1129–1131.

Fehér, E., Major, B., Bélafi-Bakó, K., Gubicza, L., 2007. On the background of enhanced stability and reusability of enzymes in ionic liquids. Biochem. Soc. Trans. 35, 1624–1627.

Forsyth, S.A., MacFarlane, D.R., Thomson, R.J., von Itzstein, M., 2002. Rapid, clean, and mild *O*-acetylation of alcohols and carbohydrates in an ionic liquid. Chem. Commun. 714–715.

Frank, H.S., Evans, M.W., 1945. Free volume and entropy in condensed systems III Entropy in binary liquid mixtures; partial molal entropy in dilute solutions; structure and thermodynamics in aqueous electrolytes. J. Chem. Phys. 13, 507–532.

Frank, H.S., Wen, W.-Y., 1957. Ion-solvent interaction. Structural aspects of ion-solvent interaction in aqueous solutions: a suggested picture of water structure. Discuss. Faraday Soc. 24, 133–140.

Frank, H.S., 1963. Single ion activities and ion-solvent interaction in dilute aqueous solutions. J. Phys. Chem. 67, 1554–1558.

Franks, F., 1973. Water, a comprehensive treatise. In: Aqueous Solutions of Simple Electrolytes, vol. 3. Plenum Press, New York.

Fujita, K., MacFarlane, D.R., Forsyth, M., 2005. Protein solubilising and stabilising ionic liquids. Chem. Commun. 4804–4806.

Fujita, K., Forsyth, M., MacFarlane, D.R., Reid, R.W., Elliott, G.D., 2006. Unexpected improvement in stability and utility of cytochrome *c* by solution in biocompatible ionic liquids. Biotechnol. Bioeng. 94, 1209–1213.

Fujita, K., MacFarlane, D.R., Forsyth, M., Yoshizawa-Fujita, M., Murata, K., Nakamura, N., Ohno, H., 2007. Solubility and stability of cytochrome *c* in hydrated ionic liquids: effect of oxo acid residues and kosmotropicity. Biomacromolecules 8, 2080–2086.

Garcia, S., Lourenco, N.M.T., Lousa, D., Sequeira, A.F., Mimoso, P., Cabral, J.M.S., Afonso, C.A.M., Barreiros, S., 2004. A comparative study of biocatalysis in non-conventional solvents: ionic liquids, supercritical fluids and organic media. Green Chem. 6, 466–470.

Gill, I., 2001. Bio-doped nanocomposite polymers: sol-gel bioencapsulates. Chem. Mater. 13, 3404–3421.

Gorke, J.T., Srienc, F., Kazlauskas, R.J., 2008. Hydrolase-catalyzed biotransformations in deep eutectic solvents. Chem. Commun. 1235–1237.

Guo, Z., Xu, X., 2005. New opportunity for enzymatic modification of fats and oils with industrial potentials. Org. Biomol. Chem. 3, 2615–2619.

Guo, Z., Xu, X., 2006. Lipase-catalyzed glycerolysis of fats and oils in ionic liquids: a further study on the reaction system. Green Chem. 8, 54–62.

Guo, Z., Chen, B., Murillo, R.L., Tan, T., Xu, X., 2006. Functional dependency of structures of ionic liquids: do substituents govern the selectivity of enzymatic glycerolysis? Org. Biomol. Chem. 4, 2772–2776.

Guo, Z., Lue, B.-M., Thomasen, K., Meyer, A.S., Xu, X., 2007. Predictions of flavonoid solubility in ionic liquids by COSMO-RS: experimental verification, structural elucidation, and solvation characterization. Green Chem. 9, 1362–1373.

Guo, Z., Kahveci, D., Özçelik, B., Xu, X., 2009. Improving enzymatic production of diglycerides by engineering binary ionic liquid medium system. New Biotechnol. 26, 37–43.

Ha, S.H., Lee, S.H., Dang, D.T., Kwon, M.S., Chang, W.-J., Yu, Y.J., Byun, I.S., Koo, Y.-M., 2008. Enhanced stability of *Candida antarctica* lipase B in ionic liquids. Korean J. Chem. Eng. 25, 291–294.

Habeeb, A.F.S.A., 1967. Preparation of enzymically active, water-insoluble derivatives of trypsin. Arch. Biochem. Biophys. 119, 264–268.

Haerens, K., Matthijs, E., Chmielarz, A., Van der Bruggen, B., 2009. The use of ionic liquids based on choline chloride for metal deposition: a green alternative? J. Environ. Manage. 90, 3245–3252.

Halling, P.J., 1992. Salt hydrates for water activity control with biocatalysts in organic media. Biotechnol. Tech. 6, 271–276.

Hara, P., Hanefeld, U., Kanerva, L.T., 2009. Immobilised *Burkholderia cepacia* lipase in dry organic solvents and ionic liquids: a comparison. Green Chem. 11, 250–256.

Hara, P., Mikkola, J.-P., Murzin, D.Y., Kanerva, L.T., 2010. Supported ionic liquids in *Burkholderia cepacia* lipase-catalyzed asymmetric acylation. J. Mol. Catal. B: Enzym. 67, 129–134.

Henderson, W.A., 2006. Glyme-lithium salt phase behavior. J. Phys. Chem. B 110, 13177–13183.

Henderson, W.A., 2007. Crystallization kinetics of glyme-LiX and PEO-LiX polymer electrolytes. Macromolecules 40, 4963–4971.

Hernández-Fernández, F.J., de los Ríos, A.P., Rubio, M., Gómez, D., Víllora, G., 2007. Enhancement of activity and selectivity in lipase-catalyzed transesterification in ionic liquids by the use of additives. J. Chem. Tech. Biotechnol. 82, 882–887.

Hernández-Fernández, F.J., de los Ríos, A.P., Tomás-Alonso, F., Gómez, D., Víllora, G., 2009. Stability of hydrolase enzymes in ionic liquids. Can. J. Chem. Eng. 87, 910–914.

von Hippel, P.H., Schleich, T., 1969. Ion effects on the solution structure of biological macromolecules. Acc. Chem. Res. 2, 257–265.

von Hippel, P.H., Wong, K.-Y., 1965. On the conformational stability of globular proteins. J. Biol. Chem. 240, 3909–3923.

Hilhorst, R., Spruijt, R., Laane, C., Veeger, C., 1984. Rules for the regulation of enzyme activity in reserved micelles as illustrated by the conversion of apolar steroids by 20 β-hydroxysteroid dehydrogenase. Eur. J. Biochem. 144, 459–466.

Hiroto, M., Matsushima, A., Kodera, Y., Shibata, Y., Inada, Y., 1992. Chemical modification of lipase with a comb-shaped synthetic copolymer of polyoxyethylene allyl methyl diether and maleic anhydride. Biotechnol. Lett. 14, 559–564.

Hofmeister, F., 1888. Zur Lehre von der Wirkung der Salze. Arch. Exp. Pathol. Pharmakol. 24, 247–260.

Honig, B., Nicholls, A., 1995. Classical electrostatics in biology and chemistry. Science 268, 1144–1149.

Hou, Y., Gu, Y., Zhang, S., Yang, F., Ding, H., Shan, Y., 2008. Novel binary eutectic mixtures based on imidazole. J. Mol. Liq. 143, 154–159.

Hu, Y., Yang, J., Jia, R., Ding, Y., Li, S., Huang, H., 2014. Chemical modification with functionalized ionic liquids: a novel method to improve the enzymatic properties of *Candida rugosa* lipase. Bioprocess Biosyst. Eng. 37, 1617–1626.

Husum, T.L., Jorgensen, C.T., Christensen, M.W., Kirk, O., 2001. Enzyme catalysed synthesis in ambient temperature ionic liquids. Biocatal. Biotransform. 19, 331–338.

Inada, Y., Takahashi, K., Yoshimoto, T., Ajima, A., Matsushima, A., Saito, Y., 1986. Application of polyethylene glycol-modified enzymes in biotechnological processes: organic solvent-soluble enzymes. Trends Biotechnol. 4, 190–194.

Irimescu, R., Kato, K., 2004. Investigation of ionic liquids as reaction media for enzymatic enantioselective acylation of amines. J. Mol. Catal. B: Enzym. 30, 189–194.

Itoh, T., Nishimura, Y., Kashiwagi, M., Onaka, M., 2003. Efficient lipase-catalyzed enantioselective acylation in an ionic liquid solvent system. In: Rogers, R.D., Seddon, K.R. (Eds.), Ionic Liquids as Green Solvents: Progress and Prospects. American Chemical Society, Washington, DC, pp. 251–261.

Itoh, T., Han, S., Matsushita, Y., Hayase, S., 2004. Enhanced enantioselectivity and remarkable acceleration on the lipase-catalyzed transesterification using novel ionic liquids. Green Chem. 6, 437–439.

Itoh, T., Matsushita, Y., Abe, Y., Han, S., Wada, S., Hayase, S., Kawatsura, M., Takai, S., Morimoto, M., Hirose, Y., 2006. Increased enantioselectivity and remarkable acceleration of lipase-catalyzed transesterification by using an imidazolium PEG-alkyl sulfate ionic liquid. Chem. Eur. J. 12, 9228–9237.

Jansen, E.F., Olson, A.C., 1969. Properties and enzymatic activities of papain insolubilized with glutaraldehyde. Arch. Biochem. Biophys. 129, 221–227.

Jenkins, H.D.B., Marcus, Y., 1995. Viscosity *B*-coefficients of ions in solution. Chem. Rev. 95, 2695–2724.

Jia, R., Hu, Y., Liu, L., Jiang, L., Huang, H., 2013a. Chemical modification for improving activity and stability of lipase B from *Candida antarctica* with imidazolium-functional ionic liquids. Org. Biomol. Chem. 11, 7192–7198.

Jia, R., Hu, Y., Liu, L., Jiang, L., Zou, B., Huang, H., 2013b. Enhancing catalytic performance of porcine pancreatic lipase by covalent modification using functional ionic liquids. ACS Catal. 3, 1976–1983.

Jia, R., Hu, Y., Huang, H., 2014. Improving the catalytic performance of porcine pancreatic lipase in the presence of [MMIm][$MeSO_4$] with the modification of functional ionic liquids. Process Biochem. 49, 668–672.

Jiang, Y., Guo, C., Xia, H., Mahmood, I., Liu, C., Liu, H., 2009. Magnetic nanoparticles supported ionic liquids for lipase immobilization: enzyme activity in catalyzing esterification. J. Mol. Catal. B: Enzym. 58, 103–109.

Jiang, Y., Guo, C., Gao, H., Xia, H., Mahmood, I., Liu, C., Liu, H., 2012. Lipase immobilization on ionic liquid-modified magnetic nanoparticles: Ionic liquids controlled esters hydrolysis at oil–water interface. AIChE J. 58, 1203–1211.

Jones, G., Dole, M., 1929. The viscosity of aqueous solutions of strong electrolytes with special reference to barium chloride. J. Am. Chem. Soc. 51, 2950–2964.

Kaar, J.L., Jesionowski, A.M., Berberich, J.A., Moulton, R., Russell, A.J., 2003. Impact of ionic liquid physical properties on lipase activity and stability. J. Am. Chem. Soc. 125, 4125–4131.

Kaftzik, N., Wasserscheid, P., Kragl, U., 2002. Use of ionic liquids to increase the yield and enzyme stability in the β-galactosidase catalysed synthesis of *N*-acetyllactosamine. Org. Proc. Res. Dev. 6, 553–557.

Kaftzik, N., Neumann, S., Kula, M.-R., Kragl, U., 2003. Enzymatic condensation reactions in ionic liquids. In: Rogers, R.D., Seddon, K.R. (Eds.), Ionic Liquids as Green Solvents: Progress and Prospects. American Chemical Society, Washington, DC, pp. 206–211.

Kahveci, D., Guo, Z., Özçelik, B., Xu, X., 2009. Lipase-catalyzed glycerolysis in ionic liquids directed towards diglyceride synthesis. Process Biochem. 44, 1358–1365.

Kahveci, D., Guo, Z., Özçelik, B., Xu, X., 2010. Optimisation of enzymatic synthesis of diacylglycerols in binary medium systems containing ionic liquids. Food Chem. 119, 880–885.

Kamlet, M.J., Doherty, R.M., Abraham, M.H., Marcus, Y., Taft, R.W., 1988. Linear solvation energy relationship. 46. An improved equation for correlation and prediction of octanol/water partition coefficients of organic nonelectrolytes (including strong hydrogen bond donor solutes). J. Phys. Chem. 92, 5244–5255.

Karajanagi, S.S., Vertegel, A.A., Kane, R.S., Dordick, J.S., 2004. Structure and function of enzymes adsorbed onto single-walled carbon nanotubes. Langmuir 20, 11594–11599.

Kay, R.L., Vituccio, T., Zawoyski, C., Evans, D.F., 1966. Viscosity B coefficients for the tetraalkylammonium halides. J. Phys. Chem. 70, 2336–2341.

Kim, H.S., Koo, Y.-M., 2012. Effects of physicochemical properties of ionic liquids on butyl acetate synthesis using *Candida antarctica* lipase B. Korean J. Chem. Eng. 29, 1610–1614.

Kim, H.S., Ha, S.H., Sethaphong, L., Koo, Y.-M., Yingling, Y.G., 2014. The relationship between enhanced enzyme activity and structural dynamics in ionic liquids: a combined computational and experimental study. Phys. Chem. Chem. Phys. 16, 2944–2953.

Klähn, M., Lim, G.S., Seduraman, A., Wu, P., 2011a. On the different roles of anions and cations in the solvation of enzymes in ionic liquids. Phys. Chem. Chem. Phys. 13, 1649–1662.

Klähn, M., Lim, G.S., Wu, P., 2011b. How ion properties determine the stability of a lipase enzyme in ionic liquids: a molecular dynamics study. Phys. Chem. Chem. Phys. 13, 18647–18660.

Kodera, Y., Tanaka, H., Matsushima, A., Inada, Y., 1992. Chemical modification of L-asparaginase with a comb-shaped copolymer of polyethylene glycol derivative and maleic anhydride. Biochem. Biophys. Res. Commun. 184, 144–148.

Kotlewska, A.J., van Rantwijk, F., Sheldon, R.A., Arends, I.W.C.E., 2011. Epoxidation and Baeyer–Villiger oxidation using hydrogen peroxide and a lipase dissolved in ionic liquids. Green Chem. 13, 2154–2160.

Kragl, U., Eckstein, M., Kaftzik, N., 2002. Enzyme catalysis in ionic liquids. Curr. Opin. Biotechnol. 13, 565–571.

Kunz, W., Henle, J., Ninham, B.W., 2004. 'Zur Lehre von der Wirkung der Salze' (about the science of the effect of salts: Franz Hofmeister's historical papers). Curr. Opin. Colloid Interface Sci. 9, 19–37.

Laane, C., Boeren, S., Vos, K., 1985. On optimizing organic solvents in multi-liquid-phase biocatalysis. Trends Biotechnol. 3, 251–252.

Laane, C., Boeren, S., Vos, K., Veeger, C., 1987. Rules for optimization of biocatalysis in organic solvents. Biotechnol. Bioeng. 30, 81–87.

Lai, J.-Q., Hu, Z.-L., Sheldon, R.A., Yang, Z., 2012. Catalytic performance of cross-linked enzyme aggregates of *Penicillium expansum* lipase and their use as catalyst for biodiesel production. Process Biochem. 47, 2058–2063.

Lang, M., Kamrat, T., Nidetzky, B., 2006. Influence of ionic liquid cosolvent on transgalactosylation reactions catalyzed by thermostable β-glycosylhydrolase CelB from *Pyrococcus furiosus*. Biotechnol. Bioeng. 95, 1093–1100.

Laszlo, J.A., Compton, D.L., 2001. α-Chymotrypsin catalysis in imidazolium-based ionic liquids. Biotechnol. Bioeng. 75, 181–186.

Lau, R.M., van Rantwijk, F., Seddon, K.R., Sheldon, R.A., 2000. Lipase-catalyzed reactions in ionic liquids. Org. Lett. 2, 4189–4191.

Lau, R.M., Sorgedrager, M.J., Carrea, G., van Rantwijk, F., Secundo, F., Sheldon, R.A., 2004. Dissolution of *Candida antarctica* lipase B in ionic liquids: effects on structure and activity. Green Chem. 6, 483–487.

Lee, J.K., Kim, M.-J., 2002. Ionic liquid-coated enzyme for biocatalysis in organic solvent. J. Org. Chem. 67, 6845–6847.

Lee, S.H., Lee, S.B., 2009. Octanol/water partition coefficients of ionic liquids. J. Chem. Tech. Biotechnol. 84, 202–207.

Lee, S.-M., Chang, W.-J., Choi, A.-R., Koo, Y.-M., 2005. Influence of ionic liquids on the growth of *Escherichia coli*. Korean J. Chem. Eng. 22, 687–690.

Lee, S.H., Ha, S.H., Lee, S.B., Koo, Y.-M., 2006. Adverse effect of chloride impurities on lipase-catalyzed transesterifications in ionic liquids. Biotechnol. Lett. 28, 1335–1339.

Lee, S.H., Doan, T.T.N., Ha, S.H., Koo, Y.-M., 2007a. Using ionic liquids to stabilize lipase within sol–gel derived silica. J. Mol. Catal. B: Enzym. 45, 57–61.

Lee, S.H., Doan, T.T.N., Ha, S.H., Chang, W.-J., Koo, Y.-M., 2007b. Influence of ionic liquids as additives on sol-gel immobilized lipase. J. Mol. Catal. B: Enzym. 47, 129–134.

Lee, S.H., Koo, Y.-M., Ha, S.H., 2008. Influence of ionic liquids under controlled water activity and low halide content on lipase activity. Korean J. Chem. Eng. 25, 1456–1462.

Linert, W., Jameson, R.F., Taha, A., 1993. Donor numbers of anions in solution: the use of solvatochromic Lewis acid–base indicators. J. Chem. Soc. Dalton Trans. 3181–3186.

Liu, Q., Janssen, M.H.A., van Rantwijk, F., Sheldon, R.A., 2005. Room-temperature ionic liquids that dissolve carbohydrates in high concentrations. Green Chem. 7, 39–42.

Liu, Y., Huang, L., Dong, S., 2007. Electrochemical catalysis and thermal stability characterization of laccase–carbon nanotubes-ionic liquid nanocomposite modified graphite electrode. Biosens. Bioelectron. 23, 35–41.

Liu, Y., Chen, D., Yan, Y., 2013. Effect of ionic liquids, organic solvents and supercritical CO_2 pretreatment on the conformation and catalytic properties of *Candida rugosa* lipase. J. Mol. Catal. B: Enzym. 90, 123–127.

Lopez-Serrano, P., Cao, L., van Rantwijk, F., Sheldon, R.A., 2002. Cross-linked enzyme aggregates with enhanced activity: application to lipases. Biotechnol. Lett. 24, 1379–1383.

Lou, W.-Y., Zong, M.-H., 2006. Efficient kinetic resolution of (*R*,*S*)-1-trimethylsilylethanol via lipase-mediated enantioselective acylation in ionic liquids. Chirality 18, 814–821.

Lou, W.-Y., Zong, M.-H., Wu, H., Xu, R., Wang, J.-F., 2005. Markedly improving lipase-mediated asymmetric ammonolysis of D,L-*p*-hydroxyphenylglycine methyl ester by using an ionic liquid as the reaction medium. Green Chem. 7, 500–506.

Lourenço, N.M.T., Barreiros, S., Afonso, C.A.M., 2007. Enzymatic resolution of Indinavir precursor in ionic liquids with reuse of biocatalyst and media by product sublimation. Green Chem. 9, 734–736.

Lozano, P., De Diego, T., Carrié, D., Vaultier, M., Iborra, J.L., 2001a. Over-stabilization of *Candida antarctica* lipase B by ionic liquids in ester synthesis. Biotechnol. Lett. 23, 1529–1533.

Lozano, P., de Diego, T., Guegan, J.-P., Vaultier, M., Iborra, J.L., 2001b. Stabilization of α-chymotrypsin by ionic liquids in transesterification reactions. Biotechnol. Bioeng. 75, 563–569.

Lozano, P., de Diego, T., Carrié, D., Vaultier, M., Iborra, J.L., 2003a. Enzymatic catalysis in ionic liquids and supercritical carbon dioxide. In: Rogers, R.D., Seddon, K.R. (Eds.), Ionic Liquids as Green Solvents: Progress and Prospects. American Chemical Society, Washington, DC, pp. 239–250.

Lozano, P., de Diego, T., Carrie, D., Vaultier, M., Iborra, J.L., 2003b. Enzymatic ester synthesis in ionic liquids. J. Mol. Catal. B: Enzym. 21, 9–13.

Lozano, P., De Diego, T., Gmouh, S., Vaultier, M., Iborra, J.L., 2005. Dynamic structure–function relationships in enzyme stabilization by ionic liquids. Biocatal. Biotransform. 23, 169–176.

Lozano, P., Piamtongkam, R., Kohns, K., De Diego, T., Vaultier, M., Iborra, J.L., 2007a. Ionic liquids improve citronellyl ester synthesis catalyzed by immobilized *Candida antarctica* lipase B in solvent-free media. Green Chem. 9, 780–784.

Lozano, P., De Diego, T., Sauer, T., Vaultier, M., Gmouh, S., Iborra, J.L., 2007b. On the importance of the supporting material for activity of immobilized *Candida antarctica* lipase B in ionic liquid/hexane and ionic liquid/supercritical carbon dioxide biphasic media. J. Supercrit. Fluids 40, 93–100.

Lozano, P., Bernal, J.M., Piamtongkam, R., Fetzer, D., Vaultier, M., 2010. One-phase ionic liquid reaction medium for biocatalytic production of biodiesel. ChemSusChem 3, 1359–1363.

Lozano, P., García-Verdugo, E., Bernal, J.M., Izquierdo, D.F., Burguete, M.I., Sánchez-Gómez, G., Luis, S.V., 2012a. Immobilised lipase on structured supports containing covalently attached ionic liquids for the continuous synthesis of biodiesel in $scCO_2$. ChemSusChem 5, 790–798.

Lozano, P., Bernal, J.M., Navarro, A., 2012b. A clean enzymatic process for producing flavour esters by direct esterification in switchable ionic liquid/solid phases. Green Chem. 14, 3026–3033.

Lozano, p., Bernal, J.M., Sánchez-Gómez, G., López-López, G., Vaultier, M., 2013. How to produce biodiesel easily using a green biocatalytic approach in sponge-like ionic liquids. Energy Environ. Sci. 6, 1328–1338.

Lue, B.-M., Guo, Z., Xu, X., 2010. Effect of room temperature ionic liquid structure on the enzymatic acylation of flavonoids. Process Biochem. 45, 1375–1382.

MacFarlane, D.R., Pringle, J.M., Johansson, K.M., Forsyth, S.A., Forsyth, M., 2006. Lewis base ionic liquids. Chem. Commun. 1905–1917.

Mann, J.P., McCluskey, A., Atkin, R., 2009. Activity and thermal stability of lysozyme in alkylammonium formate ionic liquids—influence of cation modification. Green Chem. 11, 785–792.

Marcus, Y., 1994. Viscosity *B*-coefficients, structural entropies and heat capacities, and the effects of ions on the structure of water. J. Solution Chem. 23, 831–848.

Margolin, A.L., Navia, M.A., 2001. Protein crystals as novel catalytic materials. Angew. Chem. Int. Ed. 40, 2204–2222.

Maruyama, T., Nagasawa, S., Goto, M., 2002. Poly(ethylene glycol)-lipase complex that is catalytically active for alcoholysis reactions in ionic liquids. Biotechnol. Lett. 24, 1341–1345.

Maruyama, T., Yamamura, H., Kotani, T., Kamiya, N., Goto, M., 2004. Poly(ethylene glycol)-lipase complexes that are highly active and enantioselective in ionic liquids. Org. Biomol. Chem. 2, 1239–1244.

Mateo, C., Palomo, J.M., van Langen, L.M., van Rantwijk, F., Sheldon, R.A., 2004. A new, mild cross-linking methodology to prepare cross-linked enzyme aggregates. Biotechnol. Bioeng. 86, 273–276.

Migliorini, M.V., Donato, R.K., Benvegnu, M.A., Goncalves, R.S., Schrekker, H.S., 2008. Imidazolium ionic liquids as bifunctional materials (morphology controller and pre-catalyst) for the preparation of xerogel silica's. J. Sol-Gel Sci. Technol. 48, 272–276.

Minton, A.P., 1981. Excluded volume as a determinant of macromolecular structure and reactivity. Biopolymers 20, 2093–2120.

Monhemi, H., Housaindokht, M.R., Bozorgmehr, M.R., Googheri, M.S.S., 2012. Enzyme is stabilized by a protection layer of ionic liquids in supercritical CO_2: Insights from molecular dynamic simulation. J. Supercrit. Fluids 69, 1–7.

Moniruzzaman, M., Kamiya, N., Nakashima, K., Goto, M., 2008. Water-in-ionic liquid microemulsions as a new medium for enzymatic reactions. Green Chem. 10, 497–500.

Moniruzzaman, M., Kamiya, N., Goto, M., 2009. Biocatalysis in water-in-ionic liquid microemulsions: a case study with horseradish peroxidase. Langmuir 25, 977–982.

Moniruzzaman, M., Nakashima, K., Kamiya, N., Goto, M., 2010. Recent advances of enzymatic reactions in ionic liquids. Biochem. Eng. J. 48, 295–314.

Muldoon, M.J., Gordon, G.M., Dunkin, I.R., 2001. Investigation of solvent-solute interaction in room temperature ionic liquids using solvatochromic dyes. J. Chem. Soc. Perkin Trans. 2, 433–435.

Mutschler, J., Rausis, T., Bourgeois, J.-M., Bastian, C., Zufferey, D., Mohrenz, I.V., Fischer, F., 2009. Ionic liquid-coated immobilized lipase for the synthesis of methylglucose fatty acid esters. Green Chem. 11, 1793–1800.

Myers, J.K., Pace, C.N., Scholtz, J.M., 1995. Denaturant m values and heat capacity changes: relation to changes in accessible surface areas of protein unfolding. Protein Sci. 4, 2138–2148.

Naik, P.U., Nara, S.J., Harjani, J.R., Salunkhe, M.M., 2007. Ionic liquid anchored substrate for enzyme catalysed kinetic resolution. J. Mol. Catal. B: Enzym. 44, 93–98.

Nakashima, K., Okada, J., Maruyama, T., Kamiya, N., Goto, M., 2006a. Activation of lipase in ionic liquids by modification with comb-shaped poly(ethylene glycol). Sci. Technol. Adv. Mater. 7, 692–698.

Nakashima, K., Maruyama, T., Kamiya, N., Goto, M., 2006b. Homogeneous enzymatic reactions in ionic liquids with poly(ethylene glycol)-modified subtilisin. Org. Biomol. Chem. 4, 3462–3467.

Nara, S.J., Harjani, J.R., Salunkhe, M.M., 2002. Lipase-catalysed transesterification in ionic liquids and organic solvents: a comparative study. Tetrahedron Lett. 43, 2979–2982.

Narayan, V.S., Klibanov, A.M., 1993. Are water-immiscibility and apolarity of the solvent relevant to enzyme efficiency? Biotechnol. Bioeng. 41, 390–393.

Némethy, G., 1967. Hydrophobic interactions. Angew. Chem. Int. Ed. Engl. 6, 195–206.

Nightingale, E.R., 1962. Viscosity of aqueous solutions. III. Tetramethylammonium bromide and the role of the tetraalkylammonium ions. J. Phys. Chem. 66, 894–897.

Noël, M., Lozano, P., Vaultier, M., Iborra, J.L., 2004. Kinetic resolution of *rac*-2-pentanol catalyzed by *Candida antarctica* lipase B in the ionic liquid, 1-butyl-3-methylimidazolium bis [(trifluoromethyl)sulfonyl]amide. Biotechnol. Lett. 26, 301–306.

Nordwald, E.M., Kaar, J.L., 2013. Stabilization of enzymes in ionic liquids via modification of enzyme charge. Biotechnol. Bioeng. 110, 2352–2360.

Oehlke, A., Hofmann, K., Spange, S., 2006. New aspects on polarity of 1-alkyl-3-methylimidazolium salts as measured by solvatochromic probes. New J. Chem. 30, 533–536.

Ou, G., Yang, J., He, B., Yuan, Y., 2011. Buffer-mediated activation of *Candida antarctica* lipase B dissolved in hydroxyl-functionalized ionic liquids. J. Mol. Catal. B: Enzym. 68, 66–70.

Pagni, R.M., 2002. Ionic liquids as alternatives to traditional organic and inorganic solvents. In: Rogers, R.D., Seddon, K.R., Volkov, S. (Eds.), Green Industrial Applications of Ionic Liquids. Kluwer Academic Publishers, Dordrecht, pp. 105–127.

Park, S., Kazlauskas, R.J., 2001. Improved preparation and use of room-temperature ionic liquids in lipase-catalyzed enantio- and regioselective acylations. J. Org. Chem. 66, 8395–8401.

Park, S., Kazlauskas, R.J., 2003. Biocatalysis in ionic liquids - advantages beyond green technology. Curr. Opin. Biotechnol. 14, 432–437.

Partridge, J., Halling, P.J., Moore, B.D., 1998. Practical route to high activity enzyme preparations for synthesis in organic media. Chem. Commun. 841–842.

Pavlidis, I.V., Gournis, D., Papadopoulos, G.K., Stamatis, H., 2009. Lipases in water-in-ionic liquid microemulsions. Structural and activity studies. J. Mol. Catal. B: Enzym. 60, 50–56.

Pavlidis, I.V., Tzafestas, K., Stamatis, H., 2010. Water-in-ionic liquid microemulsion-based organogels as novel matrices for enzyme immobilization. Biotechnol. J. 5, 805–812.

Perutz, M.F., 1978. Electrostatic effects in proteins. Science 201, 1187–1191.

Petkovic, M., Ferguson, J.L., Gunaratne, H.Q.N., Ferreira, R., Leitão, M.C., Seddon, K.R., Rebelo, L.P.N., Pereira, C.S., 2010. Novel biocompatible cholinium-based ionic liquids - toxicity and biodegradability. Green Chem. 12, 643–649.

Qiu, Z., Texter, J., 2008. Ionic liquids in microemulsions. Curr. Opin. Coll. Interface Sci. 13, 252–262.

Quiocho, F.A., Richards, F.M., 1964. Intermolecular cross linking of a protein in the crystalline state: Carboxypeptidase-A. Proc. Natl. Acad. Sci. U.S.A. 52, 833–839.

van Rantwijk, F., Sheldon, R.A., 2007. Biocatalysis in ionic liquids. Chem. Rev. 107, 2757–2785.

van Rantwijk, F., Madeira Lau, R., Sheldon, R.A., 2003. Biocatalytic transformations in ionic liquids. Trends Biotechnol. 21, 131–138.

van Rantwijk, F., Secundo, F., Sheldon, R.A., 2006. Structure and activity of *Candida antarctica* lipase B in ionic liquids. Green Chem. 8, 282–286.

Reetz, M.T., 2006. Practical protocols for lipase immobilization via sol-gel techniques. In: Guisan, J.M. (Ed.), Immobilization of Enzymes and Cells, second ed. Humana Press, Totowa, NJ, pp. 65–76.

Reichardt, C., 1994. Solvatochromic dyes as solvent polarity indicators. Chem. Rev. 94, 2319–2358.

Reichardt, C., 2005. Polarity of ionic liquids determined empirically by means of solvatochromic pyridinium N-phenolate betaine dyes. Green Chem. 7, 339–351.

Réjasse, B., Lamare, S., Legoy, M.-D., Besson, T., 2004. Stability improvement of immobilized *Candida antarctica* lipase B in an organic medium under microwave radiation. Org. Biomol. Chem. 2, 1086–1089.

Réjasse, B., Besson, T., Legoy, M.-D., Lamare, S., 2006. Influence of microwave radiation on free *Candida antarctica* lipase B activity and stability. Org. Biomol. Chem. 4, 3703–3707.

de los Ríos, A.P., Hernández-Fernández, F.J., Martínez, F.A., Rubio, M., Víllora, G., 2007. The effect of ionic liquid media on activity, selectivity and stability of *Candida antarctica* lipase B in transesterification reactions. Biocatal. Biotransform. 25, 151–156.

de los Ríos, A.P., van Rantwijk, F., Sheldon, R.A., 2012. Effective resolution of 1-phenyl ethanol by *Candida antarctica* lipase B catalysed acylation with vinyl acetate in protic ionic liquids (PILs). Green Chem. 14, 1584–1588.

Robinson, D.R., Jencks, W.P., 1965. The effect of compounds of the urea-guanidinium class on the activity coefficient of acetyltetraglycine ethyl ester and related compounds. J. Am. Chem. Soc. 87, 2462–2470.

Ropel, L., Belvèze, L.S., Aki, S.N.V.K., Stadtherr, M.A., Brennecke, J.F., 2005. Octanol–water partition coefficients of imidazolium-based ionic liquids. Green Chem. 7, 83–90.

Roy, I., Gupta, M.N., 2004. Preparation of highly active α-chymotrypsin for catalysis in organic media. Bioorg. Med. Chem. Lett. 14, 2191–2193.

Ryu, K., Dordick, J.S., 1992. How do organic solvents affect peroxidase structure and function? Biochemistry 31, 2588–2598.

Safavi, A., Maleki, N., Farjami, F., 2010. Phase behavior and characterization of ionic liquids based microemulsions. Colloids Surf. A 355, 61–66.

Samoilov, O.Y., 1953. Hydration of ions in aqueous solution. Bull. Acad. Sci. USSR, Div. Chem. Sci. 2, 219–225.

Samoilov, O.Y., 1957. A new approach to the study of hydration of ions in aqueous solutions. Discuss. Faraday Soc. 24, 141–146.

Sangster, J., 1989. Octanol-water partition coefficients of simple organic compounds. J. Phys. Chem. Ref. Data 18, 1111–1229.

Sarma, T.S., Ahluwalia, J.C., 1973. Experimental studies on the structure of aqueous solutions of hydrophobic solutes. Chem. Soc. Rev. 2, 203–232.

Schellman, J.A., 2003. Protein stability in mixed solvents: a balance of contact interaction and excluded volume. Biophys. J. 85, 108–125.

Schoevaart, R., Wolbers, M.W., Golubovic, M., Ottens, M., Kieboom, A.P.G., van Rantwijk, F., van der Wielen, L.A.M., Sheldon, R.A., 2004. Preparation, optimization, and structures of cross-linked enzyme aggregates (CLEAs). Biotechnol. Bioeng. 87, 754–762.

Schöfer, S.H., Kaftzik, N., Wasserscheid, P., Kragl, U., 2001. Enzyme catalysis in ionic liquids: lipase catalysed kinetic resolution of 1-phenylethanol with improved enantioselectivity. Chem. Commun. 425–426.

Schröder, U., Wadhawan, J.D., Compton, R.G., Marken, F., Suarez, P.A.Z., Consorti, C.S., de Souza, R.F., Dupont, J., 2000. Water-induced accelerated ion diffusion: voltammetric studies in 1-methyl-3-[2,6-(S)-dimethylocten-2-yl]imidazolium tetrafluoroborate, 1-butyl-3-methylimidazolium tetrafluoroborate and hexafluorophosphate ionic liquids. New J. Chem. 24, 1009–1015.

Seddon, K.R., Stark, A., Torres, M.-J., 2000. Influence of chloride, water, and organic solvents on the physical properties of ionic liquids. Pure Appl. Chem. 72, 2275–2287.

Shah, S., Gupta, M.N., 2007. Kinetic resolution of (±)-1-phenylethanol in [Bmim][PF_6] using high activity preparations of lipases. Bioorg. Med. Chem. Lett. 17, 921–924.

Shah, S., Solanki, K., Gupta, M.N., 2007. Enhancement of lipase activity in non-aqueous media upon immobilization on multi-walled carbon nanotubes. Chem. Cent. J. 1, 30.

Shakeri, M., Kawakami, K., 2010. Significant changes in the transesterification activity of free and mesoporous-immobilized *Rhizopus oryzae* lipase in ionic liquids. J. Biotechnol. 145, 281–283.

Sheldon, R.A., Schoevaart, R., Van Langen, L.M., 2005. Cross-linked enzyme aggregates (CLEAs): a novel and versatile method for enzyme immobilization (a review). Biocatal. Biotransform. 23, 141–147.

Sheldon, R.A., Schoevaart, R., van Langen, L.M., 2006. Cross-linked enzyme aggregates. In: Guisan, J.M. (Ed.), Immobilization of Enzymes and Cells, second ed. Humana Press, Totowa, NJ, pp. 31–45.

Sheldon, R.A., Sorgedrager, M., Janssen, M.H.A., 2007. Use of cross-linked enzyme aggregates (CLEAs) for performing biotransformations. Chim. Oggi 25, 62–67.

Sheldon, R.A., 2007. Cross-linked enzyme aggregates (CLEAs): stable and recyclable biocatalysts. Biochem. Soc. Trans. 35, 1583–1587.

Shen, Z.-L., Zhou, W.-J., Liu, Y.-T., Ji, S.-J., Loh, T.-P., 2008. One-pot chemoenzymatic syntheses of enantiomerically-enriched *O*-acetyl cyanohydrins from aldehydes in ionic liquid. Green Chem. 10, 283–286.

Shi, F., Zhang, Q., Li, D., Deng, Y., 2005. Silica-gel-confined ionic liquids: a new attempt for the development of supported nanoliquid catalysis. Chem. Eur. J. 11, 5279–5288.

Solanki, K., Gupta, M.N., 2008. Optimising biocatalyst design for obtaining high transesterification activity by α-chymotrypsin in non-aqueous media. Chem. Cent. J. 2, 2.

de Souza, R.L., de Faria, E.L.P., Figueiredo, R.T., dos Santos Freitas, L., Iglesias, M., Mattedi, S., Zanin, G.M., dos Santos, O.A.A., Coutinho, J.A.P., Lima, Á.S., Soares, C.M.F., 2013. Protic ionic liquid as additive on lipase immobilization using silica sol–gel. Enzym. Microb. Technol. 52, 141–150.

Spickermann, C., Thar, J., Lehmann, S.B.C., Zahn, S., Hunger, J., Buchner, R., Hunt, P.A., Welton, T., Kirchner, B., 2008. Why are ionic liquid ions mainly associated in water? A Car–Parrinello study of 1-ethyl-3-methyl-imidazolium chloride water mixture. J. Chem. Phys. 129, 104505.

St. Clair, N.L., Navia, M.A., 1992. Cross-linked enzyme crystals as robust biocatalysts. J. Am. Chem. Soc. 114, 7314–7316.

Stamatis, H., Xenakis, A., Kolisis, F.N., 1999. Bioorganic reactions in microemulsions: the case of lipases. Biotechnol. Adv. 17, 293–318.

Sureshkumara, M., Lee, C.-K., 2009. Biocatalytic reactions in hydrophobic ionic liquids. J. Mol. Catal. B: Enzym. 60, 1–12.

Tamaki, K., Ohara, Y., Isomura, Y., 1973. Viscosity *B* coefficients for some alkyl sulfates in aqueous solutions. Bull. Chem. Soc. Jpn. 46, 1551–1552.

Tanaka, H., Oka, Y., 2005. Chaotropic ions and multivalent ions activate sperm in the viviparous fish guppy *Poecilia reticulata*. Biochim. Biophys. Acta Gen. Subj. 1724, 173–180.

Tanford, C., 1978. The hydrophobic effect and the organization of living matter. Science 200, 1012–1018.

Tang, S., Baker, G.A., Zhao, H., 2012. Ether- and alcohol-functionalized task-specific ionic liquids: attractive properties and applications. Chem. Soc. Rev. 41, 4030–4066.

Theppakorn, T., Kanasawud, P., Halling, P.J., 2004. Activity of immobilized papain dehydrated by *n*-propanol in low-water media. Biotechnol. Lett. 26, 133–136.

Toral, A.R., de los Ríos, A.P., Hernández, F.J., Janssen, M.H.A., Schoevaart, R., van Rantwijk, F., Sheldon, R.A., 2007. Cross-linked *Candida antarctica* lipase B is active in denaturing ionic liquids. Enzym. Microb. Technol. 40, 1095–1099.

Tsangaris, J.M., Martin, R.B., 1965. Viscosities of aqueous solutions of dipolar ions. Arch. Biochem. Biophys. 112, 267–272.

Turner, M.B., Spear, S.K., Huddleston, J.G., Holbrey, J.D., Rogers, R.D., 2003. Ionic liquid salt-induced inactivation and unfolding of cellulase from *Trichoderma reesei*. Green Chem. 5, 443–447.

Ulbert, O., Frater, T., Belafi-Bako, K., Gubicza, L., 2004. Enhanced enantioselectivity of *Candida rugosa* lipase in ionic liquids as compared to organic solvents. J. Mol. Catal. B: Enzym. 31, 39–45.

Ursoiu, A., Paul, C., Kurtán, T., Péter, F., 2012. Sol-gel entrapped *Candida antarctica* lipase B — A biocatalyst with excellent stability for kinetic resolution of secondary alcohols. Molecules 17, 13045–13061.

Ventura, S.P.M., Santos, L.D.F., Saraiva, J.A., Coutinho, J.A.P., 2012. Ionic liquids microemulsions: the key to *Candida antarctica* lipase B superactivity. Green Chem. 14, 1620–1625.

Veronese, F.M., Largajolli, R., Boccú, E., Benassi, C.A., Schiavon, O., 1985. Surface modification of proteins activation of monomethoxy-polyethylene glycols by phenylchloroformates and modification of ribonuclease and superoxide dismutase. Appl. Biochem. Biotechnol. 11, 141–152.

Vioux, A., Viau, L., Volland, S., Le Bideau, J., 2010. Use of ionic liquids in sol-gel; ionogels and applications. Comptes Rendus Chim 13, 242–255.

Wang, H., Jing, Y., Wang, X., Yao, Y., Jia, Y., 2011. Ionic liquid analogous formed from magnesium chloride hexahydrate and its physico-chemical properties. J. Mol. Liq. 163, 77–82.

Wasserscheid, P., Welton, T., 2008. Ionic Liquids in Synthesis, second ed. Wiley-VCH, Weinheim.

Wiggins, P.M., 1997. Hydrophobic hydration, hydrophobic forces and protein folding. Physica A 238, 113–128.

Wiggins, P.M., 2003. Methods for the Separation of Isomers. US Patent 6638360.

Woodward, C.A., Kaufman, E.N., 1996. Enzymatic catalysis in organic solvents: polyethylene glycol modified hydrogenase retains sulfhydrogenase activity in toluene. Biotechnol. Bioeng. 52, 423–428.

Xue, L., Zhao, Y., Yu, L., Sun, Y., Yan, K., Li, Y., Huang, X., Qu, Y., 2013. Choline acetate enhanced the catalytic performance of *Candida rogusa* lipase in AOT reverse micelles. Colloids Surf. B 105, 81–86.

Yang, Z., Pan, W., 2005. Ionic liquids: Green solvents for nonaqueous biocatalysis. Enzym. Microb. Technol. 37, 19–28.

Yang, Z., Yue, Y.-J., Huang, W.-C., Zhuang, X.-M., Chen, Z.-T., Xing, M., 2009. Importance of the ionic nature of ionic liquids in affecting enzyme performance. J. Biochem. 145, 355–364.

Yang, Z., Zhang, K.-P., Huang, Y., Wang, Z., 2010. Both hydrolytic and transesterification activities of *Penicillium expansum* lipase are significantly enhanced in ionic liquid [BMIm][PF_6]. J. Mol. Catal. B: Enzym. 63, 23–30.

Yang, Z., 2009. Hofmeister effects: an explanation for the impact of ionic liquids on biocatalysis. J. Biotechnol. 144, 12–22.

Yu, H.W., Chen, H., Wang, X., Yang, Y.Y., Ching, C.B., 2006. Cross-linked enzyme aggregates (CLEAs) with controlled particles: application to *Candida rugosa* lipase. J. Mol. Catal. B: Enzym. 43, 124–127.

Zaks, A., Klibanov, A.M., 1988. Enzymatic catalysis in nonaqueous solvents. J. Biol. Chem. 263, 3194–3201.

Zarcula, C., Corîci, L., Croitoru, R., Ursoiu, A., Peter, F., 2010. Preparation and properties of xerogels obtained by ionic liquid incorporation during the immobilization of lipase by the sol–gel method. J. Mol. Catal. B: Enzym. 65, 79–86.

Zech, O., Thomaier, S., Kolodziejski, A., Touraud, D., Grillo, I., Kunz, W., 2010. Ionic liquids in microemulsions - a concept to extend the conventional thermal stability range of microemulsions. Chem. Eur. J. 16, 783–786.

Zhang, J., Ma, Y., Shi, F., Liu, L., Deng, Y., 2009. Room temperature ionic liquids as templates in the synthesis of mesoporous silica via a sol–gel method. Microporous Mesoporous Mater. 119, 97–103.

Zhao, H., Song, Z., 2007. Nuclear magnetic relaxation of water in ionic-liquid solutions: determining the kosmotropicity of ionic liquids and its relationship with the enzyme enantioselectivity. J. Chem. Tech. Biotechnol. 82, 304–312.

Zhao, H., Song, Z., 2010. Migration of reactive trace compounds from Novozym® 435 into organic solvents and ionic liquids. Biochem. Eng. J. 49, 113–118.

Zhao, H., Olubajo, O., Song, Z., Sims, A.L., Person, T.E., Lawal, R.A., Holley, L.A., 2006a. Effect of kosmotropicity of ionic liquids on the enzyme stability in aqueous solutions. Bioorg. Chem. 34, 15–25.

Zhao, H., Campbell, S., Jackson, L., Song, Z., Olubajo, O., 2006b. Hofmeister series of ionic liquids: kosmotropic effect of ionic liquids on the enzymatic hydrolysis of enantiomeric phenylalanine methyl ester. Tetrahedron: Asymmetry 17, 377–383.

Zhao, H., Jackson, L., Song, Z., Olubajo, O., 2006c. Enhancing protease enantioselectivity by ionic liquids based on chiral- or ω-amino acids. Tetrahedron: Asymmetry 17, 1549–1553.

Zhao, H., Campbell, S., Solomon, J., Song, Z., Olubajo, O., 2006d. Improving the enzyme catalytic efficiency using ionic liquids with kosmotropic anions. Chin. J. Chem. 24, 580–584.

Zhao, H., Jackson, L., Song, Z., Olubajo, O., 2006e. Using ionic liquid [EMIM][CH_3COO] as an enzyme-'friendly' co-solvent for resolution of amino acids. Tetrahedron: Asymmetry 17, 2491–2498.

Zhao, H., Baker, G.A., Song, Z., Olubajo, O., Crittle, T., Peters, D., 2008. Designing enzyme-compatible ionic liquids that can dissolve carbohydrates. Green Chem. 10, 696–705.

Zhao, H., Baker, G.A., Song, Z., Olubajo, O., Zanders, L., Campbell, S.M., 2009a. Effect of ionic liquid properties on lipase stabilization under microwave irradiation. J. Mol. Catal. B: Enzym. 57, 149–157.

Zhao, H., Jones, C.L., Cowins, J.V., 2009b. Lipase dissolution and stabilization in ether-functionalized ionic liquids. Green Chem. 11, 1128–1138.

Zhao, H., Jones, C.L., Baker, G.A., Xia, S., Song, Z., Olubajo, O., Person, V.N., 2009c. Regenerating cellulose from ionic liquids for an accelerated enzymatic hydrolysis. J. Biotechnol. 139, 47–54.

Zhao, H., Song, Z., Olubajo, O., Cowins, J.V., 2010a. New ether-functionalized ionic liquids for lipase-catalyzed synthesis of biodiesel. Appl. Biochem. Biotechnol. 162, 13–23.

Zhao, H., Baker, G.A., Cowins, J.V., 2010b. Fast enzymatic saccharification of switchgrass after pretreatment with ionic liquids. Biotechnol. Prog. 26, 127–133.

Zhao, H., Baker, G.A., Holmes, S., 2011. New eutectic ionic liquids for lipase activation and enzymatic preparation of biodiesel. Org. Biomol. Chem. 9, 1908–1916.

Zhao, H., Zhang, C., Crittle, T.D., 2013. Choline-based deep eutectic solvents for enzymatic preparation of biodiesel from soybean oil. J. Mol. Catal. B: Enzym. 85–86, 243–247.

Zhao, H., 2005. Effect of ions and other compatible solutes on enzyme activity, and its implication for biocatalysis using ionic liquids. J. Mol. Catal. B: Enzym. 37, 16–25.

Zhao, H., 2006a. Are ionic liquids kosmotropic or chaotropic? An evaluation of available thermodynamic parameters for quantifying the ion kosmotropicity of ionic liquids. J. Chem. Technol. Biotechnol. 81, 877–891.

Zhao, H., 2006b. Viscosity *B*-coefficients and standard partial molar volumes of amino acids, and their roles in interpreting the protein (enzyme) stabilization. Biophys. Chem. 122, 157–183.

Zhou, Y., Schattka, J.H., Antonietti, M., 2004. Room-temperature ionic liquids as template to monolithic mesoporous silica with wormlike pores via a sol-gel nanocasting technique. Nano Lett. 4, 477–481.

Zhou, H., Chen, J., Ye, L., Lin, H., Yuan, Y., 2011. Enhanced performance of lipase-catalyzed kinetic resolution of secondary alcohols in monoether-functionalized ionic liquids. Biores. Technol. 102, 5562–5566.

Zou, B., Hu, Y., Jiang, L., Jia, R., Huang, H., 2013. Mesoporous material SBA-15 modified by amino acid ionic liquid to immobilize lipase via ionic bonding and cross-linking method. Ind. Eng. Chem. Res. 52, 2844–2851.

Chapter 5

Rational Design of Ionic Liquids for Lipid Processing

Zheng Guo
Department of Engineering, Aarhus University, Aarhus, Denmark

5.1 INTRODUCTION

As a class of neoteric solvents, ionic liquids (ILs) have a unique molecular feature, which are composed solely of ions, typically cations and anions as a pair (Wilkes and Zaworotko, 1992; Welton, 1999; Wasserscheid and Keim, 2000). ILs have attracted considerable interest as unconventional solvents for process, catalysis, reaction, and electrochemical applications (Brennecke and Maginn, 2001; Earle and Seddon, 2000; Wilkes, 2002; Guo et al., 2007a; van Rantwijk and Sheldon, 2007). Many of the interests are rested on the concept of ILs as "designer solvents" (Plechkova and Seddon, 2008; Huddleston and Rogers, 1998; Baudequin et al., 2005; Seddon, 1997). In principle, both cation and anion can be designed and synthesized. More importantly the polarity and hydrophilicity/lipophilicity can be readily adjusted by a judicious selection of cation/anion and ILs have thus been referred to as "designer solvents" (Smiglak et al., 2006, 2007; Niedermeyer et al., 2013). This provides tremendous opportunities for design of a processing medium with the desired physical and chemical attributes (Freemantle, 1998); however, a practical implementation of this concept represents a big challenge (McLeese et al., 2010; Abe et al., 2010). The main limitation is the scarce available experimental data (Zhang et al., 2006). Hence a substantial amount of experimental work has to be carried out to determine and screen thermophysical properties of ILs and their mixtures for a preferable property.

Design of an IL system for a specific application can typically be categorized into two different levels (Scheme 5.1) (Smiglak et al., 2007; McLeese et al., 2010; Palomar et al., 2008, 2009). Level I is the selection of an IL from the existing IL pool or a new combination from existing cations/anions for a specific topic, which is a very basic level of application. The advanced Level II is based on the description of the physical and chemical

Ionic Liquids in Lipid Processing and Analysis. http://dx.doi.org/10.1016/B978-1-63067-047-4.00005-2

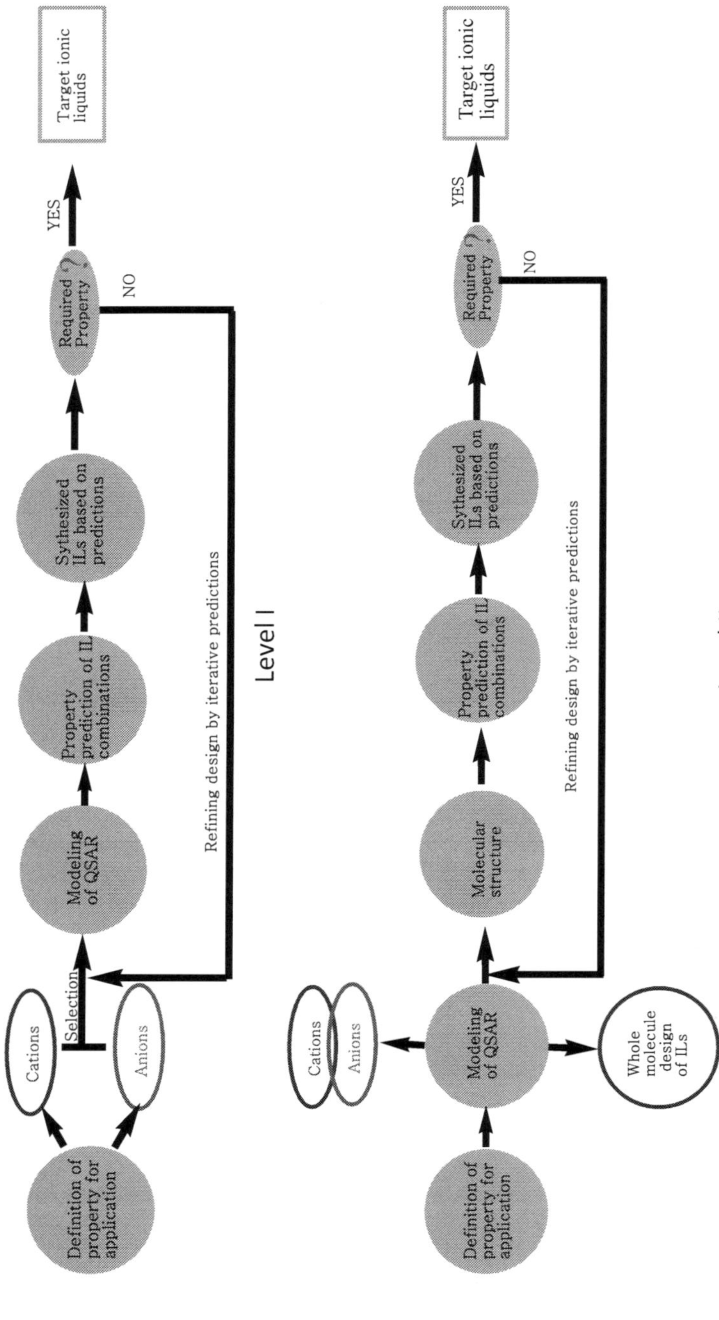

SCHEME 5.1 Two levels of a priori design of ionic liquids (ILs). *QSAR*, quantitative structure–activity relationship.

properties for a targeted application, the structure of a new IL, that exhibits the desired properties and is designed and synthesized; the latter is a more advanced predictive molecular design. In the first case, the availability of data for the property of ILs is crucial, which is currently limited; for the second case, a predictive model that fully establishes the link between the desired physicochemical property of IL and its constituent ions and fragments for a priori design is required. Nevertheless, in both cases development and application of a predictive model is of great interest (Niedermeyer et al., 2013; Palomar et al., 2007; Guo et al., 2007b; Deetlefs et al., 2006; Kato and Gmehling, 2005).

5.1.1 Motivation and Conditions for Design of Task-Specific ILs

The rational design of ILs requires a deep understanding of (1) the molecular characteristics that govern the physical property of ILs, such as melting point, density, and so on; and (2) interaction or solvation with the molecules of process (substrate), such as polarity, solvation effects. The methodologies to attain these understandings constitute the basis for design of task-specific ILs (Table 5.1). First, the design task should be describable in a qualitative and quantitative way. The types and the range of selection/design of ILs can be narrowed. Second, which is also of most importance, a model to quantitatively and correctly describe the structural attributes of ILs with respect to various physicochemical properties should be established. This model can be physically founded methods such as ab initio quantum mechanics (QM) (Kresse and Hafner, 1993; Kirchner et al., 2015; Emel'yanenko et al., 2007), molecular dynamics (MD) (Borodin et al., 2009; Maginn, 2009; Canongia Lopes and Padua, 2006; Borner et al., 2012; Morrow and Maginn, 2002), conductor-like

TABLE 5.1 Rationales of Design of Ionic Liquids (ILs) for a Specific Application

Rationales	Descriptions
Quantitative Definition of design task	The task of application can be clearly defined. The properties of the ILs required to fulfill the task can be defined qualitatively and quantitatively.
Modeling of quantitative structure–activity relationship (QSAR)	Empirical or ab initio model to quantitatively and correctly describe the structural attributes of ILs toward various physicochemical properties can be established.
Verification methodology of design	The methods to experimentally or theoretically verify the effectiveness and accuracy of the design and prediction of ILs are available.

screening model for real solvents (COSMO-RS) (Diedenhofen et al., 2003; Klamt et al., 2010), or classical correlative thermodynamic models, such as UNIversal Functional-group Activity Coefficients method (UNIFAC) (Kato and Gmehling, 2005; Nebig et al., 2007; Lei et al., 2012), UNIversal QUasi-chemical Activity Coefficients (UNIQUAC) (Banerjee et al., 2005; Santiago et al., 2009; Simoni et al., 2008), and so on. The correctness and quality to describe the structure—activity relationship determine the accuracy of the prediction. Last, a method to validate the prediction should be available or developed. These methods can be experimental or theoretical, among which the verification from other existing models will be a support. Of course, the best verification is that the measured values of the property of the ILs conceived of by model prediction has a good agreement with the predicted value.

ILs have found a variety of applications in different fields (Wasserscheid and Keim, 2000; van Rantwijk and Sheldon, 2007; Plechkova and Seddon, 2008; Huddleston and Rogers, 1998; Han and Row, 2010). Different applications have different emphases on the properties required (Brennecke and Maginn, 2001; Weingärtner, 2008; Wilkes, 2004; Huddleston et al., 2001). For instance, for electrochemistry, diffusivity, and electric conductivity will be the properties of interest (Wilkes et al., 1982; Liu et al., 2010); for food/drug process, compatible with food elements, toxicity and chemical stability will be a concern (El Seoud et al., 2007; Marrucho et al., 2014). Indeed, theoretical applications to predict IL properties have highlighted growth 2000s; however, none of the models have general applications in the prediction of all IL properties in high quality. Instead, some models have better accuracy than others in predicting some specific properties; for example, the QM model can better predict molecular structure and H-bonding interactions (Kirchner et al., 2015; Emel'yanenko et al., 2007); MD shows good prediction of melting point, heat capacity and diffusivity etc. of ILs (Morrow and Maginn, 2002; Tsuzuki et al., 2009); COSMO-RS is generally suitable for estimation of vapor pressure, and better estimation of activity coefficients and phase equilibria for the more demanding systems, and so on (Klamt et al., 2010); and UNIFAC (trained with experimental data) achieves good quality in prediction of activity coefficients and phase equilibria data for the simple molecules (Kato and Gmehling, 2005). Therefore, for different applications, different models may be selected and conditions for the prediction can also be varied.

Table 5.2 lists the general conditions needed for performing a theoretical prediction of IL property for a task-specific application. To make accurate predictions, a quantitative property value has to be provided. For some application, additional conditions will be required for consideration when the prediction is performed. With these conditions some irrelevant structures of ILs can be ruled out, and the range of screening will thus be narrowed. Upon a predictive model being established, experimental data are always needed as a training database for determination of the parameter constants and

TABLE 5.2 Conditions/Procedure for a Priori Design of Ionic Liquids (ILs)

Conditions	Descriptions
Definition of application objectives	The objectives of the application can be defined in a quantitative manner. The chemicophysical properties of ILs to attain the objectives have been delineated quantitatively.
Conditions pertaining to specific application	For a specific application, such as using for solvent enzymatic reactions, as food processing aid, etc., the property requirement for biocompatible with enzyme and nontoxic will be a precondition for IL design.
Availability of property data for model training	For a specific application, empirical model needs experimental data of desired property for regression of coefficient constants. For ab initio model, IL data are required for the model study and training, by which the parameterizations of the model can be achieved.
Correct modeling of quantitative structure–activity relationship (QSAR)	Establishment of a correct modeling of QSAR is the first of all conditions for a rational design of ILs. A sound theoretical basis and the parameters with clear physical meaning will be preferable.
Practicability and feasibility of synthesis	It makes sense only when a priori designed IL is attainable by means of available synthetic techniques and building blocks. The predictive design will be restrained and adjusted based on practicability and feasibility of synthesis.
Verification and model correction	A prediction or design of ILs should be able to refine in molecular structure when the property of synthetic ILs is deviated significantly from the designing. An iteration design should be performed to refine the structure for the best approach to the desired property.

parameterization of the model. The ultimate objective is to acquire a new IL that is able to undertake the desired task. Therefore, the conceived ILs should be able to synthesize with available building blocks and current synthetic techniques. The applicability of the model and the accuracy of the prediction have to be verified with the real property of synthetic ILs, and in most cases an iterative prediction/design is used.

5.1.2 Steps Toward a Priori Design of ILs

It is clear that prediction is an essential complement to experiments to acquire the property of ILs. However, real a priori designs of ILs need substantial knowledge, powerful modeling, and potent synthetic tools (Smiglak et al., 2007; Niedermeyer et al., 2013). However, many computational studies have been performed that are limited to a few ILs consisting of the most frequent cations and anions combinations (Deetlefs et al., 2006; Mondal and Balasubramanian, 2014). The related predictions are also limited to validate the effectiveness of the prediction model by comparing the calculated and experimental data (Kato and Gmehling, 2005; Nebig et al., 2007; Lei et al., 2012), or to explain the property of ILs with respect to their underlying molecular structures (Morrow and Maginn, 2002; Tsuzuki et al., 2009). In practice, the really valuable prediction is that the predicted structure could lead to a given set of properties; or to a newly conceived structure. Such a prediction should avoid/reduce the reliance on the search of new preferable ILs on chemical intuition or simple extrapolation from known compounds (Palomar et al., 2008, 2009).

Depending on the application requirement and ease of prediction, a priori design of ILs can be categorized into two levels as shown in Scheme 5.1. After many years' development of IL synthesis, a pool of cations and anions has been established, with cations represented by imidazolium, ammonium, pyridinium, pyrrolidinium, and phosphonium, and so on; and anions by halogens, acetate, dicyanamide, alkylsulfonate, alkylsulfate, tetrafluoroborate, hexafluorophosphate, bis(trifluoromethylsulfonyl)imide, and so on (Plechkova and Seddon, 2008; Zhang et al., 2006; Hough et al., 2007). The primary design of ILs for a specific application is to randomly select cations and anions from the pool, which can be categorized as the Level I design (Scheme 5.1). The selected ion-pair as input to the model is used to predict the property of the combination. Based on the best match of the predicted property with the target for design, the optimized structures will be determined for synthesis. The effectiveness and the degree of success of prediction will be ultimately validated by the property of synthesized ILs based on model selected structures. The synthesis of ILs with model prediction recommended structures will be an essential step. Examination of the property difference between the synthetic and targeted ILs will reveal the distance of the acquired property to the target. In most cases, an iteration prediction for refined design has to be performed to attain the target design. However, it should be pointed out that, due to the limited available structure of cations/anions pool, as well as insufficiency of modeling, Level I design may not generate a perfect design.

More advanced design (Level II) will rest on more accurate prediction/description of the structure–property relationship. Level II design often refers to more complicated design tasks, and thus more factors will be taken into consideration in addition to the main property requirement (Scheme 5.1). For example, to develop ILs for food or drug applications, toxicity and degradability will be considered. Some elements and functional groups with potential toxicity will be avoided, while more functional groups and complicated structures will be included when modeling prediction is performed. Advanced design may be executed in two ways: (1) selection from cations/anions pool but with more complex functional groups; and (2) an initio design based on molecular descriptors characteristic of the structure–property relationship, where, except for classic cations/anions, other ILs such as natural compounds (like amino acids)-derived ILs (Fukumoto et al., 2005; Ohno and Fukumoto, 2007; Chiappe et al., 2010a) and deep eutectic solvents (Zhang et al., 2012) will be considered for structure assessment and screening. In either of them, the modeling needs methodology to provide a more concise description of the structure–property relationship and large computation capacity to cope with complex structures and more variables and large numbers of item freedom. Similar to Level I prediction, the synthesis of the designed ILs has to be conducted and their properties are measured to verify the design and prediction. In many cases a refining design based on iterative predictions and feedback of synthetic ILs has to be implemented.

It has to be acknowledged that the Level II design scheme is an ideal situation, which has not been able to be implemented with currently available methodology, at least not completely (Palomar et al., 2008, 2009; Das and Roy, 2013; Ratti, 2014). This is because of (1) limited accuracy of quantitative structure–property relationships (QSPR) methodology and (2) limitation of synthetic methods to perform a successful synthesis of designed structures. The state-of-the-art modeling for ILs will be detailed in the following sections.

5.2 EXPERIMENTAL APPROACHES FOR CHARACTERIZATION OF PHYSICOCHEMICAL PROPERTIES OF ILs

To optimize the use of ILs and design the desirable ILs, knowledge of their physical and chemical properties is essentially important (Seddon, 1997; Morrow and Maginn, 2002). From the industrial viewpoint, a fundamental understanding of the chemical and physical properties of ILs should be known before exploiting its industrial application (Brennecke and Maginn, 2001; Plechkova and Seddon, 2008). Physical properties such as melting point, density, and viscosity are related to the mechanical and engineering components associated with a process. For academic research, physical and chemical properties are also indispensable to validate the theoretical models or select proper ILs (Zhang et al., 2006). For example, densities, viscosities, and surface tensions determine important parameters including rates of liquid–liquid phase separation, mass transfer, power requirements of mixing, and pumping. Other physical properties, such as refractive index, are related to certain chemical properties despite providing a bulk property description.

5.2.1 Measurements of Physicochemical Properties of ILs

There have been considerable studies on the measurement of physical properties of ILs and correlation of those properties with structures (Zhang et al., 2006; Tokuda et al., 2004, 2005, 2006). Table 5.3 listed the physical properties

TABLE 5.3 Physical Properties of Ionic Liquids (ILs) and the Methods for Measurement

Property	Measurement Method	Examples of ILs	References
Melting point/glass transition temperature	DSC	1,3-Dimethylimidazolium chloride; 1,3-dimethylimidazolium tetrafluoroborate; 1,3-dimethylimidazolium hexafluorophosphate	Ohno and Yoshizawa (2002)
Boiling point/ decomposition temperature	Thermogravimetric analyzer	1-Hexyl-3-methylimidazolium chloride; 1-hexyl-3-methylimidazolium hexafluorophosphate	Huddleston et al. (2001)
Viscosity	Viscometer or rheological meter	*n*-Hexyl pyridinium tetrafluoroborate; *n*-hexyl pyridinium bis((trifluoromethyl)sulfonyl)imides	Hardacre et al. (2002)
Density	Gravimetric analysis	*n*-Butyl pyridinium tetrafluoroborate; *n*-butyl pyridinium bis((trifluoromethyl)sulfonyl)imides	Noda et al. (2001)
Surface tension	Tensiometer	1-Octyl-3-methylimidazolium bromide; 1-octyl-3-methylimidazolium tetrafluoroborate; 1-octyl-3-methylimidazolium hexafluorophosphate	Law and Watson (2001)
Refractive index	Refractive index analyzer	[C_4mim][$MeSO_4$]; [C_4mim][OAc]; [C_4mim][OTf]; [C_4mim][NTf_2]	Tariq et al. (2009)

Solute partition	Liquid scintillation analyzer	1-Butyl-3-methylimidazolium hexafluorophosphate; 1-butyl-3-methylimidazolium tetrafluoroborate; 1-butyl-3-methylimidazolium bis((trifluoromethyl)sulfonyl)imides	Huddleston et al. (2001)
Self-diffusivity	Pulsed field gradient nuclear magnetic resonance	Trihexyltetradecylphosphonium bis(mandelato)borate	Filippov et al. (2013)
Heat capacity	DSC relative to an R-alumina (sapphire) sample	1-Hexyl-3-methylimidazolium bis((trifluoromethyl) sulfonyl)imides, 1-hexyl-3-methylimidazolium bromide, 1-hexyl-3-methylimidazolium tetraborate	Crosthwaite et al. (2005)
Phase equilibrium	"Cloud point" method/UV spectroscopy	1-*n*-Butyl-3-methylimidazolium hexafluorophosphate	Sahandzhieva et al. (2006)
Conductivity	Conductivity meter	Dimethyl-ethyl-propylammonium bis((trifluoromethyl) sulfonyl)imides; trimethyl-butylammonium bis((trifluoromethyl)sulfonyl)imides; triethyl-methylammonium bis((trifluoromethyl)sulfonyl) imides	MacFarlane et al. (2000)

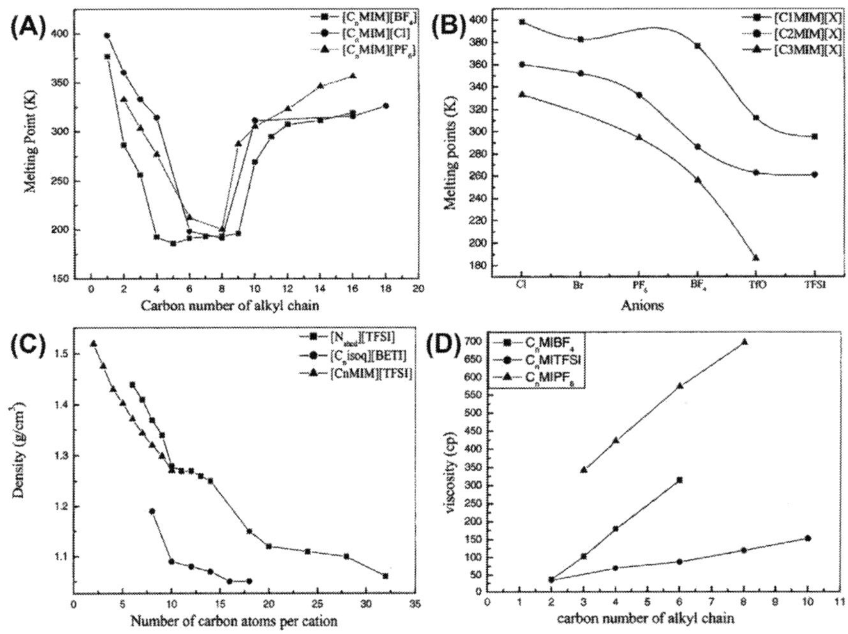

FIGURE 5.1 Variation of physical properties of ionic liquids against their structural evolutions. (A) Melting point variation with carbon number in alkyl chain for C_nMIX. (B) Melting point variation with anions. (C) Density variation of several systems with number of carbon atoms. (D) Viscosity variation of several systems with number of carbon atoms. *Reproduced from Zhang, S., Sun, N., He, X., Lu, X., Zhang, X., 2006. Physical properties of ionic liquids: database and evaluation. J. Phys. Chem. Ref. Data 35(4), 1475–1517 with permission.*

frequently investigated and the methods used for measurement. As can be seen from Table 5.3, nothing is special with ILs; the methods used for property measurement for conventional solvents are also applicable for ILs. However, unlike normal volatile solvents, the vapor pressure and flammable point are difficult to measure for ILs.

From a modeling point of view, the purpose of examination of physical and chemical properties of ILs as their structure changes is to establish the correlation between the property and molecular segments, and elucidate the intermolecular interactions governing the displayed property. Fig. 5.1 depicted the variation of some selected physical properties of ILs with the corresponding structure changes. Studies (Zhang et al., 2006) reveal that (1) the phase transition temperature of ILs is governed by van der Waals forces and electrostatic interaction force, and the impact of the two forces play different roles for different kinds of ILs; (2) the density of ILs is governed by the composing moieties, symmetry and property of cations/anions; the densities of most of the ILs are bigger than water except for pyrrolidinium dicyano-diamide and guanidinium; (3) the viscosities of ILs are governed essentially by van der Waals interactions and H-bonding. It is thus easy to understand property change against structural variation in Fig. 5.1. ILs with symmetric cation exhibit the highest melting points; the melting points appear to fall steadily with increasing alkyl chain length. When the carbon number in the substituents of 1-alkyl-3-methylimdazolium comes to 4–9, there is a platform, then the melting points increase with the carbon number as the increase in van der Waals interaction becomes dominating (Fig. 5.1A). The melting temperatures generally decrease with increasing anion radius except for PF_6 (the calculated values of Cl, Br, BF_4, PF_6, CF_3SO_3(TfO), $(CF_3SO_2)_2N$ (Tf_2N) are 2.70, 3.12, 3.44, 3.60, 3.79, and 4.39 Å, respectively) (Fig. 5.1B). It is because the larger anion radius induces weaker electrostatic interaction with imidazolium cation. However, the ILs with anions of PF_6 have strong hydrogen bonds for the sake of an F atom and their melting points are comparatively higher. Fig. 5.1C showed that the density for the three series of ILs decreases as the number of carbon atoms in the alkyl group and sum carbon number for quaternary ammonium ILs increases. It is interesting to note that the density of 1-methylimidazolium ILs decreases linearly with increasing temperature but at a rate less than that for molecular organic solvents. ILs are higher viscosity liquids in comparison to conventional solvents in one to three orders of magnitude. Alkyl chain lengthening or fluorination makes the salt more viscous, due to an increase in van der Waals interactions and hydrogen bonds (Fig. 5.1D).

5.2.2 Use of the Database of ILs (http://ilthermo.boulder.nist.gov)

For decades, there are increasing numbers of new ILs synthesized or physical properties of them measured by different research groups and appeared in various published literature (Tokuda et al., 2004, 2005, 2006; Branco et al., 2011; Pereiro et al., 2007). A comprehensive database with the collection of those data will definitely be a powerful tool for researchers to consult. In 2006, Zhang et al. published their collection of 1680 pieces of data on the physical properties for 588 available ILs, from which 276 kinds of cations and 55 kinds of anions were extracted. The presented properties include phase transition temperature, density, viscosity, surface tension, conductivity, and electrochemical window.

Since 2003, Merck KGaA and the Ionic Liquid Working Group of the Center for Environmental Research and Sustainable Technology (UFT) joined into a university–industry partnership to develop The UFT/Merck Ionic Liquids Biological Effects Database (http://www.il-eco.uft.uni-bremen.de), which is addressed to producers, developers, and downstream users of ILs in different fields of application to facilitate the selection of (eco)toxicologically favorable ILs. The aim of UFT/Merck database is "to get data about (eco)toxicological hazards of ionic liquids structures and to identify ionic liquids with optimal technical applicability combined with a reduced hazard potential for man and the environment." In latest update of the UFT/Merck database, up to 791 substances with

TABLE 5.4 Status of Ionic Liquids (IL) Database—ILThermo (v2.0)

Last Update	August 7, 2015
ILs	1345
All compounds	2765
References	2165
Pure IL systems	1022
Binary mixture systems	8723
Ternary mixture systems	2061
Data points (pure ILs)	76,110
Data points (binary mixtures)	224,971
Data points (ternary mixtures)	106,800
Total data points	407,881

(eco)toxicological data have been collected, also including several biodegradation test data to determine the biodegradability of IL structures.

In 2003, the International Union of Pure and Applied Chemistry (IUPAC) launched the Ionic Liquids Database (http://www.iupac.org/nc/home/projects/project-db), whose objective is to "Create an open-access, free, online, comprehensive database for storage and retrieval of metadata and numerical data for ionic liquids, including their syntheses, structure, properties, and uses." This led to the establishment of the Ionic Liquids Database—ILThermo (http://ilthermo.boulder.nist.gov) when the project ended in 2007. Table 5.4 showed the current status of the database, which revealed that up to 1345 types of ILs and a total of 407,881 data points are available for use. The database also covers binary and ternary systems in addition to pure IL systems. For each system, 48 physical and chemical properties are currently available for selection of search. In the search interface of the database, chemical formula (case-sensitive), CAS registry number, or name (part or full) can be used as input for the search, and the available filters include number of mixture components, reference, and property. Scheme 5.2 displays the interface of search results with the input of "imidazolium" and selection of "viscosity." In terms of the property of viscosity of imidazolium based systems, 2828 hits were available.

5.3 PREDICTIONS OF PHYSICOCHEMICAL PROPERTIES OF ILs

Achieving the full potential of these exciting new classes of solvents will require that chemical engineers bring to bear the full assortment of our unique

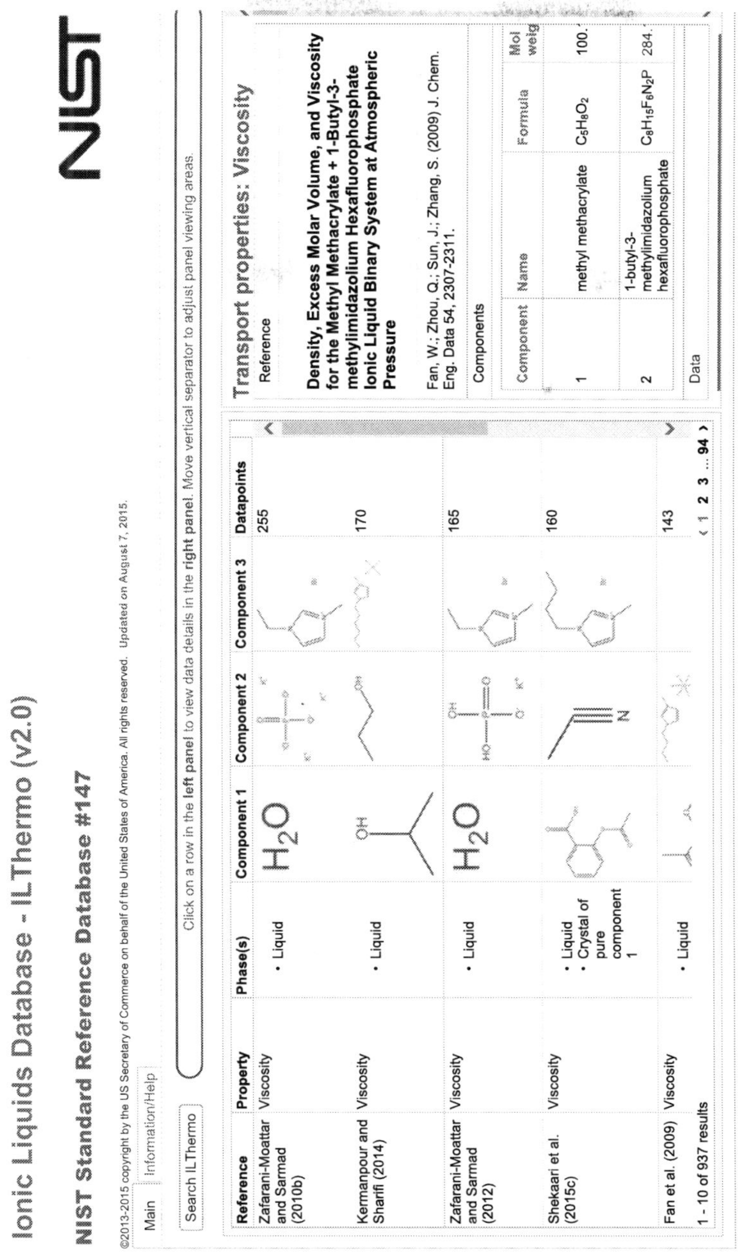

SCHEME 5.2 Search interface of Ionic Liquids (ILs) Database—ILThermo (v2.0).

mathematical and technical skills on the many problems involved in the design and development of ILs for practical applications. Depending on the reliance on or independency of the experimental data, the modeling or prediction of the properties of ILs can be categorized into empirical, semiempirical, or theoretical models. (1) Mathematical regression or fitting experimental data to property description equation are pure empirical models, among which solvatochromic dye to determine IL polarity (Reichardt, 2005; Muldoon et al., 2001; Carmichael and Seddon, 2000) and group contribution method (UNIFAC) (Kato and Gmehling, 2005; Nebig et al., 2007; Lei et al., 2012; Paduszynski and Domanska, 2013) to predict IL physical properties are representative methods. (2) Based on molecular properties using experimental quantities as descriptors the derived method such as Linear Solvation Energy Relationship (LSER) (Kamlet et al., 1983; Breitbach and Armstrong, 2008; Lee, 2005; Acree and Abraham, 2006; Sprunger et al., 2007; Anderson et al., 2002; Reichardt, 1979) (or Linear Free Energy Relationship, LFER) can be grouped as semiempirical methods. (3) The method based on structural molecule descriptors such as topological indices, quantum-chemical, geometrical, and electrostatic descriptors and so forth, are used to encode molecular features; this kind of method is ab initio modeling (Bernazzani et al., 2006; Krossing et al., 2006; Katritzky et al., 2002a,b). (4) Molecular dynamic modeling (Morrow and Maginn, 2002; Tsuzuki et al., 2009) and COSMO-RS (Palomar et al., 2007, 2008, 2009) can be seen as theoretical modeling. These representative methodologies will be detailed in the following sections.

5.3.1 Empirical/Semiempirical Approaches

Most chemical reactions take place in solution; therefore, the selection of an appropriate solvent for the reaction under study is, among other reaction parameters, of paramount importance for the success of a chemical process (Welton, 1999; Wasserscheid and Keim, 2000; Bini et al., 2008). It is known that the property of solvents could influence the position of reaction equilibria, reaction rate, and the position and intensity of spectral absorption bands (eg, UV/Vis, IR, EPR, NMR) (Reichardt, 1979, 1994, 2005; Muldoon et al., 2001; Chiappe et al., 2010b; Hallett and Welton, 2011; Kawai et al., 2004). Therefore, the chemists usually attempt to understand solvent effects on chemical processes or examine the intermolecular solute−solvent interactions in terms of the solvent *polarity*.

5.3.1.1 Scaling of IL Polarity by Reichardt's Solvatochromic Probes

According to IUPAC (Reichardt, 2005; Muller, 1994), solvent polarity is simply defined as the "overall solvation capability (or solvation power) for (1) educts and products, which influences chemical equilibria; (2) reactants and activated complexes ("transition states"), which determines reaction rates; and (3) ions or molecules in their ground and first excited state, which is responsible for light absorptions in the various wavelength regions. This overall solvation capability depends on the action of all, nonspecific and specific, intermolecular solute–solvent interactions, excluding such interactions leading to definite chemical alterations of the ions or molecules of the solute" (Reichardt, 1994; Poole, 2004). So far, up to 184 different methods, including dipole moment, polarizability index, kinetic measurements, equilibrium measurement, and spectroscopic measurements, have been attempted to scale solvent polarity (Katritzky et al., 2004), among which Reichardt's solvatochromic probes (betaine dye No. 30, $E_T(30)$) method (E_T^N scale) represents one of the most referred quantitative measures of solvent polarity. This has also been applied to characterization of the polarity of ILs (Reichardt, 2005).

Scheme 5.3 displayed the molecular structure of solvatochromic standard betaine dye no. 30 (2,6-diphenyl-4-(2,4,6-triphenylpyridinium-1-yl)phenolate) used as a solvent-dependent reference process to define empirically a solvent polarity scale, $E_T(30)$ values. As illustrated in Scheme 5.3, the principle for polarity scaling is that, with increasing solvent polarity, the dipolar ground-state ($\mu_G \approx 15$ D) is more stabilized by solvation than the less dipolar Franck–Condon excited state ($\mu_G \approx 6$ D), which may be even somewhat destabilized because its solvation shell is still equal to that of the ground state according to the Franck–Condon principle. As a result, the long-wavelength solvatochromic, intramolecular CT absorption band is hypsochromically

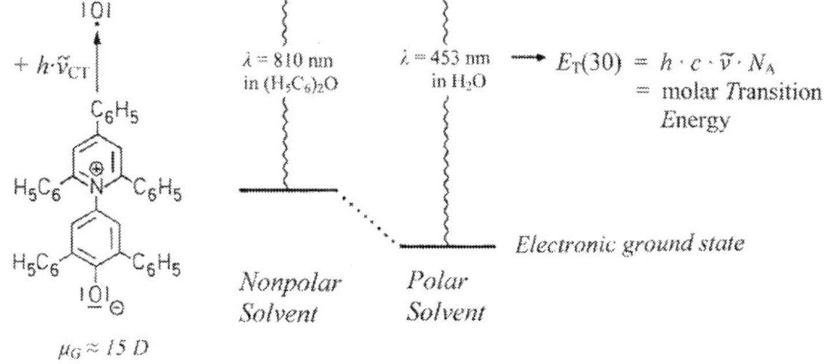

SCHEME 5.3 Qualitative illustration of the solvent influence on the intramolecular charge-transfer visible absorption of the standard betaine dye no. 30 ($\mu_G > \mu_E$) and definition of the $E_T(30)$ values as its molar electronic transition energies. *Reproduced from Reichardt, C., 2005. Polarity of ionic liquids determined empirically by means of solvatochromic pyridinium N-phenolate betaine dyes. Green Chem. 7, 339–351 with permission.*

shifted from $\lambda_{max} = 810$ nm in diphenyl ether (the least polar solvent in which the betaine dye is sufficiently soluble) to $\lambda_{max} = 453$ nm in water, which corresponds to a solvent-induced band shift of $\Delta\lambda = -357$ nm. These $E_T(30)$ values are simply defined as the molar transition energies (in kcal/mol; 1 kcal = 4.184 kJ) of the standard betaine dye no. 30, measured in solvents of different polarity at room temperature (25°C) and normal pressure (1 bar), according to (Eq. [1]).

$$E_T(30)\ \text{kcal/mol} = hc\widetilde{\upsilon}_{max}N_A = \left(2.8591 \times 10^{-3}\right)\widetilde{\upsilon}_{max}\ \text{cm}^{-1} = \frac{28,591}{\lambda_{max}}\ \text{nm} \quad [1]$$

where $\widetilde{\upsilon}_{max}$ is the wavenumber and λ_{max} the wavelength of the maximum of the long-wavelength, solvatochromic, intramolecular CT absorption band of the standard betaine dye, and h, c, and N_A are Planck's constant, the speed of light, and Avogadro's constant, respectively.

The $E_T(30)$ scale ranges from 63.1 kcal/mol for water, the most polar solvent, to 30.7 kcal/mol for tetramethylsilane (TMS), the least polar solvent, for which $E_T(30)$ values are experimentally available. $E_T(30)$ values can also be normalized as dimensionless E_T^N scale by using water ($E_T^N = 1.00$) and TMS ($E_T^N = 0.00$) as reference solvents to fix the scale, according to (Eq. [2]).

$$\begin{aligned} E_T^N &= [E_T(\text{solvent}) - E_T(\text{TMS})]/[E_T(\text{water}) - E_T(\text{TMS})] \\ &= [E_T(\text{solvent}) - 30.7]/32.4 \end{aligned} \quad [2]$$

A graphical representation of the E_T^N solvent polarity scale is given in Fig. 5.5, with a selection of 14 common molecular solvents above the line and eight groups of ILs below the line. In going from left to right above the line, with increasing solvent polarity, we first find the group of apolar non-HBD (aprotic) solvents (eg, cyclohexene, benzene, THF, dichloro-methane), followed by the dipolar non-HBD (aprotic) solvents (eg, acetone, DMF, DMSO), and eventually the group of dipolar HBD (protic) solvents (eg, cyclohexanol, 1-propanol, ethanol, phenol, methanol, ethane-1,2-diol, and 2,2,2-trifluoroethanol) (Fig. 5.2).

In the spectrum of solvent polarity, ILs are positioned among middle to high polar solvents (Fig. 5.2). More specifically, the polarities of different types of ILs are summarized in Table 5.5.

Actually $E_T(30)$ scale can be well correlated with multiple interacting solvent effects model Kamlet, Abboud, and Taft linear solvation energy relationship $XYZ = XYZ_0 + a\alpha + b\beta + s\pi^*$ (Eq. [9]), where α, β, and π^* represent hydrogen bond acidity, hydrogen bond basicity, and dipolarity/polarizability effects, respectively. a, b, and s are the corresponding regression coefficients) given by (Eqs. [3] and [4]) (Marcus, 1993):

$$E_T(30)\ \text{kcal/mol} = 31.2 + 11.5\pi^* + 15.2\alpha \quad [3]$$

$$E_T^N = 0.01 + 0.36\pi^* + 0.47\alpha \quad [4]$$

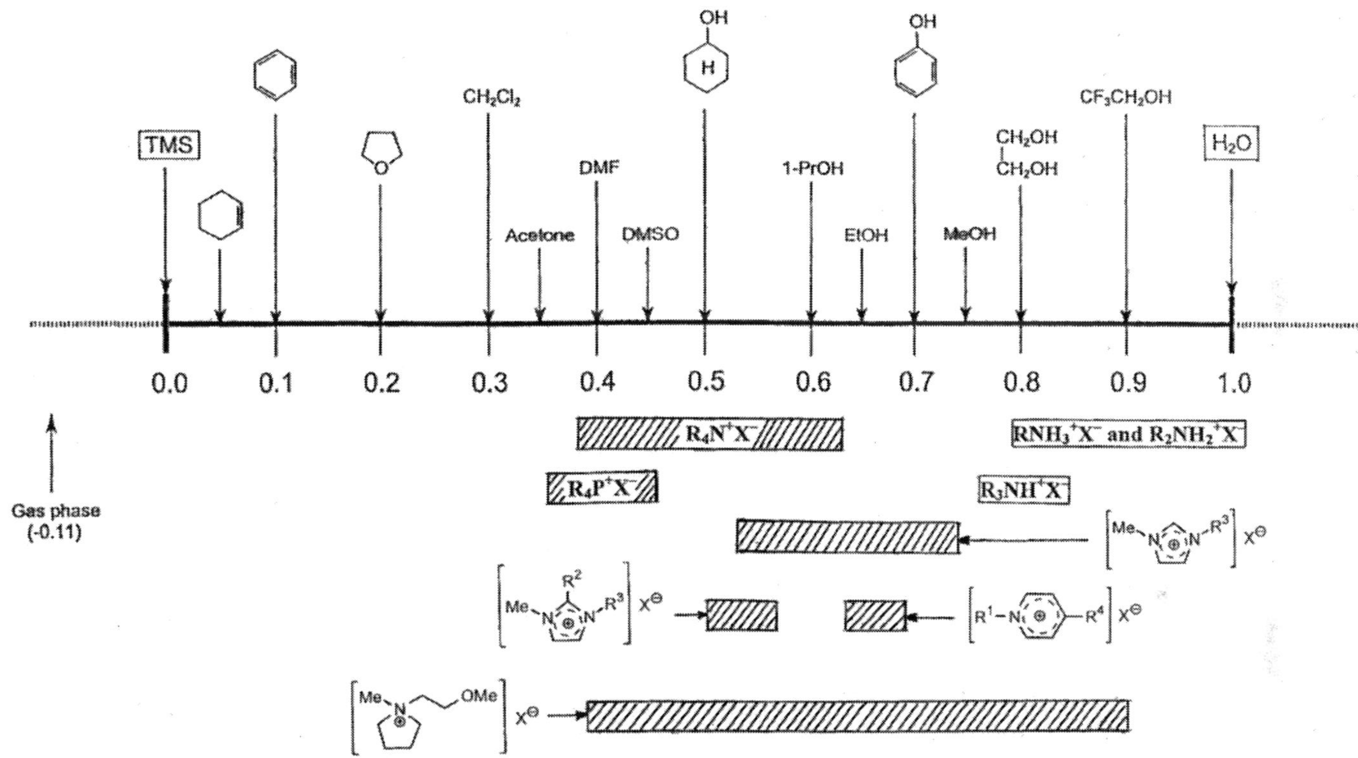

FIGURE 5.2 Normalized solvent polarity scale E_T^N [cf. Eq. [2] for its definition] with $E_T^N = 0.00$ for tetramethylsilane (TMS) and $E_T^N = 1.00$ for water as arbitrarily fixed points, with ordering of 14 selected solvents and the inclusion of eight groups of ionic liquids on this scale. The gas-phase E_T^N value is an extrapolated value and not directly measurable. *Reproduced from Reichardt, C., 2005. Polarity of ionic liquids determined empirically by means of solvatochromic pyridinium N-phenolate betaine dyes. Green Chem. 7, 339–351 with permission.*

TABLE 5.5 Polarity of Ionic Liquids (ILs) (NARTLIS) According to Their $E_T(30)$ (Bold-Faced) and E_T^N Values (in Parentheses)

ILs	Range	Mean Value	Molecular Liquids with Corresponding $E_T(30)$ Values
Primary and secondary alkylammonium salts, $[RNH_3]^+X^-$ and $[R_2NH_2]^+X^-$	**57–66** (0.81–1.1)	**61** (0.94)	H_2O, 63 $F_3C{-}CH_2OH$, 60
Tertiary trialkylammonium salt, $[R_3NH]^+X^-$	**57** (0.81)	**57** (0.81)	$HOCH_2{-}CH_2OH$, 56 $H{-}CONH_2$, 56
Quaternary tetraalkylammonium salts, $[R_4N]^+X^-$	**43–51** (0.38–0.63)	**47** (0.50)	$H_3C{-}CN$, 46 $H_3C{-}SO{-}CH_3$, 45
Quaternary tetraalkylphosphonium salts, $[R_4P]^+X^-$	**42–45** (0.35–0.44)	**43** (0.38)	$H_3C{-}CO{-}CH_3$, 42 $H{-}CON(CH_3)_2$, 43
1-Methyl-3-alkylimidazolium salts, $[R_3mim]^+X^-$	**48–55** (0.53–0.75)	**52** (0.66)	$H_3C{-}CH_2{-}OH$, 52 $H_3C{-}CONHCH_3$, 52
1-Methyl-2,3-dialkylimidazolium salts, $[R_2,R_3mim]^+X^-$	**47–49** (0.50–0.56)	**48** (0.53)	2-Propanol, 48 1-Heptanol, 48.5
1-Alkyl- and 1,4-dialkylpyridinium salts	**51–53** (0.63–0.69)	**52** (0.66)	$H_3C{-}CH_2OH$, 52 $H_3C{-}CONHCH_3$, 52
1-Methyl-1-(2-methoxyethyl) pyrrolidinium salts	**43–60** (0.38–0.90)	**53** (0.69)	$H_3C{-}CH_2OH$, 52 $H_3C{-}CONHCH_3$, 52

These multiparameter correlations clearly demonstrate that the $E_T(30)$ respectively, E_T^N values measure preferably the solvent's dipolarity/polarizability (given by π^*) and HBD acidity (given by α).

5.3.1.2 *Prediction of Thermophysical Properties and Phase Behaviors of ILs by Group Contribution Methods*

Since 1975 Fredenslund and coworkers first proposed UNIQUAC UNIFAC (Fredenslund et al., 1975). The UNIFAC model as a group contribution method has been well developed and elaborated as an empirical or semi-empirical method "for predicting liquid-phase activity coefficients provides the process design engineer with a useful tool for calculating vapor–liquid equilibrium compositions in the frequently encountered situation where no binary (or higher) experimental information is available" (Larsen et al., 1987).

The UNIFAC correlation attempts to break down the problem of predicting interactions between molecules by describing molecular interactions based upon the functional groups attached to the molecule. Although it is theoretically possible to make prediction by ab initio methods such as molecular dynamics (Kresse and Hafner, 1993; Kirchner et al., 2015; Emel'yanenko et al., 2007; Borodin et al., 2009; Maginn, 2009; Canongia Lopes and Padua, 2006; Borner et al., 2012; Morrow and Maginn, 2002) or COSMO-RS (Diedenhofen et al., 2003; Klamt et al., 2010), as a simplified model the concept of UNIFAC still has its unique value and has also been extended to the IL area (Lei et al., 2012; Gardas and Coutinho, 2008a; Paduszynski and Domanska, 2013). The group contribution methods have attempted to predict the properties of pure ILs such as melting point, density, viscosity, heating capacity, conductivity, and critical property, as well as the phase behaviors in mixed systems (Xiong et al., 2012; Gardas and Coutinho, 2009).

5.3.1.3 Prediction of the Properties of Pure ILs

It is known that the working principle of UNIFAC is that the property of a compound is calculated by summing up the contributions of certain defined groups of atoms, considering at the same time the number frequency of each group occurring in the molecule (Fredenslund et al., 1975; Larsen et al., 1987). Hence, the molecules have to be decomposed or broken down into the group or fragments with defined interaction parameters. Different from conventional solvent molecules as neutral molecules, ILs consist of solely ions with more substituents and groups attached to the ions. Valderrama et al. (2008) and Valderrama and Robles (2007) extended the conventional group contribution method to an IL system, where the authors adjusted the group contributions to critical temperature, pressure, volume, and acentric factors. They successfully predicted the critical properties and densities of 200 ILs with lower levels of average absolute deviations (5.9%).

Ye and Shreeve (2007), based on the known volumes of molecular/group species, derived volume parameters of groups/fragments for ILs. They developed a method for rapid and accurate estimation of the density of 59 room-temperature ILs based on the group additivity method. Gardas and Coutinho (2009) further extended this concept to the prediction of a broad range of thermophysical properties of ILs, such as density, surface tension, viscosity, speed of sound, and liquid heat capacity. In the prediction of respective property, different approaches have been employed; for example, Jenkins' approach for density (Gardas and Coutinho, 2008a), Orrick–Erbar method for viscosity (Gardas and Coutinho, 2008b), Vogel–Tammann–Fulcher (VTF) equation (Ghatee et al., 2010) for viscosity and electrical conductivity, Stokes–Einstein relation (Gardas and Coutinho, 2009) for self-diffusion coefficient, and linear correlation for refractive index and thermal conductivity (Gardas and Coutinho, 2009), among others. The group contribution parameters were derived from the regression of the equation with experimental data of ILs. In this prediction, the molecular volume is an important parameter that simply is treated as the sum of the volumes of cation and anion.

Valderrama and Rojas (2010) proposed a mass connectivity index based molecular parameter for prediction of IL properties. The molecular connectivity concept refers to an algorithm to encode bond contributions to a molecular branching index (Valderrama et al., 2011a,b). A simpler mass connectivity index, designated by λ, which considers both the type of connections between groups and the mass of such groups, is defined in Table 5.6.

TABLE 5.6 Groups Considered in this Study for Defining the Mass Connectivity Index

No	Groups without rings	M_i
1	$-CH_3$	15.04
2	$-CH_2-$	14.03
3	$>CH-$	13.02
4	$>C<$	12.01
5	$=CH_2$	14.03
6	$=CH-$	13.02
7	$=C<$	12.01
8	$-OH$	17.01
9	$-O-$	16.00
10	$>C=O$	28.01
11	$-COO-$	44.01
12	$-NH_2$	16.02
13	$-NH-$	15.02
14	$>N-$	14.01
15	$=N-$	14.01
16	$-CN$	26.02
17	$-NO_2$	46.01
18	$-F$	19.00
19	$-Cl$	35.45
20	$-Br$	79.90

TABLE 5.6 Groups Considered in this Study for Defining the Mass Connectivity Index—cont'd

No	Groups without rings	M_i
21	–I	126.90
22	–P	30.97
23	–B	10.81
24	–S–	32.07
25	$-SO_2$	64.07
26	–Al	26.98
27	–Sb	121.75
No	**Groups with rings**	M_i
28	$-CH_2-$	14.03
29	=CH–	13.02
30	>C<	12.01
31	=C<	12.01
32	–OH	17.01
33	–NH–	15.02
34	>N–	14.01
35	=N–	14.01

Data adapted from Valderrama, J.O., Rojas, R.E., 2010. Mass connectivity index, a new molecular parameter for the estimation of ionic liquid properties. Fluid Phase Equilib. 297, 107–112.

The mass connectivity index of a molecule is defined as the sum of the inverse of the mass connectivity interactions, calculated as the square root of the product of the mass of groups immediately connected in a molecule: $\lambda = \sum \left({}^{1}\!/\!\sqrt{m_i m_j} \right)_{ij}$. In this equation, m_i and m_j are the mass of neighboring groups i and j in a molecule. In summing up the connections, $m_i m_j$ is different from $m_j m_i$. The use of this equation can be demonstrated by an example of IL trimethylsulfonium tetrafluoroborate ($(CH_3)_3S + BF_4^-$). According to Table 5.6, $-CH_3$ is group 1, –S– is group 2, –B is group 3, and –F is group 4. The mass groups are $m_1 = 15.035$, $m_2 = 32.066$, $m_3 = 10.811$, and $m_4 = 18.998$, respectively. The number of connections between the groups in a molecule of [S111] [BF_4] are shown in Table 5.7.

TABLE 5.7 Group Connections for [S111] [BF_4] for the Calculation of the MCI

	$-CH_3$	–S–	–B	–F
$-CH_3$	–	3	–	–
–S–	3	–	1	–
–B	–	1	–	4
–F	–	–	4	–

The MCI (λ) for trimethylsulfonium tetrafluoroborate can be calculated by (Eq. [5]) and the value is 0.8389.

$$\lambda = \sum \left(\frac{1}{\sqrt{m_i m_j}} \right)_{ij} = \frac{3}{\sqrt{m_1 m_2}} + \frac{3}{\sqrt{m_2 m_1}} + \frac{1}{\sqrt{m_2 m_3}} + \frac{1}{\sqrt{m_3 m_2}} + \frac{4}{\sqrt{m_3 m_4}} + \frac{4}{\sqrt{m_4 m_3}} \quad [5]$$

The influence of molecular structure, represented by the connectivity index on the property such as density (Eq. [6]) and heat capacity (Eq. [7]) is included in the temperature functionality for each of the properties.

$$\rho = \rho_0 + a\lambda(T - T_0) \quad [6]$$

$$C_p = C_{p_0} + \left[c(T - T_0) + d\left(T^2 - T_0^2\right)\right] \quad [7]$$

The constant a for the density and c and d for the heat capacity are determined from experimental data taken from the literature. The prediction from this method gave fairly good results: for density with 479 data points yielding % Av. absolute deviation, 0.3%, and heat capacity with 541 data points producing % Av. absolute deviation, 0.9% (Valderrama et al., 2011a). Later the same authors further elaborated their predictions with incorporation of both molecular volume and mass connectivity index (Valderrama et al., 2011b).

Albert and Müller (2014) used first-order groups as molecular descriptors to describe the ions, which allows for a wide range of applications and high accuracy at the same time. With experimental heat capacity data for 106 ILs and thermal conductivity data for 39 ILs, the prediction achieved a mean absolute deviation of 5.4% for the heat capacity and of 8.1% for the thermal conductivity.

5.3.1.4 Prediction of Phase Behaviors in IL-Mixed Systems

Original and modified UNIFAC models have been proven a powerful tool in prediction of thermodynamic property, such as activity coefficients, which are expressed as a function of a combinatorial contribution ln γ_i^C, essentially due to differences in size and shape of the molecules; and a residual contribution, that is, ln γ_i^R, essentially due to energetic interactions (temperature). Lei et al. (2012) extended this concept to imidazolium-based IL systems and developed a decomposing method of ILs into groups by treating the core structure of ILs, the skeletons of the cation and the anion as a whole; for example, [bmim][BF_4] is decomposed as one CH_3 group, three CH_2 groups, and one [MIM][BF_4] group. Based on this concept, the authors developed UNIFAC parameter matrix as illustrated in Fig. 5.3, with 45 main groups and 77 subgroups for ILs.

The authors used this developed model for Prediction of Activity Coefficients at Infinite Dilution in mixed IL systems, Prediction of Vapor–Liquid Equilibria (VLE) of Systems with ILs at Finite Concentration, Screening of Suitable ILs as Entrainer in Separation Processes, and others, and good predictive results were obtained (Lei et al., 2009). Kato et al. (2004) used group contribution UNIFAC to predict the VLE of binary systems containing the ILs, and the relative mean deviations between the measured and predicted pressures obtained using the UNIQUAC model are achieved as 4.7%.

It should be pointed out, current UNIFAC methods used for prediction of IL thermodynamic properties are far from universally applicable, and it differs from case to case in terms of definition of group contribution parameters. Many of them have limited application range and some correction parameters have to be introduced for a more accurate prediction. A more universal group contribution method is expected to develop in the near future and serves design of ILs for chemical reaction, process, and separation.

5.3.2 Fast Screening of ILs: QSPR and Linear Free Energy Descriptors

QSPR model is a function to predict the property of a molecule (responsive variables, predictors) by associating with its chemical structure (structural variable, descriptors) (Das and Roy, 2013; Katritzky et al., 2002a; Valderrama and Robles, 2007; Ye and Shreeve, 2007). This function can be a pure theoretical description or a regression from experimental quantity (Abraham et al., 2004). The application of theoretical QSPR, represented by MD modeling in IL prediction will be addressed in the following session. A simple but frequently used empirical or semiempirical model to describe solute–solvent interaction is the LSER model, first proposed by Kamlet and Taft (1976), Taft and Kamlet (1976), Kamlet et al. (1977), and IUPAC (1997). This model is based on multivariate linear regression of solvatochromic parameters (Marcus, 1993). Another multivariate linear regression model is the Abraham LFER model based on chromatographic parameters (Acree and Abraham, 2006; Sprunger et al., 2007; Anderson et al., 2002).

Although Reichardt's dye and Nile Red as well as fluorescent probes and the Rohrschneider-McReynolds gas–liquid chromatography method have been

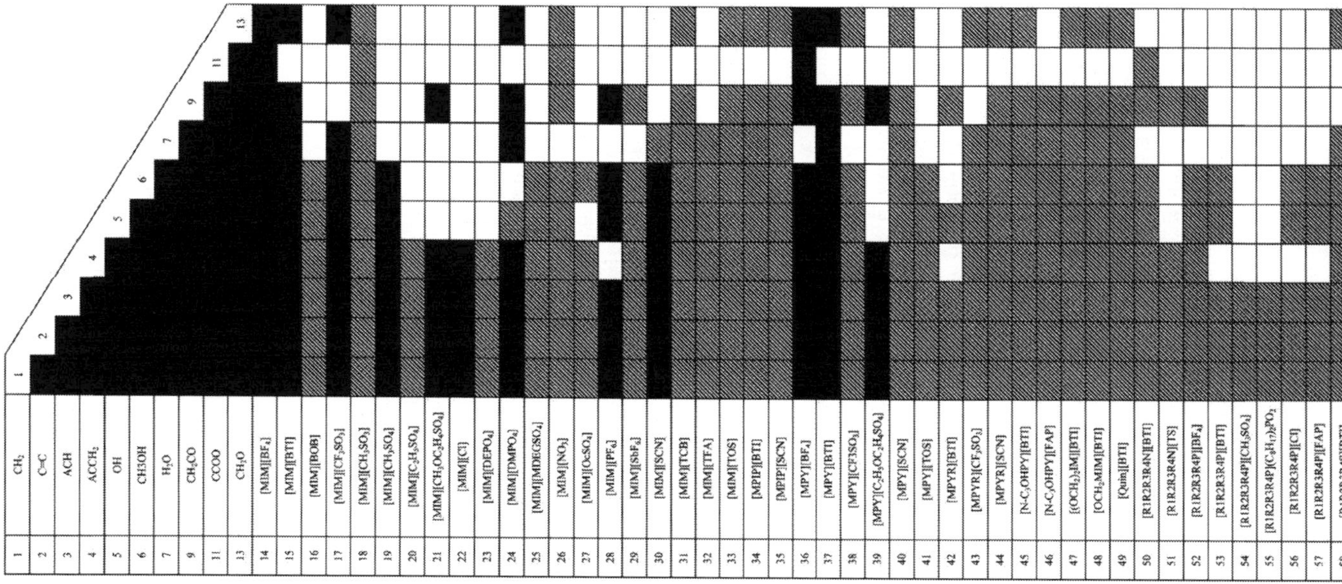

FIGURE 5.3 UNIversal Functional-group Activity Coefficients method parameter matrix for imidazolium-based ILs. Parameters published in Lei et al. (2009), Parameters published in Lei et al. (2012), No parameters available. *Reproduced from Lei, Z., Dai, C., Liu, X., Xiao, L., Chen, B., 2012. Extension of the UNIFAC model for ionic liquids. Ind. Eng. Chem. Res. 51(37), 12135–12144.*

used to characterize ILs (Reichardt, 2005; Acree and Abraham, 2006; Sprunger et al., 2007; Anderson et al., 2002). The motivation of developing multiparameter models to characterize the interaction between solute–solvents is to solve the problem of inefficiency of the single-parameter polarity approach to analyze and explain the multifacet role/property of a solvent when used as media for reaction or processing. This is especially true for ILs, which are among the most complex solvents where dispersive, π–π, n–π, hydrogen bonding, dipolar, ionic/charge–charge interactions are simultaneously presented (Anderson et al., 2002). An interesting progress in developing understanding of the physical meaning of the descriptors of LFERs is that two sets of molecular descriptors, the five experimental Abraham, and the five COSMOments of Klamt's COSMO-RS, are shown to exhibit a large overlap as far as their chemical content, which have been compared to a data set of 470 compounds (Zissimos et al., 2002). This somewhat provides experimental basis for using COSMO-RS descriptors to analyze interactions in IL involved systems.

In the original equation of the Kamlet–Taft model (Kamlet et al., 1983), a variety of parameters have been included (Eq. [8]):

$$XYZ = XYZ_0 + s(\pi^* + d\delta) + a\alpha + b\beta + h\delta_H + e\xi \qquad [8]$$

where *XYZ* is some property of solvent and XYZ_0 is a constant item. The π^* scale is an index of solvent dipolarity/polarizability, which measures the ability of the solvent to stabilize a charge or a dipole by virtue of its dielectric effect. δ parameter is a polarizability correction term. The α scale of solvent HBD (hydrogen-bond donor) acidities describes the ability of the solvent to donate a proton in a solvent-to-solute hydrogen bond. The β scale of hydrogen-bond acceptor (HBA) basicities provides a measure of the solvent's ability to accept a proton (donate an electron pair) in a solute-to-solvent hydrogen bond. δ_H term is the *Hildebrand solubility parameter*, a measure of the solvent–solvent interactions that are interrupted in creating a cavity for the solute (the cavity term). The ξ parameter is a coordinate covalency measure. The *s*, *d*, *a*, *b*, *h*, and *e* coefficients in the equation measure the relative susceptibilities of *XYZ* to the indicated solvent property scales. Actually in many cases the δ, δ_H, and ξ terms are constants, therefore, the Kamlet–Taft equation can also be simplified as (Eq. [9]):

$$XYZ = XYZ_0 + a\alpha + b\beta + s\pi^* \qquad [9]$$

where α, β, and π^* represent hydrogen bond acidity, hydrogen bond basicity, and dipolarity/polarizability effects, respectively. *a*, *b*, and *s* are the corresponding regression coefficients. The π^* values can be determined by measuring the bathochromical shift of UV/vis spectrum of nonprotonic indicator (b = 0) *N,N′*-diethyl-4-nitroaniline or Reichardt's $E_T(30)$ betaine. In HBD solvents ($a \neq 0$), the spectrum of the betaine is shifted further hypsochromically. The magnitudes of the enhanced shifts are linear with the *a* values and the *a* values can thus be determined by regression.

The determination of *β value* involves the UV/vis spectrum of 3,5-dinitroaniline *α* protonic indicator whose absorption maximum is shifted bathochromically with increasing solvent dipolarity and with type-B hydrogen bonding by the amine protons to HBA solvents ($d = 0$, $b \neq 0$).

Lee et al. (2008) investigated the polarity and Kamlet–Taft parameters of pyridinium-based ILs in the range 25–65°C by using three dye probes: Reichardt's dye, *N,N*-diethyl-4-nitroaniline, and 4-nitroaniline. Based on Kamlet–Taft parameters, they concluded that alkylpyridinium bis(trifluoromethylsulfonyl)imides are more polar than DMSO, DMF, and acetone polar solvents but less polar than imidazolium-based ILs. They also found out that substituting a methyl group on the cation decreases polarity (*a* and *β*). Substituting a methyl group in the C(2) position on the cation leads to low polarity and reduces the dependence of temperature on polarity (similar for π^*).

In the attempt to characterize ILs by Kamlet–Taft parameters, Welton's group (Crowhurst et al., 2003) characterized imidazolium-based ILs with varied anions and found that π^* is high for all the ILs studied and varies with both anion and cation; *a* is generally moderate and depends mainly on the cation; *β* is also moderate and depends mainly on the anion. In another study, the same group examined a variety of dye probes and observed the dye selection can have a dramatic influence on the resulting measurement (Ab Rani et al., 2011). They found, in general, π^* values vary with the cation such that morpholinium > imidazolium > pyridinium > pyrrolidinium > phosphonium ILs sharing the same anion. The *β* value is dominated by the anion of the IL, with the cation having a secondary effect. The Kamlet–Taft *a* values of the ILs did not agree with measurements of their hydrogen-bond donor abilities derived from GC measurements (Table 5.8).

5.3.3 Molecular Dynamic Modeling

Theoretically, ILs can be modeled/simulated based on structural molecular/atomistic descriptors, such as topological indices, and quantum-chemical, geometrical, and electrostatic descriptors to encode molecular features (Kresse and Hafner, 1993; Kirchner et al., 2015; Emel'yanenko et al., 2007). This methodology depends on no or little experimental data to take advantage of full, realistic representation of the molecules, which could be carried out at an ab initio level. Quantum mechanics in theory can provide the most accurate and reliable means for computing the structure and energetics of a molecule, however, in practice it is difficult to obtain highly accurate intermolecular energies, and it is not yet possible to carry out a fully quantum mechanical calculation of ILs in the condensed phase and extract the kind of thermodynamic and transport properties we are interested in (Emel'yanenko et al., 2007; Kroon et al., 2007). As a quantitative approach, an atomistic simulation involves the construction of a model that faithfully reproduces the geometric and energetic features of the real system (Maginn, 2007, 2009; Morrow and Maginn, 2002). MD and Monte Carlo (MC) have been used to model ILs by

TABLE 5.8 E_T^N and Kamlet–Taft Values for a Selection of Ionic Liquids and Solvents

Solvent	E_T^N	π^*	a	β	References
Water	1.00	1.33	1.12	0.14	Reichardt (1994) and Persson (1986)
Methanol	0.762	0.73	1.05	0.61	Reichardt (1994) and Persson (1986)
Acetone	0.350	0.704	0.202	0.539	Persson (1986)
Hexane	0.009	−0.12	0.07	0.04	Reichardt (1994)
$[EtNH_3][NO_3]$	0.954	1.12	1.10	0.46	Shetty et al. (1987) and Poole et al. (1989)
$[bmim][BF_4]$	0.670	1.047	0.627	0.376	Muldoon et al. (2001)
$[bmim][PF_6]$	0.669	1.032	0.634	0.207	Muldoon et al. (2001)
$[bmim][N(Tf)_2]$	0.644	0.984	0.617	0.243	Muldoon et al. (2001)
[bmim][TfO]	0.656	1.006	0.625	0.464	Muldoon et al. (2001)

generating atomistic trajectory deterministically evolved as a function of time (MD) (Maginn, 2007); or sampling states from the appropriate ensemble probability distribution (MC) (Shah et al., 2002). By subjecting the resulting trajectory to statistical mechanical analysis, all the relevant thermodynamic/transport properties of the model system can be determined (Maginn, 2007, 2009). The prediction quality is governed by accuracy of the model matching reality and generated trajectories.

According to Maginn (2007), there are two key elements behind effective simulation of ILs by MD: (1) a good model (force field) and (2) a proper method to conduct simulation (conformational sampling, balance between computation quality, and expense). This force field typically has the following functional form (Eq. [10]) for the configurational potential energy:

$$U_{\text{tot}} = \sum_{\text{bonds}} k_b(r-r_0)^2 + \sum_{\text{angles}} k_\theta(\theta-\theta_0)^2 + \sum_{\text{dihedrals}} k_X[1+\cos(n_0X-\delta_0)] + \sum_{\text{impropers}} k_\psi(\psi-\psi_0)^2 + \sum_{i=1}^{N-1}\sum_{j=i+1}^{N}\left\{4\in_{ij}\left[\left(\frac{\sigma_{ij}}{r_{ij}}\right)^{12}-\left(\frac{\sigma_{ij}}{r_{ij}}\right)^{6}\right]+\frac{q_iq_j}{r_{ij}}\right\} \quad [10]$$

The simulation snapshots of Fig. 5.4, especially those rendered under the red/green polar/nonpolar convention, provide a powerful visual insight into the nature and evolution of the observed structures as the length of the nonpolar

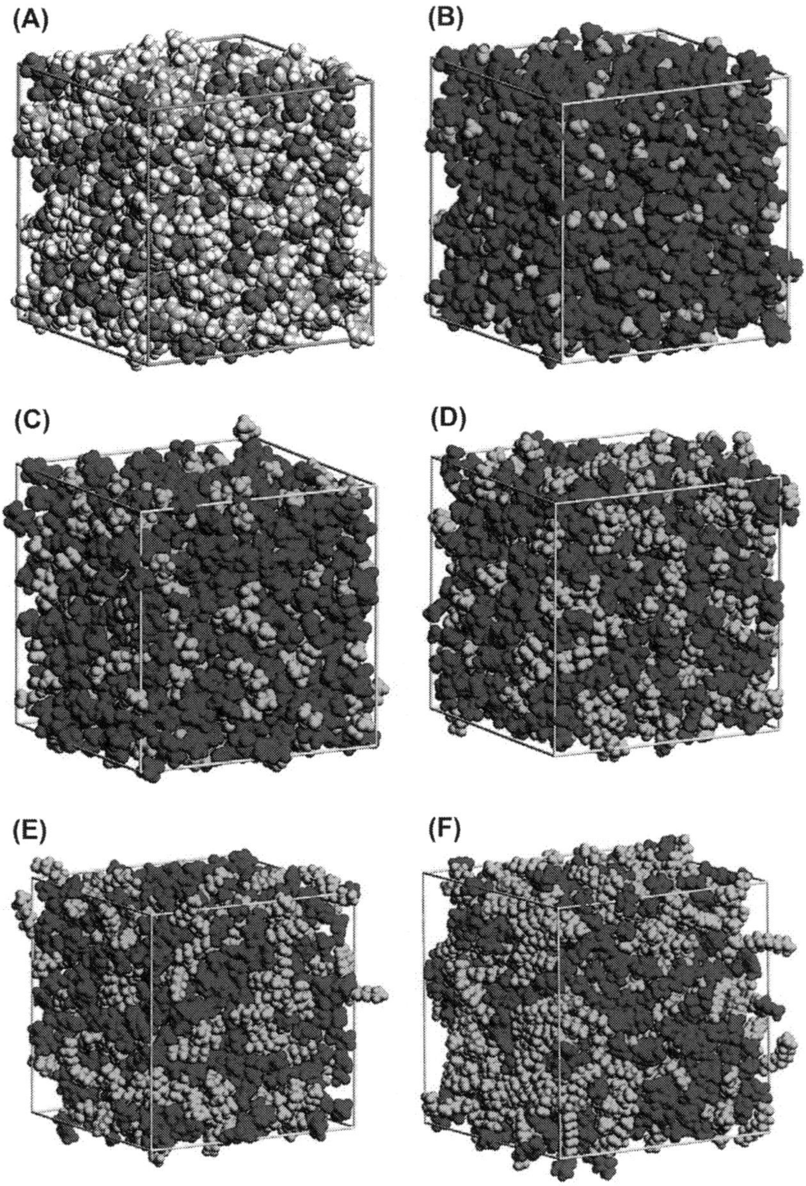

FIGURE 5.4 Snapshots of simulation boxes containing 700 ions of $[C_n mim][PF_6]$. The application of a coloring code enables clear identification of the charged and nonpolar domains that form in ILs. The lengths of the box sides are given: (A) $[C_2mim][PF_6]$ CPK coloring; (B) $[C_2mim][PF_6]$ same configuration as in (A) with red (dark gray in print versions)/green (light gray in print versions) (charged/nonpolar) coloring; (C) $[C_4mim][PF_6]$ $l = 49.8$ Å; (D) $[C_6mim][PF_6]$ $l = 52.8$ Å; (E) $[C_8mim][PF_6]$ $l = 54.8$ Å; (F) $[C_{12}mim][PF_6]$ $l = 59.1$ Å. *Reproduced from Canongia Lopes, J.N.A., Padua, A.A.H., 2006. Nanostructural organization in ionic liquids. J. Phys. Chem. B. 110, 3330–3335 with permission.*

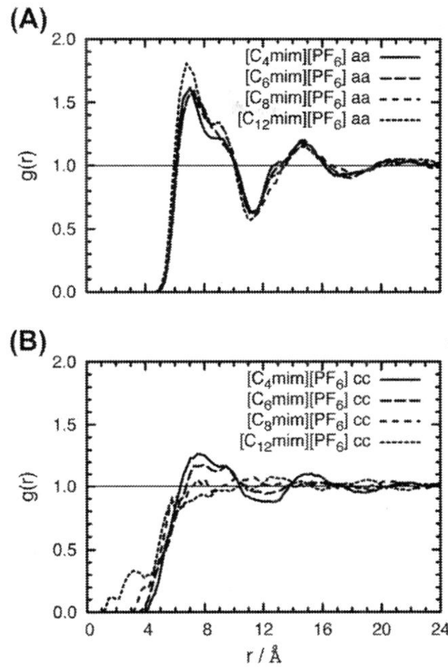

FIGURE 5.5 Center-of-mass radial distribution functions between anion–anion (aa) (A) and cation–cation (cc) (B) pairs in $[C_n\text{mim}][PF_6]$. Pairs denoted as ii (C) and ia (D) correspond to (the centers of mass of) the imidazolium rings in the cations relative to each other and relative to the anion, respectively. *Reproduced from Canongia Lopes, J.N.A., Padua, A.A.H., 2006. Nanostructural organization in ionic liquids. J. Phys. Chem. B. 110, 3330–3335 with permission.*

chain is increased. However, a more complete analysis can be performed if the radial distribution functions (and corresponding static structure factors) are considered (Fig. 5.5).

Properties computed by an MD force field include molar volumes, average energies, liquid structure via radial distribution functions, and mean square displacements as a function of time. Properties computed by MC simulations include the liquid density, cohesive energy density (and thus enthalpy of vaporization), isothermal compressibility, and volumetric expansion coefficient (Shah et al., 2002; Shah and Maginn, 2004). The liquid structure was also determined from radial distribution functions.

Success of molecular simulation for accurately predicting various properties depends on the quality of the inter- and intramolecular potential functions (ie, the force fields) (Borodin, 2009; Liu et al., 2004). Several force fields, including AMBER (Cornell et al., 1995), CHARMM (MacKerell et al., 1998), and OPLS (Jorgensen et al., 1996), have been developed and applied successfully in a variety of systems. The group of Lynden-Bell (Hanke et al., 2001) developed both the all-atom and united-atom force fields. Morrow and Maginn (2002, 2003) reported molecular dynamics of $[\text{bmim}][PF_6]$ by a force field from CHARMM. de Andrade et al. (2002a,b) presented a force field mainly using the parameterization strategies from AMBER. Canongia Lopes et al. (2004) proposed a model for a series of dialkylimidazolium cations based on the OPLS-AA/AMBER framework and discussed the parameterization by the torsion energy profiles in detail. Borodin (2009) developed a many-body polarizable force field applicable to a variety of ILs to achieve an accurate description/prediction of density, heat of vaporization, self-diffusion coefficients, ionic conductivity, and viscosity of small organic molecules and ILs. It should be pointed out that, despite these efforts, there are a huge number of cation and anion types for which force-field parameters do not yet exist. Moreover, many of the force fields that have been proposed have not been subjected to a rigorous validation procedure.

The second key element in the improvement of MD simulation is the difficulty in making conformational sampling (Maginn, 2009). If natural relaxation times are longer than the practical timescale accessible to conventional MD (typically 10−100 ns) then MD cannot generate equilibrium conformations. Two approaches have been attempted to tackle this challenge. The first is to *coarse grain* the system so that nonessential degrees of freedom are removed (Wang et al., 2007). This permits faster simulations and longer time steps, thereby enabling larger systems to be simulated for longer times. Another approach is the *accelerated dynamics method* used to overcome some of the limitations of standard MD (Voter, 1997).

However, it deserves to be pointed out that, although the ab initio simulation methodologies and computing power have converged to enable sophisticated and accurate molecular simulations to be carried out on ILs, it is still unlikely that simulations would have been able to address many of the interesting questions that are being tackled today, especially for design of ILs for a specific application. According to Maginn (2009), the success of the ab initio simulation in exploiting IL applications depend largely on its advancements in two important areas. The first area is in property prediction. Simulations have already been shown to be capable of making quantitative predictions of thermodynamic and transport properties of ILs, and this will continue be an important area in the future (Maginn, 2007). The second important area of application is in providing qualitative insight into the nature of ILs (Lynden-Bell et al., 2007). The new simulation should look into the presence of nano-sized polar/nonpolar domains and organization of ILs at interfaces to understand nanoscale phenomena of ILs (Goebel et al., 2009) such that simulations may help to suggest new application areas. To achieve better simulations the existing problems in ab initio simulation has to be solved, which include (1) developing accurate force fields and validating them in a much larger range of compounds; (2) improving the methods to conduct simulations; the attempted approaches include coarse graining computation (Wang et al., 2007) and accelerated dynamics methods (Voter, 1997); (3) developing new MC simulation methods in the area of ILs (Shah et al., 2002; Shah and Maginn, 2004).

5.3.4 Quantum Chemistry Modeling and COSMO-RS

Density functional theory (DFT; conceptually roots in the Thomas-Fermi model) is among the most popular and versatile methods used for quantum chemical calculation (Vignale and Rasolt, 1987). With this theory, the properties of a molecule or an ion can be determined by using the function of electron density (Yang, 1991). The most successful application in using quantum chemical calculation for prediction of the properties and phase behaviors of ILs and their mixture with other organic solvents is COSMO-RS, which was first proposed by A. Klamt in 1995. The COSMO-RS model was further refined and parameterized by Klamt et al. in 1998.

According to Klamt (1995), COSMO-RS is a quantum chemistry-based statistical thermodynamics model for the prediction of thermodynamic properties of fluids and liquid mixtures, which integrates concepts from quantum chemistry, dielectric continuum models, electrostatic surface interactions, and statistical thermodynamics. The COSMO-RS theory has been described in detail elsewhere (Eckert and Klamt, 2002; Klamt and Eckert, 2000). In brief, starting from the surface polarization charge densities σ from DFT COSMO calculations, COSMO-RS considers all interactions, especially electrostatic interaction and H-bonding, in a liquid system as contact interactions of the molecular surfaces, which are written as pair interactions of the respective polarization change densities σ and σ' of the contacting surface. Then, the chemical potential of a surface segment with screening charge density σ in an ensemble can be described by the normalized distribution function $p_S(\sigma')$ given by (Eq. [11]):

$$\mu_s(\sigma) = -\frac{RT}{a_{\text{eff}}} \ln\left[\int p_s(\sigma')\exp\left(\frac{1}{RT}(\mu_s(\sigma') - E_{\text{misfit}}(\sigma') - E_{\text{HB}}(\sigma'))d\sigma'\right)\right] \quad [11]$$

where $\mu_s(\sigma)$ is a measure for the affinity of the system S to a surface polarity σ; E_{misfit} represents the electrostatic contact interaction energy; E_{HB} represents the energy contribution from H-bonding interaction; and α_{eff} is the effective contact area between two surface segments. Not being a function of individual surface contacts, E_{vdw} is not included in (Eq. [11]) but added to the reference energy in solution a posteriori. The chemical potential of compound X_i in system S is then available from integration of the σ-potential over the surface of X_i. The capability of COSMO-RS to predict the chemical potential μ_s^X of any solute X in any pure or mixed solvent S at variable temperature T enables the calculation of any thermodynamic liquid—liquid equilibrium (LLE) calculation and prediction.

The course of a COSMO-RS calculation can be illustrated by a flow chart as shown in Fig. 5.6. It always starts from quantum chemical COSMO calculation (using DMol^3, Turbomole, or similar) resulting in a COSMO file containing geometry, energy, and surface screening charge densities of the

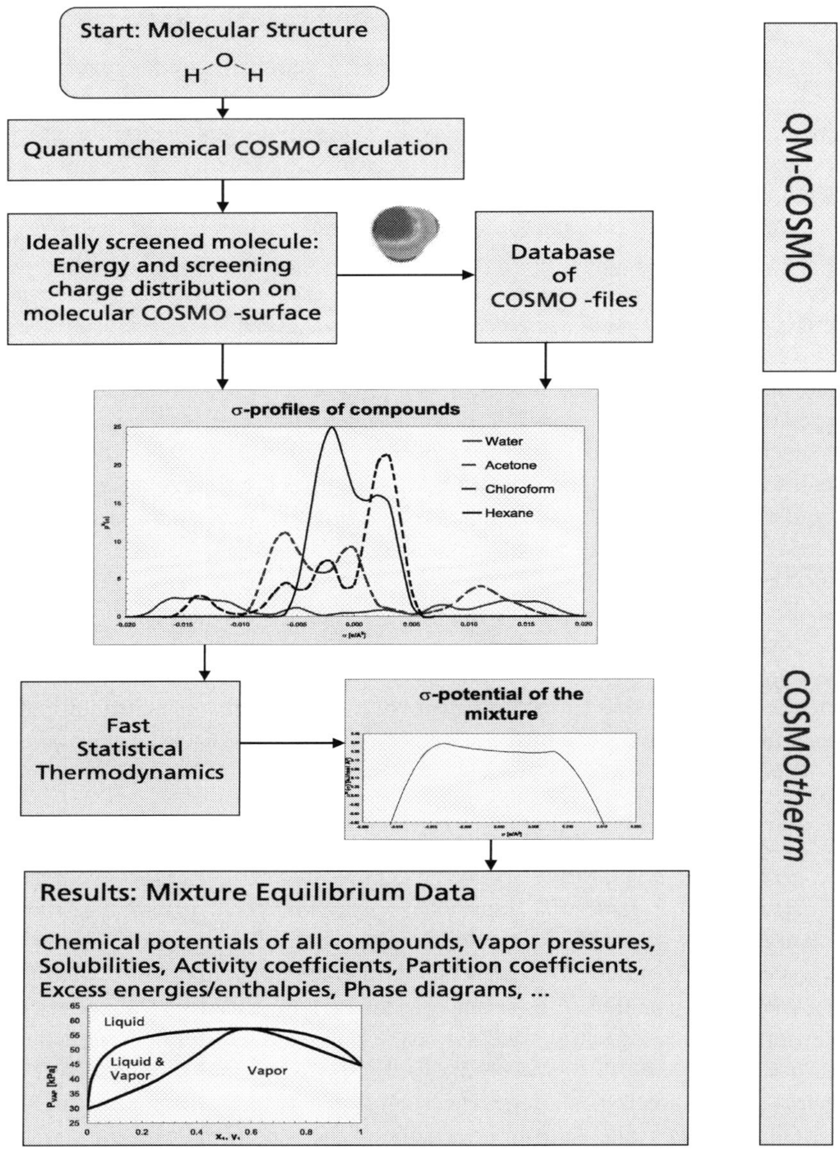

FIGURE 5.6 Flow chart of conductor-like screening model for real solvents (COSMO-RS) calculation.

considered molecule. For standard molecules the COSMO file can alternatively be taken from a database of precalculated COSMO files. The prediction of thermodynamic properties can be performed with the COSMOtherm software, which is commercially available.

The very characteristic descriptor of a molecule in COSMO-RS is σ-profile, reducing the 3D molecular surface charge distribution to a histogram, which contains the main chemical information necessary to predict its thermodynamic properties and interaction with other molecules (Eckert and Klamt, 2002; Diedenhofen and Klamt, 2010). Fig. 5.7 shows the surface polarization charge density and σ-profiles of some representative solvents (A), cations (B), and anions (C) of ILs (Palomar et al., 2010). As can be seen, from deep blue to deep red indicates an evolution from strong positive charge to negative charge (Fig. 5.7).

COSMO-RS depends on an extremely small number of adjustable parameters (7 basic parameters and 2 for each element; these element-specific parameters are the radius and the vdw energy per area, as well as 9 τ_{vdw} values), but is able to predict all kinds of equilibrium thermodynamic properties or derived quantities; for example, vapor pressure, free energy of solvation, activity coefficients, partition coefficients, solubility and solid–liquid equilibria, LLE and VLE, phase diagrams, azeotropes, miscibility gaps, excess enthalpies and excess free energies, pK_A of acids and bases, various QSPR models, reaction constants and liquid extraction equilibria, and so on.

Diedenhofen et al. (2003) presented the first example of using COSMO-RS to predict activity coefficients of 38 solutes (hydrocarbons, alcohols, and organic polar compounds) in [C_2mim][NTf_2], [C_2C_1mim][NTf_2], and [C_1C_4py][BF_4] at infinite dilution, which indicated that, except for underestimation for the activity coefficients in [C_1C_4py][BF_4], the same accuracy as in normal organic solvents can be achieved. Similarly Banerjee and Khanna (2006) predicted the activity coefficients of hydrocarbons and alcohols in phosphonium-based ILs in infinite dilution with average deviations between 8% and 16%.

Zhang et al. (2008) performed a COSMO-RS screening study of Henry's law constants of CO_2 in 408 ILs to help find an improved CO_2 capture capability for ILs that are based on the [$(C_2F_5)_3PF_3$] anion. Marsh et al. (2002, 2004) predict the LLE of [C_4mim][PF_6]-alcohols using both the *meta*-file and electroneutral ion mixture approach and found out the use of the electroneutral ion mixtures improved the results markedly, leading to a reasonable description of the composition dependence and a remarkably good prediction. Banerjee et al. (2006) used a COSMO-RS based ion-pair model for the VLE prediction of 13 ILs with benzene, cyclohexane, acetone, 2-propanol, water, and tetrahydrofuran; and a good agreement with the experimental values with a root mean square deviation of the vapor pressure of about 6%.

As a powerful tool, the capability of COSMO-RS is not limited to the prediction of the thermodynamic properties of a given molecule, rather to screen the right molecule structure from a large of pool of possible structure; or for a more advanced application design of a molecular structure based on property inputs. The latter application depends on several factors, while the former application has been approved through a few case studies. Based on

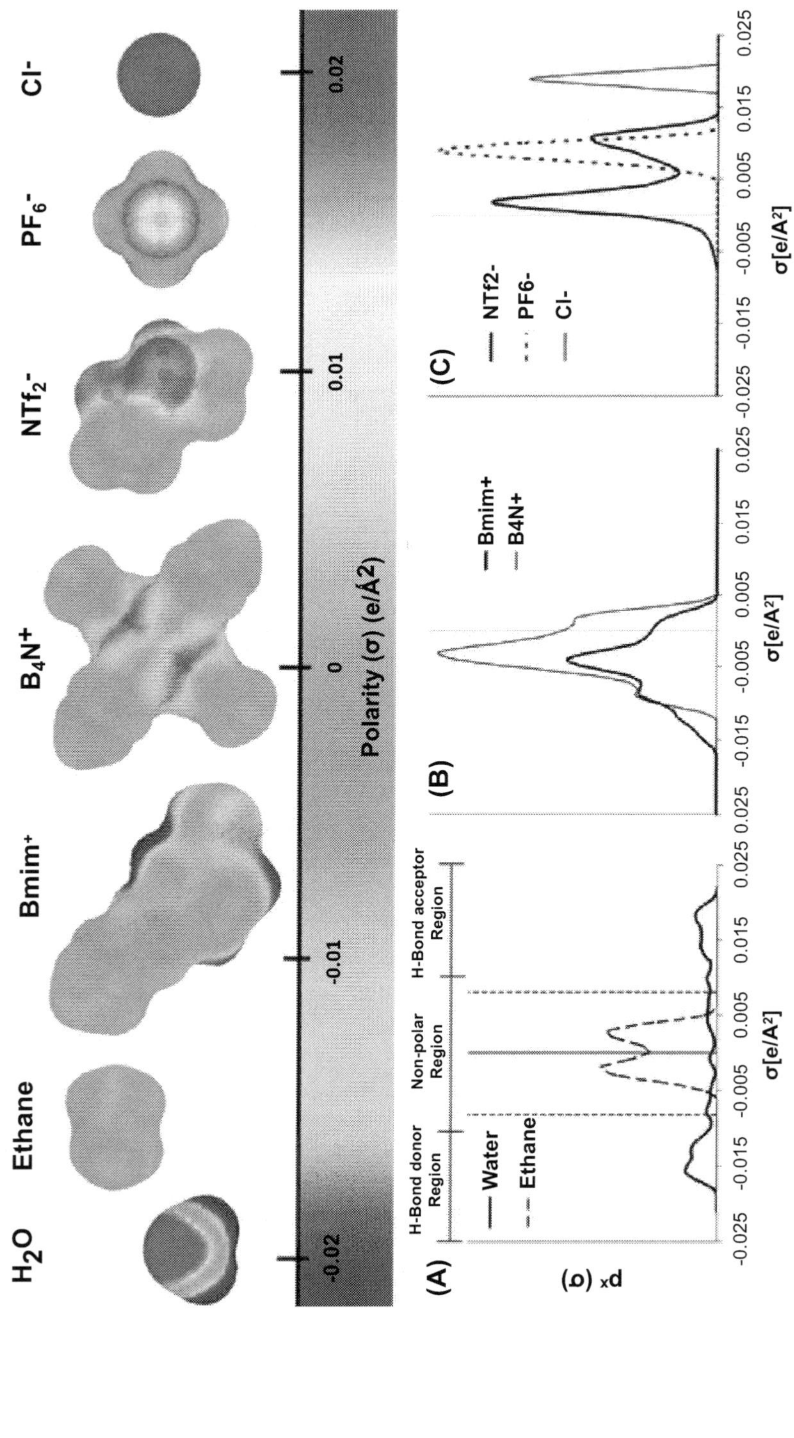

FIGURE 5.7 Surface polarization charge density and s-profile of some representative organic solvents (A), cations (B), and anions (C) of ionic liquids.

COSMO-RS predictions, Jork et al. (2005) achieved optimization of ILs as entrainers in the distillative separation of tetrahydrofurane/water and methylcyclohexane/toluene. Lei et al. (2007) screened imidazolium-based ILs with different anion pairs based on COSMO-RS prediction for extraction of amphetamine and nikethamide from urine and found out [C_2mim] + [OAc] is the desired IL structure. Maiti et al. (2008) used predicted solubilities for the solvent screening to dissolve 1,3,5-triamino-2,4,6-trinitrobenzene crystal. For a small evaluation set of imidazolium cations with [F], [Cl], and [Br] as anion, an excellent agreement with the experimental data has been obtained.

To screen suitable IL candidates for dissolving flavonoids for design of enzymatic reaction system for lipophilization of flavonoids, Guo et al. (2007b) conducted COSMO-RS based prediction of the solubilities of esculin in 1888 possible IL structures (combinations of 59 cations and 32 anions). The cations cover the most important types of possible cations, such as imidazolium, pyridinium, pyrrolidinium, ammonium, phosphonium, sulfonium, guanidinium, isoquinolinium, and isouronium; and the anions represent most investigated groups. They found out that H-bonding interaction is the most dominant interaction for ILs (followed by misfit and van der Waals interactions) to determine the solubility of flavonoids, and the anionic part has greater effect on the overall H-bonding capability of the ILs (Fig. 5.8). Moreover, they based on the basicity of anions, categorized into three groups, corresponding to the classification of the solubility of flavonoids. Importantly the authors concluded that COSMO σ-moment descriptors, which roughly denote the characteristic properties of the ILs, might be of general value to have a fast estimation for the solubilities of flavonoids as well as those compounds with massive moieties as H-bonding donors. Based on this work, a strategy by combination of a strong polar anion and a moderate to strong hydrophobic cation to be an ideal IL is determined.

Similarly, Kahlen et al. (2010) examined the potential of a COSMO-RS-based screening of ILs (combinations of the 32 anions and 71 cations) with respect to their dissolving power for cellulose, where the activity coefficient of the solute (cellotriose as a model molecule) at infinite dilution is considered a qualitative measure for the dissolving power of the solvent. Based on the prediction, some new ILs (with chloride, acetate, decanoate, and bis(2,4,4-trimethylpentyl)phosphinate as anions) as potential candidates for cellulose dissolution were suggested.

Grabda et al. (2014) applied COSMO-RS to the problem of extracting rare earth metal salts $NdCl_3$ and $DyCl_3$ from an aqueous solution using IL extraction solvents. A comprehensive screening with 4400 different ILs was conducted. The most promising candidates with the highest selectivity in the extraction process were validated by experiment. Alevizou and Voutsas (2014) evaluate the accuracy of the COSMO-RS model in solubility predictions of cinnamic acid derivatives in ILs, organic solvents, and in binary solvents composed of an IL and an organic solvent.

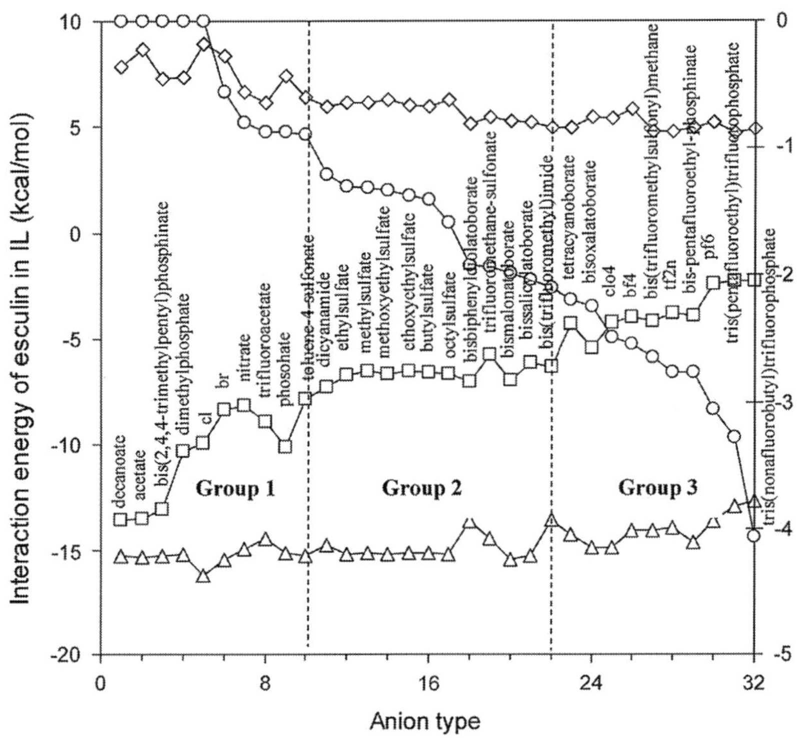

FIGURE 5.8 Conductor-like screening model for real solvents-derived descriptors used to characterize the solvation interactions of esculin and BMIM-based ionic liquids (ILs) with different anions. (○) Predicted solubility, (◇) misfit interaction energy, (□) H-bonding interaction energy, and (Δ) van der Waals interaction energy in respective ILs at 298.15 K. All interactions are calculated at indefinite dilution.

5.4 RATIONAL DESIGN OF ILs FOR LIPIDS PROCESSING

Lipids represents a family of structurally diverse compounds ubiquitously existing in plants, animals, and microorganisms (IUPAC-IUB, 1977). It is difficult to specify many different molecular species on this planet; however, just in the human body it is estimated to be about 180,000 according to lipidomics analysis (Brugger, 2014). Even though the lipids from plant sources, which is the current the main stream for food and other industrial applications, might be simple, the number can be added up to thousands with various structures (Fahy et al., 2005). In terms of molecular structures, lipids can be categorized as fatty acyls and glycerolipids containing linear chains; glycerophospholipids, sphingolipids, and saccharolipids containing polar heads; and sterol lipids, phenol lipids, and polyketides with conjugated cyclo structures (Schmelzer et al., 2007; IUPAC-IUB, 1977). Therefore, when these lipids are either used as starting parent molecules for modification or as building blocks for synthesis, specifically in design of ILs for carrying on these processes, the similarity, difference, and compatibility of the ILs with the lipids are the key factor to be considered.

As illustrated in Scheme 5.1, a priori design of an IL typically proceeds through a few steps. In principle, a priori design is possible for any task-specific application. However, in reality the success of designing/discovering the best IL candidate for such an application requires substantial background knowledge and research/experimental experience in the area of interest. It is also the case in lipid processing and biorefinery areas. Herein we attempt to demonstrate how ILs could be designed for a specific application, and how a processing system was developed.

Case 1: Screening, discovery and design of ILs as processing media for enzymatic production of partial glycerides.

Partial glycerides, especially mono- and diglycerides (DG), are major ingredients in the food emulsifier market. The current production mainly depends on chemical glycerolysis of oils and fats at elevated temperature (200–220°C). The main technical challenges include low conversion of triglycerides (TG) and inapplicability of high-temperature process to unsaturated oils, which resulted from incompatibility between hydrophobic substrate TG and hydrophilic substrate glycerol (Bornscheuer, 1995). Guo and Xu (2005) reported the first IL-based reaction system for enzymatic glycerolysis and achieved remarkably high conversion (nearly 100%) and selectivity (90%) toward production of monoglycerides (MG). A type of ammonium IL, cocosalkyl pentaethoxi methylammonium methosulfate (Ammoeng 100), was identified to be able to create a compatible system for both substrates and induce the shift of reaction equilibrium to the formation of MG.

Based on this observation, they used the COSMO-RS model to screen/predict another type of ammonium IL Ammoeng 102, by comparison of the similarity of σ-profiles between 1-monoolein, cation of Ammoeng 100, and cation of Ammoeng 102, as a potential medium for enzymatic glycerolysis for MG production (Guo et al., 2006). Excellent conversion and selectivity for Ammoeng 102 mediated reaction were also obtained as for Ammoeng 100. Moreover, the COSMO-RS model analysis of the interactions between substrates, products, and ILs with respect to misfit interaction, van der Waals interaction, and chemical potential, concluded that the hydrophobic substituents in the cation contribute to the increase of glyceride solubility and the strong polar anion and polyethoxyl group to improve reaction selectivity (Chen et al., 2008). This inspired another idea that an IL reaction facilitating DG production might be created (Kahveci et al., 2009a) (Fig. 5.9).

The enzymatic glycerolysis reactions in trioctylmethylammonium trifluoroacetate ([TOMA Tf_2N]) and Ammoeng 120 produced comparable yield of DG to those most efficient conventional systems but with less by-products (Kahveci et al., 2009b). By applying a binary IL system that consisted of one IL with better DG production selectivity and another IL being able to achieve higher conversion of TG, the combination of TOMA. Tf_2N/Ammoeng 102 was

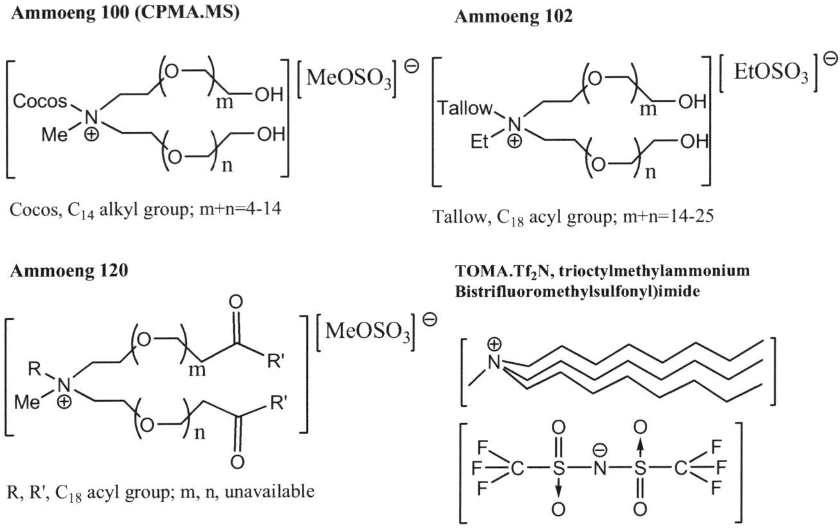

FIGURE 5.9 Molecular structures of the ionic liquids used for enzymatic productions of mono-/diglycerides (Guo and Xu, 2005; Guo et al., 2006; Chen et al., 2008; Kahveci et al., 2009a,b; Kahveci et al., 2010).

found to be an efficient system of DG production; 85% (mol%) of oil conversion and up to 90% (mol%) of total DG yield (73%, wt%) were obtained under optimized reaction conditions (Kahveci et al., 2010).

This case study gives a vivid example that how a series of lipid processes were evolved and developed by combining experimental study with modeling screening and prediction, and how modeling assisted interaction analysis and understanding could lead to exploration of related areas and extending the use of ILs. This work also demonstrated that binary (or multiple) IL systems can be a very potential area to create the possibility that a single IL cannot bring in (Kahveci et al., 2009b, 2010).

Case 2: Deconstruction of lignocellulosic biomass with ILs for biorefinery.

Our current chemical industries depend heavily on fossil organic matter petroleum oil, natural gas, and coal, which are exhausting and unsustainable (Ragauskas et al., 2006; Clark et al., 2009). As alternative sources of biofuels and biochemicals from edible components of food crops such as sucrose, starch, and vegetable oils, it is competitive with the increasing need for food by the rapidly growing world population (Food and Agricultural Organization of the United Nations, 2008). Lignocellulosic biomass is the most abundant plant material on our planet and therefore available in much higher quantities (also due to higher yields per area of land) and at lower cost than starch- and sucrose-based materials (Chheda et al., 2007). A great potential value from lignocellulosic biomass for sustainability of the modern chemical industry and energy supply can thus be anticipated (Binder and Raines, 2009).

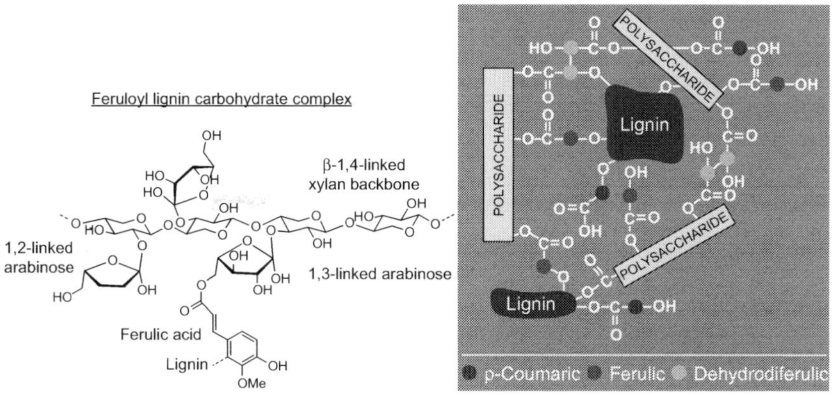

FIGURE 5.10 Grass lignin-carbohydrate complexes with phenoleic acid (left) and network structure (right). *Reproduced from Iiyama, K., Lam, T., Stone, B., 1994. Covalent cross-links in the cell wall. Plant Physiol. 104(2), 315–320 with permission.*

Lignocellulose is composed of three biopolymers—cellulose, hemicellulose, and lignin—making up approximately 90% of the dry matter (Brandt et al., 2013). In principle, cellulose and hemicellulose can be depolymerized into hexose and/or pentose, which can be used for microbial fermentation (Kumar et al., 2009). A schematic representation of the network structure of grass lignocellulose is depicted in Fig. 5.10 (Iiyama et al., 1994), in which a highly structured, multiply chemically bonded network can be seen. Therefore, one of the key technical challenges is how to deconstruct the biomass, condition the treated biomass and the hydrolysates (detoxification and neutralization), and convert (hydrolysis/depolymerization of the polysaccharides) them into fermentable sugars. Indeed, sustainable utilization of lignocellulosic biomass for bioenergy production (bioethanol) is an extremely interesting topic. Due to ionic nature and designer property for solvation, ILs hold the promise to deconstruct lignocellulosic biomass to facilitate its conversion to fuels and chemicals in so-called integrated biorefinery (Clark et al., 2009). Welton and his coworkers (Brandt et al., 2013) reviewed the progress of deconstruction of lignocellulosic biomass with ILs.

Like any other technology evolution, the discovery of ILs to dissolve cellulose was also trial-and-error when it was found that 1-ethylpyridiniumchloride ((C_2Pyr)Cl) can effectively dissolve cellulose (Gränacher, 1934). With more identified ILs added to the list, it has been observed that the IL anion plays an important role in determining an IL's ability to dissolve cellulose (Swatloski et al., 2002; Wu et al., 2004; Fukaya et al., 2006, 2008). The ILs identified to date contain anions that can form strong hydrogen bonds with hydroxyl groups (eg, chloride, carboxylates (acetate, formate, propionate, lactate), dialkyl phosphates, dialkyl and trialkylphosphonates, and amino acid anions). The dissolving power of these relevant ILs has been typically attributed to strong

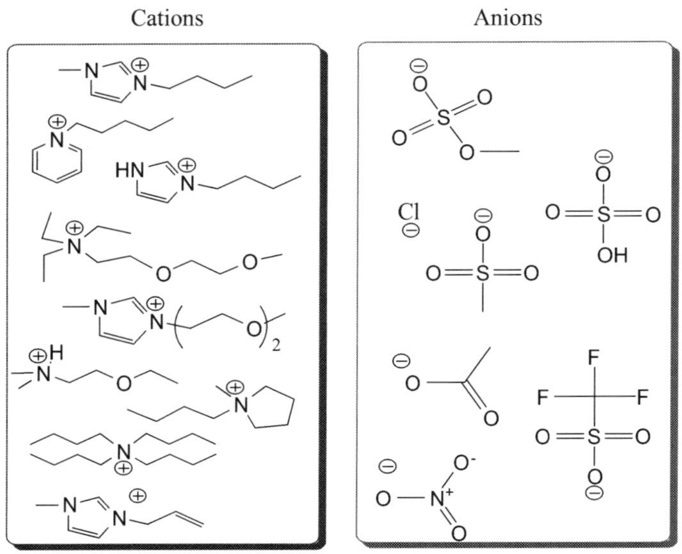

FIGURE 5.11 Typical types of cations and anions for constituting ionic liquids that are widely used for dissolving cellulose.

hydrogen-bonding interactions between the anions and equatorial hydroxyl groups on the cellulose.

On the basis of experimental results, theoretical understanding of the IL-mediated deconstruction of lognocellulosic biomass and process has been deepened with the introduction of modeling and polarity descriptors including COSMO-RS prediction (Kahlen et al., 2010; Casas et al., 2012) and Kamlet–Taft solvent parameters (Brandt et al., 2011; Froschauer et al., 2012). The common structural characteristics of optimum ILs for cellulose dissolution are generalized: (1) ILs should have a relatively small, noncoordinating cations and small hydrogen-bonding anions; (2) ILs should have high hydrogen-bond basicity and low hydrogen-bond acidity; (3) IL cations and anions should not be decorated with long aliphatic alkyl chains. Typical types of cations and anions for constituting ILs that are widely used for dissolving cellulose are shown in Fig. 5.11.

As an ionic solvent, ILs have the ability to decrystallize the cellulose portion of lignocellulosic biomass and simultaneously disrupt the lignin and hemicellulose network. The dissolved cellulose can be modified in solution or regenerated (reprecipitated) by adding water or other solvents. The ordering of the regenerated cellulose is reduced compared to the initial state, which results in significantly accelerated hydrolysis with cellulases compared to native cellulose. Of course, after IL-based deconstruction treatment, there are still substantial processes for lignocellulose toward fermentable sugars, which have been intensively discussed elsewhere (Brandt et al., 2013).

5.5 CONCLUDING REMARKS AND PERSPECTIVES

The main purpose of the development of modeling is to assist or improve experiments, not to replace experimental practice, which it may never be able to do. As any other modeling or theory does, the importance of IL modeling and prediction at the early stage of IL technology development is to give a rough estimation of the property of a perceived structure of ILs assigned for a specific application. At this stage the prediction quality cannot be high because (1) the chemical and physical data of ILs is scarcely available; (2) the knowledge and structure–property relationship have not been established; (3) to a large extent, the modeling for ILs derived from conventional solvents/salts has not been well defined or optimized.

Since the modern ILs with anions of hexafluorophosphate and tetrafluoroborate were obtained by Wilkes and Zaworotko in 1992, more than two decades have passed. Thousands of ILs have been added to the IL library with hundreds of thousands of data points of physicochemical properties documented. It can be anticipated that, with development of new techniques/instruments for measurement, more and more physicochemical data will be added to the IL database. On one hand, IL modeling benefits from the increasing database for refining and parameter optimization; on the other hand, the practical requirement for the quality of prediction for IL modeling has also been enhanced.

As reviewed in this context, empirical and semiempirical approaches can offer a rough categorization of polarity of ILs. UNIFAC, after adapted for modification, still showed its effectiveness for prediction of phase behaviors of pure and IL-solvent mixtures; however, UNIFAC parameters have to be continuously added to an ever-enlarging IL library. As a simplified QSPR model, Linear Free Energy Descriptors offer a readily available tool to analyze complex interactions within ILs and the solutes dissolved therein. Therefore, the derived descriptors are very useful for fast screening of ILs for a specific purpose; for example, screening of ILs by using Kamlet–Taft solvent parameters for dissolving cellulose (Brandt et al., 2013). Based on structural molecular/atomistic descriptors, dynamic molecular modeling can certainly provide more accurate prediction of many properties. However, it needs professional background knowledge of QM, force-field and simulation methods, as well as computation skills; this is not an easy job for common users.

COSMO-RS is a well-established platform for prediction of thermodynamic properties and the COSMOtherm software also provides a user-friendly operation interface. Therefore, this program has been widely used in different studies. With the expanding applications of ILs, more and more molecular aggregation and clustering behaviors and interactions with micromolecules (eg, proteins) will be interesting areas for COSMO-RS development.

As previously discussed, a priori design of ILs for task-specific application is still very challenging. Empirical and semiempirical models have their limitations to fulfill this task. Ab initio MD can be simplified and commercialized as user-friendly software; this will be a promising future perspective, alternative to COSMO-RS, contributing to the exploitation of application potential of ILs in chemical and biological engineering, including lipid processing.

ACKNOWLEDGMENT

A critical review and comments to this chapter by Dr Andreas Klamt (COSMOlogic GmbH & Co. KG, Leverkusen, Germany) is gratefully acknowledged.

REFERENCES

Ab Rani, M.A., Brandt, A., Crowhurst, L., Dolan, A., Hassan, N.H., Hallett, J.P., Hunt, P.A., Lui, M.Y., Niedermeyer, H., Perez-Arlandis, J.M., Schrems, M., To, T.Q., Welton, T., Wilding, R., 2011. Understanding the polarity of ionic liquids. Phys. Chem. Chem. Phys. 13, 16831–16840.

Abe, Y., Yoshiyama, K., Yagi, Y., Hayase, S., Kawatsura, M., Itoh, T., 2010. A rational design of phosphonium salt type ionic liquids for ionic liquid coated-lipase catalyzed reaction. Green Chem. 12, 1976–1980.

Abraham, M.H., Ibrahim, A., Zissimos, A.M., 2004. Determination of sets of solute descriptors from chromatographic measurements. J. Chromatogr. A 1037, 29–47.

Acree Jr., W.E., Abraham, M.H., 2006. The analysis of solvation in ionic liquids and organic solvents using the Abraham linear free energy relationship. J. Chem. Technol. Biotechnol. 81, 1441–1446.

Albert, J., Müller, K., 2014. A group contribution method for the thermal properties of ionic liquids. Ind. Eng. Chem. Res. 53 (44), 17522–17526.

Alevizou, E., Voutsas, E.C., 2014. Evaluation of COSMO-RS model in binary and ternary mixtures of natural antioxidants, ionic liquids and organic solvents. Fluid Phase Equilib. 369, 55–67.

Anderson, J.L., Ding, J., Welton, T., Armstrong, D.W., 2002. Characterizing ionic liquids on the basis of multiple solvation interactions. J. Am. Chem. Soc. 124 (47), 14247–14254.

de Andrade, J., Boes, E.S., Stassen, H., 2002a. A force field for liquid state simulations on room temperature molten salts: 1-ethyl-3-methylimidazolium tetrachloroaluminate. J. Phys. Chem. B 106, 3546–3548.

de Andrade, J., Boes, E.S., Stassen, H., 2002b. Computational study of room temperature molten salts composed by 1-Alkyl-3-methylimidazolium cations-force-field proposal and validation. J. Phys. Chem. B 106, 13344–13351.

Banerjee, T., Khanna, A., 2006. Infinite dilution activity coefficients for trihexyltetradecyl phosphonium ionic liquids: measurements and COSMO-RS prediction. J. Chem. Eng. Data 51, 2170–2177.

Banerjee, T., Singh, M.K., Sahoo, R.K., Khanna, A., 2005. Volume, surface and UNIQUAC interaction parameters for imidazolium based ionic liquids via Polarizable Continuum Model. Fluid Phase Equilib. 234, 64–76.

Banerjee, T., Singh, M.K., Khanna, A., 2006. Prediction of binary VLE for imidazolium based ionic liquid systems using COSMO-RS. Ind. Eng. Chem. Res. 45, 3207–3219.

Baudequin, C., Bregeon, D., Levillain, J., Guillen, F., Plaquevent, J.-C., Gaumont, A.-C., 2005. Chiral ionic liquids, a renewal for the chemistry of chiral solvents? Design, synthesis and applications for chiral recognition and asymmetric synthesis. Tetrahedron: Asymmetry 16, 3921–3945.

Bernazzani, L., Duce, C., Micheli, A., Mollica, V., Sperduti, A., Starita, A., Tiné, M.R., 2006. Predicting physical–chemical properties of compounds from molecular structures by recursive neural networks. J. Chem. Inf. Model 46, 2030–2042.

Binder, J.B., Raines, R.T., 2009. Simple chemical transformation of lignocellulosic biomass into furans for fuels and chemicals. J. Am. Chem. Soc. 131 (5), 1979–1985.

Bini, R., Chiappe, C., Mestre, V.L., Pomellic, C.S., Welton, T., 2008. A rationalization of the solvent effect on the Diels–Alder reaction in ionic liquids using multiparameter linear solvation energy relationships. Org. Biomol. Chem. 6, 2522–2529.

Borner, A., Li, Z., Levin, D., 2012. Modeling of an ionic liquid electrospray using molecular dynamics with constraints. J. Chem. Phys. 136, 124507.

Bornscheuer, U.T., 1995. Lipase-catalyzed syntheses of monoacylglycerols. Enzym. Microb. Technol. 17, 578–586.

Borodin, O., Smith, G.D., Kim, H., April 9, 2009. Viscosity of a room temperature ionic liquid: predictions from nonequilibrium and equilibrium molecular dynamics simulations. J. Phys. Chem. B 113 (14), 4771–4774.

Borodin, O., 2009. Polarizable force field development and molecular dynamics simulations of ionic liquids. J. Phys. Chem. B 113, 11463–11478.

Branco, L.C., Carrera, G.V.S.M., Aires-de-Sousa, J., Martin, I.L., Frade, R., Afonso, C.A.M., 2011. Physico-chemical properties of task-specific ionic liquids. In: Kokorin, A. (Ed.), Ionic Liquids: Theory, Properties, New Approaches. InTech.

Brandt, A., Ray, M.J., To, T.Q., Leak, D.J., Murphy, R.J., Welton, T., 2011. Ionic liquid pretreatment of lignocellulosic biomass with ionic liquid–water mixtures. Green Chem. 13, 2489–2499.

Brandt, A., Gräsvik, J., Hallett, J.P., Welton, T., 2013. Deconstruction of lignocellulosic biomass with ionic liquids. Green Chem. 15, 550–583.

Breitbach, Z.S., Armstrong, D.W., 2008. Characterization of phosphonium ionic liquids through a linear solvation energy relationship and their use as GLC stationary phases. Anal. Bioanal. Chem. 390, 1605–1617.

Brennecke, F., Maginn, E.J., 2001. Ionic liquids: innovative fluids for chemical processing. AIChE J. 47, 2384–2389.

Brugger, B., 2014. Lipidomics: analysis of the lipid composition of cells and subcellular organelles by electrospray ionization mass spectrometry. Annu. Rev. Biochem. 83, 79–98.

Canongia Lopes, J.N.A., Padua, A.A.H., 2006. Nanostructural organization in ionic liquids. J. Phys. Chem. B 110, 3330–3335.

Canongia Lopes, J.N.C., Deschamps, J., Padua, A.A.H., 2004. Modeling ionic liquids using a systematic all-atom force field. J. Phys. Chem. B 108, 2038–2047.

Carmichael, A.J., Seddon, K.R., 2000. Polarity study of some 1-alkyl-3-methylimidazolium ambient-temperature ionic liquids with the solvatochromic dye, Nile Red. J. Phys. Org. Chem. 13, 591–595.

Casas, A., Palomar, J., Alonso, M.V., Oliet, M., Omar, S., Rodriguez, F., 2012. Comparison of lignin and cellulose solubilities in ionic liquids by COSMO-RS analysis and experimental validation. Ind. Crop Prod. 37, 155–163.

Chen, B., Guo, Z., Tan, T., Xu, X., 2008. Structures of ionic liquids dictate the conversion and selectivity of enzymatic glycerolysis: theoretical characterization by COSMO-RS. Biotechnol. Bioeng. 99 (1), 18–29.

Chheda, J.N., Huber, G.W., Dumesic, J.A., 2007. Liquid-phase catalytic processing of biomass-derived oxygenated hydrocarbons to fuels and chemicals. Angew. Chem. Int. Ed. 46, 7164–7183.

Chiappe, C., Marra, A., Mele, A., 2010a. Synthesis and applications of ionic liquids derived from natural sugars. Top. Curr. Chem. 295, 177–195.

Chiappe, C., Malvaldi, M., Pomelli, C.S., 2010b. The solvent effect on the Diels–Alder reaction in ionic liquids: multiparameter linear solvation energy relationships and theoretical analysis. Green Chem. 12, 1330–1339.

Clark, J.H., Deswarte, F.E.I., Farmer, T.J., 2009. The integration of green chemistry into future biorefineries. Biofuels Bioprod. Biorefin. 3, 72–90.

Cornell, W.D., Cieplak, P., Bayly, C.I., Gould, I.R., Merz, K.M., Ferguson, D.M., Spellmeyer, D.C., Fox, T., Caldwell, J.W., Kollman, P.A., 1995. A second generation force field for the simulation of proteins, nucleic acids, and organic molecules. J. Am. Chem. Soc. 117, 5179–5197.

Crosthwaite, J.M., Muldoon, M.J., Dixon, J.K., Anderson, J.L., Brennecke, J.F., 2005. Phase transition and decomposition temperatures, heat capacities and viscosities of pyridinium ionic liquids. J. Chem. Thermodyn. 37, 559–568.

Crowhurst, L., Mawdsley, P.R., Perez-Arlandis, J.M., Salter, P.A., Welton, T., 2003. Solvent-solute interactions in ionic liquids. Phys. Chem. Chem. Phys. 5, 2790–2794.

Das, R.N., Roy, K., 2013. Advances in QSPR/QSTR models of ionic liquids for the design of greener solvents of the future. Mol. Divers. 17 (1), 151–196.

Deetlefs, M., Seddon, K.R., Shara, M., 2006. Predicting physical properties of ionic liquids. Phys. Chem. Chem. Phys. 8, 642–649.

Diedenhofen, M., Klamt, A., 2010. COSMO-RS as a tool for property prediction of IL mixtures—a review. Fluid Phase Equilib. 294, 31–38.

Diedenhofen, M., Eckert, F., Klamt, A., 2003. Prediction of infinite dilution activity coefficients of organic compounds in ionic liquids using COSMO-RS. J. Chem. Eng. Data 48 (3), 475–479.

Earle, M.J., Seddon, K.R., 2000. Ionic liquids. Green solvents for the future. Pure Appl. Chem. 72 (7), 1391–1398.

Eckert, F., Klamt, A., 2002. Fast solvent screening via quantum chemistry: COSMO-RS approach. AIChE J. 369–385.

El Seoud, O.A., Koschella, A., Fidale, L.C., Dorn, S., Heinze, T., 2007. Applications of ionic liquids in carbohydrate chemistry: a window of opportunities. Biomacromolecules 8 (9), 2629–2647.

Emel'yanenko, V.N., Verevkin, S.P., Heintz, A., 2007. The gaseous enthalpy of formation of the ionic liquid 1-butyl-3-methylimidazolium dicyanamide from combustion calorimetry, vapor pressure measurements, and ab initio calculations. J. Am. Chem. Soc. 129, 3930–3937.

Fahy, E., Subramaniam, S., Brown, H.A., Glass, C.K., Merrill Jr., A.H., Murphy, R.C., Raetz, C.R., Russell, D.W., Seyama, Y., Shaw, W., et al., 2005. A comprehensive classification system for lipids. J. Lipid Res. 46, 839–862.

Filippov, A., Shah, F.U., Taher, M., Glavatskih, S., Antzutkin, O.N., 2013. NMR self-diffusion study of a phosphonium bis(mandelato)borate ionic liquid. Phys. Chem. Chem. Phys. 15 (23), 9281–9287.

Food and Agricultural Organisation of the United Nations, 2008. The State of Food and Agriculture 2008. Rome. Available at: http://www.fao.org/.

Fredenslund, A., Jones, R.L., Prausnitz, J.M., 1975. Group-contribution estimation of activity coefficients in nonideal liquid mixtures. AIChE J. 21, 1086–1099.

Freemantle, M., 1998. Designer solvents: ionic liquids may boost clean technology development. Chem. Eng. News 76, 32–37.

Froschauer, C., Hummel, M., Laus, G., Schottenberger, H., Sixta, H., Weber, H.K., Zuckerstätter, G., 2012. Dialkyl phosphate-related ionic liquids as selective solvents for xylan. Biomacromolecules 13, 1973–1980.

Fukaya, Y., Sugimoto, A., Ohno, H., 2006. Superior solubility of polysaccharides in low viscosity, polar, and halogen-free 1,3-dialkylimidazolium formates. Biomacromolecules 7, 3295–3297.

Fukaya, Y., Hayashi, K., Wada, M., Ohno, H., 2008. Cellulose dissolution with polar ionic liquids under mild conditions: required factors for anions. Green Chem. 10, 44–46.

Fukumoto, K., Yoshizawa, M., Ohno, H., 2005. Room temperature ionic liquids from 20 natural amino acids. J. Am. Chem. Soc. 127 (8), 2398–2399.

Gardas, R.L., Coutinho, J.A.P., 2008a. Extension of the Ye and Shreeve group contribution method for density estimation of ionic liquids in a wide range of temperatures and pressures. Fluid Phase Equilib. 263, 26–32.

Gardas, R.L., Coutinho, J.A.P., 2008b. A group contribution method for viscosity estimation of ionic liquids. Fluid Phase Equilib. 266, 195–201.

Gardas, R.L., Coutinho, J.A.P., 2009. Group contribution methods for the prediction of thermophysical and transport properties of ionic liquids. AIChE J. 55 (5), 1274–1290.

Ghatee, M.H., Zare, M., Zolghadr, A.R., Moosavi, F., 2010. Temperature dependence of viscosity and relation with the surface tension of ionic liquids. Fluid Phase Equilib. 291, 188–194.

Goebel, R., Hesemann, P., Weber, J., Moeller, E., Friedrich, A., Beuermann, S., Taubert, A., 2009. Surprisingly high, bulk liquid-like mobility of silica-confined ionic liquids. Phys. Chem. Chem. Phys. 11, 3653–3662.

Grabda, M., Panigrahi, M., Oleszek, S., Kozak, D., Eckert, F., Shibata, E., Nakamura, T., 2014. COSMO-RS screening for efficient ionic liquid extraction solvents for $NdCl_3$ and $DyCl_3$. Fluid Phase Equilib. 383 (15), 134–143.

Gränacher, C., 1934. Cellulose Solution. U.S. Patent 1943176.

Guo, Z., Xu, X., 2005. New opportunity for enzymatic modification of fats and oils with industrial potentials. Org. Biomol. Chem. 3 (14), 2615–2619.

Guo, Z., Chen, B., Murillo, R.L., Tan, T., Xu, X., 2006. Functional dependency of structures of ionic liquids: do substituents govern the selectivity of enzymatic glycerolysis? Org. Biomol. Chem. 4, 2772–2776.

Guo, Z., Lue, B.-M., Xu, X., 2007a. Ionic liquids: neoteric 'green' media for lipid processing. Inform 18 (2), 78–82.

Guo, Z., Lue, B.-M., Thomsen, K., Meyer, A.S., Xu, X., 2007b. Predictions of flavonoid solubility in ionic liquids by COSMO-RS: experimental verification, structural elucidation, and solvation characterization. Green Chem. 9, 1362–1373.

Hallett, J.P., Welton, T., 2011. Room-temperature ionic liquids: solvents for synthesis and catalysis. 2. Chem. Rev. 111, 3508–3576.

Han, D., Row, K.H., 2010. Recent applications of ionic liquids in separation technology. Molecules 15, 2405–2426.

Hanke, C.G., Price, S.L., Lynden-Bell, R.M., 2001. Intermolecular potentials for simulations of liquid imidazolium salts. Mol. Phys. 99, 801–809.

Hardacre, C., Holbrey, J.D., Katdare, S.P., Seddon, K.R., 2002. Alternating copolymerisation of styrene and carbon monoxide in ionic liquids. Green Chem. 4, 143–146.

Hough, W.L., Smiglak, M., Rodríguez, H., Swatloski, R.P., Spear, S.K., Daly, D.T., Pernak, J., Grisel, J.E., Carliss, R.D., Soutullo, M.D., Davis Jr., J.H., Rogers, R.D., 2007. The third evolution of ionic liquids: active pharmaceutical ingredients. New J. Chem. 31, 1429–1436.

Huddleston, J.G., Rogers, R.D., 1998. Room temperature ionic liquids as novel media for 'clean' liquid–liquid extraction. Chem. Commun. 1765–1766.

Huddleston, J.G., Visser, A.E., Matthew Reichert, W., Willauer, H.D., Broker, G.A., Rogers, R.D., 2001. Characterization and comparison of hydrophilic and hydrophobic room temperature ionic liquids incorporating the imidazolium cation. Green Chem. 3, 156–164.

Iiyama, K., Lam, T., Stone, B., 1994. Covalent cross-links in the cell wall. Plant Physiol. 104 (2), 315–320.

IUPAC, 1997. Compendium of Chemical Terminology (the "Gold Book"). Compiled by A. D. McNaught and A. Wilkinson, second ed. Blackwell Scientific Publications, Oxford.

IUPAC-IUB Commission on Biochemical Nomenclature (CBN). The nomenclature of lipids (Recommendations 1976). 1977. Eur. J. Biochem. 79: 11–21; 1977. Hoppe-Seylers, Z. Physiol. Chem. 358: 617–631; 1977. Lipids 12: 455–468; 1977. Mol. Cell. Biochem. 17: 157–171; 1978. Chem. Phys. Lipids 21: 159–173; 1978. J. Lipid Res. 19: 114–128; 1978. Biochem. J. 171: 21–35. http://www.chem.qmul.ac.uk/iupac/lipid/.

Jorgensen, W.L., Maxwell, D.S., TiradoRives, J., 1996. Development and testing of the OPLS all-atom force field on conformational energetics and properties of organic liquids. J. Am. Chem. Soc. 118, 11225–11236.

Jork, C., Kristen, C., Pieraccini, D., Stark, A., Chiappe, C., Beste, Y.A., Arlt, W., 2005. Tailor-made ionic liquids. J. Chem. Thermodyn. 37, 537–558.

Kahlen, J., Masuch, K., Leonhard, K., 2010. Modelling cellulose solubilities in ionic liquids using COSMO-RS. Green Chem. 12, 2172–2181.

Kahveci, D., Guo, Z., Özçelik, B., Xu, X., 2009a. Lipase-catalyzed glycerolysis in ionic liquids directed towards diglyceride synthesis. Process Biochem. 44, 1358–1365.

Kahveci, D., Guo, Z., Özçelik, B., Xu, X., 2009b. Improving enzymatic production of diglycerides by engineering binary ionic liquid medium system. New Biotechnol. 26, 37–43.

Kahveci, D., Guo, Z., Özçelik, B., Xu, X., 2010. Optimisation of enzymatic synthesis of diacylglycerols in binary medium systems containing ionic liquids. Food Chem. 119, 880–885.

Kamlet, M.J., Taft, R.W., 1976. The solvatochromic comparison method. I. The beta-scale of solvent hydrogen-bond acceptor (HBA) basicities. J. Am. Chem. Soc. 98 (2), 377–383.

Kamlet, M.J., Abboud, J.L., Taft, R.W., 1977. The solvatochromic comparison method. 6. The "π"* scale of solvent polarities. J. Am. Chem. Soc. 99 (18), 6027–6038.

Kamlet, M.J., Abboud, J.L.M., Abraham, M.H., Taft, R.W., 1983. Linear solvation energy relationships. 23. A comprehensive collection of the solvatochromic parameters, π^*, α, and β, and some methods for simplifying the generalized solvatochromic equation. J. Org. Chem. 48 (17), 2877–2887.

Kato, R., Gmehling, J., 2005. Systems with ionic liquids: measurement of VLE and γ^{∞} data and prediction of their thermodynamic behavior using original UNIFAC, mod. UNIFAC(Do) and COSMO-RS(Ol). J. Chem. Thermodyn. 37, 603–619.

Kato, R., Krummen, M., Gmehling, J., 2004. Measurement and correlation of vapor–liquid equilibria and excess enthalpies of binary systems containing ionic liquids and hydrocarbons. Fluid Phase Equilib. 224, 47–54.

Katritzky, A.R., Lomaka, A., Petrukhin, R., Jain, R., Karelson, M., Visser, A.E., Rogers, R.D., 2002a. QSPR correlation of the melting Point for pyridinium bromides, potential ionic liquids. J. Chem. Inf. Comput. Sci. 42, 71–74.

Katritzky, A.R., Jain, R., Lomaka, A., Petrukhin, R., Karelson, M., Visser, A.E., Rogers, R.D., 2002b. Correlation of the melting points of potential ionic liquids (imidazolium bromides and benzimidazolium bromides) using the CODESSA program. J. Chem. Inf. Comput. Sci. 42, 225–231.

Katritzky, A.R., Fara, D.C., Yang, H., Tämm, K., 2004. Quantitative measures of solvent polarity. Chem. Rev. 104 (1), 175–198.

Kawai, A., Hidemori, T., Shibuya, K., 2004. Polarity of room-temperature ionic liquid as examined by EPR spectroscopy. Chem. Lett. 33 (11), 1464–1465.

Kirchner, B., Hollóczki, O., Canongia Lopes, J.N., Padua, A.A.H., 2015. Multiresolution calculation of ionic liquids. WIREs Comput. Mol. Sci. 5, 202–214.

Klamt, A., Eckert, F., 2000. COSMO-RS: a novel and efficient method for the a priori prediction of thermophysical data of liquids. Fluid Phase Equilib. 172, 43–72.

Klamt, A., Jonas, V., Bürger, T., Lohrenz, J.C.W., 1998. Refinement and parametrization of COSMO-RS. J. Phys. Chem. A 102 (26), 5074–5085.

Klamt, A., Eckert, F., Arlt, W., 2010. COSMO-RS: an alternative to simulation for calculating thermodynamic properties of liquid mixtures. Annu. Rev. Chem. Biomol. Eng. 1, 101–122.

Klamt, A., 1995. Conductor-like screening model for real solvents: a new approach to the quantitative calculation of solvation phenomena. J. Phys. Chem. 99 (7), 2224–2235.

Kresse, G., Hafner, J., 1993. Ab initio molecular dynamics for liquid metals. Phys. Rev. B 47, 558–561.

Kroon, M.C., Buijs, W., Peters, C.J., Witkamp, G.-J., 2007. Quantum chemical aided prediction of the thermal decomposition mechanisms and temperatures of ionic liquids. Thermochim. Acta 465, 40–47.

Krossing, I., Slattery, J.M., Daguenet, C., Dyson, P.J., Oleinikova, A., Weingärtner, H., 2006. Why are ionic liquids liquid? A simple explanation based on lattice and solvation energies. J. Am. Chem. Soc. 128, 13427–13434.

Kumar, P., Barrett, D.M., Delwiche, M.J., Stroeve, P., 2009. Methods for pretreatment of lignocellulosic biomass for efficient hydrolysis and biofuel production. Ind. Eng. Chem. Res. 48 (8), 3713–3729.

Larsen, B.L., Rasmussen, P., Fredenslund, A., 1987. A modified UNIFAC group-contribution model for prediction of phase equilibria and heats of mixing. Ind. Eng. Chem. Res. 26 (11), 2274–2286.

Law, G., Watson, P.R., 2001. Surface tension measurements of N-alkylimidazolium ionic liquids. Langmuir 17 (20), 6138–6141.

Lee, J.-M., Ruckes, S., Prausnitz, J.M., 2008. Solvent polarities and Kamlet-Taft parameters for ionic liquids containing a pyridinium cation. J. Phys. Chem. B 112, 1473–1476.

Lee, S.B., 2005. Analysis of solvation in ionic liquids using a new linear solvation energy relationship. J. Chem. Technol. Biotechnol. 80, 133–137.

Lei, Z., Chen, B., Li, C., 2007. COSMO-RS modeling on the extraction of stimulant drugs from urine sample by the double actions of supercritical carbon dioxide and ionic liquid. Chem. Eng. Sci. 62, 3940–3950.

Lei, Z., Zhang, J., Li, Q., Chen, B., 2009. UNIFAC model for ionic liquids. Ind. Eng. Chem. Res. 48, 2697–2704.

Lei, Z., Dai, C., Liu, X., Xiao, L., Chen, B., 2012. Extension of the UNIFAC model for ionic liquids. Ind. Eng. Chem. Res. 51 (37), 12135–12144.

Liu, Z., Huang, S., Wang, W., 2004. A refined force field for molecular simulation of imidazolium-based ionic liquids,. J. Phys. Chem. B 108, 12978–12989.

Liu, H., Liu, Y., Li, J., 2010. Ionic liquids in surface electrochemistry. Phys. Chem. Chem. Phys. 12, 1685–1697.

Lynden-Bell, R.M., Del Popolo, M.G., Youngs, T.G.A., Kohanoff, J., Hanke, C.G., Harper, J.B., Pinilla, C.G., 2007. Simulations of ionic liquids, solutions, and surfaces. Acc. Chem. Res. 40, 1138–1145.

MacFarlane, D.R., Sun, J., Golding, J., Meakin, P., Forsyth, M., 2000. High conductivity molten salts based on the imide ion. Electrochim. Acta 45, 1271–1278.

MacKerell, A.D., Bashford, D., Bellott, M., Dunbrack, R.L., Evanseck, J.D., Field, M.J., Fischer, S., Gao, J., Guo, H., Ha, S., Joseph-McCarthy, D., Kuchnir, L., Kuczera, K., Lau, F.T.K., Mattos, C., Michnick, S., Ngo, T., Nguyen, D.T., Prodhom, B., Reiher, W.E., Roux, B., Schlenkrich, M., Smith, J.C., Stote, R., Straub, J., Watanabe, M., Wiorkiewicz-Kuczera, J., Yin, D., Karplus, M., 1998. All-atom empirical potential for molecular modeling and dynamics studies of proteins. J. Phys. Chem. B 102, 3586–3616.

Maginn, E.J., 2007. Atomistic simulation of the thermodynamic and transport properties of ionic liquids. Acc. Chem. Res. 40, 1200–1207.

Maginn, E.J., 2009. Molecular simulation of ionic liquids: current status and future opportunities. J. Phys. Condens. Matter 21, 373101.

Maiti, A., Pagoria, P.F., Gash, A.E., Han, T.Y., Orme, C.A., Gee, R.H., Fried, L.E., 2008. Solvent screening for a hard-to-dissolve molecular crystal. Phys. Chem. Chem. Phys. 10, 5050–5056.

Marcus, Y., 1993. The properties of organic liquids that are relevant to their use as solvating solvents. Chem. Soc. Rev. 22, 409–416.

Marrucho, I.M., Branco, L.C., Rebelo, L.P.N., 2014. Ionic liquids in pharmaceutical applications. Annu. Rev. Chem. Biomol. Eng. 5, 527–546.

Marsh, K.N., Deev, A., Wu, A.C.-T., Tran, E., Klamt, A., 2002. Room temperature ionic liquids as replacements for conventional solvents – a review. Korean J. Chem. Eng. 19, 357–362.

Marsh, K.N., Boxall, J.A., Lichtenthaler, R., 2004. Fluid Phase Equilib. 219, 93–98.

McLeese, S.E., Eslick, J.C., Hoffmann, N.J., Scurto, A.M., Camarda, K.V., 2010. Design of ionic liquids via computational molecular design. Comput. Chem. Eng. 34, 1476–1480.

Mondal, A., Balasubramanian, S., 2014. Quantitative prediction of physical properties of imidazolium based room temperature ionic liquids through determination of condensed phase site charges: a refined force field. J. Phys. Chem. B 118 (12), 3409–3422.

Morrow, T.I., Maginn, E.J., 2002. Molecular dynamics study of the ionic liquid 1-n-butyl-3-methylimidazolium hexafluorophosphate. J. Phys. Chem. B 106 (49), 12807–12813.

Morrow, T.I., Maginn, E.J., 2003. Molecular dynamics study of the ionic liquid 1-n-butyl-3-methylimidazolium hexafluorophosphate. J. Phys. Chem. B 107, 9160.

Muldoon, M.J., Gordon, C.M., Dunkin, I.R., 2001. Investigations of solvent–solute interactions in room temperature ionic liquids using solvatochromic dyes. J. Chem. Soc. Perkin Trans. 2, 433–435.

Muller, P., 1994. Glossary of terms used in physical organic chemistry. Pure Appl. Chem. 66, 1077–1184.

Nebig, S., Bölts, R., Gmehling, J., 2007. Measurement of vapor-liquid equilibria (VLE) and excess enthalpies (HE) of binary systems with 1-alkyl-3-methylimidazolium bis(trifluoromethylsulfonyl)imide and prediction of these properties and γ^{∞} using modified UNIFAC (Dortmund). Fluid Phase Equilib. 258, 168–178.

Niedermeyer, H., Ashworth, C., Brandt, A., Welton, T., Hunt, P.A., 2013. A step towards the a priori design of ionic liquids. Phys. Chem. Chem. Phys. 15 (27), 11566–11578.

Noda, A., Hayamizu, K., Watanabe, M., 2001. Pulsed-gradient spin–echo ^{1}H and ^{19}F NMR ionic diffusion coefficient, viscosity, and ionic conductivity of non-chloroaluminate room-temperature ionic liquids. J. Phys. Chem. B 105 (20), 4603–4610.

Ohno, H., Fukumoto, K., 2007. Amino acid ionic liquids. Acc. Chem. Res. 40 (11), 1122–1129.

Ohno, H., Yoshizawa, M., 2002. Ion conductive characteristics of ionic liquids prepared by neutralization of alkylimidazoles. Solid State Ionics 154–155, 303–309.

Paduszynski, K., Domanska, U., 2013. Extension of modified UNIFAC (Dortmund) matrix to piperidinium ionic liquids. Fluid Phase Equilib. 353, 115–120.

Palomar, J., Ferro, V.R., Torrecilla, J.S., Rodríguez, F., 2007. Density and molar volume predictions using COSMO-RS for ionic liquids. An approach to solvent design. Ind. Eng. Chem. Res. 46 (18), 6041–6048.

Palomar, J., Torrecilla, J.S., Ferro, V.R., Rodríguez, F., 2008. Development of an a priori ionic liquid design tool. 1. Integration of a novel COSMO-RS molecular descriptor on neural networks. Ind. Eng. Chem. Res. 47 (13), 4523–4532.

Palomar, J., Torrecilla, J.S., Ferro, V.R., Rodríguez, F., 2009. Development of an a priori ionic liquid design tool. 2. Ionic liquid selection through the prediction of COSMO-RS molecular descriptor by inverse neural network. Ind. Eng. Chem. Res. 48 (4), 2257–2265.

Palomar, J., Torrecilla, J.S., Lemus, J., Ferro, V.R., Rodriguez, F., 2010. A COSMO-RS based guide to analyze/quantify the polarity of ionic liquids and their mixtures with organic cosolvents. Phys. Chem. Chem. Phys. 12, 1991–2000.

Pereiro, A.B., Legido, J.L., Rodriguez, A., 2007. Physical properties of ionic liquids based on 1-alkyl-3-methylimidazolium cation and hexafluorophosphate as anion and temperature dependence. J. Chem. Thermodyn. 39, 1168–1175.

Persson, I., 1986. Solvation and complex formation in strongly solvating solvents. Pure Appl. Chem. 58, 1153–1161.

Plechkova, N.V., Seddon, K.R., 2008. Applications of ionic liquids in the chemical industry. Chem. Soc. Rev. 37, 123–150.

Poole, S.K., Shetty, P.H., Poole, C.F., 1989. Chromatographic and spectroscopic studies of the solvent properties of a new series of room-temperature liquid tetraalkylammonium sulfonates. Anal. Chim. Acta 218, 241–264.

Poole, C.F., 2004. Chromatographic and spectroscopic methods for the determination of solvent properties of room temperature ionic liquids. J. Chromatogr. A 1037, 49–82.

Ragauskas, A.J., Williams, C.K., Davison, B.H., Britovsek, G., Cairney, J., Eckert, C.A., Frederick, W.J., Hallett, J.P., Leak, D.J., Liotta, C.L., Mielenz, J.R., Murphy, R., Templer, R., Tschaplinski, T., 2006. The path forward for biofuels and biomaterials. Science 311, 484–489.

van Rantwijk, F., Sheldon, R.A., 2007. Biocatalysis in ionic liquids. Chem. Rev. 107, 2757–2785.

Ratti, R., 2014. Ionic liquids: synthesis and applications in catalysis. Adv. Chem. 2014. http://dx.doi.org/10.1155/2014/729842. Article ID 729842, 16 pages.

Reichardt, C., 1979. Empirical parameters of solvent polarity as linear free-energy relationships. Angew. Chem. Int. Ed. Engl. 18, 98–110.

Reichardt, C., 1994. Solvatochromic dyes as solvent polarity indicators. Chem. Rev. 94 (8), 2319–2358.

Reichardt, C., 2005. Polarity of ionic liquids determined empirically by means of solvatochromic pyridinium N-phenolate betaine dyes. Green Chem. 7, 339–351.

Sahandzhieva, K., Tuma, D., Breyer, S., Kamps, Á.P.-S., Maurer, G., 2006. Liquid–liquid equilibrium in mixtures of the ionic liquid 1-n-Butyl-3-methylimidazolium hexafluorophosphate and an alkanol. J. Chem. Eng. Data 51 (5), 1516–1525.

Santiago, R.S., Santos, G.R., Aznar, M., 2009. UNIQUAC correlation of liquid–liquid equilibrium in systems involving ionic liquids: the DFT–PCM approach. Fluid Phase Equilib. 278, 54–61.

Schmelzer, K., Fahy, E., Subramaniam, S., Dennis, E.A., 2007. The lipid maps initiative in lipidomics. Methods Enzymol. 432, 171–183.

Seddon, K.R., 1997. Ionic liquids for clean technology. J. Chem. Technol. Biotechnol. 68, 351–356.

Shah, J.K., Maginn, E.J., 2004. A Monte Carlo simulation study of the ionic liquid 1-n-butyl-3-methylimidazolium hexafluorophosphate: liquid structure, volumetric properties and infinite dilution solution thermodynamics of CO_2. Fluid Phase Equilib. 222, 195–203.

Shah, J.K., Brennecke, J.F., Maginn, E.J., 2002. Thermodynamic properties of the ionic liquid 1-n-butyl-3-methylimidazolium hexafluorophosphate from Monte Carlo simulations. Green Chem. 4, 112–118.

Shetty, P.H., Youngberg, P.J., Kersten, B.R., Poole, C.F., 1987. Solvent properties of liquid organic salts used as mobile phases in microcolumn reversed-phase liquid chromatography. J. Chromatogr. A 411, 61–79.

Simoni, L.D., Lin, Y., Brennecke, J.F., Stadtherr, M.A., 2008. Modeling liquid–liquid equilibrium of ionic liquid systems with NRTL, electrolyte-NRTL, and UNIQUAC. Ind. Eng. Chem. Res. 47 (1), 256–272.

Smiglak, M., Mathew Reichert, W., Holbrey, J.D., Wilkes, J.S., Sun, L., Thrasher, J.S., Kirichenko, K., Singh, S., Katritzky, A.R., Rogers, R.D., 2006. Combustible ionic liquids by design: is laboratory safety another ionic liquid myth? Chem. Commun. 2554–2556.

Smiglak, M., Metlen, A., Rogers, R.D., 2007. The second evolution of ionic liquids: from solvents and separations to advanced materials—energetic examples from the ionic liquid cookbook. Acc. Chem. Res. 40 (11), 1182–1192.

Sprunger, L., Clark, M., Acree Jr., W.E., Abraham, M.H., 2007. Characterization of room-temperature ionic liquids by the Abraham model with cation-specific and anion-specific equation coefficients. J. Chem. Inf. Model. 47, 1123–1129.

Swatloski, R.P., Spear, S.K., Holbrey, J.D., Rogers, R.D., 2002. Dissolution of cellose with ionic liquids. J. Am. Chem. Soc. 124, 4974–4975.

Taft, R.W., Kamlet, M.J., 1976. The solvatochromic comparison method. 2. The alpha-scale of solvent hydrogen-bond donor (HBD) acidities. J. Am. Chem. Soc. 98 (10), 2886–2894.

Tariq, M., Forte, P.A.S., Costa Gomes, M.F., Canongia Lopes, J.N., Rebelo, L.P.N., 2009. Densities and refractive indices of imidazolium- and phosphonium-based ionic liquids: effect of temperature, alkyl chain length, and anion. J. Chem. Thermodyn. 41, 790–798.

Tokuda, H., Hayamizu, K., Ishii, K., Susan, M.A.B.H., Watanabe, M., 2004. Physicochemical properties and structures of room temperature ionic liquids. 1. Variation of anionic species. J. Phys. Chem. B 108 (42), 16593–16600.

Tokuda, H., Hayamizu, K., Ishii, K., Susan, M.A.B.H., Watanabe, M., 2005. Physicochemical properties and structures of room temperature ionic liquids. 2. Variation of alkyl chain length in imidazolium cation. J. Phys. Chem. B 109 (13), 6103–6110.

Tokuda, H., Hayamizu, K., Ishii, K., Susan, M.A.B.H., Watanabe, M., 2006. Physicochemical properties and structures of room-temperature ionic liquids. 3. Variation of cationic structures. J. Phys. Chem. B 110 (6), 2833–2839.

Tsuzuki, S., Shinoda, W., Saito, H., Mikami, M., Tokuda, H., Watanabe, M., 2009. Molecular dynamics simulations of ionic liquids: cation and anion dependence of self-diffusion coefficients of ions. J. Phys. Chem. B 113 (31), 10641–10649.

Valderrama, J.O., Robles, P.A., 2007. Critical properties, normal boiling temperatures, and acentric factors of fifty ionic liquids. Ind. Eng. Chem. Res. 46 (4), 1338–1344.

Valderrama, J.O., Rojas, R.E., 2010. Mass connectivity index, a new molecular parameter for the estimation of ionic liquid properties. Fluid Phase Equilib. 297, 107–112.

Valderrama, J.O., Sanga, W.W., Lazzús, J.A., 2008. Critical properties, normal boiling temperature, and acentric factor of another 200 ionic liquids. Ind. Eng. Chem. Res. 47 (4), 1318–1330.

Valderrama, J.O., Toro, A., Rojas, R.E., 2011a. Prediction of the heat capacity of ionic liquids using the mass connectivity index and a group contribution method. J. Chem. Thermodyn. 43, 1068–1073.

Valderrama, J.O., Martinez, G., Rojas, R.E., 2011b. Predictive model for the heat capacity of ionic liquids using the mass connectivity index. Thermochimica Acta 513, 83–87.

Vignale, G., Rasolt, M., 1987. Density-functional theory in strong magnetic fields. Phys. Rev. Lett. 59 (20), 2360–2363.

Voter, A.F., 1997. Hyperdynamics: accelerated molecular dynamics of infrequent events. Phys. Rev. Lett. 78, 3908–3911.

Wang, Y., Jiang, W., Yan, T., Voth, G.A., 2007. Understanding ionic liquids through atomistic and coarse-grained molecular dynamics simulations. Acc. Chem. Res. 40, 1193–1199.

Wasserscheid, P., Keim, W., 2000. Ionic liquids - new 'solutions' for transition metal catalysis. Angew. Chem. Int. Ed. 39 (21), 3773–3789.

Weingärtner, H., 2008. Understanding ionic liquids at the molecular level: facts, problems, and controversies. Angew. Chem. Int. Ed. 47, 654–670.

Welton, T., 1999. Room-temperature ionic liquids. Solvents for synthesis and catalysis. Chem. Rev. 99, 2071–2084.

Wilkes, J.S., Zaworotko, M.J., 1992. Chem. Commun. 965–967.

Wilkes, J.S., Levisky, J.A., Wilson, R.A., Hussey, C.L., 1982. Dialkylimidazolium chloroaluminate melts: a new class of room-temperature ionic liquids for electrochemistry, spectroscopy and synthesis. Inorg. Chem. 21 (3), 1263–1264.

Wilkes, J.S., 2002. Green industrial applications of ionic liquids. In: Rogers, R.D., Seddon, K.R., Volkov, S. (Eds.), NATO Science Series II: Mathematics, Physics and Chemistry, vol. 92. Kluwer, Dordrecht, pp. 295–320.

Wilkes, J.S., 2004. Properties of ionic liquid solvents for catalysis. J. Mol. Catal. A: Chem. 214, 11–17.

Wu, J., Zhang, J., Zhang, H., He, J.S., Ren, Q., Guo, M., 2004. Homogeneous acetylation of cellulose in a new ionic liquid. Biomacromolecules 5, 266–268.

Xiong, Y., Ding, J., Peng, C., Liu, H., 2012. Group contribution methods for prediction of thermophysical properties and phase behavior of ionic liquids. CIESC J. 63 (3), 667–675.

Yang, W., 1991. Direct calculation of electron density in density-functional theory. Phys. Rev. Lett. 66 (11), 1438–1441.

Ye, C., Shreeve, J.M., 2007. Rapid and accurate estimation of densities of room-temperature ionic liquids and salts. J. Phys. Chem. A 111 (8), 1456–1461.

Zhang, S., Sun, N., He, X., Lu, X., Zhang, X., 2006. Physical properties of ionic liquids: database and evaluation. J. Phys. Chem. Ref. Data 35 (4), 1475–1517.

Zhang, X., Liu, Z., Wang, W., 2008. Screening of ionic liquids to capture CO_2 by COSMO-RS and experiments. AIChE J. 54, 2717–2728.

Zhang, Q., De Oliveira Vigier, K., Royer, S., Jérôme, F., 2012. Deep eutectic solvents: syntheses, properties and applications. Chem. Soc. Rev. 41, 7108–7146.

Zissimos, A.M., Abraham, M.H., Klamt, A., Eckert, F., Wood, J., 2002. A comparison between the two general sets of linear free energy descriptors of Abraham and Klamt. J. Chem. Inf. Comput. Sci. 42, 1320–1331.

Chapter 6

Synthesis and Properties of Lipid-Inspired Ionic Liquids

Arsalan Mirjafari
Florida Gulf Coast University, Department of Chemistry and Physics, Fort Myers, Florida, United States

Richard A. O'Brien, James H. Davis, Jr.
University of South Alabama, Department of Chemistry, Mobile, Alabama, United States

6.1 INTRODUCTION

6.1.1 Lipid-Inspired Materials

Bilayer lipid membranes constitute the domain component of cellular membranes and they contain characterized hydrophilic polar head groups connected to the hydrophobic long tails. Lipids usually form lamellar bilayer structures over the majority of their phase diagram and they form vesicles or liposomes in the dilute aqueous solutions. The idea behind the formation of vesicles is simple and straightforward. Long-chain amphiphilic molecules, which have polar heads and nonpolar tails, are dissolved in aqueous media within the certain concentration range. Planar bilayer lipid membranes are generated in which the hydrophobic heads become exposed to the aqueous solution. As a result, they can form an aggregated structure where all the hydrophobic tails are directed toward the core of aggregates (Israelachvili et al., 1976).

In the last few decades, many studies have been performed to reveal the liposomes' structures as well as their biological importance. The formation of vesicles has turned out to be an archetype of self-assembly, extending well beyond lipid building block usages and leading to outstanding novel possibilities and applications in materials science (Antonietti and Föster, 2003). Biologically-inspired materials containing bilayer membranes is a growing technology since they are the perfect model to study the cellular membranes, and also demonstrate a great potential for helping to establish specific beneficial processes and achieve selectivity within a biological setting as drug delivery (Peetla et al., 2009) and biosensor for detecting biological agents (Cooper, 2002).

Ionic Liquids in Lipid Processing and Analysis. http://dx.doi.org/10.1016/B978-1-63067-047-4.00006-4

6.1.2 Ionic Liquids: Synthesis and Applications

The term ionic liquids (ILs) is used to describe organic salts that are liquid below 100°C due to their poorly coordinated anions and cations. ILs frequently share a number of interesting properties that distinguish them from volatile organic solvents such as negligibly low vapor pressure, excellent thermal/chemical/radiochemical stability, low flammability, and excellent solubility toward a wide variety of substances ranging from asphalt to biopolymers (eg, cellulose, silk and chitin). More importantly, the designer nature of ILs has driven their exploration and exploitation in many fields of science and engineering (Wasserscheid and Welton, 2008).

Advances in their development have transitioned them from the first generation, where they were merely regarded as solvents, to the next generation of task-specific ionic liquids (TSILs), where functionalization of the cations and/or anions provided specific properties (eg, physical, chemical, magnetic, optical, and biological) designed to accomplish specific applications (Davis, 2004). Generally, TSILs or functionalized ILs are a unique class of ILs, which possess a potential spectrum of utility extending far beyond conventional ILs. This definitely shows that it was possible to design salts that remain liquid at room temperature while incorporating functional groups into the structure of cations and/or anions. This concept has accounted for the wide popularity of ILs across different disciplines in science, technology, and engineering. In this concept, TSILs have been utilized in a multitude of diverse applications from synthesis/catalysis chemistry (Hallett and Welton, 2011) to energy, material, and biological applications (Smiglak et al., 2014; Tang et al., 2012). Advances in the synthesis and functionalization of ILs have generated excitement and a certain expectation that these ionic fluids have been exploited in academia and industry. The ease of synthesis along with the unique physicochemical properties of TSILs make them ideal candidates for translation from the laboratory setting into the industrial arena. Since BASF announced the BASIL process as the first industrial process involving an IL in 2002, many other processes using ILs have been followed by the chemical industry (Plechkova and Seddon, 2008).

6.2 LIPID-INSPIRED IONIC LIQUIDS: A NEW CLASS OF LIPIDOID MATERIALS

ILs containing long chain appendages ($\geq C_8$) are finding a wide spectrum of applications as solvents for metal nanoparticle stabilization (Yang et al., 2014), energy storage devices (Armand et al., 2009), phase transfer catalysis (O'Brien et al., 2013), and lubricants (Zhou et al., 2009). In other words, their function is strongly related to their fluidity, and that is quantified by reference to the melting point, T_m, in which increased fluidity corresponds to a lower T_m value.

Due to the presence of long carbon chains, lipid-inspired ionic liquids (LIILs) display mesostructural ordering with segregated regions of polar and nonpolar domains. This self-assembly is driven by the combination of Coulombic, van der Waals, and H-bonding interactions. Molecular modeling suggests that the nanostructure of the ILs can be efficiently adjusted by changing the relative proportions of the polar and nonpolar portions of the species; transitioning from polar domains with nonpolar alkyl tails, to a nonpolar region with disperse ionic domains (Smith et al., 2013). These applications depend remarkably upon how the ILs with long tails behave in the liquid state.

The biggest challenge is designing the ILs with attached long alkyl side chains to the cations, while keeping their melting points under ambient temperature. In fact, as the alkyl chain length increases above the seven carbons, the T_m of salts increase dramatically and are above room temperature for chain lengths above 12 carbon atoms with many common anions (Murray et al., 2010). As the alkyl side chain length increases, it becomes the dominant structural feature of the cation; increasing the chain length then increases the dispersion force interaction between the chains while no longer providing asymmetry, resulting in higher melting points, which increase in the same way as adding methylene groups to straight chain alkanes. The development for the formulation of poorly water-soluble or water-insoluble ILs has remained an important and challenging issue.

LIILs, first described in our recent publications (Kwan et al., 2013; Mirjafari et al., 2012, 2014; Murray et al., 2010, 2013), are a novel class of biomaterials that utilize structural features similar to natural lipids to allow the incorporation of the lipophilic structural elements while keeping their melting points below room temperature. In the first generation of lipidic ILs, naturally-occurring fatty acids were used to synthesize the 1-alkyl-3-methylimidazolium ions, paired with bistriflimide (NTf_2) anion, to show the dramatic melting point decrease when a *cis*-double bond was introduced into the side alkyl chain (Murray et al., 2010). In the second generation, we have demonstrated similar effects with synthetic asymmetric structural features such as cyclopropyl group—branching—that introduce "kinks" in the alkyl chain structures (Kwan et al., 2013). We also utilized the "click" chemistry to prepare the synthetic analogs of the naturally-derived side chains and asymmetric features are introduced through the inclusion of sulfur atoms and branches in the alkyl chain, both easily facilitated through the use of thiol-ene reaction to form the third generation of these functional fluids. This synthetic motif has also allowed us to examine the effect of the location of the sulfur atom and branches in the alkyl chain on the thermophysical properties of the species providing valuable structure/property information (Mirjafari et al., 2014; O'Brien et al., 2011). The synthesis and physicochemical properties of the first generation of lipidic ILs has been fully discussed in a review by our group (Mirjafari et al., 2012). In this chapter, we will primarily focus on the second and third generations of lipid-inspired materials.

6.2.1 Fatty Acid-Derived Ionic Liquids

The examples of first and second generations of lipidic ILs, ILs **1–8**, along with their melting points are illustrated in Scheme 6.1. Each of the ILs has a C_{18} alkyl appendage identical to that in a natural fatty acid. ILs **1–6** were prepared in a three-step process starting from fatty-alcohol mesylates, 1-methylimidazole or 1,2-dimethylimidazole, NaI, and $KNTf_2$ (Mirjafari et al., 2012). Two additional steps are required to synthesize ILs **7** and **8** from their corresponding fatty-alcohol mesylates (Kwan et al., 2013).

The T_m values reported in Scheme 6.1 were determined by differential scanning calorimetry (DSC). Like most lipid-like materials, the new ILs showed rich phase behavior below their melting points. The reported T_m values are the highest temperature phase changes exhibited by the ILs in a liquid state. ILs **1** and **2** feature fully saturated C_{18} side chains (identical to stearic acid). Due to the strong van der Waals interaction between the saturated long hydrocarbon chains, their T_m values are considerably higher (IL **1** = 53.5°C; IL **2** = 70.0°C). However, as is apparent from the data in Scheme 6.1, inclusion of double bond(s) in the side chain of imidazolium-type lipidic ILs (ILs **1–6**) resulted in profound decreases in their T_m relative to ILs **1** and **2**, with their linear saturated side-chains. Oleyl-based ILs **3** and **4** are the unsaturated analogs of ILs **1** and **2**, and they have a *cis*-unsaturated bond between C9 and C10 positions of the side chains. Likewise, IL **5**, elaidyl-based IL, is a *trans* counterpart of **3** with C_{18} side chains, which only differs on the basis of the C9–C10 double bond's configuration. IL **6** is the analog of linoleic acid, which contains two double bonds at the positions of C9–C10 and C12–C13,

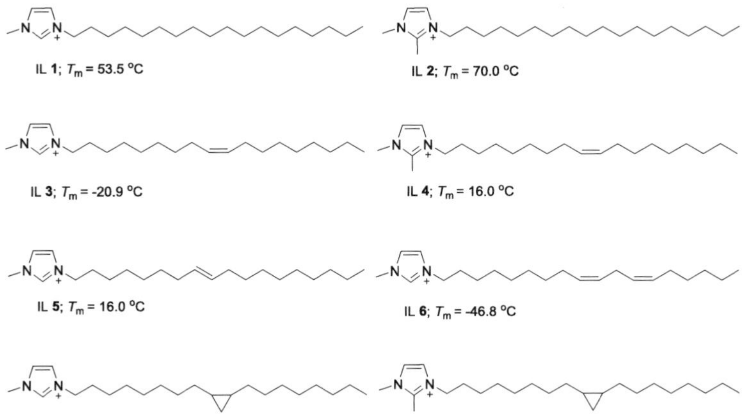

SCHEME 6.1 Structures and melting points of C_{18}-bearing LIILs (the paired anion is NTf_2^-).

respectively. Indeed, the introduction of *cis* double bonds (kinks) into the long aliphatic tails of these salts leads to a radical decrease in their T_m values when compared to their saturated and *trans* counterparts. This decrease consequently leads to an increase in fluidity that closely parallels that observed with their molecular relatives.

However, the *ene*-induced reductions in T_m are accompanied by an increase in the chemical fragility of the LIILs. In other words, at elevated temperatures and in the presence of air, *ene*-bearing ILs are subjected to slow oxidative cleavage of its alkyl side chain at the site of unsaturation similar to the process of fat rancidification. To address this issue, therefore, we embarked to synthesize branched LIILs by inclusion of a branching element (eg, cyclopropyl moiety) at the C9−C10 position of the side chains, which has the packing-disruptive module in the long alkyl appendages. Indeed, branching is known to influence the physical properties of organic molecules including lipids. Among the lipid structural features found in nature that tend to depress melting points are *cisoid* cyclopropyl moieties. Such moieties occur most commonly in lipids of nonmammalian sources that range from microorganisms to fruiting plants like *Litchi chinensis*, which contain relatively large amounts of cyclopropanated components (Gaydou et al., 1993). The overall depressions in T_m brought about by the inclusion of a cyclopropyl group in the side chains of cations ($\Delta T_m = 62.1°C$ in **1** vs **7**; $\Delta T_m = 52.0°C$ in **2** vs **8**) are substantial even when compared to those brought by chain-included olefins. While not as efficient as olefins in bringing about this effect, ILs incorporating side-chain cyclopropanated modules are likely to be less sensitive toward oxidative degradation.

For the imidazolium-based ILs, the hydrogen bonding between the cation and the anion is mainly determined by the most acidic hydrogen at C2 position of imidazolium ring. As such, the methylation at C2 position disrupts the H-bonding interaction between cation and anion, leading to an increase in the melting points and viscosity of ILs (Izgorodina and MacFarlane, 2011). The effect of C2-alkylation of an imidazolium ring by raising the melting points of ILs can be observed in ILs **1**, **3**, and **7** versus **2**, **4**, and **8**, respectively. Additionally, in continuation with our effort to make varied LIILs, we synthesized C2-substituted LIILs via a cyclization−alkylation reaction, which comprised reacting a diamine with natural saturated fatty acids (eg, palmitic, stearic, and arachidic acids), followed by the metathesis reaction using $KNTf_2$. This three-step procedure leads to the low-yielding formation of a series of new imidazolium-based LIILs containing C_{16}, C_{18}, C_{20} saturated chains at C2 position of imidazolium ring (Scheme 6.2). It is worthy to mention that due to the high temperature of reaction, C2-unsaturated lipidic ILs were not formed in this process.

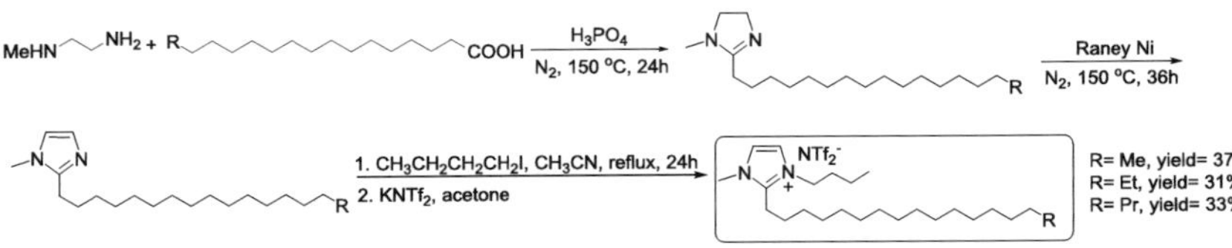

SCHEME 6.2 Multistep synthesis of C2-substituted C_{16}, C_{18}, and C_{20} saturated LIILs.

6.2.2 Thiol-ene "Click" Chemistry Mediated Synthesis of Lipid-Inspired Ionic Liquids

6.2.2.1 Thiol-ene "Click" Chemistry

In 2001, Sharpless et al. reported a new concept for conducting organic reactions, which was based upon the premise that organic synthesis should take advantage of the long history of development and progress during the 20th century and focus attention on highly chemo/regio/stereo-selective, simple, and orthogonal reactions with no side product formation, and that yields heteroatom-linked compounds under a variety of mild conditions. Several efficient reactions, which are capable of producing a wide catalog of functional synthetic molecules and organic materials, have been categorized accordingly under the term of click reactions. Following Sharpless's visionary characterization of several idealized reactions as click reactions, the materials science and synthetic chemistry communities have pursued numerous routes toward the identification and implementation of these click reactions (Kolb et al., 2001; Kolb and Sharpless, 2003; Moses and Moorhouse, 2007).

The thiol-ene click reaction, highly efficient reactions of thiols with reactive carbon—carbon double bonds, or simply *enes*, is an efficient coupling between thiols and alkenes, which was well known in the early 1900s. The radical-based additions of thiols across alkenes carry many of the attributes of click reactions. The thiol-ene reaction has received much attention for the rapid development in its applicability across multiple areas of chemistry including small molecule organic chemistry, polymer/copolymer and material functionalization, cellular microarray technology, dendrimer synthesis, glycopolymer synthesis, biocompatible network materials, microfluidic devices, electro-optical materials, surface pattering, nanoimprinting, micellular drug delivery and vaccine development (Hoyle and Bowman, 2010; Kade et al., 2010).

6.2.2.2 Synthesis of Thioether-Functionalized Lipid-Inspired Ionic Liquids via Thiol-ene Click Chemistry

The continued success of TSILs in a wide spectrum of applications ranging from catalyst/solvent to medicine imposes a challenge to develop practical syntheses of these functionalized fluids. In turn, this challenge has focused on the greater need to elaborate efficient and orthogonal approaches, like click chemistry, for the preparation of functional materials. Underlying both of these synthetic goals is the additional desire to develop methodologies that fit into the focus area of green chemistry (Sheldon, 2012). For the first time, thiol-ene click chemistry was employed using *ene*-functionalized ILs to assess its potential utility in TSIL synthesis (O'Brien et al., 2011). In this strategy, a series of novel TSILs with highly polar polyhydroxylated appendages containing sulfur moieties was prepared via the radical-based thiol-ene addition. TSILs with multiple hydroxyl and thioether functional groups were readily prepared by the radical addition of 1-thioglycerol and salts of imidazolium/quaternary ammonium cations bearing tethered allyl and/or vinyl groups. 1-Thioglycerol was used for initial study in order to demonstrate the incorporation of a desirable functional group (—OH) into an IL by way of the orthogonal reaction of *ene*-bearing IL ions. The reactions proceed quickly and cleanly, and give acceptable yields of products without optimization (O'Brien et al., 2011).

Alkene-induced LIILs (ILs **3**—**6**) possess very low T_m values; however, the presence of double bond(s) increases the chemical fragility of the IL much like unsaturated lipids, which are subject to oxidative cleavage of the side chains at the site of unsaturation. While side-chain cleavage appears likely to be advantageous from the standpoint of IL biodegradability, this might complicate their use in certain practical applications. Accordingly, we embarked to synthesize other types of lipid-inspired ILs, ones in which the packing-disruptive module in the long alkyl appendage is something other than an olefinic moiety, while allowing the highly lipophilic character of the original materials to be retained. After considering several options, we elected to investigate the selective substitution of side chain methylene groups by sulfur atoms because C—S bonds are typically 10% longer than C—C bonds (184 pm vs 153 pm) and C—S—C bond angles are compressed relative to C—C—C angles (~99° vs ~109°), kinks are created that are potentially alkene-mimicking as packing disruptors. Furthermore, while thioethers can undergo slow aerobic oxidation (eg, to sulfoxides), this reaction does not culminate in molecular cleavage.

To evaluate our hypothesis, we used the thiol-ene reaction with different *ene*-functionalized ILs by reacting them photochemically with alkyl thiols, yielding a novel series of C_{16}, C_{18}, and C_{20} lipidic ILs, incorporating the thioether linkage at various positions within the alkyl chain (Scheme 6.3) (O'Brien et al., 2014). A library of the 1-methyl-3-thialkaylimidazolium

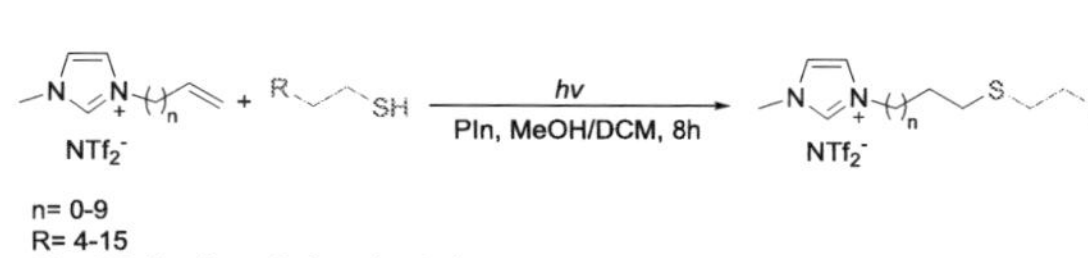

SCHEME 6.3 Synthesis of 1-methyl-3-thialkaylimidazolium bistriflimide-based lipidic ILs via thiol-ene click reaction.

bistriflimide-based LIILs, which contain **16**, **18**, and **20** equivalent carbons in their linear alkyl side chains with a sulfur heteroatom that is strategically introduced from C3 to C12, was displayed in Schemes 6.4A–C. ILs **11–20**, **23–31**, and **34–43** were exclusively formed in anti-Markovnikov orientation and near quantitative yields. They were simply purified by washing with hexanes, in which none of the starting or product salts were soluble and no further purifications, such as chromatographic methods, were required to obtain the desired salts. These lipidic ILs were categorized into three sets based upon their chain length (C_{16}, C_{18}, and C_{20}). Each set was designed to be compared to two previously reported, set-specific reference compounds, one an analog with a fully saturated, all-carbon side-chain and the other

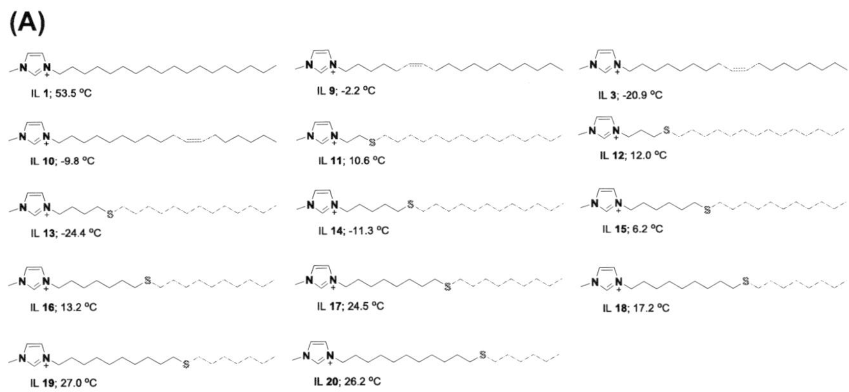

SCHEME 6.4A Structures of LIILs containing 18 equivalent carbons in length and their T_m values (paired anion is NTf_2^-).

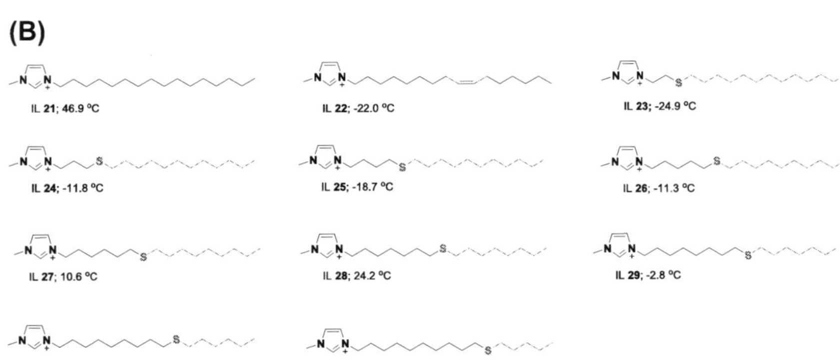

SCHEME 6.4B Structures of LIILs containing 16 equivalent carbons in length and their T_m values (paired anion is NTf_2^-).

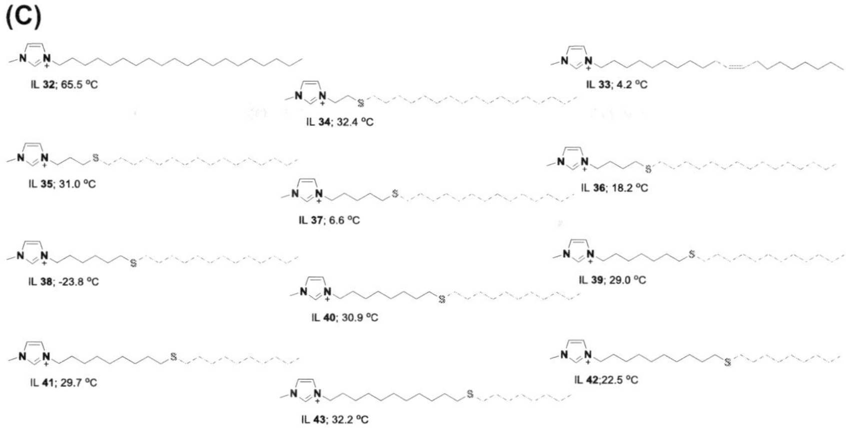

SCHEME 6.4C Structures of LIILs containing 20 equivalent carbons in length and their T_m values (paired anion is NTf_2^-).

bearing a *cis*-alkene module. The anion of each of the new and reference ILs is bis(trifluoromethane)sulfonimide (NTf_2^-). The ILs that comprise set C_{18} are alike in incorporating 18-atom side chains, but differ in the placement of the double bonds (reference compounds) or sulfur atoms in each. Sets C_{16} and C_{20} were similarly formatted with respect to their side-chains, save for differing in the overall number of chain atoms in each (O'Brien et al., 2014).

The new ILs were evaluated by DSC, and the resulting T_m data compiled to create Schemes 6.4–6.6. Note that T_m as reported here is the temperature at which the phase transition to a liquid, isotropic, or anisotropic occurs, regardless of the nature of the preceding state. As hoped, the T_m of each new IL was notably lower than the corresponding all-carbon reference compound with the same number of side-chain backbone atoms (C18: Scheme 6.4A; C16: Scheme 6.4B; C18: Scheme 6.4C). For instance, IL **1**, with its saturated all-carbon side chain, has a T_m value of 53.5°C but the formal replacement of its C11 methylene unit with an S atom provides IL **19**, which has a T_m of 27.0°C ($\Delta T_m = 26.5$°C). Likewise, while reference IL 32 has a T_m of 62.5°C, its S11 counterpart 42 has a T_m of 22.5°C ($\Delta T_m = 40.0$°C) (O'Brien et al., 2014).

It has been observed for several homologous series of *N*-methyl-*N*′-alkylimidazolium salts that T_m decreases as the alkyl side chain increases in length from one to about seven carbons, after which further elongation leads to a sharp T_m increase (Mirjafari et al., 2012). Rothenberg's group has proposed that the former domain constitutes a *symmetry-breaking region*, the impact of which is eventually outweighed by that of the cumulative interchain dispersion forces of progressively longer alkyl groups (Lopez-Martin et al., 2007). We subsequently showed that the lower T_m values in imidazolium-type ILs are observed regardless of whether the site of unsaturation was within the symmetry breaking region or beyond it, although the magnitude of the effect was greater in the latter circumstance. In order to determine if the *S-effect* operative in thioether-functionalized ILs is likewise present when a sulfur atom is in the symmetry-breaking region of the alkyl appendage, we prepared new lipidic ILs containing thioether linkages, which in each side chain incorporates sulfur in the C3−C12 positions from the imidazolium ring.

It is apparent from the foregoing data that strategically placed sulfur atoms can—as thioether linkages—have a significant depressive effect on the T_m values of lipidic compounds in which they are incorporated, although in no case is the degree of T_m depression (relative to saturated and unsaturated reference compounds) as great as that evoked by unsaturation. However, the depressive effect (eg, ΔT_m relative to the appropriate saturated standard) manifested by these formal, single S-atom replacements of a two-atom functional unit was in all cases at least 45% of that wrought by the olefin module, and as great as 72%. It is also noteworthy that the effect is general, being operative in ILs and fatty acids alike, suggesting it to be a versatile tool for engineering T_m and potentially other fluid properties (eg, viscosity) in a range of lipidoid materials.

As the sulfur is moved away from the cationic head-group, the melting point decreases initially, reaching a minimum, and then begins to increase as the sulfur approaches the end of the chain. This is consistent with the behavior observed for the short chain, molecular thioethers and ethers alike. The magnitude of melting point depression relative to the saturated homolog is observed to have a strong dependence on the position of sulfur in the side chains. The sulfur position corresponding to the lowest melting point for a homologous series shifts further down the chain as the chain length is increased, indicating that the maximum effect takes place near the center of the ion. Thus, regardless of the chemical identity of the symmetry breaking moiety (unsaturation or sulfur atom) the maximum melting point depression seems to occur when the moiety is placed near the center of the structure (O'Brien et al., 2014).

An initial theoretical validation of the concept was provided by comparing the results of high-level DFT calculations on a truncated thioether chain segment model, $CH_3CH_2SCH_2CH_3$, with those from the all-carbon backbone analog $CH_3CH_2CH_2CH_2CH_3$. Strikingly, computations on the former predicted that the room-temperature population of the two degenerate *gauche* conformers of the C−S−C−C bond ought to collectively comprise about 63% of the total conformer population around the linkage. This considerably exceeds the collective *gauche* populations at the central C−C−C−C linkage of the all-hydrocarbon counterpart, and comports with earlier computational results as well as crystallographic analyses of preferred thioether conformations. Furthermore, on examining the computed structures it is visually apparent that the *gauche* conformation of the C−S−C−C bond produces a distinct kinking of the side chain in which it is incorporated (O'Brien et al., 2014).

6.2.2.3 *Synthesis of Alkyl-Branched Thioether-Functionalized Lipid-Inspired Ionic Liquids: The Remarkable Solvent Effect on the Pathway of Thiol-ene Reaction*

Recently, our group has reported that the nature of solvent plays a crucial role in the outcome of the thiol-ene reaction to synthesize a novel family of LIILs (Mirjafari et al., 2014). In the presence of a polar protic solvent, such as MeOH, the regioselectivity of the thiol-ene reaction is reversed to yield the Markovnikov-oriented addition in appendages containing a sulfur heteroatom at different positions of their lipidic tails. This offers considerable promise as a robust, efficient, and orthogonal method to construct structurally diverse imidazolium-type lipidic ILs where their C_{18} linear/branched cationic tails contain a sulfur heteroatom in different positions (Mirjafari et al., 2014). In this study, asymmetric features are introduced through the inclusion of a sulfur atom along with methyl branching, easily facilitated through the use of thiol-ene click chemistry. Furthermore, an important fact to note is that the highly branched structures of natural fatty acids (ie, tuberculostearic, phytanic, and pristanic acids) offer the potential of preserving low melting points by introducing asymmetry along the side chain while eliminating the possibility of oxidative decomposition.

With the opportunity to synthesize linear/branched lipidic ILs via a simple solvent selection, we endeavored to create a new library of branched LIILs, which were synthesized via two similar synthetic pathways (Scheme 6.5, Paths 1 and 2) and grouped into three sets (linear, internally methyl-branched, and externally methyl-branched). Particularly, Path 1 is the photochemical addition of a variety of alkylthiols across *ene*-bearing ILs. On the other hand, Path 2 is the addition of thioether-based ILs across the variety of alkenes (Scheme 6.5). Unexpectedly, we discovered that the nature of solvent remarkably influences the pathway and regioselectivity of thiol-ene photocoupling. In particular, use of $MeOH/CH_2Cl_2$ (1:1 ratio) as solvents yields the anti-Markovnikov adducts (Scheme 6.5, linear ILs), while MeOH as the lone solvent leads to the exclusive formation of Markovnikov-oriented methyl-branched thioether-type ILs (Scheme 6.5, internally/externally-branched ILs), which could be explained on the basis of addition of sulfur-centered radical cation across *enes*.

The linear ILs are alike in incorporating linear 18 atom side chains, but differ in the positions of sulfur atom (Scheme 6.4a, ILs **12**, **14**, **16**, and **17**), which have been covered in the last section. On the other hand, branched ILs are similarly formatted with respect to their side-chains and contain internal (Scheme 6.6, ILs **44**–**46**) or external (Scheme 6.6, **47**–**50**) methyl branches toward the cationic ring and located adjacent to sulfur atom, which were prepared exclusively in a Markovnikov fashion. The resulting LIILs of both Paths 1 and 2 were formed in quantitative yields (88–100%) and were simply purified by washing with hexanes. No chromatographic separation was required. It is noteworthy that the use of 0.5 equiv./alkene of 2,2-dimethoxy-2-phenylacetophenone, as a photoinitiator, is necessary for 100% addition of thiols across *enes* that can be recognized by disappearance of the olefinic peaks in ^{1}H and ^{13}C NMR spectra.

As stated previously, the T_m value of each linear thioether-type IL was remarkably lower than that of the corresponding all-carbon reference compound with the same number of side-chain backbone atoms. As previously mentioned, sulfur atoms have profound depressive effects on the T_m values of lipidic compounds. Besides sulfur's role, side-chain branching is an important structural feature of the new ILs, which has an important impact upon T_m and proves to be similarly effective. As shown in Scheme 6.6, methyl moieties embedded in the different positions of lipidic aliphatic side-chains of ILs **44–50** through the Markovnikov-oriented addition are offered to be significantly effective in decreasing the T_m values of such materials relative to linear analogs with the same number of carbons. For instance, stearyl-based IL **1**, with a saturated C_{18} side chain, has a T_m value of 53.5°C. The replacement of its methylene unit with a sulfur atom at the position of 6 yields demonstrably the lower T_m of −11.3°C for IL **14** ($\Delta T_m = 64.8$°C). Notably, Markovnikov-oriented inclusion of a sulfur and a methyl branch at the positions of 6 and 7 forms IL **48** with T_m value of −27.7°C, which is considerably lower than both (**14** vs **48**: $\Delta T_m = 16.4$°C; **1** vs **48**: $\Delta T_m = 81.2$°C).

Theoretically, olefinic hydrothiolation proceeds through two competitive pathways either in Markovnikov or anti-Markovnikov fashion. Typically, the insertion of thiyl radicals across double bonds shows the dominant anti-Markovnikov regioselectivity because of enhanced stability of carbon-centered radical intermediate. Generally, the mechanism of photo-induced thiol-ene reaction is a radical step-growth involving a radical-mediated addition of a thiol across an external *ene* to form a thioether as a single product in regioselective manner. The mechanism develops along a step-growth pathway and proceeds via a typical chain process with initiation, propagation, and termination steps. Initiation begins with the reaction of thiol with an appropriate photoinitiator resulting in the photolysis of weak S–H bond and generation of thiyl radical, RS•. Propagation is a two-step process wherein a thiyl radical adds across an *ene*, generating a carbon-centered radical intermediate, which can then undergo a chain-transfer step by abstracting a hydrogen atom from another thiol. The step-growth path results in formation of a thioether product with predominantly anti-Markovnikov orientation along with a new thiyl radical, which then initiates another propagation step. It is worth noting that the regioselectivity is explained by the enhanced stability of the carbon-centered radical intermediate upon addition onto the less substituted alkene carbon (Hoyle and Bowman, 2010).

Intriguingly, we found the solvent polarity exerts a great influence on the regioselectivity of thiol-ene reaction in our systems. We observed that the

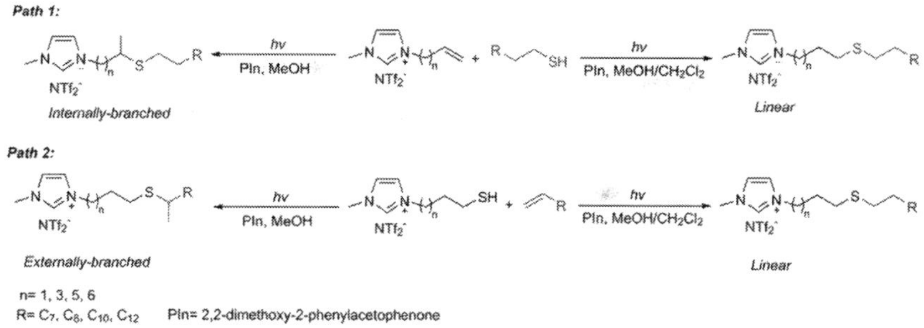

SCHEME 6.5 Formation of branched lipidic ILs via Markovnikov-mediated photoaddition of thiols across *enes*.

predominant formation of Markovnikov-oriented adducts can be triggered simply by changing the solvent from 1:1 mixture of $MeOH/CH_2Cl_2$ to pure MeOH (Scheme 6.7). Based upon the literature precedence (Monney et al., 2013), the Markovnikov regioselectivity of thiol-ene might be explained in terms of formation of a relatively stable sulfur-centered radical cation intermediate (Scheme 6.7, I), which is followed by the loss of proton to form the S-centered radical II. Afterward, H^+ adds to the less substituted end of C=C bond to form a stable carbocation III, which reacts with S-centered radical II, followed by the electron donation from the photoinitiator's radical anion, $[PIn]^-$, which leads to the formation of the branched S-radical cation IV. The hydrogen abstraction gives the final thiol-ene adducts in Markovnikov fashion. In fact, methanol as a polar protic solvent induced a significant reversal of regioselectivity in comparison to the same reaction in $MeOH/CH_2Cl_2$ due to the increasing stabilization of released H^+ via hydrogen bonding. In addition, the polar transition state for the electrophilic addition of proton across alkene and also intermediate carbocation III can be stabilized by the improved solvation in a protic solvent (eg, methanol).

The unusual property of S-radical cations is their propensity to form two-center, three-electron bonds with nonbonding electrons of other sulfur atom $[S\therefore S]^+$ or heteroatoms providing a pair of electrons $[S\therefore X]^+$ (Monney et al., 2013). The formation of intermediate I can be justified by the low ionization potential of sulfur atom, which is considerably below 10 eV. Therefore, an electron can be removed photochemically from the unshared pair of electron with more p character by the photoinitiator, which leads to form S-radical cations. Recently, the Glass group has reported on the formation of stabilized sulfur radical cations by interacting to aromatic π-system and formation of $[S\therefore\pi]^+$ complex (Monney et al., 2013). In this study, we believe the main stabilizing factor is the interaction of sulfur radical cation I with π-system of imidazolium ring, which can be explained in terms of formation of relatively stable intermediate (Scheme 6.7).

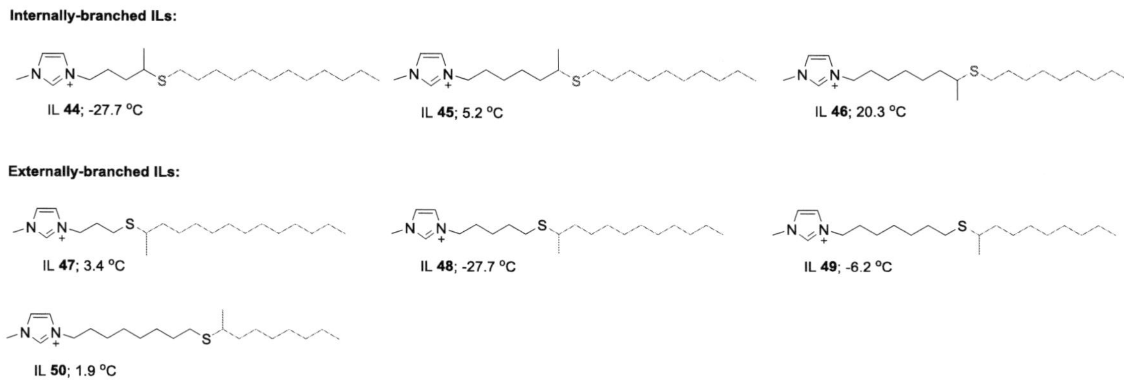

SCHEME 6.6 A library of internally- and externally-branched new LIILs along with their T_m values (paired anion is NTf_2^-).

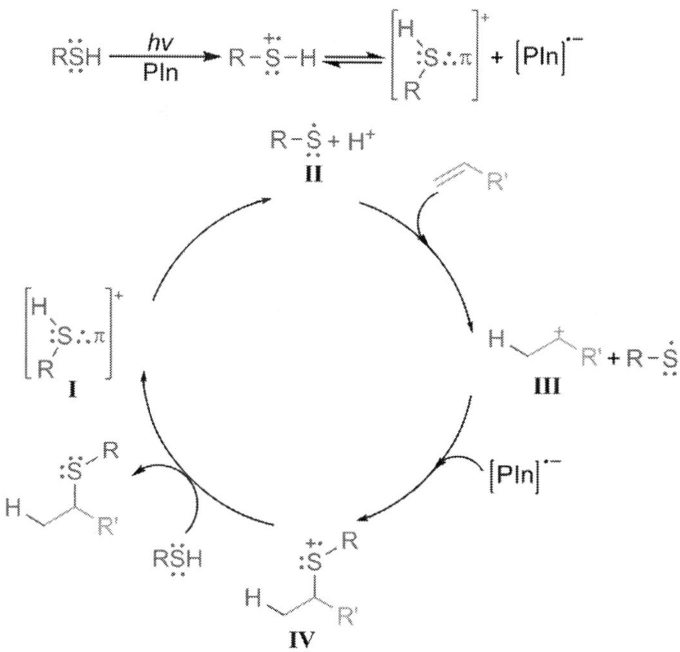

SCHEME 6.7 Proposed mechanism for the Markovnikov-oriented thiol-ene reaction through S-radical cations formation.

6.2.2.4 Thioether-Functionalized Quaternary Ammonium Salts as Phase Transfer Catalysts

In our continuing efforts to explore the synthetic utility of the thiol-ene click chemistry, we employed an earlier *ene*-bearing substrate (O'Brien et al., 2011) and expanded the range of thiols incorporated into the IL framework. Using this strategy, a series of novel tetra(4-thiaalkyl)ammonium bromides containing sulfur moieties were prepared via the radical-based thiol-ene addition of tetraallylammonium bromide with a library of alkylthiols with chain lengths ranging from C_2—C_{18} (Scheme 6.8) (O'Brien et al., 2013).

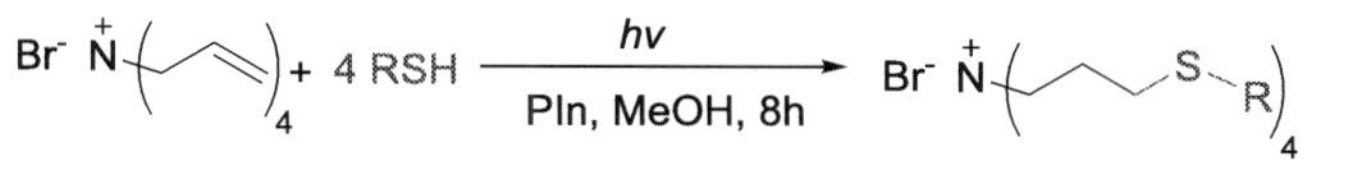

15 examples, Yield: 51-98%

R= ethyl, propoyl, isopropyl, *n*-butyl, isobutyl, *t*-butyl, anteisobutyl, hexyl, heptyl, octyl, tetradecyl, octadecyl, $CH_2CH_2(CF_2)_7CF_3$, cyclohexyl, benzyl.
PIn= 2,2-dimethoxy-2-phenylacetophenone

SCHEME 6.8 Preparation of tetra(4-thiaalkyl)ammonium salts via thiol-ene reaction.

In order to expand our understanding of these novel quaternary ammonium salts, we chose phase-transfer catalysis (PTC) since it offers both a potential application and a tool for evaluating behavior. Because the mechanism of the PTC cycle is generally well understood (Naik and Doraiswamy, 1998), the activity of these salts compared to their traditional analogs provides a means of characterizing them. A commonly used reaction for screening the activity of phase transfer catalysts is the nucleophilic substitution of cyanide for bromide (Chandler et al., 1998). In this work, we have chosen as a model reaction the cyanide displacement on 1-bromopentane and the results indicate that the quaternary ammonium salts containing sulfur yield slower reaction rates than those without. However, the reaction rate is significantly increased compared to that without a catalyst present. With no catalyst, approximately 3% conversion of the 1-bromopentane was achieved in a 24h period.

One of the unusual features of these thiaquat salts is their melting points as compared with their tetraalkylammonium bromide counterparts. For the tetradodecylammonium bromide, the melting point as measured by the transition onset from the modulated DSC scan was 86.3°C. This value is in good agreement with a previously reported melting point (89°C) (Xenopolous et al., 1992), as are the two lower temperature solid/solid mesophase transitions. For the tetra(4-thiadodecyl)ammonium bromide, the melting point is significantly higher, 234°C from the modulated scan and 238°C from the traditional scan. This range is consistent with visual determination of the melting point of 234–235°C. The compound is observed to decompose upon melting, and samples were observed to change color from white to brown after scans at temperatures higher than 210°C, even when the melting point was not reached. This decomposition temperature is significantly higher than that of the tetraalkyl analog, which exhibits an endothermic degradation beginning at around 160°C.

The higher melting point of tetra(4-thiadodecyl)ammonium bromide relative to tetradodecylammonium bromide is indicative of greater cation–anion interactions in the sulfur-containing salt. It is unclear whether the phase transition, which certainly disrupts solid-phase interactions between ion pairs, additionally disrupts internal ion–pair interactions or interactions between a bromide of one ion pair and a hydrogen alpha to the sulfur on another ion pair. Regardless, the higher melting point is consistent with other measurements and calculations in this work indicating enhanced interactions between the ions, including the predictions made using quantum-based computational methods and the experimentally-observed reduction of reaction rate. Furthermore, the greater thermal stability may be due to the sulfur's proximity to the central nitrogen, electronically or sterically preventing degradation reactions.

6.3 CONCLUSION AND PROSPECTIVE

ILs with long-chain appendages and low melting points are of great interest in the soft material field. These LIILs have structural features similar to natural lipids, specifically, long alkyl chains with *cis*-unsaturations, which lowers their melting points relative to homologs with saturated alkyl chains. Since 2012, our group has presented different alternate structural features such as sulfur inclusion and branched substituents (eg, cyclopropyl or methyl) to bring about depressions in the T_m values of the LIILs, which are likely to be more resistant to aerobic degradation than their olefinic counterparts. Specifically, we have used thiol-ene chemistry for the synthesis of a novel class of lipid-inspired ILs with linear and branched long alkyl appendages containing a sulfur heteroatom at different positions of their lipidic tails. The clear and explainable relationships between these T_m values and structures of ILs are apparent. The strategically placed thioether groups have significant depressive effects on the T_m values of lipidic ILs in which they are incorporated. The magnitude of melting point depression relative to the saturated analog is observed to have a strong dependence on the position of the sulfur in the side chain. Additionally, a novel class of methyl branched thioether-type LIILs, which are simultaneously lipophilic and low-melting salts, were synthesized via exclusive Markovnikov-oriented thiol-ene addition using a polar protic solvent such as MeOH due to the marked tendency of sulfur-centered radical cation intermediates to become stabilized by coordination with π-system of imidazolium ring.

In anticipation of future developments, an extension of these studies has broad implications for the applications of ILs to investigations of specially tailored synthetic soft materials. The present data shows that the LIILs containing long chain appendages with low melting points may have significant applications for IL use in enzymatic catalysis, lubricants, heat-transfer fluids and gas storage/separation, solvents for nonpolar biomolecules such as cholesterol, and models for lipid bilayers.

ACKNOWLEDGMENTS

This material is based upon work supported by the National Science Foundation under Grant Number 1133101.

REFERENCES

Antonietti, M., Föster, S., 2003. Vesicles and liposomes: a self-assembly principle beyond lipids. Adv. Mater. 15, 1323–1333.

Armand, M., Endres, F., MacFarlane, D.R., Ohno, H., Scrosati, B., 2009. Ionic–liquid materials for the electrochemical challenge in future. Nat. Mater. 8, 621–629.

Chandler, K., Culp, C.W., Lamb, D.R., Liotta, C.L., Eckert, C.A., 1998. Phase-transfer catalysis in supercritical carbon dioxide: kinetic and mechanistic investigations of cyanide displacement on benzyl chloride. Ind. Eng. Chem. Res. 37, 3252–3259.

Cooper, M.A., 2002. Optical biosensors in drug discovery. Nat. Rev. Drug Discov. 1, 515–528.

Davis Jr., J.H., 2004. Task-specific ionic liquid. Chem. Lett. 33, 1072–1077.

Gaydou, E.M., Ralaimanarivo, A., Bianchini, J.P., 1993. Cyclopropanoic fatty acids of litchi (*Litchi chinensis*) seed oil. A reinvestigation. J. Agric. Food Chem. 41, 886–890.

Hallett, J.P., Welton, T.T., 2011. Room-temperature ionic liquids. Solvents for synthesis and catalysis. Chem. Rev. 111, 3508–3576.

Hoyle, C.E., Bowman, C.N., 2010. Thiol–ene click chemistry. Angew. Chem. Int. Ed. 49, 1540–1573.

Israelachvili, J.N., Mitchell, D.J., Ninham, B.W., 1976. Theory of self-assembly of hydrocarbon amphiphiles into micelles and bilayers. J. Chem. Soc. Faraday Trans. 2 (72), 1525–1568.

Izgorodina, E.I., MacFarlane, D.R., 2011. Nature of hydrogen bonding in charged hydrogen-bonded complexes and imidazolium-based ionic liquids. J. Phys. Chem. B 115, 14659–14667.

Kade, M.J., Burke, D.J., Hawker, C.J., 2010. The power of thiol-ene chemistry. J. Polym. Sci. Polym. Chem. 48, 743–750.

Kolb, H.C., Finn, M.G., Sharpless, K.B., 2001. Click chemistry: diverse chemical function from few good reactions. Angew. Chem. Int. Ed. 40, 2004–2021.

Kolb, H.C., Sharpless, K.B., 2003. The growing impact of click chemistry on drug discovery. Drug Discov. Today 8, 1128–1137.

Kwan, M.-L., Mirjafari, A., McCabe, J.R., O'Brien, R.A., Essi IV, D.F., Baum, L., West, K.N., Davis Jr., J.H., 2013. Synthesis and thermophysical properties of ionic liquids: cyclopropyl moieties versus olefins as T_m-reducing elements in lipid-inspired ionic liquids. Tetrahedron Lett. 54, 12–14.

Lopez-Martin, I., Burello, E., Davey, P.N., Seddon, K.R., Rothenberg, G., 2007. Anion and cation effects on imidazolium salt melting points: a descriptor modelling study. Chem. Phys. Chem. 8, 690–695.

Mirjafari, A., O'Brien, R.A., Murray, S.M., Mattson, K.L., Mobarrez, N., West, K.N., Davis Jr., J.H., 2012. Lipid-inspired ionic liquids containing long-chain appendages: novel class of biomaterials with attractive properties and applications. In: ACS Symp. Ser., vol. 1117, pp. 199–216.

Mirjafari, A., O'Brien, R.A., West, K.N., Davis Jr., J.H., 2014. Synthesis of new lipid-inspired ionic liquids by thiol-ene chemistry: profound solvent effect on reaction pathway. Chem. Eur. J. 20, 7576–7580.

Monney, N.P.A., Bally, T., Bhagvathy, G.S., Glass, R.S., 2013. Spectroscopic evidence for a new type of bonding between a thioether radical cation and a phenyl group. Org. Lett. 15, 4932–4935.

Moses, J.E., Moorhouse, A.D., 2007. The growing applications of click chemistry. Chem. Soc. Rev. 36, 1249–1262.

Murray, S.M., O'Brien, R.A., Mattson, K.M., Ceccarelli, C., Sykora, R.E., West, K.N., Davis Jr., J.H., 2010. The fluid-mosaic model, homeoviscous adaptation, and ionic liquids: dramatic lowering of the melting point by side-chain unsaturation. Angew. Chem. Int. Ed. 49, 2755–2758.

Murray, S.M., Zimlich, T.K., Mirjafari, A., O'Brien, R.A., Davis Jr., J.H., West, K.N., 2013. Thermophysical properties of imidazolium-based lipidic ionic liquids. J. Chem. Eng. Data 58, 1516–1522.

Naik, S.D., Doraiswamy, L.K., 1998. Phase transfer catalysis: chemistry and engineering. J. AIChE 44, 612–646.

O'Brien, R.A., Mirjafari, A., Jajam, V., Capley, E.N., Stenson, A.C., West, K.N., Davis Jr., J.H., 2011. Functionalized ionic liquids with highly polar polyhydroxylated appendages and their rapid synthesis via thiol-ene click chemistry. Tetrahedron Lett. 52, 5173–5175.

O'Brien, R.A., West, C.W., Hollingsworth, B.E., Stenson, A.C., Henderson, C.B., Mirjafari, A., Mobarrez, N., West, K.N., Mattson, K.L., Salter, E.A., Wierzbicki, A., Davis Jr., J.H., 2013. A simple and rapid route to novel tetra(4-thiaalkyl)ammonium bromides. RSC Adv. 3, 24612–24617.

O'Brien, R.A., Mirjafari, A., Mattson, K.M., Murray, S.M., Mobarrez, N., Salter, E.A., Wierzbicki, A., Davis Jr., J.H., West, K.N., 2014. The effect of the sulfur position on the melting points of lipidic 1-methyl-3-thiaalkylimidazolium ionic liquids. J. Phys. Chem. B 118, 10232–10239.

Peetla, C., Stine, A., Labhasetwar, V., 2009. Biophysical interactions with model lipid membranes: applications in drug discovery and drug delivery. Mol. Pharm. 6, 1264–1276.

Plechkova, N.V., Seddon, K.R., 2008. Applications of ionic liquids in the chemical industry. Chem. Soc. Rev. 37, 123–150.

Sheldon, R.A., 2012. Fundamentals of green chemistry: efficiency in reaction design. Chem. Soc. Rev. 41, 1437–1451.

Smiglak, M., Pringle, J.M., Lu, X., Han, L., Zhang, S., Gao, H., Macfarlane, D.R., Rogers, R.D., 2014. Ionic liquids for energy, materials, and medicine. Chem. Commun. 50, 9228–9250.

Smith, A.M., Lovelock, K.R., Gosvami, N.N., Licence, P., Dolan, A., Welton, T., Perkin, S., 2013. Monolayer to bilayer structural transition in confined pyrrolidinium-based ionic liquids. J. Phys. Chem. Lett. 4, 378–382.

Tang, S., Baker, G.A., Zhao, H., 2012. Ether- and alcohol-functionlized task-specific ionic liquids: attractive properties and applications. Chem. Soc. Rev. 41, 4030–4066.

Wasserscheid, P., Welton, T., 2008. Ionic Liquids in Synthesis, second ed. Wiley-VCH Verlag GmbH, Weinheim.

Xenopolous, A., Cheng, J., Yasuniwa, M., Wunderlich, B., 1992. Mesophases of alkylammonium salts. First order transitions. Mol. Cryst. Liq. Cryst. 214, 63–79.

Yang, M., Campbell, P.S., Santini, C.C., Mudring, A.-V., 2014. Small nickel nanoparticle arrays from long chain imidazolium ionic liquids. Nanoscale 6, 3367–3375.

Zhou, F., Liang, Y., Liu, W., 2009. Ionic liquid lubricants: designed chemistry for engineering applications. Chem. Soc. Rev. 38, 2590–2599.

Chapter 7

New Opportunities from Ionic Liquid for Chemical and Biochemical Processes of Lipids

Bianca Pérez, Wei Wei
Aarhus University, Department of Engineering, Aarhus, Denmark

7.1 INTRODUCTION

Ionic liquids (ILs) have been used to replace hazardous, volatile, and flammable organic solvents in different academy and industry applications. IL's tunable properties such as viscosity, polarity, and solubility allow their customization according to specific demands through selection of a combination of anions and cations (Moniruzzaman et al., 2010a,b,c). The countless number of cations and anions available to obtain IL permit the preparation of IL, which can even be immiscible with each other (Werner et al., 2010). Their attractive miscibility properties including electric conductivity, polarity, and nucleophilicity, mainly originate from the interplay of noncovalent interactions between the cation/anion of IL. The asymmetry of the ions prevents the salt from packing and consequently, crystallizing, leading to IL's low melting points. Although ILs are generally more expensive than organic solvents, they can be tuned to achieve its purpose and lower their cost. For example, Ku and Lu were able to develop cheap quaternary ammonium salts for the preparation of electrolytes with low volatility (Ku and Lu, 2011). Regardless of being highly tunable, IL's chemical and physical properties generally fall in a specific range (Table 7.1) (Werner et al., 2010). However, IL's countless possible combinations of cation/anion implies that significant efforts have to be made to identify/synthesize superior IL properties for a specific application. Nonetheless, ILs offer new opportunities in the industry and academic research for a wide variety of applications including synthetic chemistry (Welton, 1999; Wasserscheid and Welton, 2008), energy

Ionic Liquids in Lipid Processing and Analysis. http://dx.doi.org/10.1016/B978-1-63067-047-4.00007-6

TABLE 7.1 Chemical/Physical Property Range of Ionic Liquids (Werner et al., 2010)

Chemical/Physical Property	Typical Range of Most Ionic Liquids
Melting point/glass transition	0–60°C
Density	1.1–1.6 g/L[a]
Viscosity	40–800 mPa s[a]
Thermal stability	230–300°C[b]
Surface tension	30–50 mN/m
Heat capacity	300–400 J/mol per Kelvin
Water miscibility	Ionic liquid can be soluble or not in water
Hydrolytic stability	Hydrolysis of the heterocyclic cations can occur under extreme conditions
Base stability	All 1,3-dialkylimidazolium ionic liquids are subject to deprotonation
Corrosion	Most ionic liquids are corrosive versus Cu
Toxicity	Increasing lipophilicity leads to higher toxicity
Price	25–250 € kg^{-1}

[a]Estimation made for a production scale of 1000 kg and for a purity >98%.
[b]Thermalgravimetric analysis at 10 K/min.

conversion (MacFarlane et al., 2014), chemical engineering (Visser et al., 2001; Anderson and Armstrong, 2005), and biotechnology (van Rantwijk and Sheldon, 2007; Roosen et al., 2008).

The wide panel of opportunities offered by IL also comes with challenges (see Table 7.2). For instance, although ILs offer higher selectivity, faster reaction rates, and higher enzyme stability in biocatalysis, they also generate difficulties when it comes to product isolation and controlling pH (Park and Kazlauskas, 2003). Regardless, IL's tunable properties continuously encourage scientific efforts when it comes to the development of novel technologies. For example, IL properties have promoted the search for new materials that improve the way energy is stored, generated, or delivered (MacFarlane et al., 2014). Furthermore, IL has encouraged the development of alternative chemical and biochemical processes in academy and industry research. Although chemical processes were originally designed to control reactions, perform extraction, and process materials while avoiding chemical manufacture waste and volatile organic compound emission to the environment, the

TABLE 7.2 Ionic Liquid Opportunities and Challenges for Industrial and Academic Research

Research Field	Ionic Liquid Applications	Opportunities	Challenges
Biotechnology	Biocatalysis, biotransformation; lipase production and activation; lipids processing; synthesis of antioxidants; synthesis of sugar/carbohydrates; production of biodiesel and oleochemicals	Higher selectivity; enzyme stability; and faster reaction rates	Product isolation; controlling pH; and purification of ionic liquids
Synthetic chemistry	Organic synthesis; nanomaterial synthesis; polymerization; microwave chemistry; multiphasic reactions and extractions	Simplified catalyst recovery; no atmospheric solvent emission; can be designed to control chemistry	Expensive; lack of corrosion and toxicity data
Energy conversion	Fuel cells; batteries; heat storage; solar panels; metal plating	Enhance safety and stability	Efficiency lower than that reported for traditional solvents; less viscous ionic liquids are needed in electrochemical devices
Analysis	Matrices for mass spectrometry; gas chromatography columns; stationary phase for high-performance liquid chromatography	Tunable selectivity and high solubilizing capacity; chiral separation; withstand high temperature and high vacuum	Design of hydrophobic task-specific ionic liquids
Chemical engineering	Extraction and recovery; separation; coating; lubricants; plasticizers	Higher operational safety due to ionic liquid negligible vapor pressure	More difficult and costly to prepare than organic solvents
Pharmacy	Drug delivery methodologies; pharmaceutical solvents; active pharmaceutical ingredients	Avoid polymorphism; increase water solubility; and alter bioavailability	Only applicable for ready ionizable drugs; manufacture; scale-up; purification; toxicity

latter still comprises one-third of industrial emission worldwide. Traditional solvents used such as water, methanol, ethanol, dichloromethane, acetonitrile, toluene, and benzene possess a relatively narrow liquid region (range between the freezing point and the boiling point of a substance; eg, water liquid region is 100°C) from 75 to 200°C (Brennecke and Maginn, 2001). Thus, the chemical industry is under pressure to replace toxic/hazardous and volatile organic solvents that cause negative environmental impact by novel solvents and methodologies that "green" industrial processes (Sheldon, 2001). This has generated enormous scientific effort to develop novel "green" reaction media. In this context, ILs, perceived as nonvolatile liquids, offered novel strategies for the design of chemical and biochemical processes for industrial applications.

In addition to replacing traditional volatile solvents in chemical and biochemical processes, ILs also offer a wide range of possibilities to overcome current challenges in the pharmaceutical industry. ILs can be designed to not only be biologically active but also improve bioavailability of drugs (Shamshina and Rogers, 2014); that is, they can increase the solubility of a drug in both lipophilic and hydrophilic environments (Ferraz et al., 2011; Hough et al., 2007). This demonstrates IL prospective use for the development of novel drug delivery systems for therapeutic purposes. For instances, ILs could potentially provide novel methodologies for the delivery of lipophilic active ingredients such as essential fatty acids, which present poor aqueous solubility.

Despite increasing IL use to improve drug bioavailability, ILs have not been yet fully explored for pharmaceutical industry applications. This chapter focuses on research carried out so far, which shows the novel strategies and opportunities offered by IL for the development of drug delivery systems for lipophilic and hydrophilic active pharmaceutical ingredients (APIs).

7.2 IONIC LIQUIDS FOR DRUG DELIVERY PURPOSES

Since the first report in the literature of ILs containing 1-alkyl-3-methylimidazolium salts by Wilkes et al. in 1982, ILs have rapidly evolved from being nonvolatile compounds with thermal stability into helping tune biologically APIs (Hough and Rogers, 2007). Three generations of ILs have been identified so far (Fig. 7.1): (1) the first generation contains unique characteristics including thermal stability and low or nonvolatility, (2) the second generation generates new materials for multiple applications based on the tunable physical and chemical properties of IL, and (3) the third generation tunes API using IL to maintain the positive features of a drug while suppressing the undesired properties associated with the use of such (Hough et al., 2007). The later generation has had a significant impact in the

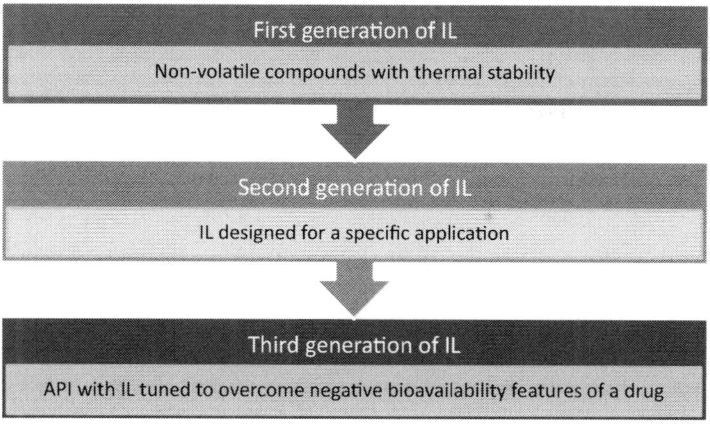

FIGURE 7.1 Evolution of ionic liquids.

pharmaceutical industry since IL has not only been shown to improve drug solubility but also biological activity (Suresh et al., 2012).

7.2.1 Synthesis and Characterization of Ionic Liquids from Active Pharmaceutical Ingredients

Nitrogen-containing heterocyclic ILs are frequently used to prepare API-IL. The preparation of API-IL is based on the chemical modification of a drug with ionizable functional groups to eliminate their undesirable physical or chemical properties. So far several cations and anions (**1**–**15**) have been used for the preparation of API-IL (Fig. 7.2) (Shamshina et al., 2013; Hough-Troutman et al., 2009). Also, different synthetic pathways have demonstrated to be effective for the preparation of API-IL: (1) metathesis reaction (Cole et al., 2011; Frizzo et al., 2013), (2) ion exchange strategies (Ferraz et al., 2012), and (3) solvent-free conditions (Bica et al., 2010; Bica and Rogers, 2010). The most common procedure for the preparation of API-IL is metathesis reactions where an anion halide reacts with an alkaline form. An example constituted the one reported by Cole et al. (2011), who carried out the preparation of a series of ILs that contained a cationic imidazolium or pyridinium and an anionic beta-lactam antibiotic. They promoted counterion exchange by the slow addition of an aqueous solution containing [Na^+ (**14**)] to a solution of [(**3**)Br^-] in chloroform. Although metathesis reactions are mostly used, solvent-free synthesis offered unique advantages over metathesis reactions (Bica et al., 2010; Bica and Rogers, 2010). Solvent-free reactions prevent the use of solvents and generate API-IL that do not contain metal, halides, or solvent traces as desired for pharmaceutical industry purposes.

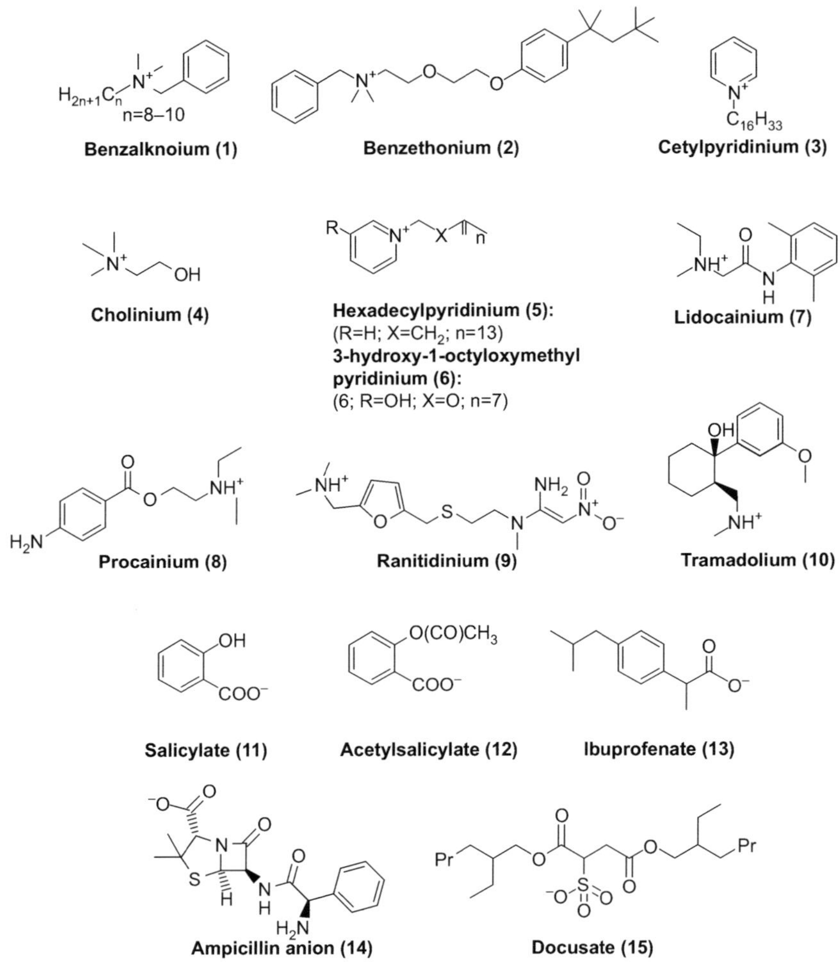

FIGURE 7.2 Examples of anions and cations **1–15** used for preparation of API-IL.

In this type of reaction, a stoichiometric mixture of the base of an ion and the acid of the counterion is melted to yield a flowing liquid.

IL chemical and physical properties including melting point, thermal stability, solubility, viscosity, and density can be characterized by different techniques:

1. Differential scanning calorimetry provides information regarding thermal properties including melting point (thermal properties of API-IL reported so far are displayed in Table 7.3); that is, generally, an IL melts below 100°C though some ILs do not present melting point but glass-transition, or glass-transition and melting point (Ngo et al., 2000).

TABLE 7.3 Thermal Properties of API-IL Reported as of 2015

API-IL	Cation	Anion	Thermal Properties	References
Benzalkonium acesulfamate	Benzalkonium (antibacterial)	Acesulfamate (noncaloric sugar)	Form: white solid m.p.: 90°C T (glass transition): N.R.	Hough-Troutman et al. (2009)
Benzalkonium saccharinate	Benzalkonium (antibacterial)	Saccharinate (noncaloric sugar)	Form: white solid m.p.: 74°C T (glass transition): N.R.	Hough-Troutman et al. (2009)
Benzalkonium salicylate	Benzalkonium (antibacterial)	Salicylate (antiinflammatory, analgesic, antipyretic)	Form: colorless solid m.p.: 98°C T (glass transition): 51°C	Bica et al. (2010)
Benzethonium acetylsalicylate	Benzethonium (antibacterial)	Acetylsalicylate (antiinflammatory, analgesic, antipyretic)	Form: colorless glass m.p.: N.R. T (glass transition): 14°C	Bica et al. (2010)
Benzethonium saccharinate	Benzethonium (antibacterial)	Saccharinate (noncaloric sugar)	Form: colorless oil m.p.: liquid at RT. T (glass transition): −4°C	Dean et al. (2009)
Benzethonium salicylate	Benzethonium (antibacterial)	Salicylate (antiinflammatory, analgesic, antipyretic)	Form: colorless glass m.p.: N.R. T (glass transition): −14°C	Bica et al. (2010)
Cetylpyridinium acetylsalicylate	Cetylpyridinium (antibacterial)	Acetylsalicylate (antiinflammatory, analgesic, antipyretic)	Form: colorless solid m.p.: N.R. T (glass transition): 3°C	Bica et al. (2010)

Continued

TABLE 7.3 Thermal Properties of API-IL Reported as of 2015—cont'd

API-IL	Cation	Anion	Thermal Properties	References
Cetylpyridinium ampicillin	Cetylpyridinium (antibacterial)	Ampicillin (antibacterial)	Form: pale yellow solid m.p.: 86°C T (glass transition): −20°C	Ferraz et al. (2012)
Cetylpyridinium salicylate	Cetylpyridinium (antibacterial)	Salicylate (antiinflammatory, analgesic, antipyretic)	Form: colorless waxy solid m.p.: 74°C T (glass transition): N.R.	Bica et al. (2010)
Cholinium ampicillin	Cholinium	Ampicillin (antibacterial)	Form: pale yellow solid m.p.: 58°C T (glass transition): −20°C	Ferraz et al. (2012)
Didecyldimethylammonium acesulfamate	Didecyldimethylammonium (antibacterial)	Acesulfamate (noncaloric sugar)	Form: viscous liquid m.p.: liquid at RT T (glass transition): −53°C	Hough-Troutman et al. (2009)
Didecyldimethylammonium ibuprofen	Didecyldimethylammonium (antibacterial)	Ibuprofenate (antiinflammatory)	Form: yellow liquid m.p.: liquid at RT T (glass transition): −73°C	Hough et al. (2007)
Didecyldimethylammonium saccharinate	Didecyldimethylammonium (antibacterial)	Saccharinate (noncaloric sugar)	Form: viscous liquid m.p.: 16°C T (glass transition): −33°C	Hough-Troutman et al. (2009)
1-ethyl-3-methylimidazolium ampicillin	1-Ethyl-3-methylimidazolium	Ampicillin (antibacterial)	Form: pale yellow solid m.p.: 72°C T (glass transition): −18°C	Ferraz et al. (2012)
Hexadecylpyridinium acesulfamate	Hexadecylpyridinium (antibacterial)	Acesulfamate (noncaloric sugar)	Form: white solid m.p.: 57°C T (glass transition): −11°C	Hough-Troutman et al. (2009)

Hexadecylpyridinium saccharinate	Hexadecylpyridinium (antibacterial)	Saccharinate (noncaloric sugar)	Form: white solid m.p.: 66°C T (glass transition): N.R.	Hough-Troutman et al. (2009)
3-hydroxy-1-octyloxymethylpyridinium acesulfamate	3-Hydroxy-1-octyloxymethylpyridinium (antimicrobial)	Acesulfamate (noncaloric sugar)	Form: colorless crystal m.p.: 79–81°C T (glass transition): N.R.	Hough-Troutman et al. (2009)
3-hydroxy-1-octyloxymethylpyridinium saccharinate	3-Hydroxy-1-octyloxymethylpyridinium (antimicrobial)	Saccharinate (noncaloric sugar)	Form: colorless crystal m.p.: 95–98°C T (glass transition): N.R.	Hough-Troutman et al. (2009)
Lidocainium acetylsalicylate	Lidocainium (pain reliever)	Acetylsalicylate (antiinflammatory, analgesic, antipyretic)	Form: colorless viscous liquid m.p.: 61°C T (glass transition): N.R.	Bica et al. (2010)
Lidocainium docusate	Lidocainium (pain reliever)	Docusate (emollient)	Form: colorless gel m.p.: liquid at RT T (glass transition): −29°C	Hough et al. (2007)
Lidocainium salicylate	Lidocainium (pain reliever)	Salicylate (antiinflammatory, analgesic, antipyretic)	Form: colorless viscous liquid m.p.: liquid at RT T (glass transition): 20°C	Bica et al. (2010)
Procainamidium salicylate	Procainamidium (antiarrhythmic)	Salicylate (antiinflammatory, analgesic, antipyretic)	Form: colorless viscous liquid m.p.: liquid at RT T (glass transition): 20°C	Bica et al. (2010)
Procainium salicylate	Procainium (local anesthetic)	Salicylate (antiinflammatory, analgesic, antipyretic)	Form: colorless viscous liquid m.p.: liquid at RT T (glass transition): 14°C	Bica et al. (2010)

Continued

TABLE 7.3 Thermal Properties of API-IL Reported as of 2015—cont'd

API-IL	Cation	Anion	Thermal Properties	References
Propantheline acesulfamate	Propantheline (antimuscarinic)	Acesulfamate (noncaloric sugar)	Form: colorless oil m.p.: liquid at RT T (glass transition): −20°C	Dean et al. (2009)
Propantheline *p*-toluenesulfonate	Propantheline (antimuscarinic)	*p*-Toluenesulfonate	Form: colorless oil m.p.: liquid at RT T (glass transition): 7°C	Dean et al. (2009)
Ranitidine docusate	Ranitidine (histamine H_2-receptor antagonist)	Docusate (emollient)	Form: dark red gel m.p.: liquid at RT T (glass transition): −12°C	Hough et al. (2007)
Tetraethylammonium ampicillin	Tetraethylammonium	Ampicillin (antibacterial)	Form: pale yellow solid m.p.: 79°C T (glass transition): −19°C	Ferraz et al. (2012)
Tetrabutylphosphonium salicylate	Tetrabutylphosphonium (antimicrobial)	Salicylate (antiinflammatory, analgesic, antipyretic)	Form: colorless crystal m.p.: 57°C T (glass transition): −56°C	Bica et al. (2010)
Tramadolium acetylsalicylate	Tramadolium (pain reliever)	Acetylsalicylate (antiinflammatory, analgesic, antipyretic)	Form: colorless viscous liquid m.p.: liquid at RT T (glass transition): −14°C	Bica et al. (2010)
Trihexyltetradecylphosphonium ampicillin	Trihexyltetradecylphosphonium	Ampicillin (antibacterial)	Form: yellow viscous liquid m.p.: liquid at RT T (glass transition): N.R.	Ferraz et al. (2012)

N.R., not reported.

2. Thermalgravimetric analysis (TGA) helps analyzing the thermal stability of a substance. Generally, thermal stability depends on the strength of the heteroatom–carbon bond and the heteroatom–hydrogen bond (Ngo et al., 2000).
3. UV–visible spectroscopy can allow the characterization of IL solubility (Cole et al., 2011). ILs can be designed to be miscible or immiscible in water or other organic solvents depending on the anions and cations used.
4. Viscosity measurements are carried out using a viscometer (Wilkes et al., 2007). IL viscosity shows correlation with compounds' propensity to form hydrogen bonds and van der Waals interactions (Frizzo et al., 2013).
5. Several methodologies are available to either measure or predict the density of an IL (Wilkes et al., 2007). IL density seems to decrease when the alkyl chain of the cation increases (Jacquemin, 2006). One model used to predict the density of IL is the cohesion factor in the cubic equations of state, which demonstrated to be effective in predicting the density of six commercially available ILs (Rabari et al., 2014).

IL's chemical and physical property differences compared to traditional pharmaceutical active salts might be explained by molecular dynamic simulations and nuclear magnetic resonance studies, which suggested IL to form nanostructure organizations in aqueous medium instead of dissolving in water as single anions (Bhargava and Klein, 2009; Zhao et al., 2008). This offers new opportunities for the development of more efficient delivery modes making it possible, for instance, to deliver an API in a more localized way, which could also avoid toxicity and metabolite of the drugs in other part of the body (Stoimenovski et al., 2010).

7.2.2 Ionic Liquid Toxicity and Bioavailability for Pharmaceutical Industry Purposes

The lack of information available regarding IL toxicity has hampered the development of API-IL. Although some IL are considered "green solvents," some others are extremely toxic (Rogers and Seddon, 2003). Nonetheless, their tunable properties offered the possibility to redesign them by adjusting the cation/anion combination in order to generate biocompatible and nontoxic ILs for pharmaceutical industry purposes. Until now IL toxicity studies based on imidazolium, pyridinium, or quaternary ammonion-IL suggest that the longer the alkyl chain in the imidazole or pyridine ring, the higher the toxicity of IL (Couling et al., 2006). The latter could be explained by the fact that increased lipophilicity might promote membrane disruption since higher amounts of compound would incorporate inside the biological membrane.

Couling et al. (2006) developed quantitative structure–property relationship models to assess structural features in the IL that might be responsible for a material toxicity to aquatic organisms. Accordingly, this model not only predicted that lipophilicity correlates with the length of the alkyl chain, but also that toxicity increases with the number of nitrogen atoms in a cation aromatic ring. They found that toxicity might increase following this trend: ammonium < pyridinium < imidazolium < triazolium < tetrazolium. In addition, results showed that toxicity might decrease with (1) ring methylation, and (2) the number of negatively charge atoms in the cation; for example, the oxygen in choline. Despite the fact that studies did not focus on the anion, results suggest that toxicity might increase to some extent with the number of positive charged atoms on the anion. Further work carried out by Ranke et al. demonstrated that increasing alkyl chain length of 3-alkoxymethyl-1-methylimidazolium and 3-alkyl-1-methylimidazolium IL cations only increases the cytotoxicity up to an alkyl chain length of 12 carbon atoms. However, correlations found are limited to hydrophilic anions nonreactive under physiological conditions, and is sufficient to the solubility of the IL in the cytotoxicity test medium (Ranke et al., 2007).

Besides toxicity, bioavailability is also an important factor taken into account for the design and evaluation of new methodologies for drug delivery. Highly ionic drugs have difficulty crossing membranes and reaching their corresponding target. However, ions pair can overcome bioavailability issues and enhance the transport of ionic drugs through membranes, which make ILs beneficial for drug delivery purposes. Thus, chemically modifying drugs using IL constitutes a highly promising approach with potential applications in the pharmaceutical industry. In addition, the tunable character of IL through the selection of a proper counterion can drastically affect API solubility either in hydrophilic or hydrophobic environment, enhancing respective drug delivery into the body. An example regarding improvement of solubility in a lipophilic environment is lidocaine docusate (**7.15**), which joins together a hydrophilic cation (lidocaine) with a hydrophobic anion (docusate) to form a compound with reduced or controlled water solubility, consequently elongating the resident time of the drug in the skin (Hough et al., 2007).

7.3 FORMULATIONS OF ACTIVE PHARMACEUTICAL INGREDIENTS WITH IONIC LIQUIDS FOR DRUG DELIVERY

The drug delivery market is estimated to be worth up to $224 billion dollars by 2017 (http://www.marketsandmarkets.com/PressReleases/drug-delivery-technologies.asp). Most active drugs do not make it to formulation for being barely soluble or insoluble in water medium; around 40% of the new biologically active drugs have poor or no solubility in water (Moniruzzaman et al., 2010a,b,c). Methods to deliver these drugs include using excipients such as ethanol, dimethylsufoxide, and other surfactants. However, to obtain significant enhanced delivery effects, high amounts of the excipients is generally needed, making the formulation rather toxic. Thus drug bioavailability is a huge obstacle in the pharmaceutical industry.

Nowadays, pharmaceutical companies mainly rely on solid API because drugs converted into salts are generally more stable and water soluble (Stahl and Wermuth, 2008; Paulekuhn et al., 2007). However, crystalline materials can exhibit polymorphism and different polymorphs may have different physicochemical properties, which can drastically change drug bioavailability. A well-known example of polymorphism constituted the million-dollar loss caused by the polymorphism of ritonavir, an HIV-protease inhibitor, whose polymorph proved to be poorly soluble in the original system developed to administrate ritonavir (Shamshina et al., 2013). A tunable, drug-specific strategy of interest relies on the attractive properties of IL. IL could be designed to enhance desired features of a pharmaceutical compound and at the same time to (1) chemically stabilize pharmaceutical compounds, (2) convert solid active drugs into liquids with benign counter-ion, (3) avoid polymorphism of crystalline materials, (4) increase water solubility, and (5) improve transdermal penetration (Sekhon, 2011). In fact, the use of IL has already generated new technologies for drug delivery such as liquefaction of API, development of a prodrug with IL, and modification of a drug's environment to improve its bioavailability.

7.3.1 Liquefaction of Drugs with Ionic Liquids

Ranitidine docusate (**9.15**) constituted one of the first examples of API-IL where a drug is liquefied to avoid its polymorphic conversion (Hough et al., 2007). Rogers's research group synthesized ranitidine docusate (**9.15**) through a metathesis reaction between ranitidine hydrochloride (an antagonist of histamine H_2-receptor) and sodium docusate (an emollient). They were able to demonstrate that a polymorphic API could be converted into a room-temperature IL. In this work, they also synthesized didecyldimethylammonium ibuprofen (**16.13**) and lidocaine docusate (**7.15**). The former demonstrated that the anion could be the API rather than the cation. This compound was synthesized using as starting materials didecyldimethylammonium bromide (an antibacterial) and sodium ibuprofen (an antiinflammatory) (Scheme 7.1). In the case of lidocaine docusate, also liquid at room temperature, Hough et al. showed that the API-IL generated a longer antinociceptive effect than the original drug (lidocaine hydrochloride), suggesting an alternative mechanism of action.

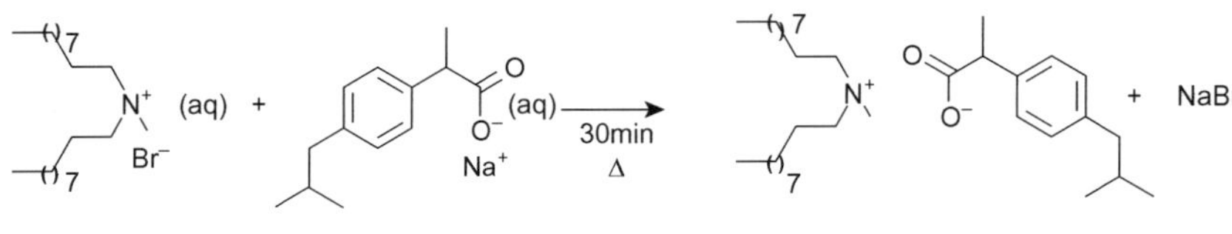

(16) Br⁻ (13) Na⁺ (16.13)

SCHEME 7.1 Preparation of didecyldimethylammonium ibuprofen (**16.13**).

Studies about the bioavailability of lidocaine docusate (**7.15**) are currently ongoing (in 2015).

Development of API-IL has demonstrated multiple times to be a promising strategy to overcome polymorphism, solubility, and drug delivery problems; increase efficacy; and add functionality (Rogers et al., 2010, 2012a,b). Ferraz et al. (2012) prepared several API-IL using ampicillin **14** as an anion in combination with the following cations: 1-ethyl-3-methylimidazolium **17**, 1-hydroxy-ethyl-3-methylimidazolium **18**, choline **4**, tetraethylammonium **19**, cetylpyridinium **3**, and trihexyltetradecylphosphonium **20**, since these cations can significantly influence ampicillin solubility and membrane permeability. All cations except for **20** were selected based on their low toxicity. Ferraz et al. were able to demonstrate that API-IL properties such as thermal stability, solubility, and melting point could significantly depend on the cation used to yield the IL. API-IL (**4.14**) provided the most interesting results for medicinal purposes since it presented low melting and toxicity and high water solubility and biocompatibility.

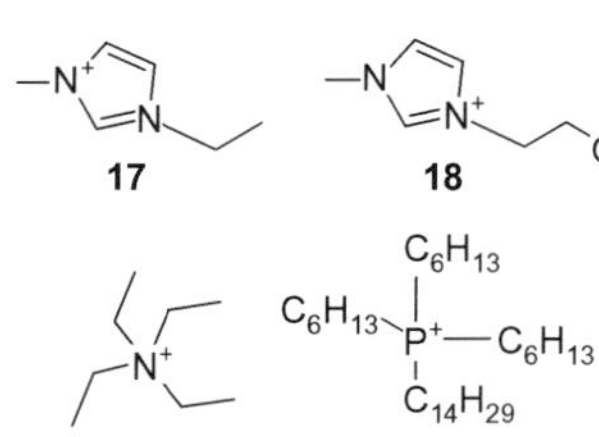

Aiming to improve solubility of aspirin and circumvent gastrointestinal irritation by current aspirin formulation, Bica et al. (2010) successfully prepared a series of ILs of aspirin. They obtained a series of salts using acetylsalycilate or sacylate in combination with different antibacterial and antimicrobial cations such as tetrabutylphosphonium **21**, cetylpyridinium **3**, benzethonium **2**, benzalkonium **1**, hexetidinium **22**, and also tramadolium **10** (analgesic), lidocainium **7** (a local anesthetic), procainium **8**, and procainiumamide **23** (an antiarrhythmic). Though several dual functioning liquid salts were synthesized and characterized, they present poor stability when exposed to moisture during long-term storage, which potentially limits their use as therapeutic agents.

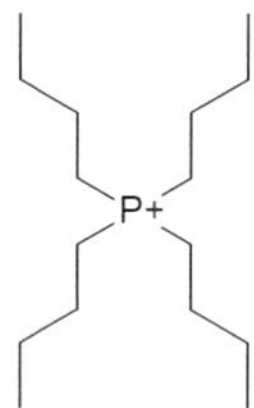

21

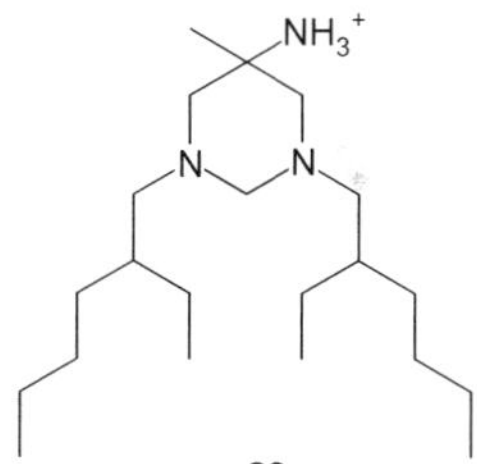

22

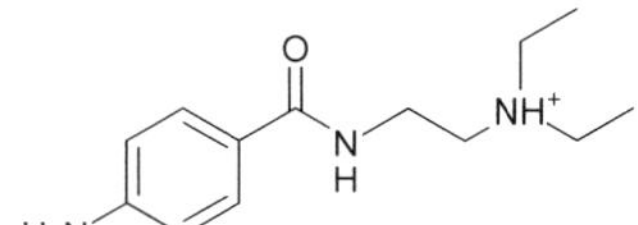

23

Encouraged by the liquefaction approach, Suresh and coworkers used betulinic acid **24** as the biologically active component and synthesized different ionic derivatives that presented not only improved water solubility but also enhanced biological activity (Suresh et al., 2012). They investigated the antitumor activity of four betulinic acid ionic derivatives (Fig. 7.3) against melanoma (A375), neuroblastoma (SH-SY5Y), breast adenocarcinoma (MCF7), and epidermoid carcinoma (A431). Betulinic acid IL **25–28** showed higher water solubility and in vitro activity against different cell lines than their pattern drug, betulinic acid. In fact, [choline][betulinate] **28** was 3–4 times more active than betulinic acid and 100 times more water soluble. Although ionic derivatives presented melting points higher than 100°C, which do not allow them to classify as API-IL, results demonstrate the potential the use of the liquefaction approach to obtain API with improved bioavailability (Zhao et al., 2012).

7.3.1.1 Anticrystal Engineering Approach

An approach that aimed to not only liquefy drugs but also narrow the search of cations and anions for the preparation of active IL is the anticrystal engineering approach introduced by Dean et al. (2009). To illustrate the concept, Dean et al. prepared a series of API-IL using different anions **2**, **4**, **29–33** and cations **34–36** (Fig. 7.4) selected from the API salts and the Generally Recognized As Safe list. Results suggested that anions capable of forming anion dimers are not useful building blocks for the preparation of IL. For instance, they found that anion **29** has a tendency to promote anion–anion interactions that led to melting points higher than 100°C. In addition, results showed that cations and anions containing hydrogen bond donors and acceptors yield crystalline salts. In fact, choline phenytoin (**4.30**),

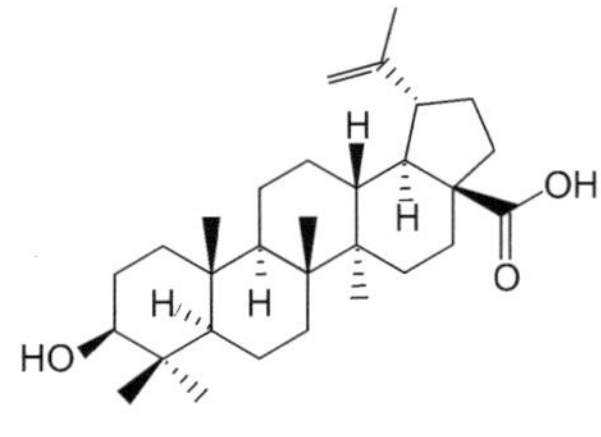

24

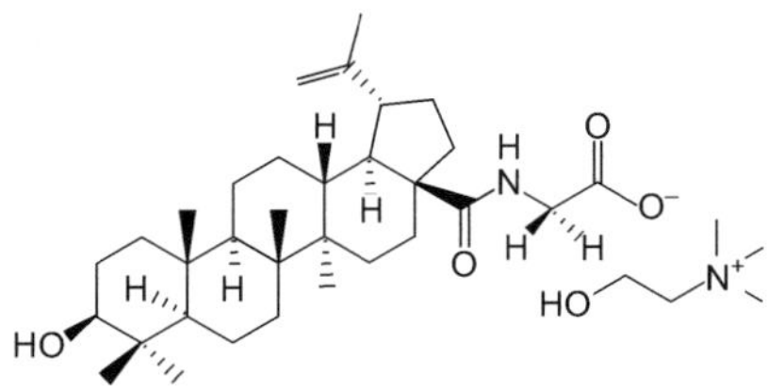

Cholinium salt of betulinic acid-glycine (25)

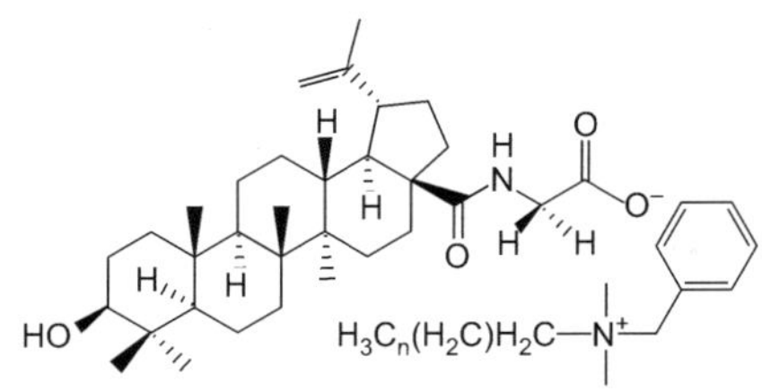

Benzalkonium salt of betulinic acid-glycine n=10, 12, 14 (26a-c)

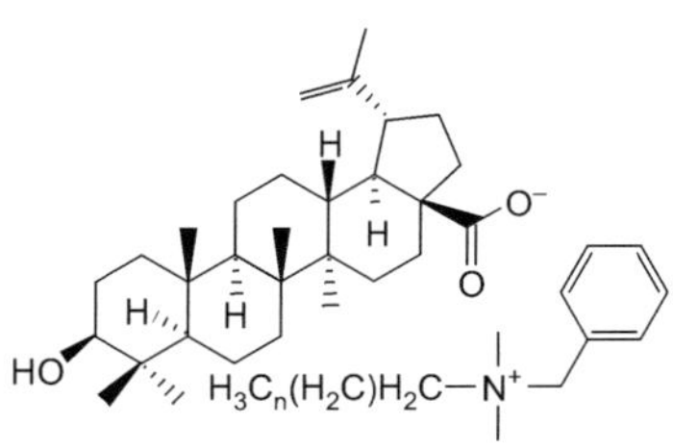

[benzalkonium][betulinate] (27)

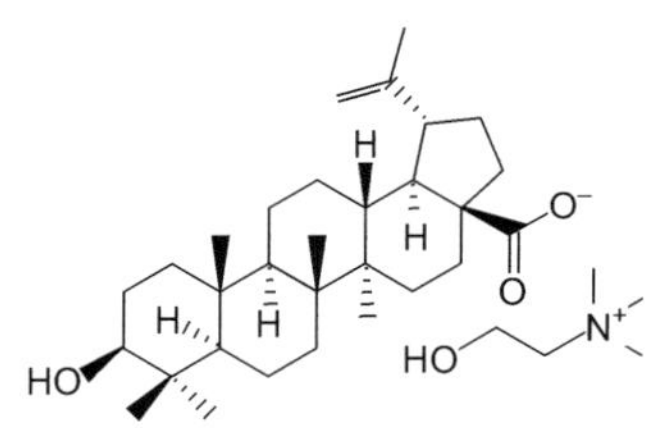

[cholinium][betulinate] (28)

FIGURE 7.3 Betulinic acid and ionic derivatives (Suresh et al., 2012).

which presented more hydrogen bond interactions, was the crystalline salt with the highest melting point. Nonetheless, there were some salts that lacked cation/anion capable of forming hydrogen bond interactions and were still crystalline salt. However, Dean et al., using the anticrystal engineering approach, were able to identify three different API-IL liquids at room temperature, namely, propantheline acesulfamate (**35.33**), propantheline *p*-toluenesulfonate (**35.31**), and benzethonium saccharinate (**2.32**).

7.3.1.2 Oligomeric Active Pharmaceutical Ingredient–Ionic Liquid

A nonstoichiometric approach used to liquefy drugs using IL is the preparation of oligomeric API-IL. Oligomeric API-ILs contain hydrogen-bonded cation/anion and nonionized original pattern drugs (Bica et al., 2010). In

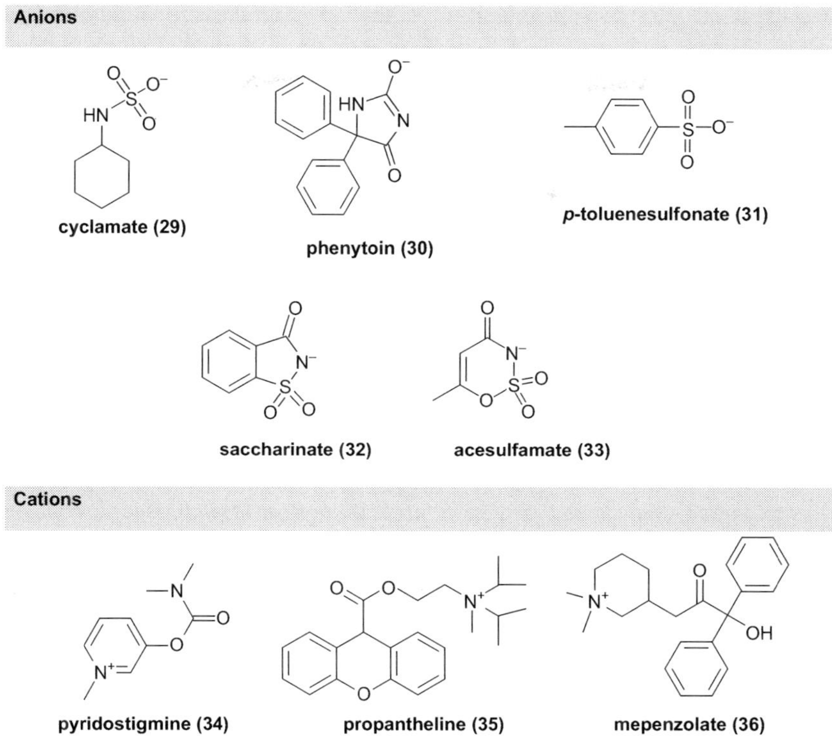

FIGURE 7.4 Most cations and anions used to prepare IL by Dean et al. to illustrate the anti-crystal engineering concept.

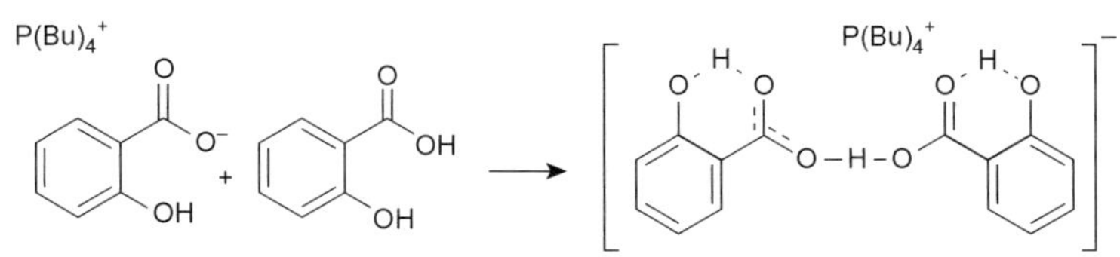

SCHEME 7.2 Oligomeric API-IL concept based on salicylic acid/salicylate proposed by Bica et al.

oligomeric API-IL, a proton is shared between two or more cationic and anionic molecules. This leads to permanent delocalization of the proton (see Scheme 7.2) and spread of the molecule charge over a larger area, preventing the salt from crystallizing. However, since API-IL are partially to fully ionized liquids, it is debatable to apply the concept of IL unless the non-covalently bonded species behave as a single ion. Bica et al. synthesized and characterized tetrabutylphosphonium salicylate (**21.11**) using salicylic acid and an aqueous solution of tetrabutylphosphonium hydroxide, which yielded a solid with a melting point of 57°C. However, the addition of 0.1 eq excess of salicylic acid led to a decrease in the melting point to 48°C; and 0.4 eq excess of salicylic acid led to a complete elimination of the melting point, which yielded a flowing liquid. The fast proton exchange could explain the melting behavior observed. Furthermore, TGA showed that increasing the excess of salicylic acid led to a reduction of the thermal stability of the obtained material. In addition, Bica et al. observed a drop in conductivity and an increase in viscosity, which could suggest that the liquefaction of tetrabutylphosphonium salicylate (**21.11**) promote a loss of the ionic property of the material. Regardless, the oligomeric API-IL approach proved to be a valuable tool to expand the liquid range of solid salts.

7.3.2 Prodrug with Ionic Liquids

The prodrug approach has also proven to be highly useful since prodrugs account for almost one-third of all small molecular weight compounds approved by 2008 (Rautio, 2010). API can be covalently modified with a prodrug capable of releasing the desired active ingredient through a metabolic process once they enter the body (Fig. 7.5). Prodrug approach offered differing advantages; for instance, it (1) prevents premature metabolism of the drug, (2) controls release rate of the drug, and/or (3) reduces drug toxicity until the drug reaches its target into the body. A further advantage of preparing a prodrug using IL is that the counter ions can be designed to provide a second function to the original drug generating, for instance, a dual action prodrug and/or an enhancer for transdermal penetration (Stoimenovski and MacFarlane, 2011). In this context, ILs offer new opportunities for the development of novel prodrugs for overcoming the negative features associated with an API, although prodrugs with IL strategy are only useful for readily ionizable pharmaceutical ingredients (Cojacaru et al., 2013).

Cojacaru et al. (2013) combined prodrug strategy with IL by adding functional groups that are charged or ionizable. The pharmaceutical compound selected was acetaminophen, an inexpensive and highly used drug whose phenolic group is not ionizable but can be functionalized to prepare an API-IL by esterification. However, since this study only aimed at proof of

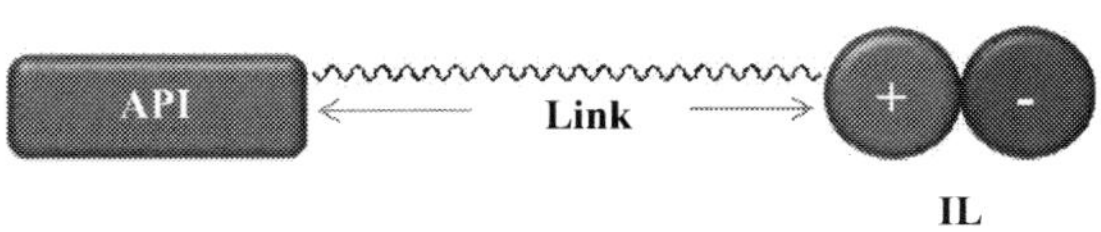

FIGURE 7.5 Prodrug with IL approach.

concept, they disregarded the potential suitability of the resulting API-IL for therapeutic use. Cojacaru et al. used cloroacetyl chloride as the linker for esterification of acetaminophen and docusate anion as the counter ion. The resultant chlorosubstituted compound was alkylated with four different IL precursors including N-methylimidazole, N-methylpyrrole, pyridine, and tributylphosphine. In total, four chloride salts (**37**−**40**) and docusate IL salts (**41**−**44**) were synthesized (see Scheme 7.3). To study if the resultant prodrug cleavage once delivered into the body, they selected chloride salt **38** and docusate IL salt **42** for analysis in three different physiological environments (pH = 7.4; pH = 6.8; pH = 1.2). Results demonstrated that in this case higher water solubility is not a requirement for the proper prodrug delivery into the body since prodrug cleavage was observed for both compounds **38** and **42**. Contrary to prodrug **38**, which presented a full release after 210 min at the three different physiological pHs, IL prodrug **42** showed that its hydrolysis depends on pH with the faster release rate observed at pH 1.2 with only 88% drug release after 210 min. Results for IL prodrug **42** suggest that the use of IL prodrugs might allow controlled release of the desired drug.

Ionizable drugs with functional groups such as carboxylic acid or amine might facilitate the design of IL prodrugs by incorporating active drugs either as the cation, anion, or both. Results so far offer new windows into the development of novel drug delivery systems with potential use in the pharmaceutical industry.

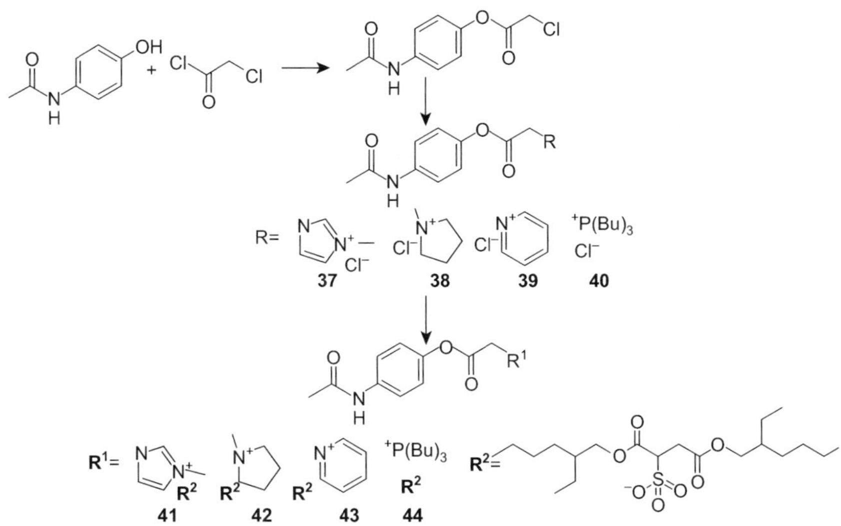

SCHEME 7.3 Synthetic pathway developed by Cojacaru et al. to obtain chloride salts **37**−**40** and IL salts **41**−**44**.

7.3.3 Solvation of Active Pharmaceutical Ingredient with Ionic Liquids

The tunable properties of IL can also facilitate the design of solvents able to solvate specific API for drug delivery; that is, IL could prevent charged API from returning to its less soluble original neutral state once they enter into the body (Mizuuchi et al., 2008; Jaitely et al., 2008; Mooney et al., 1981a,b). This could omit the need to alter the chemical structure of a drug for a more efficient delivery into the site of action. Also, this could potentially enhance the concentration of drugs for delivery and/or enable the preparation of a stable solution for drug delivery purposes (Shamshina et al., 2013). An example of IL that met the structural requirements to solubilize a poorly water soluble API constitute IL synthesized and characterized by McCrary et al. (2013). They structurally modified IL to yield hydrophilicity in one ion and lipophilicity in the other to solubilize amphiphilic drugs, namely, amphotericin B **45** and itraconazole **46**. Accordingly, they solubilized compounds **45** and **46** in IL by tuning the cation hydrophobicity of acetate-based IL and anion hydrophobicity of polyethylene glycol-based IL, correspondingly. Consequently, API **45**–**46** solubilized in respectively tuned IL displayed enhanced water solubility demonstrating that IL can be used as excipients.

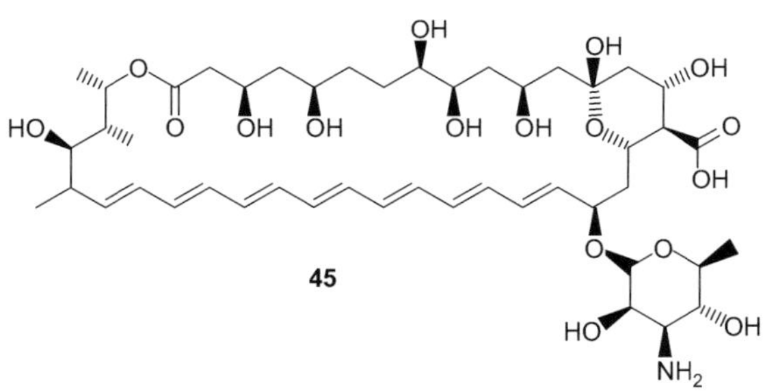

45

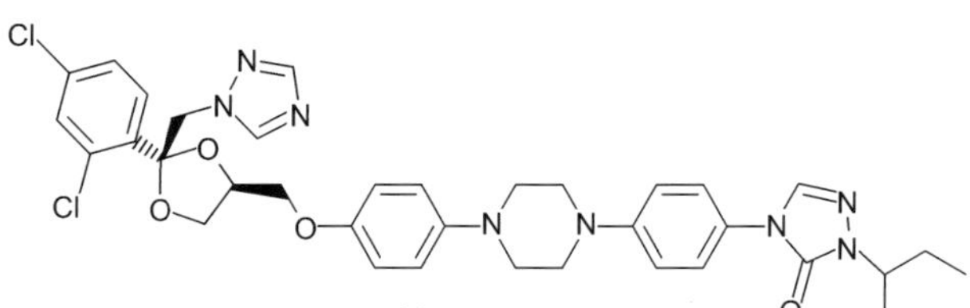

46

7.3.3.1 Microemulsions with Ionic Liquids

Another example of using the solvation strategy to delivery API involves microemulsions. Microemulsions are thermodynamically stable micro-heterogeneous and macroscopically homogeneous mixtures containing a polar phase and an apolar phase stabilized by a surfactant, with or without the need of a cosurfactant (Fanun, 2008). Microemulsion size, shape, and formation mechanisms can be studied by freeze-fracture electron microscopy, dynamic light scattering, conductivity study, UV−vis spectroscopy, and small-angle neutron scattering, among other techniques (Moniruzzaman et al., 2010a,b,c). Regardless of being called microemulsions, their droplet size is in the magnitude of ~10 nm. Microemulsions' high industry and academic research interest relied on their low interfacial tension, high thermodynamic stability, large interfacial area, and ability to solubilize immiscible liquids. Microemulsions have been used in a wide range of fields, including biotechnology. For instance, they can stabilize catalytic enzymes (Moniruzzaman et al., 2010a,b,c). Also, they can aid drug delivery in food, pharmaceutical applications, and in the petroleum industry, by accommodating insoluble or sparely soluble drugs. Microemulsions used as drug carriers cannot include TX-100 as a surfactant or cyclohexane, toluene, benzene, and dodecane as the nonpolar phase because these components are not acceptable for pharmaceutical purpose (Moniruzzaman et al., 2010a,b,c). In this context, IL, thermodynamically stable, nonvolatile, tunable solvents can potentially replace undesirable volatile organic solvents to develop drug carriers for therapeutic use (Kunz et al., 2012).

Moniruzzaman's group studied IL microemulsions capable of increasing the solubility of acyclovir **47**, a sparingly soluble drug, and increasing its topical and transdermal administration (Moniruzzaman et al., 2010a,b,c). The microemulsions contained isopropyl myterates as the apolar phase, a blend of nontoxic surfactants, namely, Tween 80 (polyoxyethylene sorbitan monooleate) and Span 20 (sorbitan laurate), and the IL dimethylimidazolium dimethylphosphate as a pseudophase since low surfactant content favors drug transdermal delivery. Accordingly, the ratio 3:2 of Tween 80 to Span 20 showed the best solubility of the drug and enhancement of acyclovir skin permeation. The drug permeability was determined using the Yucatan micropig porcine skin (Fujii et al., 1997). Results demonstrated that the use of IL microemulsions significantly increased acyclovir delivery. In addition, the carriers presented low cytotoxicity in vitro based on the reconstructed human epidermal model LabCyte™ EPI-MODEL12 (Katoh et al., 2009). Based on results, IL nonaqueous microemulsions can be used as carriers for sparingly soluble drugs for transdermal or topical administration. Moreover, Moniruzzaman et al. (2010a,b,c) demonstrated that IL containing anions capable of establishing hydrogen bond interactions (eg, $(CH_3CO)_2\ PO_2^-$ and CH_3COO^-) formed microemulsion droplets more effectively since they promote the interaction between the two surfactants (Tween 80 and Span 20). In addition to acyclovir, they also investigated the solubility of methotrexate **48** and 1-[(5-(*p*-nitrophenyl) furfurylidene) amino] hydantoin sodium **49**, other sparingly soluble drugs in most conventional solvents. Results demonstrated a higher degree of solubilization of the tested drugs, which suggests that microemulsions with IL offer new opportunities to develop drug carriers for compounds with low solubility.

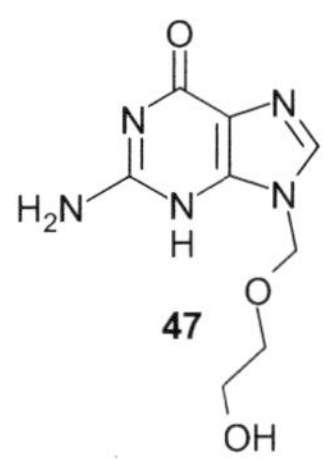

47

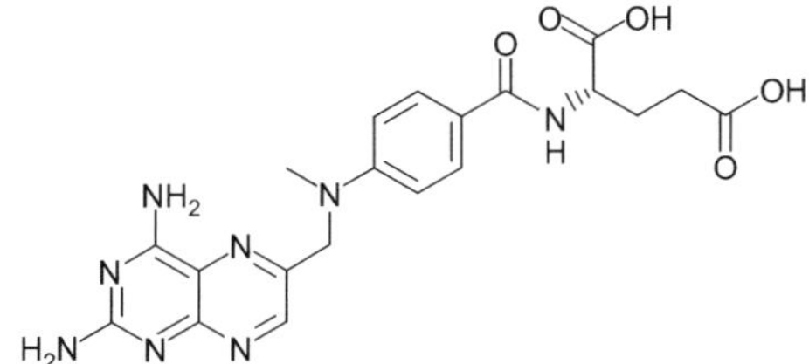

48

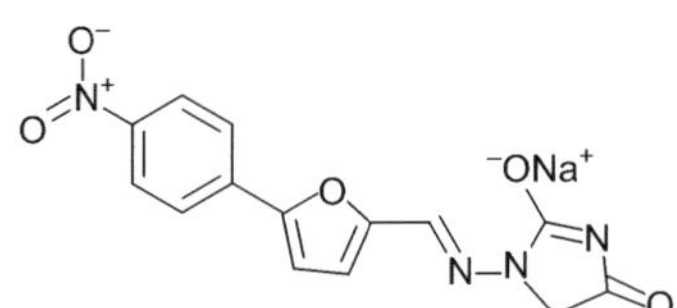

49

An interesting tool that is of valuable use to determine compound solubility in a specific IL is COSMO-RS (Zhou et al., 2014). This tool relies on density functional theory calculations of the IL and the solute molecules. However, the unknown toxicity of most ILs currently available hampers the use of IL solvents as an approach to enhance drug efficacy in the body. Developing strategies for the identification of nontoxic ILs could promote the use of ILs as future excipients. Nonetheless, the potential of ILs for drug delivery purposes remains undeniable.

7.4 SUMMARY AND CONCLUSIONS

ILs constitute a novel generation of room-temperature liquids that have demonstrated not only to be capable of offering "greener" or alternative chemical processes but have also shown to be potential biological ingredients for therapeutic use. Although additional testing is required to fully understand IL behavior once it enters into the body, the wide range of possibilities/applications that ILs offer to the pharmaceutical industry for the development of drug delivery systems is undeniable. Furthermore, ILs have been shown to overcome bioavailability challenges, offering novel strategies with potential therapeutic use to solubilize biologically active ingredients in hydrophilic and hydrophobic environments (Ferraz et al., 2011; Hough et al., 2007). On one hand, ILs can help hydrophobic compounds to overcome their poor or nonaqueous solubility, and on the other hand, ILs can also assist hydrophilic compounds, which though soluble in aqueous environment are poorly absorbed due to reduced membrane permeability. The former IL ability is of particular interest for lipids, since these IL systems could potentially offer novel methodologies for the delivery of essential fatty acids/biologically active lipids into the body. The lack of knowledge regarding IL toxicity might be hampering this process but the tunable properties of IL continue to offer pathways to overcome toxicity, allowing the potential use of these liquids as edible agents. The rapid evolution of ILs demonstrates the prospective daily use of ILs in academic research and different industrial applications.

REFERENCES

Anderson, J.L., Armstrong, D.W., 2005. Anal. Chem. 77, 6453–6462.

Bhargava, B.L., Klein, M.L., 2009. Initial stages of aggregation in aqueous solutions of ionic liquids: molecular dynamics studies. J. Phys. Chem. B 113, 9499–9505.

Bica, K., Rijksen, C., Nieuwenhuyzen, M., Rogers, R.D., 2010. In search of pure liquid salt forms of aspirin: ionic liquid approaches with acetylsalicylic acid and salicylic acid. Phys. Chem. Chem. Phys. 12, 2011–2017.

Bica, K., Rogers, R.D., 2010. Confused ionic liquid ions-a "liquification" and dosage strategy for pharmaceutically active salts. Chem. Commun. 46, 1215–1217.

Brennecke, J.F., Maginn, E.J., 2001. Ionic liquids: innovative fluids for chemical processing. AIChE J. 47, 2384–2389.

Cole, M.R., Li, M., El-Zahab, B., Janes, M.E., Hayes, D., Warner, I.M., 2011. Design, synthesis, and biological evaluation of β-lactam antibiotic-based imidazolium- and pyridinium-type ionic liquids. Chem. Biol. Drug Des. 78, 33–41.

Cojocaru, O.A., Bica, K., Gurau, G., Narita, A., McCrary, P.D., Shamshina, J.L., Barbera, P.S., Rogers, R.D., 2013. Prodrug ionic liquids: functionalizing neutral active pharmaceutical ingredients to take advantage of the ionic liquid form. Med. Chem. Commun. 4, 559.

Couling, D.J., Bernot, R.J., Docherty, K.M., Dixona, J.K., Maginn, E.J., 2006. Assessing the factors responsible for ionic liquid toxicity to aquatic organisms via quantitative structure–property relationship modeling. Green Chem. 8, 82–90.

Dean, P.M., Turanjaninm, J., Yoshizawa-Fujita, M., MacFarlane, D.R., Scott, J.L., 2009. Exploring an anti-crystal engineering approach to the preparation of pharmaceutically active ionic liquids. Cryst. Growth Des. 9, 1137–1145.

Fanun, M., 2008. Microemulsions: Properties and Applications (Surfactant Sciences). CRC Press.

Ferraz, R., Branco, L.C., Marrucho, I.M., Araújo, J., Rebelo, M., Ponte, L.P.N., Prudêncio, M.N., Noronha, C., Petrovski, J.P.Z., 2012. Development of novel ionic liquids based on ampicillin. Med. Chem. Commun. 3, 494–497.

Ferraz, R., Branco, L.C., Prudencio, C., Noronha, J.P., Petrovski, Z., 2011. ChemMedChem. 6, 975–985.

Frizzo, C.P., Tier, A.Z., Moreira, D.N., Gindri, I.M., Buriol, L., Martins, M.A.P., 2013. Pharmaceutical salts: solids to liquids by using ionic liquid design. INTECH Open Access Publisher.

Fujii, M., Yamanouchi, S., Hori, N., Iwanaga, N., Kawaguchi, N., Matsumoto, M., 1997. Evaluation of Yucatan micropig skin for use as an in vitro model for skin permeation study. Biol. Pharm. Bull. 20, 249–254.

Hough-troutman, W.L., Smiglak, M., Griffin, S., Reichert, W.M., Mirska, I., Jodynis-liebert, J., Adamska, T., Nawrot, J., Stasiewicz, M., Rogers, R.D., Pernak, J., 2009. Ionic liquids with dual biological function: sweet and anti-microbial, hydrophobic quaternary ammonium-based salts. New J. Chem. 33, 26–33.

Hough, W.L., Rogers, R.D., 2007. Ionic liquids then and now: from solvents to materials to active pharmaceutical ingredients. Bull. Chem. Soc. Jpn. 80, 2262–2269.

Hough, W.L., Smiglak, M., Rodríguez, H., Swatloski, R.P., Spear, S.K., Daly, D.T., Pernak, J., Grisel, J.E., Carliss, R.D., Soutullo, M.D., Davis Jr., J.H., Rogers, R.D., 2007. The third evolution of ionic liquids: active pharmaceutical ingredients. New J. Chem. 31, 1429–1436.

Jaitely, V., Karatas, A., Florence, A.T., 2008. Water-immiscible room temperature ionic liquids (RTILs) as drug reservoirs for controlled release. Int. J. Pharm. 354, 168–173.

Jacquemin, J., Husson, P., Padua, A.A.H., Majer, V., 2006. Density and viscosity of several pure and water-saturated ionic liquids. Green Chem. 8, 172–180.

Katoh, M., Hamajima, F., Ogasawara, T., Hata, K., 2009. Assessment of human epidermal model LabCyte EPI-MODEL for in vitro skin irritation testing according to European Centre for the Validation of Alternative Methods (ECVAM)—validated protocol. J. Toxicol. Sci. 34, 327–334.

Ku, S., Lu, S., 2011. Inexpensive room temperature ionic liquids for low volatility electrolytes of dye-sensitized solar cells. Int. J. Electrochem. Sci. 6, 5219–5227.

Kunz, W., Zemb, T., Harrar, A., 2012. Curr. Opin. Colloid Interface Sci. 2012 (17), 205–211.

MacFarlane, D.R., Tachikawa, N., Forsyth, M., Pringle, J.M., Howlett, P.C., Elliott, G.D., Davis, J.H., Watanabe, M., Simon, P., Angellg, C.A., 2014. Energy applications of ionic liquids. Energy Environ. Sci. 7, 232–250.

McCrary, P.D., Beasley, P.A., Gurau, G., Narita, A., Barber, P.S., Cojocarua, O.A., Rogers, R.D., 2013. Drug specific, tuning of an ionic liquid's hydrophilic–lipophilic balance to improve water solubility of poorly soluble active pharmaceutical ingredients. New J. Chem. 37, 2196–2202.

Mizuuchi, H., Jaitely, V., Murdan, S., 2008. Room temperature ionic liquids and their mixtures: potential pharmaceutical solvents. Eur. J. Pharm. Sci. 33, 326–331.

Moniruzzaman, M., Tamuraa, M., Taharaa, Y., Kamiya, N., Gotoa, M., 2010a. Ionic liquid-in-oil microemulsion as a potential carrier of sparingly soluble drug: characterization and cytotoxicity evaluation. Int. J. Pharm. 400, 243–250.

Moniruzzaman, M., Kamiya, N., Goto, M., 2010b. Activation and stabilization of enzymes in ionic liquids. Org. Biomol. Chem. 8, 2887–2899.

Moniruzzaman, M., Kamiya, N., Goto, M., 2010c. Ionic liquid based microemulsion with pharmaceutically accepted components: formulation and potential applications. J. Colloid Interf. Sci. 352, 136–142.

Mooney, K.G., Mintun, M.A., Himmelstein, K.J., Stella, V.J., 1981a. Dissolution kinetics of carboxylic acids I: effect of pH under unbuffered conditions. J. Pharm. Sci. 70, 13–21.

Mooney, K.G., Mintun, M.A., Himmelstein, K.J., Stella, V.J., 1981b. Dissolution kinetics of carboxylic acids II: effect of buffers. J. Pharm. Sci. 70, 22–32.

Ngo, H.L., LeCompte, K., Hargens, L., McEwen, A.B., 2000. Thermal properties of imidazolium ionic liquids. Thermochim. Acta 357–358, 97–102.

Park, S., Kazlauskas, R.J., 2003. Biocatalysis in ionic liquids – advantages beyond green technology. Curr. Opin. Biotech. 14, 432–437.

Paulekuhn, G.S., Dressman, J.B., Saal, C., 2007. Trends in active pharmaceutical ingredient salt selection based on analysis of the Orange Book database. J. Med. Chem. 50, 6665–6672.

Rabari, D., Patel, N., Joshipura, M., Banerjee, T., 2014. Densities of six commercial ionic liquids: experiments and prediction using a cohesion based cubic equation of state. J. Chem. Eng. Data. 59, 571–578.

Ranke, J., Müller, A., Bottin-Weber, U., Stock, F., Stolte, S., Arning, J., Störmann, R., Jastorff, B., 2007. Lipophilicity parameters for ionic liquid cations and their correlation to in vitro cytotoxicity. Ecotoxicol. Environ. Saf. 67, 430–438.

Rautio, J., 2010. Prodrug Strategies in Drug Design, in Prodrugs and Targeted Delivery: Towards Better ADME Properties. Wiley-VCH Verlag GmbH & Co. KgaA, Weinheim.

Rogers, R.D., Seddon, K.R., 2003. Ionic Liquids–Solvents of the future? Science 302, 792–793.

Rogers, R.D., Daly, D., MacFarlane, D., Scott, J.L., Seddon, K.R., Gurau, G., Bica, K., Turanjanin, J., Dean, P.M., 2012a. Dual functioning ionic liquids and salts thereof. PCT Int. Appl. WO 2010/078300 (2010), US2012/0046244 A1.

Rogers, R.D., Daly, D., Swatloski, R., Hough, W., Davis, J., Smiglak, M., Pernak, J., Spear, S., 2012b. Multifunctional Ionic Liquid Compositions for Overcoming Polymorphism and Imparting Improved Properties for Active Pharmaceutical, Biological, Nutritional and Energetic Ingredients. U.S. Patent No. 8,232,265; PCT/US2006/039454; WO 2007/044693 A2 (2007); MX 301158.

Rogers, R.D., Rijksen, C., Daly, D.T., Caldwell, K., Caldwell, G., Hough-Troutman, W.L., Bica, K., 2010. Compounds comprising two or more biologically functional ions and method of treating Parkinson's disease. PCT Int. Appl. WO 2010078258 A1 20100708.

Roosen, C., Müller, P., Greiner, L., 2008. Ionic liquids in biotechnology: applications and perspectives for biotransformations. Appl. Microbiol. Biotechnol. 81, 607–614.

van Rantwijk, F., Sheldon, R.A., 2007. Biocatalysis in ionic liquids. Chem. Rev. 107, 2757–2785.

Sekhon, B.S., 2011. Ionic liquids: pharmaceutical and biotechnological applications. Asian J. Phar. Biol. Res. 1, 395–411.

Shamshina, J.L., Barber, P.B., Rogers, R.D., 2013. Ionic liquids in drug delivery. Expert Opin. Drug Deliv. 10, 1367–1381.

Shamshina, J.L., Rogers, R.D., 2014. Overcoming the problems of solid state drug formulations with ionic liquids. Ther. Deliv. 5, 489–491.

Sheldon, R., 2001. Catalytic reactions in ionic liquids. Chem. Commun. 2399–2407.

Stahl, P.H., Wermuth, C.G., 2008. Handbook of Pharmaceutical Salts; Properties, Selection, and Use. Verlag Helvetica ChimicaActa, Zurich.

Stoimenovski, J., MacFarlane, D.R., 2011. Enhanced membrane transport of pharmaceutically active protic ionic liquids. Chem. Commun. 47, 11429–11431.

Stoimenovski, J., MacFarlane, D.R., Bica, K., Rogers, R.D., 2010. Crystalline vs. ionic liquid salt forms of active pharmaceutical ingredients: a position paper. Pharm. Res. 27, 521–526.

Suresh, C., Zhao, H., Gumbs, A., Chetty, C.S., Bose, H.S., 2012. New ionic derivatives of betulinic acid as highly potent anti-cancer agents. Bioorg. Me. Chem. Lett. 22, 1734–1738.

Visser, A.E., Swatloski, R.P., Griffin, S.T., Hartman, D.H., Rogers, R.D., 2001. Liquid/liquid extraction of metal ions in room temperature ionic liquids. Sep. Sci. Technol. 36, 785–804.

Wasserscheid, P., Welton, T., 2008. Ionic Liquids in Synthesis, second ed. WILEY-VCH.

Welton, T., 1999. Chem. Rev. 99, 2071–2084.

Werner, S., Haumann, M., Wasserscheid, P., 2010. Ionic liquids in chemical engineering. Annu. Rev. Chem. Biomol. Eng. 1, 203–230.

Wilkes, J.S., Levisky, J.A., Wilson, R.A., Hussey, C.L., 1982. Dialkylimidazolium chloroaluminate melts: a new class of room-temperature ionic liquids for electrochemistry, spectroscopy and synthesis. Inorg. Chem. 21, 1263–1264.

Wilkes, J.S., Wasserscheid, P., Welton, T., 2007. Ionic Liquids in Synthesis. Wiley-VCH, Weinheim.

Zhao, Y., Gao, S., Wang, J., Tang, J., 2008. Aggregation of ionic liquids [C(n)mim]Br ($n = 4, 6, 8, 10, 12$) in D2O: a NMR study. J. Phys. Chem. B 112, 2031–2039.

Zhao, H., Holmes, S.S., Baker, G.A., Challa, S., Bose, H.S., Song, Z., 2012. Ionic derivatives of betulinic acid as novel HIV-1 protease inhibitors. J. Enzym. Inhib. Med. Chem. 27, 715–721.

Zhou, T., Qi, Z.W., Sundmacher, K., 2014. Model-based method for the screening of solvents for chemical reactions. Chem. Eng. Sci. 115, 177–185.

Chapter 8

Ionic Liquids in Acylglycerol Synthesis and Modification

Derya Kahaveci
Yeditepe University, Department of Food Engineering, Istanbul, Turkey

Nanjing Zhong
Guangdong Pharmaceutical University, School of Food Science, Zhongshan, China

Xuebing Xu
Wilmar Global Research and Development Center, Shanghai, China

8.1 INTRODUCTION

Acylglycerols, which are the esters between glycerol and free fatty acids, include monoacylglycerols (MAG), diacylglycerols (DAG), and triacylglycerols (TAG). Natural oils and fats are primarily composed of TAG (more than 95% in general). Therefore, the processing manipulation of oils and fats to produce MAG and DAG has been the target in studies as well as in industry. There are synthetic studies or industrial practice to produce TAG from different materials. The use of ionic liquids (ILs) in TAG synthesis has not received substantial attention. For these concerns, this chapter is focused on the processing of MAG and DAG.

The processing of MAG and DAG can be conducted either by chemical or enzymatic approaches. For the majority of commercial operations, chemical production of partial acylglycerols is commonly used. However, a huge amount of research work has been dedicated to the use of enzymes as catalysts. In particular, ILs create a special system between glycerol and oils and fats in a low-temperature multiphase system when enzymes are used. Therefore, attention has been given to the enzyme-catalyzed processes in this chapter, where ILs play a special role in the reaction system for the formation of MAG and DAG.

8.2 ENZYMATIC PRODUCTION OF MONOACYLGLYCEROLS ASSISTED BY IONIC LIQUIDS

MAG are important nonionic emulsifiers with both hydrophilic and hydrophobic groups. They have been recognized as food-grade additives by the

Ionic Liquids in Lipid Processing and Analysis. http://dx.doi.org/10.1016/B978-1-63067-047-4.00008-8

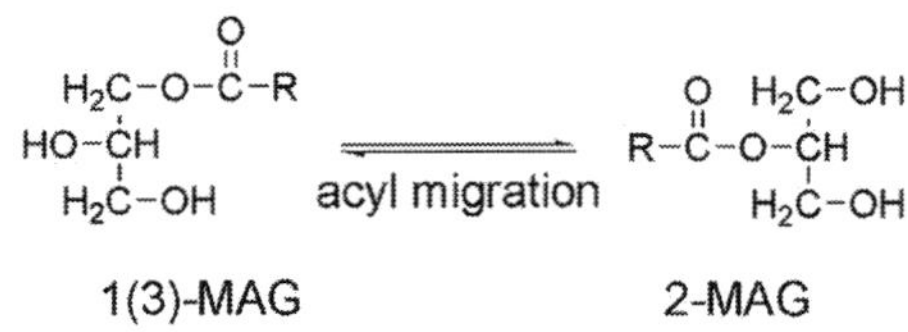

SCHEME 8.1 MAG isomers and the thermodynamic equilibrium between the isomers. *R*, fatty acid chain; *MAG*, monoacylglycerols.

Food and Drug Administration of the United States and European Union, and are widely used in food, cosmetic, and pharmaceutical industries (Damstrup et al., 2006a; Zhong et al., 2009). In addition, pure MAG are also used as intermediates or building blocks in organic synthesis. MAG consist of regioisomeric (*sn*-2) and enantiomeric (*sn*-1 and *sn*-3) isomers (Scheme 8.1), of which 1(3)-MAG and 2-MAG exist in a thermodynamic equilibrium, and 1(3)-MAG are predominant in the MAG mixtures (Zhong et al., 2014). Due to their wide applications, production of MAG is therefore of great importance.

Enzymatic reactions for MAG production were advantageous over the traditional chemical methods, in terms of lower reaction temperatures, less by-product generation, better quality of product, higher selectivity for the targeted products, and so on. Considering the low-temperature enzymatic reactions and the immiscible reactants for MAG production, cosolvents were usually used to accelerate the enzymatic reactions. In addition, some reaction media can also help to increase the targeted product selectivity, through shifting the reaction equilibrium toward the desired products formation side. Of these, ILs are believed to be promising in selective catalysis, because the structures and in turn the properties of ILs are tunable.

In lipid modification fields, IL CPMA·MS has been demonstrated quite successfully for MAG production through enzymatic glycerolysis, in which over 90% of MAG and nearly 100% conversion of TAG was obtained (Guo and Xu, 2005).

8.2.1 Types of Reactions for Monoacylglycerol Production

MAG can be produced through direct esterification (Scheme 8.2), interesterification (Scheme 8.3), partial hydrolysis or alcoholysis of TAG (Scheme 8.4), as well as glycerolysis of TAG (Scheme 8.5) (Zhong et al., 2014). Among all

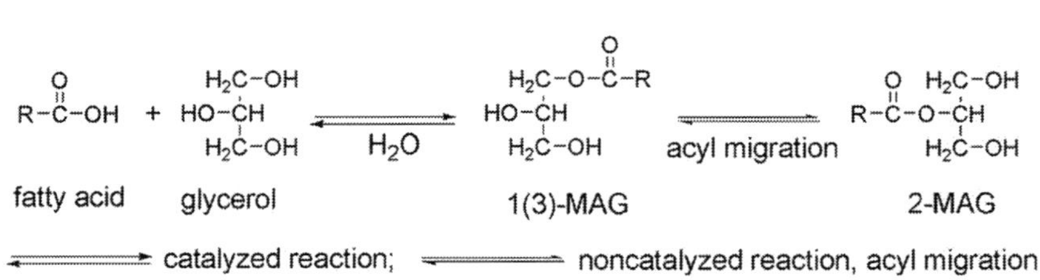

SCHEME 8.2 Reaction scheme for the production of MAG through esterification between glycerol and fatty acids.

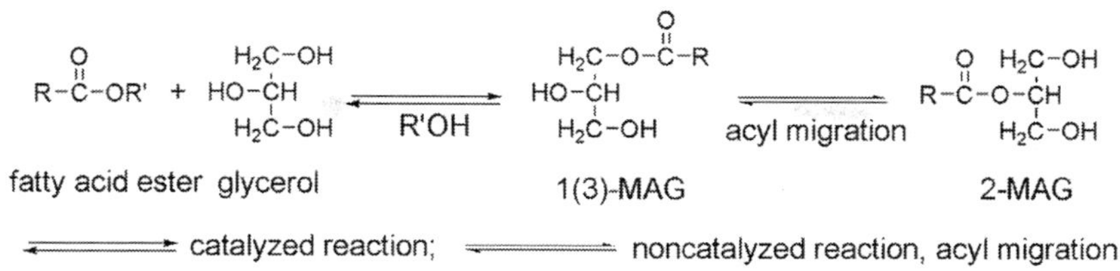

SCHEME 8.3 Reaction scheme for the production of MAG through interesterification between glycerol and fatty acid ester.

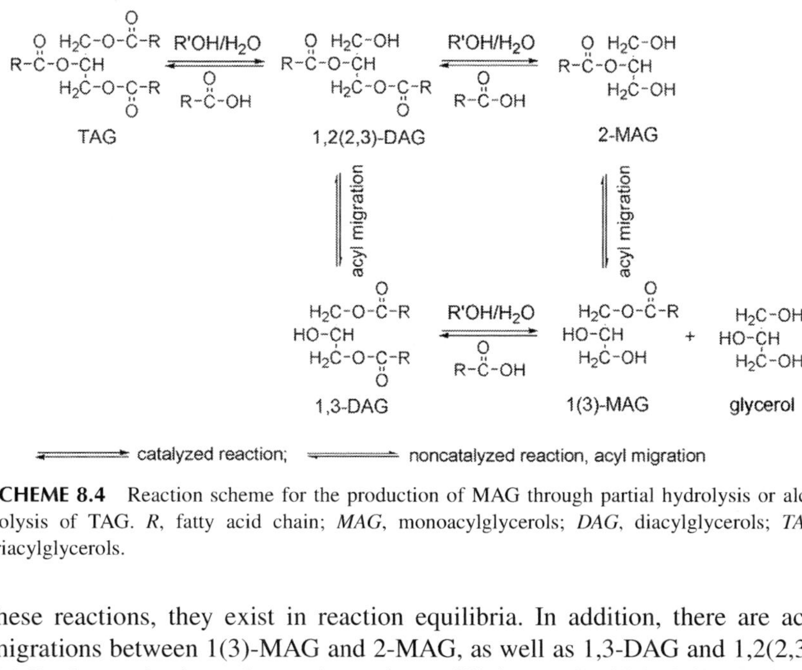

SCHEME 8.4 Reaction scheme for the production of MAG through partial hydrolysis or alcoholysis of TAG. *R*, fatty acid chain; *MAG*, monoacylglycerols; *DAG*, diacylglycerols; *TAG*, triacylglycerols.

these reactions, they exist in reaction equilibria. In addition, there are acyl migrations between 1(3)-MAG and 2-MAG, as well as 1,3-DAG and 1,2(2,3)-DAG; they exist in a thermodynamic equilibrium, of which 1(3)-MAG are predominant in the MAG mixtures and 1,3-DAG are more stable than 1,2(2,3)-DAG (Lo et al., 2008). Glycerolysis of TAG reaction is complex; Scheme 8.5A depicts the detailed reaction routes, and Scheme 8.5B is the main, overall, and simplified equation of glycerolysis of TAG, where it represents only the predominant MAG product, and it takes no account of the 2-MAG formation (Zhong et al., 2014).

Among these reaction routes for MAG production, glycerolysis of TAG is the most widely used approach, because it is more direct with high space—time efficiency: 1 mol of TAG yields 3 mol of MAG. In addition, with the rapid increase of glycerol production especially from biodiesel, conversion of glycerol to valuable products has become an urgent problem to be solved (Jérôme et al., 2008). MAG are one of such bioproducts and the glycerol derived from biodiesel production could be applied as a reactant for glycerolysis of TAG, after being purified (Feltes et al., 2013).

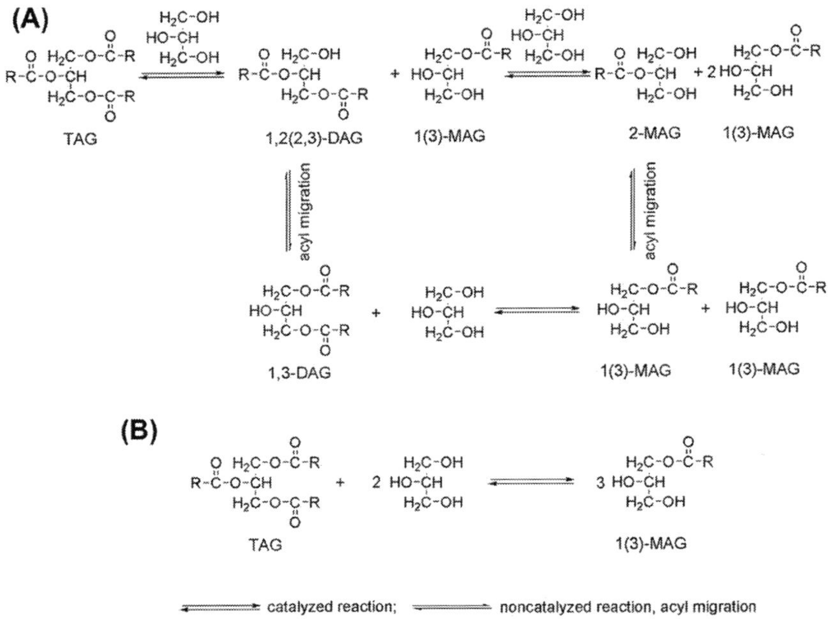

SCHEME 8.5 Reaction scheme for production of MAG through glycerolysis of TAG. (A) Detailed reaction routes, and (B) main, overall, and simplified equation for glycerolysis of TAG reaction.

Presently, the commercial production of MAG is through chemical glycerolysis of TAG at high temperatures of 200–250°C, using alkali catalysts. The chemical glycerolysis yields 40–60% MAG, depending on the glycerol excess and other reaction parameters (Guo and Xu, 2005). Commonly, molecular distillation was used to purify the MAG, since MAG at a high degree of purity exhibited better emulsifying performance (Xu, 2004). The high-temperature chemical process tends to result in dark color, burnt taste, and energy consumption. Moreover, this process is unsuitable for the modification of some functional oils with polyunsaturated fatty acids, conjugated linoleic acids, and so on, due to their heat sensitivity (Damstrup et al., 2006a; Pawongrat et al., 2007).

Enzymatic glycerolysis is believed to be a potential alternative to the chemical process, due to the much lower temperature required (lower than 80°C) as well as the high selectivity of biocatalysts (Damstrup et al., 2005; Guo and Xu, 2005). However, the miscibility of glycerol and TAG is poor at low temperatures, making the enzymatic glycerolysis reaction inefficient. Therefore, extensive studies have focused on how to create compatible systems for enzymatic glycerolysis, and *tert*-butanol as well as *tert*-pentanol have been found efficient for MAG production (Damstrup et al., 2005, 2006a,b). On the other hand, increase in the MAG selectivity through shifting the reaction equilibrium toward MAG generation is another concern in this area. To address this issue, some protocols have been developed over the years. Employment of excessive glycerol would theoretically favor the equilibrium shift toward MAG formation side, however, the MAG increase was limited. The optimal glycerol:TAG molar ratio for MAG production through enzymatic glycerolysis is usually at about 5:1; further increase in the glycerol:TAG molar ratio did not lead to the MAG increment (Damstrup et al., 2006a; Zhong et al., 2009). Even though a glycerol:TAG molar ratio of 12 was used, with MgO as catalyst, the MAG content was less than 70% (Corma et al., 1998). Another protocol is the in situ selective precipitation of MAG through temperature-programmed cooling crystallization, and over 90% of MAG can be obtained by this procedure (McNeill et al., 1990, 1991; McNeill and Yamane, 1991). Nevertheless, this protocol was only suitable for lab-scale research, the slow reaction rate as well as difficulty in the lipase recovery has restricted its application in industry. Application of traditional solvent engineering has made some progress (Rendón et al., 2001), but the success was still limited, due to the introduction of flammable and toxic solvents into the reaction systems, which made them unsuitable in industry, especially for food emulsifiers production.

Recently, a novel protocol with ILs as reaction systems was developed (Guo and Xu, 2005). The applied ILs can not only be compatible for both oil and glycerol, but also help shift the reaction equilibrium toward the MAG generation side. Remarkably, over 90% of MAG and nearly 100% conversion of TAG was obtained by this protocol.

8.2.2 Ionic Liquids Used for Monoacylglycerol Production Through Enzymatic Glycerolysis

Being recognized as "green" solvents, ILs have attracted significant attention in the science community as an alternative reaction medium. ILs have negligible vapor pressure and are capable of dissolving compounds of varying polarity (Ganske and Bornscheuer, 2005). In addition, lipase-catalyzed reactions can take place in ILs and some ILs can protect or increase the enzyme activity and stability (Tavares et al., 2013). Moreover, the most attractive characteristic of ILs may be the tunability of their properties to specific applications by altering cations, anions, and the attached substituents (Guo et al., 2006). This feature makes ILs real engineered solvents, with the judicious incorporation of some functional substituents, a broad spectrum of task-specific ILs could be theoretically obtained.

Table 8.1 summarizes the ILs used as the reaction medium for MAG production through enzymatic glycerolysis. Ammoeng 100 and Ammoeng 102 were strikingly found as good reaction media for enzymatic glycerolysis, with encouraging performance in terms of TAG conversion, MAG content, as well as desired product selectivity (MAG/DAG, mol/mol). Negligible or no reaction was observed with [BMIM]·[BF_4], [BMIM]·[PF_6], [EMIM]·[$EtSO_4$], [EMIM]·[$OctSO_4$], or [EMIM]·[TFA] as reaction medium.

TABLE 8.1 Ionic Liquids Used for Monoacylglycerol Production Through Enzymatic Glycerolysis[a]

Ionic Liquid[b]	Lipase	Conversion of Triacylglycerols (mol%)[c]	Yield of Monoacylglycerols (mol%)	Monoacylglycerols/Diacylglycerols (mol/mol)
Ammoeng 100	Novozym 435	99.07 ± 0.47	90.41 ± 1.76	33.21 ± 6.61
Ammoeng 102	Novozym 435	99.00 ± 0.49	90.25 ± 5.24	38.26 ± 6.62
Ammoeng 111	Novozym 435	42.18 ± 2.86	19.90 ± 2.45	1.02 ± 0.22
Ammoeng 112	Novozym 435	41.15 ± 0.47	31.75 ± 1.61	4.16 ± 0.29
Ammoeng 120	Novozym 435	96.61 ± 0.27	72.45 ± 1.64	3.99 ± 0.08
$[BMIM]\cdot[BF_4]$	Novozym 435	Negligible	—	—
	Lipozyme RM IM	Negligible	—	—
	Lipozyme TL IM	Negligible	—	—
$[BMIM]\cdot[PF_6]$	Novozym 435	Negligible	—	—
	Lipozyme RM IM	Negligible	—	—
	Lipozyme TL IM	Negligible	—	—
$[EMIM]\cdot[EtSO_4]$	Novozym 435	Negligible	—	—
	Lipozyme RM IM	Negligible	—	—
	Lipozyme TL IM	Negligible	—	—

[CPMA]·[BF_4]	Novozym 435	Data not shown	Data not shown	Data not shown
[CPMA]·[PF_6]	Novozym 435	Data not shown	Data not shown	Data not shown
[CPMA]·[tf_2N]	Novozym 435	Data not shown	Data not shown	Data not shown
[EMIM]·[$OctSO_4$]	Novozym 435	No reaction	–	–
[EMIM]·[TFA]	Novozym 435	No reaction	–	–
[TOMA]·[TFA]	Novozym 435	Data not shown	Data not shown	Data not shown

[a] *Reaction conditions: Unless otherwise stated, 0.5 mmol of TAG, 5 × 0.5 mmol of glycerol, 0.625 mmol of IL (2 g for Ammoeng 120), 100 mg of lipase load, 60°C, 10 h, and 600 rpm.*

[b] *Abbreviations: Ammoeng 100, Cocosaklyl pentaethoxi methyl ammonium methylsulfate; Ammoeng 102, tetraalkyl ammonium sulfate; Ammoeng 120, quaternary ammonium sulfate; Ammoeng 111, poly-[oxy(methyl-1,2-ethanediyl)]-alpha-[2-(diethylhydroxyethylammonio) ethyl]-acetate; [BMIM]·[BF_4], 1-butyl-3-methylimidazolium tetrafluoroborate; [BMIM]·[PF_6], 1-butyl-3-methylimidazolium hexafluorophosphate; [EMIM]·[$EtSO_4$], 1-ethyl-3-methylimidazolium ethylsulfate; Ammoeng 112, poly-[oxy(methyl-1,2-ethanediyl)]- alpha - [2-(diethylhydroxyethylammonio)ethyl]-dihydrogenphosphate; [CPMA]·[BF_4], Cocosaklyl pentaethoxi methyl ammonium tetrafluoroborate; [CPMA]·[PF_6], Cocosaklyl pentaethoxi methyl ammonium hexafluorophosphate; [CPMA]·[tf_2N], Cocosaklyl pentaethoxi methyl ammonium bis(trifluoromethylsulfonyl)imide; [EMIM]·[$OctSO_4$], 1-ethyl-3-methylimidazolium octylsulfate; [EMIM]·[TFA], 1-ethyl-3-methylimidazolium trifluoroacetate; [TOMA]·[TFA], trioctylmethylammonium trifluoroacetate.*

[c] *Data from Guo et al. (2006) and Chen et al. (2008).*

The structures of the ILs have a significant influence on the enzymatic glycerolysis. Simultaneous existence of long chain hydrophobic substituents and hydrophilic ethoxyl or hydroxyl moieties was found, respectively, to be essential for TAG dissolving and equilibrium shifting, which in turn increases the TAG conversion and desired product selectivity (Guo and Xu, 2005; Guo et al., 2006).

As mentioned before, one of the major problems for enzymatic glycerolysis reaction was the poor miscibility between glycerol and TAG (Zhong et al., 2010). ILs with amphiphilic structures being capable of creating a compatible system for glycerol and TAG are therefore useful to enhance the glycerolysis reaction. Ammoeng 100, with an amphiphilic structure (Fig. 8.1), gave a good glycerolysis reaction performance with a 90% MAG content and nearly 100% conversion of

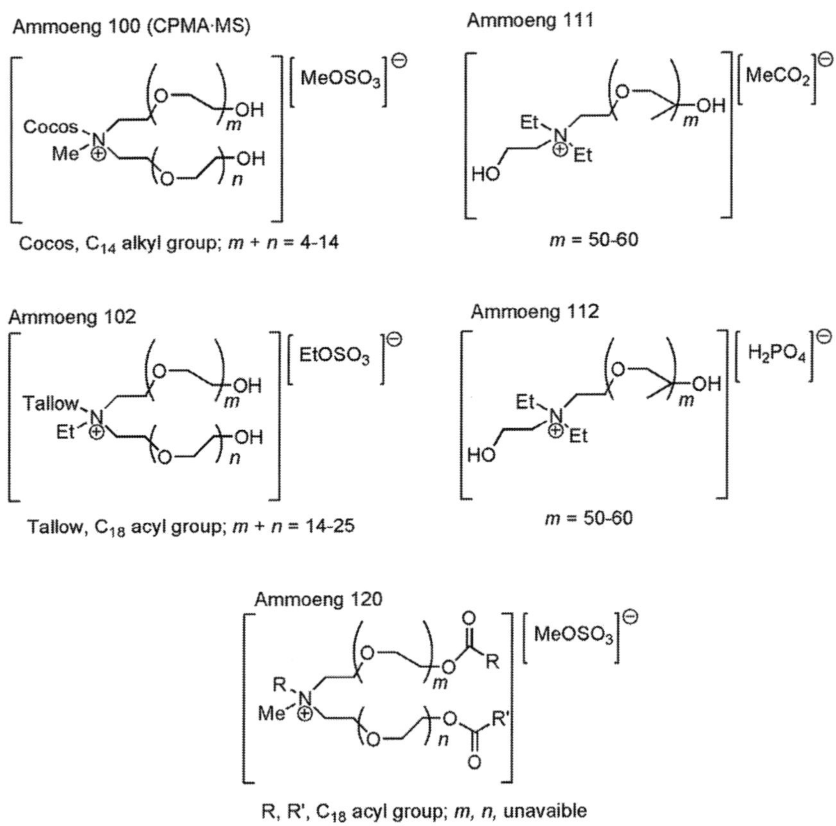

FIGURE 8.1 Molecular structures of the tetraammonium-based ILs used. *Ammoeng 100*, Cocosaklyl pentaethoxi methyl ammonium methylsulfate; *Ammoeng 102*, tetraalkyl ammonium sulfate; *Ammoeng 120*, quaternary ammonium sulfate; *Ammoeng 111*, poly-[oxy(methyl-1,2-ethanediyl)]-alpha-[2- (diethylhydroxyethylammonio) ethyl]-acetate; *Ammoeng 112*, poly-[oxy(methyl-1,2 -ethanediyl)]- alpha - [2-(diethylhydroxyethylammonio)ethyl]-dihydrogenphosphate.

TAG. Negligible reaction was observed from $[BMIM]\cdot[BF_4]$, $[BMIM]\cdot[PF_6]$, and $[EMIM]\cdot[EtSO_4]$ systems, even with other lipases, like Lipozyme TL IM or Lipozyme RM IM, as catalysts. Guo and Xu (2005) attributed this to the inefficient contact of TAG and glycerol, since they observed little oil was dissolved in these three ILs, and clear phase separation was obtained. Thus, miscibility between TAG and glycerol is quite important for the low-temperature enzymatic glycerolysis reaction to proceed.

Another concern for the enzymatic glycerolysis reaction was how to improve the desired product selectivity. Glycerolysis reaction is a multistep reversible reaction; the reaction equilibrium dictates the MAG content in the glycerolysis reaction product. Shifting the equilibrium toward the MAG formation side would undoubtedly increase the MAG content. Solvation induced by the reaction medium can shift a reaction equilibrium; as discussed previously (Zhong et al., 2014), the basic theory behind the equilibrium shifting by the solvation was as follows:

$$A + B \underset{}{\overset{K_{eq}}{\rightleftharpoons}} C + D \qquad [1]$$

$$K_{eq} = \frac{\alpha_C \cdot \alpha_D}{\alpha_C \cdot \alpha_B} = \frac{\chi_C \cdot \chi_D}{\chi_A \cdot \chi_B} \cdot \frac{\gamma_C \cdot \gamma_D}{\gamma_A \cdot \gamma_B} \qquad [2]$$

The reaction equilibrium of Eq. [1] could be correspondingly described by Eq. [2], where K_{eq} is the equilibrium constant, and α, χ, and γ represent the thermodynamic activity, mole fraction, and activity coefficient, respectively. For a given temperature, K_{eq} is constant; however, the thermodynamic activity of a component depends on its nature and specific reaction system. If there is such a reaction medium whose interaction with A, B, and D is weak, the thermodynamic activity of A, B, and D (α_A, α_B, and α_D) will not be affected much by this reaction medium. However, this reaction medium will interact with the component C (desired product) and strongly results in significant decrement in the thermodynamic activity coefficient of C (γ_C). Since K_{eq} is constant, the value of χ_C will be increased accordingly, thus increasing the content of C in the final reaction mixtures (selectivity toward the desired product C increased).

Therefore, a reaction medium that interacts with MAG strongly can theoretically shift the equilibrium toward the MAG generation side and consequently increase the MAG content.

Data from Ammoeng 100, 102, 111, 112, and 120 in Table 8.1 clearly demonstrated the influences of ILs structure on the glycerolysis reaction. On one hand, long chain hydrophobic substitutes within the ILs, such as Ammoeng 100, 111, and 120, favored the TAG dissolving and thus increased the TAG conversion, while ILs without long chain hydrophobic substitutes, like Ammoeng 111 and 112, led to much lower TAG conversion at about 41–42%. The solubility of triolein in these ILs can further explain this phenomenon, which was 0.771, 2.370, and 10.309 g/100 g, respectively, for Ammoeng 100, 102, and 120, while it was only 0.092 and 0.012 g/100 g, respectively, for Ammoeng 111

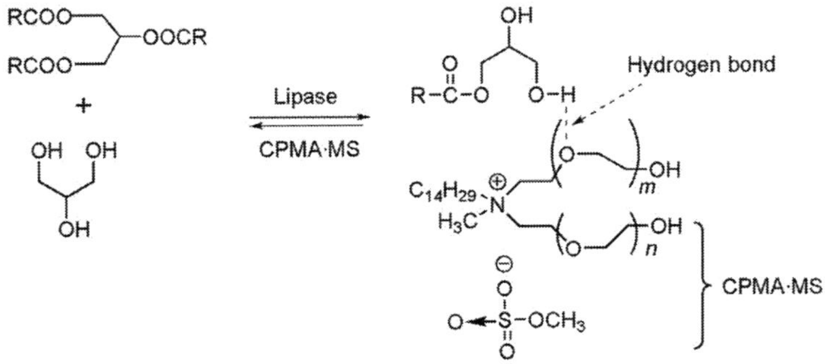

FIGURE 8.2 Lipase catalyzed glycerolysis reaction in [CPMA]·[MS] (cocosalkyl pentaethoxi methyl ammonium methosulfate) system and hydrogen bonding formation between MAG and [CPMA]·[MS].

and 112 (Chen et al., 2008). The lower solubility of TAG in Ammoeng 111 and 112 can account for the resulting lower TAG conversion in glycerolysis reaction.

On the other hand, hydrophilic ethoxyl or hydroxyl moieties within the ILs are responsible for equilibrium shifting favoring MAG formation. The hydrophilic ethoxyl or hydroxyl moieties interact with the hydrophilic MAG strongly, which result in lower thermodynamic activity of MAG (α_{MAG}) and higher α_{DAG} and α_{TAG} values (Guo and Xu, 2005). As suggested in Fig. 8.2, the moiety of polyethoxyl alcohol in Ammoeng 100 molecule could form hydrogen bonds with the hydroxyl group in the glycerol moiety of MAG, which made the hydroxyl group of MAG unreactive (Guo and Xu, 2005). Considering the structure similarity between Ammoeng 100 and 102, we can speculate that similar interaction should exist between Ammoeng 102 and MAG molecules. Performance from these two ILs systems was similar in terms of MAG content and MAG selectivity, which may confirm the speculation. Even though with a glycerol:TAG mole ratio at 1:1, MAG content could still be up to 80% (Guo et al., 2006). Though Ammoeng 120 gave a high TAG conversion being comparable to that from Ammoeng 100 and 102 systems, MAG selectivity from Ammoeng 120 was quite lower than that from Ammoeng 100 and 102 systems, with 4.0 compared to 33–38 (MAG/DAG, mol/mol). Guo et al. (2006) ascribed this lower MAG selectivity to the no free hydroxyl group within Ammoeng 120, which leads to the activity coefficient of MAG (γ_{MAG}) from Ammoeng 120 reduced not so significantly as did that from Ammoeng 100 and 102.

Calculations from conductor-like screening model for real solvents (COSMO-RS) can delineate the multiple solvation interaction among IL reaction systems at molecular level (Guo et al., 2006). COSMO-RS could give a unique 3D polarization density distribution function on the surface of a molecule X, denoting molecular interactions (so-called σ-profile). Fig. 8.3 presented the σ-profiles (A) and σ-potentials ($\mu(\sigma)$) (B) of 1-monoolein, 1,3-diolein, and cations of Ammoeng 100, 111, and 120 at 298.15 K. As can

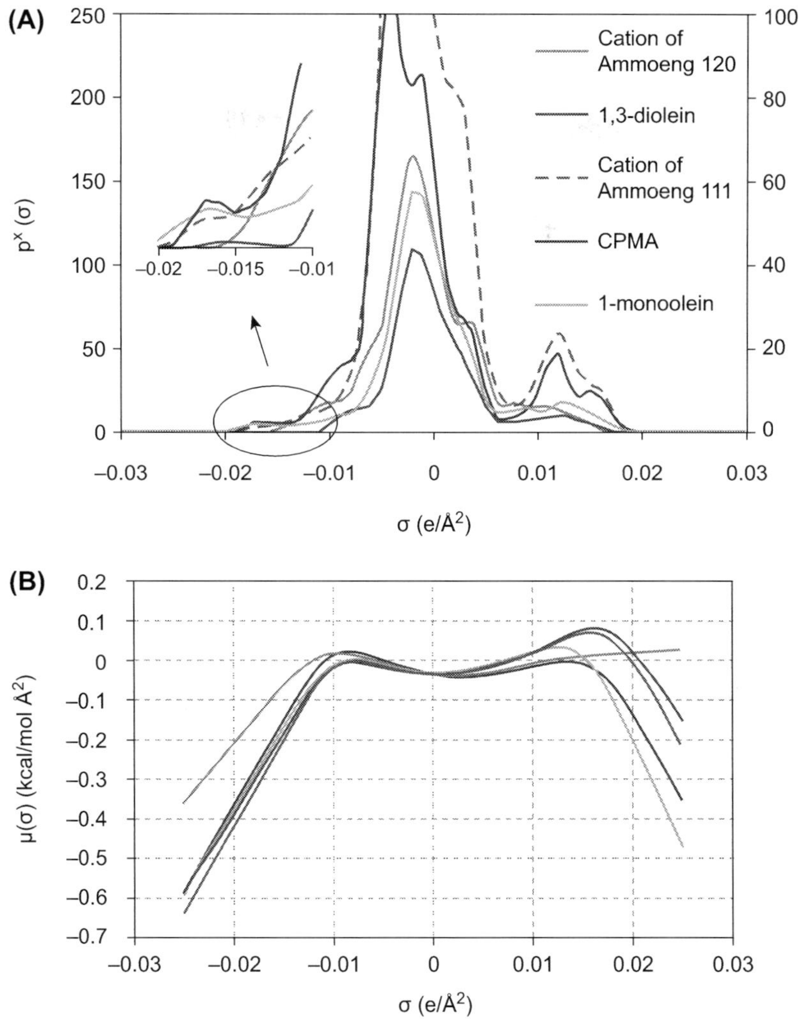

FIGURE 8.3 (A) σ-profiles of 1-monoolein, 1,3-diolein and cations of Ammoeng 100, 111, and 120; (B) σ-potentials ($\mu(\sigma)$) of 1-monoolein, 1,3-diolein and cations of Ammoeng 100, 111, and 120 at 298.15 K.

be seen, 1-monoolein gave peaks around −0.017 and 0.014 e/Å^2, corresponded to hydroxyl hydrogens and oxygens of glycerol backbone, respectively, which represented the possibility and capacity of the 1-monoolein molecule as a hydrogen bonding donor and acceptor, respectively. On the other hand, 1,3-diolein showed lower amount of surface to act as a hydrogen bonding donor or acceptor because of the relatively lower peaks below −0.01 or greater 0.01 e/Å^2. Because the structures of 1-monoolein and [CPMA]·[MS] (Ammoeng 100) are similar, both having two free hydroxyl groups and hydrophobic hydrocarbyl groups, the σ-profiles and σ-potentials of the cation of [CPMA]·[MS] and 1-monoolein are similar. The σ-potential of the cation of [CPMA]·[MS] and 1-monoolein quickly became strongly negative on both negative and positive sides (Fig. 8.3B), suggesting the strong hydrogen bonding capacity as either donor or acceptor. Contrarily, 1,3-diolein gave positive σ-potential on negative and positive sides, suggesting that it was thermodynamically unfavorable as a hydrogen bonding donor and acceptor due to the steric hindrance from two acyl groups. This can well explain the hydrogen bonding formation between MAG and [CPMA]·[MS], as suggested in Fig. 8.2.

In addition, the COSMO-RS can also calculate the solubility, Misfit interaction, van der Waals interaction, and chemical potential as well as activity coefficient. These thermodynamic properties are effective measures to delineate multiple interactions between substrates, products, and ILs, which quantitatively and qualitatively relate to the reaction evolution and profile specificity of the IL system (Guo et al., 2006; Chen et al., 2008). Of these, the infinite dilution of 1-monoolein, 1,3-diolein, and triolein in CPMA·MS was −0.154, 1.924, and 6.304, respectively. These data can help explain why the equilibrium strongly shifted to the MAG formation side (Guo et al., 2006).

It is worth mentioning that the anion part of the ILs can also influence the enzymatic glycerolysis reaction, but not so significantly as the cation part did (Chen et al., 2008). In a word, structures of ILs play an important role in dictating the enzymatic glycerolysis reaction: the hydrophobic substitutes in the cation contribute to the increase of TAG solubility and in turn the TAG conversion; while strong polar anion and polyethoxyl and free hydroxyl groups in the cation part are essential to improve the MAG selectivity. Up to 2015, Ammoeng 100 and 102 have been found to be efficient reaction media for MAG production through enzymatic glycerolysis.

8.2.3 Substrates Used for Monoacylglycerol Production Through Enzymatic Glycerolysis in ILs

Substrates for glycerolysis reaction are TAG and glycerol. The reaction protocol developed by Guo and Xu (2005), using [CPMA]·[MS] as the reaction medium and with sunflower oil as substrate through enzymatic glycerolysis, can attain high conversion of TAG and high yield of MAG. Considering that commercial oils and fats from different sources have different fatty acid profiles, it is necessary to verify the universal validity of the developed protocol. They therefore selected another four representative commercial oils and fats to test this procotol, namely rapeseed oil, palm stearin, fish oil, and CLA-TAG, rich in oleic acid (58%), palmitic acid (56%), polyunsaturated fatty acid (20:5ω3 + 22:5ω3 + 22:6ω3, 33%), and conjugated linoleic acid (80%), respectively (Guo and Xu, 2005). Results indicated that the tested oils can also give high conversion of TAG and high yield of MAG (Table 8.2), of which

TABLE 8.2 Triacylglycerols Used for Monoacylglycerol Production Through Enzymatic Glycerolysis in [CPMA]·[MS][a]

TAG Used	Temperature (°C)	Time (h)[b]	[CPMA]·[MS] Amount	Conversion of Triacylglycerols (mol%)	Yield of Monoacylglycerols (mol%)
Sunflower oil	70	24	2.2 g	98.90	89.09
Rapeseed oil	70	24	2.2 g	95.58	88.61
Palm stearin	70	24	2.2 g	97.13	90.70
Fish oil	70	24	2.2 g	99.87	88.20
CLA-TAG	70	24	2.2 g	97.90	93.30
Triolein	60	10	0.625 mmol	99.07	90.41
Tributyrin	60	10	0.625 mmol	98.94	93.40
Tricaprin	60	10	0.625 mmol	98.02	91.06

[a]*Reaction conditions: For sunflower oil, rapeseed oil, palm stearin, fish oil, and CLA-TAG, 2 mmol of triacylglycerol and 5 × 2 mmol of glycerol, 100 mg of Novozym 435, 600 rpm; for triolein, tributyrin, and tricaprin, 0.5 mmol of triacylglycerol and 5 × 0.5 mmol of glycerol, 100 mg of Novozym 435, 600 rpm.*
[b]*Reaction equilibria were essentially attained at the reaction time.*
Data from Guo and Xu (2005) and Guo et al. (2006).

CLA-TAG with high purity gave the highest yield of MAG, whereas other commercial oils consisting of diverse fatty acid profiles yield different results.

To figure out the interactions between IL and glyceride molecules with varied chain lengths and their contribution to equilibrium shifting, pure TAG including triolein, tributyrin, and tricaprin were tested, and results indicated all three pure TAG gave high conversion of TAG (>98%) and high yield of MAG (>90%) (Table 8.2). However, the MAG selectivity (MAG/DAG, mol/mol) from these three pure TAG was different: triolein 33, tributyrin 20, and tricaprin 13 (Guo et al., 2006). The authors ascribed the varied selectivity to that lipase specificity to substrates as well as different interaction of substrates and products with ILs.

In summary, the developed protocol using [CPMA]·[MS] as the reaction medium was verified to be effectively applied to different types of oils and fats. The universal applications of the protocol to different TAG profiles indicating ILs are a predominant factor to control the reaction direction.

8.2.4 Effects of Reaction Parameters

The *Candida antarctica* lipase B-catalyzed glycerolysis of sunflower oil in [CPMA]·[MS] achieved higher conversion of TAG and higher yield of MAG. This protocol has its distinct characteristics, such as lower initial reaction rate, occurrence of the induction period, a jump after the induction period, as well as bulky substrate-tolerating capacity of the [CPMA]·[MS]. Interestingly, the enzymatic reaction in Ammoeng 102 did show surprising similarity to the reaction in [CPMA]·[MS] in terms of reaction evolution, product selectivity, occurrence of induction period, as well as bulky substrate-tolerating capacity. Some of the important reaction parameters, like reaction temperature, IL dosage, substrate mole ratio, agitation speed, enzyme loading, and such, have contributed to the distinct characteristics.

8.2.4.1 Effects of Reaction Temperature on the Reaction Evolution

Table 8.3 summarizes the effects of reaction temperature on the reaction evolution in *tert*-butanol as well as [CPMA]·[MS], Ammoeng 102, and Ammoeng 120 systems. The reaction in *tert*-butanol is fast; the equilibrium was reached after 2 h reaction at 50°C without induction period. As in ILs, the overall reaction was slower than in *tert*-butanol; in addition, the initial reaction rate and length of induction period are strongly temperature-dependent (Guo and Xu, 2005). The reaction behavior in [CPMA]·[MS] and in Ammoeng 102 systems was similar. However, Ammoeng 120-mediated reaction showed some difference: no induction period was observed when temperature was ≥60°C; moreover, the length of the induction period in Ammoeng 120 was significantly shorter than that in [CPMA]·[MS] or Ammoeng 102 (Chen et al., 2008).

TABLE 8.3 Effects of Reaction Temperature on the Reaction Evolution in Different Systems[a]

Reaction Temperature (°C)	30	40	50	60	70
Induction period in *tert*-butanol (h)	Without	Without	Without	Without	Without
Time to equilibrium in *tert*-butanol (h)	Unknown	Unknown	2	Unknown	Unknown
Induction period in [CPMA]·[MS] (h)	Unknown	7	2	1.5	1.5
Time to equilibrium in [CPMA]·[MS] (h)	Unknown	>24	12	10	8
Induction period in Ammoeng 102 (h)	>6	4	2	1	Unknown
Induction period in Ammoeng 120 (h)	1	0.5	0.5	Without	Without

[a]*Reaction conditions: For tert-butanol and [CPMA]·[MS] systems, 0.5 mmol of sunflower oil, 5 × 0.5 mmol of glycerol, 2.5 g of tert-butanol and 2.2 g of [CPMA]·[MS], 100 mg of Novozym 435, 600 rpm; for Ammoeng 102 and 120 systems, 0.5 mmol of triolein, 5 × 0.5 mmol of glycerol, 0.625 mmol of Ammoeng 102 and 120, 100 mg of Novozym 435, 600 rpm.*
Data from Guo and Xu (2005) and Chen et al. (2008).

8.2.4.2 Effects of Ionic Liquid Dosage on Triacylglycerol Conversion and Monoacylglycerol Yield

Table 8.4 and Table 8.5 summarize the effects of IL ([CPMA]·[MS] and Ammoeng 102) dosage on TAG conversion and MAG yield. Table 8.4 shows that similar high conversion of TAG and high yield of MAG could be obtained when the mole ratio of oil to [CPMA]·[MS] was less than 1.2. Results from Table 8.4 also suggests that the tolerating capacity of [CPMA]·[MS] molecule to adapt MAG molecule is not unlimited, but the buffering power is higher. Herein "tolerating capacity" refers to smaller amounts of IL being able to produce more product, and "buffering power" refers to a similar high yield of MAG achieved over a wide range of mole ratio of oil to IL. The robust bulky substrate-tolerating capacity of [CPMA]·[MS] makes it more competitive in cost, since a lesser amount of IL could give a higher productivity, which is of great interest in industry (Guo and Xu, 2005).

Interestingly, like [CPMA]·[MS], Ammoeng 102 also has robust tolerating capacity for substrate (Table 8.5). Reaction performance of [CPMA]·[MS] and Ammoeng 102 was almost the same, with regard to TAG conversion and MAG yield. MAG yield declines very little and can still increase up to 90%, when triolein dosage increases from 0.4 to 1 mol for mole Ammoeng 102.

TABLE 8.4 Effects of [CPMA]·[MS] Dosage on Triacylglycerol Conversion and Monoacylglycerol Yield[a]

Mole Ratio of Oil: [CPMA]·[MS]	0.2:1	0.4:1	0.8:1	1.2:1	1.6:1	6.3:1	8.1:1
Conversion of oil (mol%)	98.60	99.07	98.60	97.29	96.63	79.37	78.65
Yield of monoacylglycerols (mol%)	91.88	90.41	88.11	86.68	80.77	66.74	54.88

[a]*Reaction conditions: Molar ratio of sunflower oil to glycerol 1:5, Novozym 435 amount 5% of oil weight, 70°C, 24 h, and 600 rpm.*
Data from Guo and Xu (2005).

Chen et al. (2008) ascribed this phenomenon to the strong buffering power of ILs for the increase of substrate concentration.

8.2.4.3 Effects of Substrate Mole Ratio on the Enzymatic Reaction

As described by (Eqs. [3]–[5]), glycerolysis of TAG is a two-step reversible reaction, in which excess glycerol would suppress the reversion of the first step (Eq. [4]) and promote the further glycerolysis of the generated DAG (Eq. [5]), leading to higher yield of MAG. Enzymatic glycerolysis in some common solvent systems, a glycerol:TAG mole ratio at 3–5/1 was usually required to ensure a high yield of MAG (Zhong et al., 2013).

$$\text{TAG} + 2\text{Gly} \overset{K_{eq}}{\rightleftharpoons} 3\text{MAG} \qquad [3]$$

$$\text{TAG} + \text{Gly} \overset{K_{eq_1}}{\rightleftharpoons} \text{MAG} + \text{DAG} \qquad [4]$$

$$\text{DAG} + \text{Gly} \overset{K_{eq_2}}{\rightleftharpoons} 2\text{MAG} \qquad [5]$$

Enzymatic glycerolysis in a [CPMA]·[MS] system showed no significant difference of TAG conversion when the glycerol:TAG mole ratio was increased from 2:1 to 6:1 (Guo and Xu, 2006). Interestingly, similar TAG conversion behavior was also observed in Ammoeng 102 systems. However, in Ammoeng 120 and 111 systems, the TAG conversions were seriously affected by glycerol dosage, of which for Ammoeng 120 it reduced from 98% at 5:1 of glycerol/TAG to 80% at 1:1, and for Ammoeng 111 from 72% to 58% (Chen et al., 2008).

As for the yield of MAG, in [CPMA]·[MS] the yield of MAG increased with the glycerol excess. Even though the [CPMA]·[MS] could induce the shift of equilibrium to the MAG generation side, by reducing the activity

TABLE 8.5 Effects of Ionic Liquids Dosage on Triacylglycerol Conversion and Monoacylglycerol Yield[a]

Mole Ratio of Oil:IL	Ammoeng 100 ([CPMA]·[MS])		Ammoeng 102	
	Conversion of Triacylglycerols (mol%)	Yield of Monoacylglycerols (mol%)	Conversion of Triacylglycerols (mol%)	Yield of Monoacylglycerols (mol%)
0.4:1	99.54 ± 0.42	90.45 ± 1.75	99.12 ± 0.44	92.03 ± 2.69
0.8:1	99.33 ± 1.38	89.75 ± 1.01	99.25 ± 0.53	91.68 ± 1.11
1:1	99.32 ± 1.45	89.23 ± 0.78	99.66 ± 1.26	90.37 ± 1.09
2:1	98.90 ± 1.04	81.24 ± 0.79	98.40 ± 0.95	79.45 ± 2.01
4:1	92.23 ± 0.86	68.40 ± 1.15	96.43 ± 0.99	68.25 ± 1.87

[a]*Reaction conditions: 0.5 mmol of triolein, 5 × 0.5 mmol of glycerol, 100 mg of Novozym 435, 60°C, and 600 rpm.*
Data from Chen et al. (2008).

coefficients of partial glycerides, more excess glycerol (more than 5:1, glycerol:TAG) was still necessary to achieve a higher yield of MAG (Guo and Xu, 2006). High yield of MAG was also obtained in Ammoeng 102 when glycerol was in excess (glycerol:TAG at 5:1).

With glycerol:TAG mole ratio at 1:1, it can evaluate the reaction selectivity based on the product MAG:DAG mole ratio. Since at the stoichiometric ratio (1:1) it should theoretically yield equimolar amounts of MAG and DAG. From Table 8.6, it can be seen that, in Ammoeng 102, even at glycerol:TAG mole ratio of 1:1, the yield of MAG can still reach up to 80% and the selectivity up to 9. Moreover, the reaction did not result in a higher yield of DAG, when glycerol:TAG mole ratio was at 1:2, in which the excess glycerol favors the formation of DAG (Guo et al., 2006). The results indicated that, like in [CPMA]·[MS], there was a strong impetus in the Ammoeng 102 system to shift the reaction equilibrium toward MAG generation side. The similar reaction behaviors from [CPMA]·[MS] and Ammoeng 102 were ascribed to the similar molecular structures (Guo et al., 2006). If we use the product MAG:DAG ratio at stoichiometric ratio of glycerol:TAG as a criterion to categorize the reaction selectivity of solvent systems for partial glyceride production, Ammoeng 102 and 112 could be classified as MAG-selective medium, and Ammoeng 120 and 111 as DAG-selective solvents (Chen et al., 2008).

8.2.4.4 Effects of Agitation Speed, Enzyme Loading, and Triacylglycerol Concentration on the Reaction Rate

The [CPMA]·[MS] can provide an environment compatible for oils and glycerol, however, this capacity was not unlimited, and phase separation occurred at higher substrate concentration. To examine the existence of external mass resistance, Guo and Xu (2006) studied the effects of agitation speed on the reaction rate in the range of 100–800 rpm. Their results indicated that the conversion of TAG increased with the agitation speed until after 600 rpm, suggesting the presence of external mass transfer limitation. Mass transfer between two phases directly relates to the interfacial area, which is dependent on the shear rates and the two-phase volume ratio. An increase in speed of agitation will result in the increasing dispersion of TAG into IL phase and subsequently a more efficient interaction with glycerol and lipase. All these would increase the reaction rate. Therefore, a proper agitation speed was required to eliminate the external mass transfer limitation in [CPMA]·[MS] system.

To further look into this system, authors investigated the effects of enzyme loading on reaction rate (Guo and Xu, 2006). The initial reaction rate showed an approximately linear increase against the enzyme concentration at the range of 10.8–32 g/L, however, when enzyme loading was over 32 g/L, the increase slowed down. If no mass transfer resistance is present, the dependence of reaction rate upon the enzyme concentration should be linear. Nevertheless, the linear increase of reaction rate against enzyme concentration in the range

TABLE 8.6 Effects of ILs as a Reaction Medium on the Selectivity of Enzymatic Glycerolysis[a]

Ionic Liquid[b]	Glycerol/Triacylglycerol (mol/mol)	Conversion of Triacylglycerols (mol%)	Yield of Monoacylglycerols (mol%)	Monoacylglycerols/ Diacylglycerols (mol/mol)
Ammoeng 102	1/1	97.86 ± 1.21	79.57 ± 3.92	9.00 ± 2.56
Ammoeng 102	1/2	95.26 ± 0.54	70.69 ± 7.06	6.88 ± 0.23
Ammoeng 120	1/1	80.49 ± 0.83	24.60 ± 1.52	0.60 ± 0.06
Ammoeng 111	1/1	27.70 ± 0.16	11.46 ± 0.47	0.75 ± 0.05
Ammoeng 112	1/1	47.28 ± 3.57	29.62 ± 1.59	1.80 ± 0.08

[a]*Reaction conditions: 0.5 mmol of triolein, corresponding glycerol with the desired glycerol: triacylglycerol ratio, 100 mg of Novozym 435, 0.625 mmol of ionic liquid (2 g for Ammoeng 120), 60°C, 10 h, and 600 rpm.*
[b]*Abbreviations as in Table 8.1.*
Data from Guo et al. (2006).

of 10.8–32 g/L did not rule out the possibility of mass transfer limitations, especially internal diffusion, on the reaction rate. The presence of mass transfer limitation was further confirmed by the TAG concentration study, in which the glycerolysis reaction rate was linearly dependent on the sunflower oil concentration (in the range of 0.19–0.50 mol/L), since the reaction rate was linearly dependent upon substrate concentrations for the mass transfer controlled reactions.

Overall, the results from the factors study, including the effects of agitation speed, enzyme loading, and TAG concentration, indicated the presence of mass transfer limitation in [CPMA]·[MS] system. Any measures to eliminate the mass transfer resistance will significantly increase the reaction rate and reduce the induction periods. This study provides a sound basis for reaction optimization.

8.2.4.5 Effects of Ionic Liquid on Lipase Activity

The structure of IL influences the lipase activity: the anion moiety of ILs could form H-bonded complexes with water molecules, and the strength of the H-bonding depends on the basicity of anions. Since lipase activity relates to the water content and water state, the anion of ILs affects the structure and consequently the activity of lipase (Lau et al., 2004).

Results from [CPMA]·[MS] system indicated that Lipozyme RM IM only achieved 15% conversion of TAG and even less conversion for Lipozyme TL IM after 24 h reaction. Fortunately, Novozym 435 exhibited a different behavior, with 99% conversion of TAG and 89% yield of MAG achieved after 6 h reaction (Guo and Xu, 2005). The good performance exhibited by Novozym 435 was attributed to its ability to keep catalytic activity even in water-depriving solvents. Authors also studied the stability of Novozym 435 in [CPMA]·[MS] system, and results showed that the Novozym 435 can catalyze the glycerolysis reaction in this system without significant activity loss, after 10 consecutive reactions with the duration of 10 h per batch and a few ethyl ether extraction processes. The good catalytic performance and the reusability of Novozym 435 in [CPMA]·[MS] system indicated the practical usability and commercial competence of this protocol.

Interestingly, Novozym 435 also exhibited good performance in Ammoeng 102 system, with almost 100% conversion of TAG and 92% yield of MAG obtained (Chen et al., 2008). In addition, Lipozyme RM IM produced 36.6% TAG conversion and 31.7% MAG yield in Ammoeng 102 system, while negligible reaction was observed for Lipozyme TL IM. Moreover, lipases AK-20 (*Pseudomonas fluorescens*) and PS-D (*Pseudomonas* sp.) were also examined in the Ammoeng 102 system and TAG conversions were found less than 10% for these two lipases. Though many publications have indicated that lipases (including *C. antarctica* B) exhibited little activity in anhydrous systems of the ILs paired with $MeSO_4^-$, $EtSO_4^-$, $MeCO_3^-$, and so forth (van Rantwijk et al., 2006), all five test lipases showed certain activity in the glycerolysis of TAG in pure tetraammonium-type ILs (including Ammoeng

100, 102, 120,111, and 112). Such behaviors were possibly attributed to the protective or stabilizing effect of the tetraammonium-type ILs on the enzyme (Chen et al., 2008).

8.3 ENZYMATIC PRODUCTION OF DIACYLGLYCEROLS ASSISTED BY IONIC LIQUIDS

DAG are natural components of various edible oils, and are surface-active molecules widely used as emulsifiers in the food industry. Commercial preparations are mixtures of MAG and DAG generated during chemical glycerolysis of TAG. Novel uses of DAG include modifiers of TAG crystal structure in fat-containing foods, and intermediates in the chemical synthesis of drug conjugates, phospholipids, and glycolipids. For many of these reactions, pure substrates are required. Therefore, lipase-catalyzed production of DAG is preferred due to the unique regioselectivity of lipases (McNeill, 1998).

There are two isomers of DAG, named as 1,3-DAG and 1,2(2,3)-DAG according to the positions of fatty acids on the glycerol backbone. The main reason for DAG oil to gain a particular interest recently is that consumption of oil rich in DAG, especially the 1,3-isomer, is proved to have positive effects on human health. DAG has two main health benefits: (1) the suppression of postprandial serum TAG elevation, and (2) the suppression of body fat accumulation (Yasukawa and Katsuragi, 2004). It was shown that energy value of DAG oil was nearly 98% of TAG oil with the similar fatty acid composition. Besides, bioavailability (Taguchi et al., 2001) and absorption rates (Murase et al., 2001) of these two oils were not different from each other. Thus, health benefits of DAG oil are considered to be the result of different metabolic fate after absorption by intestinal epithelial cells (Flickinger, 2004).

8.3.1 Lipase-Catalyzed Production of Diacylglycerols in Ionic Liquids

Although the wide possibilities obtained by modifying the chemical properties of ILs could provide in terms of yield and selectivity, the application of these solvents for DAG production has been performed only a few times. The first attempt started by using 19 different ILs as the reaction medium for production of DAG through lipase-catalyzed glycerolysis of triolein (Kahveci et al., 2009). [BMIM]·[BF_4], [BMIM]·[PF_6], and [TOMA]·[Tf_2N], those that have been widely used in many lipase-catalyzed reactions so far, resulted in 30–50% TAG conversion, while the DAG yield amounted to 30–50%. Ammoeng series of ILs (Ammoeng 100, 102, and 120) yielded a good conversion of TAG, however they differed from each other in terms of the product profiles. Ammoeng 120 produced 56% TAG conversion with 42% DAG yield, which covered 75% of the partial glycerides (MAG + DAG + FFA) generated from TAG. Contrarily, both Ammoeng 100 and 102 achieved over 90% TAG

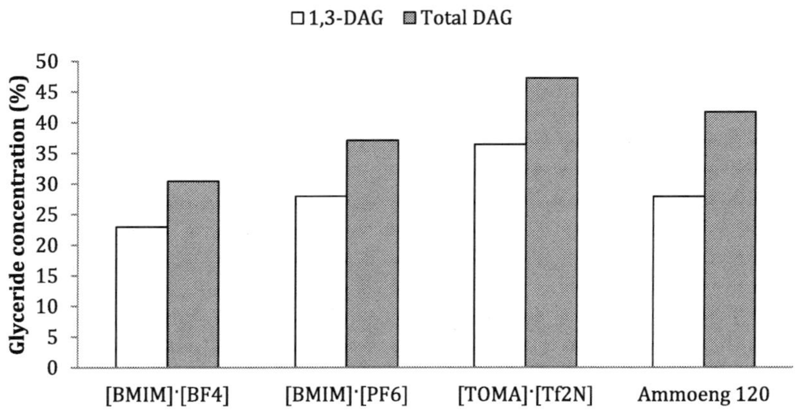

FIGURE 8.4 1,3-DAG and total DAG yield obtained by lipase-catalyzed glycerolysis in different ILs. Reaction conditions for IL-hosted reactions: 1 mmol of oil (885 mg), 0.5 mmol of glycerol (45 mg), 88.5 mg of Novozym 435, 1 g of IL, 60°C, 700 rpm and 48 h. *Data from Kahveci et al. (2009).*

conversion, but DAG occupied less than 50% of the partial glycerides. As shown in Fig. 8.4, the selectivity of the reaction was in the favorable direction as well; most of the DAG produced was in the 1,3-DAG form.

8.3.2 Effects of Reaction Parameters on Triacylglycerol Conversion and Diacylglycerol Yield

Glycerolysis of TAG in both [BMIM]·[BF_4] and [BMIM]·[PF_6] with the aim of producing MAG has been examined previously (Guo and Xu, 2005), where excessive glycerol was used and very low conversion of TAG was achieved. However, an evident conversion of TAG was observed when low glycerol dosage was employed in this work (Kahveci et al., 2009). The possible reason was suggested to be that the excessive glycerol dissolved in [BMIM]·[BF_4] and [BMIM]·[PF_6] in the previous study enlarged the polarity difference between the ILs and oil, so that the solubility of oil in these ILs was decreased. When the effect of substrate concentration was investigated (Kahveci et al., 2009), the changes of glyceride compositions with the increase of substrate concentration was remarkable for glycerolysis reactions performed in [TOMA]·[TfA] and [BMIM]·[PF_6]. Low conversion was obtained when 0.44 g oil/g IL was applied, whereas the value increased to 18% and 31%, respectively, when the concentration of oil was quadruplicated.

The effect of temperature on the glycerolysis reactions in the selected ILs were investigated at 50, 60, and 70°C (Kahveci et al., 2009). In general, TAG conversion and yield of total DAG increased with the increase of temperature. Interestingly, the ratio between 1,3-DAG and 1,2-DAG increased as the temperature increased for [BMIM]·[BF_4] and [BMIM]·[PF_6] based systems, while the change was not so regular for other IL reaction systems. In the cases of [BMIM]·[PF_6], [OMIM]·[PF_6] and [TOMA]·[TFA], both the reaction yields and rates showed improvements with the increasing temperature, which suggests that there may exist a high energy barrier for dissolution of TG in these ILs and limitation of mass transfer. High viscosities of these ILs might be the main reason.

The effect of oil-to-glycerol ratio on glycerolysis reaction was investigated at three different levels, namely 2:1, 2:1.25, and 2:1.5 (Kahveci et al., 2009). Total DAG yields and TAG conversions improved to different extents in different ILs when the glycerol dosage was increased. The most dramatic change was observed in a [TOMA]·[TFA] based system, where the total DAG yield was increased from 7% to 43% in correspondence with TAG conversion increased from 18% to 74% when the molar oil:glycerol ratio was changed from 2:1 to 2:1.5. The reaction in Ammoeng 120 also resulted in a considerable enhancement of total DAG yield and TAG conversion.

Based on the evaluation of these examinations, three IL reaction systems were selected to examine the performance of different lipases: immobilized forms of *C. antarctica* lipase B (Novozym 435), *Rhizomucor miehei* lipase (Lipozyme RM IM), and *Thermomyces lanuginosus* lipase (Lipozyme TL IM) (Kahveci et al., 2009). Lipozyme RM IM gave the highest DAG yields for [TOMA]·[Tf_2N] and Ammoeng 120; however, FFA and MAG concentrations were also high, both of which are undesired by-products of the DAG production process. Novozym 435, on the other hand, resulted in significantly lower MAG and FFA concentrations at the cost of a slightly low DAG yield, especially for [TOMA]·[Tf_2N]. The highest DAG yield was observed by the Lipozyme TL IM-catalyzed reaction in [TOMA]·[TFA], but all three lipases gave unacceptably high MAG and FFA concentrations in this IL. Different performances of different lipase preparations have been associated with the property of individual carriers; however, the effect of the IL property on enzyme performance in terms of reaction specificities was not investigated further in this work.

The property of ILs other than operation variables plays an important role to determine the conversion and selectivity of glycerolysis, even though the reaction could be remarkably improved through parameter optimization. Based on the work done by Kahveci et al. (2009), [TOMA]·[Tf_2N] and Ammoeng 120 were suggested to be preferable choices for DAG production through glycerolysis. The common structural characteristics of these two types of ILs are that both have larger hydrophobic substituents in the cation to allow oil to be dissolved at higher concentrations, which might be used as one of the indications to select an IL for DAG preparation.

8.3.3 Binary Systems for Diacylglycerol Synthesis

The use of ILs as the reaction medium for DAG synthesis through lipase-catalyzed glycerolysis led to a product with ~45% of DAG, ~30% of which

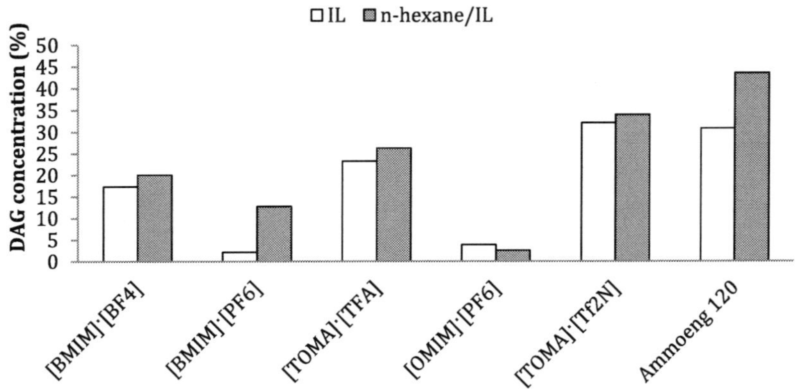

FIGURE 8.5 Comparison of DAG yield obtained in IL and *n*-hexane/IL binary systems. Reaction conditions: 2 mmol of oil (1768 mg), 1 mmol of glycerol (92 mg), 177 mg of Novozym 435, 60°C, 24 h, 700 rpm. Solvent used: 1 g of IL for IL systems, 0.5 mL: 0.5 mL for binary systems. *Data from Kahveci et al. (2010).*

was 1,3-DAG. In order to improve the yield and purity further, application of binary systems was investigated (Guo et al., 2009; Kahveci et al., 2010).

The relatively high viscosity of ILs is considered a limiting factor for various applications. The addition of organic solvents as cosolvents to ILs was considered to be practical for a dramatic reduction in the viscosity without changing the cations or anions in the system. As a common organic solvent for the food industry and also friendly to enzyme activity, *n*-hexane was directly selected as a model system for initial evaluation (Kahveci et al., 2010). [BMIM]·[BF_4], [BMIM]·[PF_6], [TOMA]·[TFA], [OMIM]·[PF_6], and [TOMA]·[Tf_2N] were mixed with *n*-hexane (1:1 by vol) and used as the glycerolysis medium catalyzed by *C. antarctica* lipase B (Novozym 435). For all the ILs tested except [OMIM]·[PF_6], the addition of *n*-hexane resulted in slight improvements in DAG yield (Fig. 8.5).

As a further step, binary IL systems were tested for the same reaction (Guo et al., 2009; Kahveci et al., 2010). Based on the previous work by Kahveci et al. (2009), cross-combination of two DAG-selective ILs ([TOMA]·[Tf_2N] and Ammoeng 120) and two MAG-selective ILs (Ammoeng 100 and 102) into binary IL systems have been examined for enzymatic production of DAG (Guo et al., 2009). [TOMA]·[Tf_2N]/Ammoeng 100 system resulted in high DAG yield and TAG conversion, while MAG concentration was relatively lowered compared to the pure Ammoeng 100 system. Addition of Ammoeng 100 or Ammoeng 102 to Ammoeng 120 did not improve TAG conversion; moreover, DAG yield was decreased in Ammoeng 120/Ammoeng 100 systems compared to pure Ammoeng 120. This observation was believed to be resulted from the high specificity of Ammoeng 100 and Ammoeng 102 for MAG production (Guo et al., 2006). Balancing the pros and cons of the evaluated binary IL systems, the authors decided to take the [TOMA]·[Tf_2N]/Ammoeng 102 system for further optimization (Kahveci et al., 2010).

Optimization of the reaction conditions for lipase-catalyzed glycerolysis in [TOMA]·[Tf_2N]/Ammoeng 102 system was performed using response surface methodology (Kahveci et al., 2010). Factors of interest were temperature, glycerol amount, and [TOMA]·[Tf_2N] concentration, whereas the responses optimized were DAG yield and TAG conversion. ANOVA showed that none of the investigated factors had a significant effect on DAG yield. The quadratic effect of glycerol amount was found to have the greatest effect on the response. DAG yield increased with increasing glycerol amount, the reason of which can be the improved availability of glycerol. In viscous and hydrophobic IL medium, it would be hard for the enzymes to interact with the highly hydrophilic glycerol. Temperature increase also had a positive effect on DAG yield. As [TOMA]·[Tf_2N] concentration in the medium decreased, in other words, Ammoeng 102 concentration increased, DAG yield increased as well. At this point, we can question the reason for using a binary IL system instead of Ammoeng 102 only, since the addition of [TOMA]·[Tf_2N] seems to have a negative effect on DAG yield. However, it should be kept in mind that Ammoeng 102 resulted in unacceptably high MAG concentration (38.8% at similar conditions; Kahveci et al., 2009), which was the reason that this IL had not been chosen for further investigation. The binary system of [TOMA]·[Tf_2N]/Ammoeng 102, on the other hand, resulted in only 3% MAG, which can be neglected. As the second response, none of the factors was found to have a significant effect on TAG conversion either. In contrary to the model generated to describe DAG yield, temperature had a slight effect on TAG conversion. Considering the cost issues and enzyme stability, a temperature in the range of 45–60°C can be chosen for the reaction. The effects of glycerol amount and [TOMA]·[Tf_2N] concentration were similar to the previous model; TAG conversion increased with increasing glycerol and decreasing [TOMA]·[Tf_2N].

The reasons for using a binary IL system consisting of [TOMA]·[Tf_2N]/Ammoeng 102 for DAG production can be summarized as (1) high DAG yield, preferably with a high proportion of 1,3-DAG isomer; (2) high TAG conversion; and (3) low MAG concentration. According to the results, these goals were generally achieved. Reactions in the mentioned system resulted in 55–75% DAG yield in the ranges used for investigated factors. This value is higher than those obtained with the other ILs used for optimization in the previous steps. Pure Ammoeng 102 had resulted in a higher DAG yield; however, MAG concentration in this IL was too high. TAG conversion of the binary system was also improved compared to the previously used ILs, again except Ammoeng 102, but as already mentioned, the binary system was preferable considering all desired parameters. Considering the optimum conditions chosen for the two models separately, it was suggested that a temperature between 55 and 60°C, glycerol amount between 1.8 and 2 mmol, and

[TOMA]·[Tf_2N] concentration less than 10% can be chosen for high DAG yield (~70%) and TG conversion (~90%) (Kahveci et al., 2010).

8.4 FUTURE PERSPECTIVES

Due to their tunable properties, ILs are promising for selective catalysis. The aforementioned developed protocols have been exemplified successfully for ILs as reaction medium to induce the selective catalysis.

The good performance from the protocols was quite advantageous over other reaction systems, in terms of MAG/DAG content, TAG conversion, and MAG/DAG selectivity. However, this protocol has not been moved to the industries, largely because of the high cost of ILs, despite the fact that ILs are recoverable and recyclable (Guo and Xu, 2006). Nevertheless, using ILs as reaction media for selective catalysis in the lipid modification field still has potential, especially with the rapid development in the field of IL synthesis. The use of ILs in industry was not a new concept; ILs have been used in electrochemical and chemical applications and also for nonsynthetic applications (Maase, 2007; Sheldon, 2001). Furthermore, ILs can be immobilized on mesoporous materials, which help to recycle them. Mesoporous materials with large pore size, like SBA-15, can serve as microreactors for glycerolysis reaction. Studies on this direction have not been seen; however, it looks interesting and promising (Zhong et al., 2014).

Another concern is the toxicity of ILs. In general, toxicity of ILs depends on both ions and the effect of the cation alkyl chain length is very pronounced, although the type of anion also exerts impact on the overall toxicity. Toxicity is an important concern for the use of ILs in food processing. As processing aids for food materials, regulatory approval is imperative. Most of the currently available ILs have not been approved for food processing. However, with the increasing attention on uses of ILs in food materials processing, the development of safer ILs is on the way. There have been a number of new claims from industry that the ILs developed can be used for food processing.

The last point needed to be addressed is the separation of ILs from other materials. As ILs have no volatility in principle, the distillation process can be used as this also commonly is used for MAG purification. However, data are needed to document the effectiveness.

REFERENCES

Chen, B., Guo, Z., Tan, T., Xu, X., 2008. Structures of ionic liquids dictate the conversion and selectivity of enzymatic glycerolysis: theoretical characterization by COSMO-RS. Biotechnol. Bioeng. 99, 18–29.

Corma, A., Iborra, S., Miquel, S., Primo, J., 1998. Catalysts for the production of fine chemicals. Production of food emulsifiers, monoglycerides, by glycerolysis of fats with solid base catalysts. J. Catal. 173, 315–321.

Damstrup, M.L., Abildskov, J., Kiil, S., Jensen, A.D., Sparsø, F.V., Xu, X., 2006a. Evaluation of binary solvent mixtures for efficient monoacylglycerol production by continuous enzymatic glycerolysis. J. Agric. Food Chem. 54, 7113–7119.

Damstrup, M.L., Jensen, T., Sparsø, F.V., Kiil, S., Jensen, A.D., Xu, X., 2005. Solvent optimization for efficient enzymatic monoacylglycerol production based on a glycerolysis reaction. J. Am. Oil Chem. Soc. 82, 559–564.

Damstrup, M.L., Jensen, T., Sparsø, F.V., Kiil, S., Jensen, A.D., Xu, X., 2006b. Production of heat-sensitive monoacylglycerols by enzymatic glycerolysis in *tert*-pentanol: process optimization by response surface methodology. J. Am. Oil Chem. Soc. 83, 27–33.

Feltes, M.M.C., Oliveira, D., Block, J.M., Ninow, J.L., 2013. The production, benefits, and applications of monoacylglycerols and diacylglycerols of nutritional interest. Food Bioprocess Technol. 6, 17–35.

Flickinger, B.D., 2004. The effect of diacylglycerols on energy expenditure and substrate utilization in humans. In: Katsuragi, Y., Yasukawa, T., Matsuo, N., Flickinger, B.D., Tokimitsu, I., Matlock, M.G. (Eds.), Diacylglycerol Oil. AOCS Press, Urbana, pp. 58–63.

Ganske, F., Bornscheuer, U.T., 2005. Lipase-catalyzed glucose fatty acid ester synthesis in ionic liquids. Org. Lett. 7, 3097–3098.

Guo, Z., Chen, B., Murillo, R.L., Tan, T., Xu, X., 2006. Functional dependency of structure of ionic liquids: do substituents govern the selectivity of enzymatic glycerolysis? Org. Biomol. Chem. 4, 2772–2776.

Guo, Z., Kahveci, D., Özçelik, B., Xu, X., 2009. Improving enzymatic production of diglycerides by engineering binary ionic liquid medium system. New Biotechnol. 26 (1/2), 37–43.

Guo, Z., Xu, X., 2005. New opportunity for enzymatic modification of fats and oils with industrial potentials. Org. Biomol. Chem. 3, 2615–2619.

Guo, Z., Xu, X., 2006. Lipase-catalyzed glycerolysis of fats and oils in ionic liquids: a further study on the reaction system. Green Chem. 8, 54–62.

Jérôme, F., Pouilloux, Y., Barrault, J., 2008. Rational design of solid catalysts for the selective use of glycerol as a natural organic building block. Chem. Sus. Chem. 1, 586–613.

Kahveci, D., Guo, Z., Özçelik, B., Xu, X., 2009. Lipase-catalyzed glycerolysis in ionic liquids directed towards diglyceride synthesis. Process. Biochem. 44, 1358–1365.

Kahveci, D., Guo, Z., Özçelik, B., Xu, X., 2010. Optimisation of enzymatic synthesis of diacylglycerols in binary medium systems containing ionic liquids. Food Chem. 119, 880–885.

Lau, M.R., Sorgedrager, M.J., Carrea, G., van Rantwijk, F., Secundo, F., Sheldon, R.A., 2004. Dissolution of *Candida antarctica* lipase B in ionic liquids: effects on structure and activity. Green Chem. 6, 483–487.

Lo, S.-K., Tan, C.-P., Long, K., Yussof, M.S.A., Lai, O.-M., 2008. Diacylglycerol oil-properties, processes and products: a review. Food Bioprocess Technol. 1, 223–233.

Maase, M., 2007. Industrial applications of ionic liquids. In: Wasserscheid, P., Welton, T. (Eds.), Ionic Liquids in Synthesis, second ed. Wiley-VCH Verlag GmbH & Co. KGaA, Weinheim, Germany, pp. 663–687.

McNeill, G.P., 1998. Enzymic processes. In: Gunstone, F.D. (Ed.), Lipid Synthesis and Manufacture. Sheffield Academic Press, Sheffield, pp. 288–320.

McNeill, G.P., Shimizu, S., Yamane, T., 1990. Solid phase enzymatic glycerolysis of beef tallow resulting in a high yield of monoglyceride. J. Am. Oil Chem. Soc. 67, 779–783.

McNeill, G.P., Shimizu, S., Yamane, T., 1991. High-yield enzymatic glycerolysis of fats and oils. J. Am. Oil Chem. Soc. 68, 1–5.

McNeill, G.P., Yamane, T., 1991. Further improvements in the yield of monoglycerides during enzymatic glycerolysis of fats and oils. J. Am. Oil Chem. Soc. 68, 6–10.

Murase, M., Mizuno, T., Omachi, T., Onizawa, K., Komine, Y., Kondo, H., Hase, T., Tokimitsu, I., 2001. Dietary diacylglycerol suppresses high fat and high sucrose diet-induced body fat accumulation in C57BL/6J mice. J. Lipid Res. 42, 372–378.

Pawongrat, R., Xu, X., H-Kittikun, A., 2007. Synthesis of monoacylglycerol rich in polyunsaturated fatty acids from tuna oil with immobilized lipase AK. Food Chem. 104, 251–258.

Rendón, X., Munguía, A.L., Castillo, E., 2001. Solvent engineering applied to lipase-catalyzed glycerolysis of triolein. J. Am. Oil Chem. Soc. 78, 1061–1066.

Sheldon, R., 2001. Catalytic reactions in ionic liquids. Chem. Commun. 23, 2399–2407.

Taguchi, H., Nagao, T., Watanabe, H., Onizawa, K., Matsuo, N., Tokimitsu, I., Itakura, H., 2001. Energy value and digestibility of dietary oil containing mainly 1,3-diacylglycerol are similar to those of triacylglycerol. Lipids 36, 379–382.

Tavares, A.P.M., Rodríguez, O., Macedo, E., 2013. New generations of ionic liquids applied to enzymatic biocatalysis. In: Kadokawa, J. (Ed.), Ionic Liquids-New Aspects for the Future. InTech, pp. 537–556.

van Rantwijk, F., Secundo, F., Sheldon, R.A., 2006. Structure and activity of *Candida antarctica* lipase B in ionic liquids. Green Chem. 8, 282–286.

Xu, X., 2004. Biocatalysis for lipid modifications. In: Dunford, N.T., Dunford, H.B. (Eds.), Nutritionally Enhanced Edible Oil and Oilseed Processing. AOCS Press, Urbana, IL, pp. 239–263.

Yasukawa, T., Katsuragi, Y., 2004. Diacylglycerols. In: Katsuragi, Y., Yasukawa, T., Matsuo, N., Flickinger, B.D., Tokimitsu, I., Matlock, M.G. (Eds.), Diacylglycerol Oil. AOCS Press, Urbana, pp. 1–15.

Zhong, N., Cheong, L.-Z., Xu, X., 2014. Strategies to obtain high content of monoacylglycerols. Eur. J. Lipid Sci. Technol. 116, 97–107.

Zhong, N., Li, L., Xu, X., Cheong, L.-Z., Li, B., Hu, S., Zhao, X., 2009. An efficient binary solvent mixture for monoacylglycerol synthesis by enzymatic glycerolysis. J. Am. Oil Chem. Soc. 86, 783–789.

Zhong, N., Li, L., Xu, X., Cheong, L.-Z., Xu, Z., Li, B., 2013. High yield of monoacylglycerols production through low-temperature chemical and enzymatic glycerolysis. Eur. J. Lipid Sci. Technol. 115, 684–690.

Zhong, N., Li, L., Xu, X., Cheong, L.-Z., Zhao, X., Li, B., 2010. Production of diacylglycerols through low-temperature chemical glycerolysis. Food Chem. 122, 228–232.

Chapter 9

Ionic Liquids in Lipid Extraction and Recovery

Michael J. Cooney
University of Hawaii at Manoa, Hawaii Natural Energy Institute, Honolulu, HI, United States

Kenneth Benjamin
South Dakota School of Mines and Technology, Department of Chemical and Biological Engineering, Rapid City, SD, United States

9.1 INTRODUCTION

9.1.1 Market Potential of Lipids

The market for lipids is a broad and expanding one that includes nutraceuticals, cosmetics, and precursors for fuels such as biodiesel. However, it can also suffer periodic fluctuations in demand, which can undermine product revenues. Biodiesel production, for example, rapidly expanded from about 5 million gallons in 2001 to 250 million gallons in 2006 (Shackleton et al., 2006), and starting in 1999, increased in the United States at an annual rate of about 200% (U.S. Energy Outlook, 2009). Starting in 2009, however, rising soybean prices and falling crude oil and diesel prices abruptly limited the competitiveness of the biodiesel industry. Regardless, the capacity for US biodiesel production remains over 1 billion gallons per year spread over about 150 plants, of which only a few have stopped producing (Gallagher, 2011). In summary, the process of biodiesel production remains one limited on a cost basis and in need of technology that reduces processing costs. A recent analysis, for example, considered the economics of producing biodiesel from microalgae using very reasonable assumptions of biomass productivity (30 gdw/m^2 per day) and lipid concentration (30% w/w) (Cooney et al., 2010). To use a cash flow, net present value calculation of the operating costs did not include the debt service cost for the capital outlays. The analysis did assume, however, all capital investments as cash (nonrecurring costs) and compared this amount with discounted cash flow returns (revenues less recurring costs) projected for the future. The cost of land was computed separately and system parameters were slightly adjusted and rounded for simplicity. The authors

Ionic Liquids in Lipid Processing and Analysis. http://dx.doi.org/10.1016/B978-1-63067-047-4.00009-X

concluded that positive net present values with reasonable rates of return will be possible only if biomass productivities and lipids yield well above 30 gdw/m^2 per day and 35% w/w, respectively, and crude oil prices above \$100 per barrel become a reality along with substantial subsidies or tax breaks. As such, an economically viable algae-to-biodiesel industry will depend upon initial government subsidies, increased price of oil, and addition to optimized biomass yields.

The nutraceutical market shows more potential. There is significant market interest in omega-3 fatty acid lipids, with the estimated global demand for omega-3 ingredients being USD 1595 million in 2010 and this is expected to be greater than USD 4000 million in 2018. The omega-3 fatty acids eicosapentaenoic acid (EPA) and docohexaenoic acid (DHA) provide significant health benefits for brain function/development and cardiovascular conditions and are considered from a nutrition standpoint to be the most important very long chain polyunsaturated fatty acids belonging to the omega-3 family (Adarme-Vega et al., 2014). Although wild fish have been the most common source of EPA and DHA for human consumption in the last decades (Adarme-Vega et al., 2014), their stocks are at great risk of being overfished (Hutchings and Reynolds, 2004). The Food and Agriculture Organization, for example, reported in 2008 that nearly 53% of marine fish stocks were close to being entirely exploited, 28% were overexploited, and 4% and were either depleted (3%) or recuperating from depletion (1%) (FAO, 2010).

There is growing consumer interest in cosmetics infused with natural and/or organic oils due to their high fatty acid content (Vermaak et al., 2011), with the major markets being Europe and North America (Shackleton et al., 2006). Natural seed oils used in cosmetics contain a range of fatty acids that contribute several beneficial properties to cosmetic and personal care products (Vermaak et al., 2011). As such, oils extracted from plant sources are used in food, energy, medicines, and for cosmetic applications. They are also used in the production of lubricants, soaps, and personal care products, as well as in the topical treatment of various conditions such as hair dandruff, muscle spasms, varicose veins, and wounds (Zimba et al., 2005). This has fueled a remarkable growth in the sales of natural organic cosmetic care products sold through retail outlets, pharmacies, and skin care clinics worldwide. The recovery of lipids from natural sources will all require the application of efficient extraction processes to remove and purify the lipids.

9.1.2 Biomass Sources of Lipids and the Process Implications of Their Extraction

The principal sources of biomass lipids include food waste collected from grease trap intercepts (Lopez et al., 2014), leftover waste recovered from animal or fish processing plants (Adeoti and Hawboldt, 2014), oil seeds (Godwin Sevara and Cooney, 2013), and high lipid bearing unicellular microorganisms (Vermaak et al., 2011). Food waste, collected from grease trap intercepts, contains free-standing fats, oils, and greases (FOG). Their extraction requires phase separation from water—a process that requires heating to achieve rapid and effective separation (U.S. Energy Outlook, 2009). The FOG is then pumped to a separate tank where it is cooled and stored until it is further processed into biodiesel (U.S. Energy Outlook, 2009).

The extraction of bio-oil from animal and fish waste is slightly more involved owing to the entrapment of the bio-oil in fat cells and tissues that must be ruptured through the application of physical, chemical, or biological treatments (Adeoti and Hawboldt, 2014). Physical treatments typically use the application of heat to rupture the fat cells, followed by the application of pressure (via pressing or centrifugation) to squeeze out the remaining bio-oil that is then stored until further processed into biodiesel (Adeoti and Hawboldt, 2014). Biological treatments apply the use of internal (autolysis) or external (hydrolysis) enzymes to break down proteins into amino acids or peptides with the bio-oil released as a by-product of the process (Adeoti and Hawboldt, 2014). Chemical treatment applies solvents under a variety of temperatures and pressures to both penetrate and break down cell and tissue barriers such that the solvent can contact the lipids. Most applications use hydrophobic organic solvents that solubilize the lipids (Cooney et al., 2009), but other emerging techniques use solvent systems that are not miscible with the lipids (Young et al., 2009). Solvent treatment is generally accelerated or enhanced by the application of high pressures (accelerated solvent extraction or supercritical solvent extraction) and/or temperatures (soxhlet digestion) that promotes solvent penetration through the internal barriers (Cooney et al., 2009). Although these systems are high energy and involve complex and costly capital equipment, they theoretically permit the manipulation of conditions (particularly with respect to supercritical conditions) that alter the physical properties of the solvent to facilitate more selective extraction.

Oil seeds present more rigid external barriers such as outer shells or hulls (Gallagher, 2011) that can limit yields when not broken or removed. The bio-oil is accumulated internally into subcellular structures called oil bodies or oleosomes (Hutchings and Reynolds, 2004) of diameters between 0.2 and 2.5 μm that comprise a matrix of triglycerides with the phosphate groups arranged in an exterior monolayer facing the cytosol (FAO, 2010). The oleosomes do not aggregate or coalesce within mature seeds owing to a layer of proteins (oleosins) that can cover their entire surface and improve their stability by means of steric hindrance and electronegative repulsion (Zimba et al., 2005). Because oils seeds are large (scale of millimeters) and possess rigid structures, the extraction of internal bio-oil can be executed on an industrial scale by mechanical presses or by using organic solvents applied to crushed or ground seeds (Global omega 3 ingredients market (EPA/DHA), 2013). Mechanical presses provide high-quality oil but, in most cases, the treatment has a low extraction rate and consumes excessive amounts of energy (Global omega 3 ingredients market (EPA/DHA), 2013). Solvent extraction can remove practically all the extractable oil although the serious environmental concerns related to the industrial use of such solvents, as well as the high temperature (ie, energy costs) and extended extraction times (ie, product degradation) often required, reduces its effectiveness (Adarme-Vega et al., 2014).

Chemical extraction of lipids from unicellular microorganisms is a complex process wherein the extent of lipid extraction is dependent upon the microorganism, the physiology of the lipids within the cells, the extent of cell wall rupture, and the chemical solvent used (Cooney et al., 2009). Although chemical solvents are the most widely used they are ineffective unless the cell walls have been ruptured or if the solvents are applied under conditions (heat or pressure) that enable rupture and penetration of the cell wall (Cartens et al., 1996). Chemical extraction of bio-oil from unicellular microorganisms is also challenged by the presence of water, which hinders migration of nonpolar lipophilic solvents into the lipid-containing cellular matrix-hindering efficient transfer of oil into the solvent phase (Samarasinghe et al., 2012), the barrier of the cell wall, and the biological matrix surrounding the triglyceride droplets (within the cells). Adding to these difficulties is the distribution of the lipids across millions of micron-sized unicellular microorganisms. As such it is difficult to design machinery that can apply the necessary sheer force at the scale of an individual cell as those that do (high pressure homogenization or ultrasonication) consume energy at levels difficult to support at a commercial scale.

The source of lipids will also ultimately affect the final choice of extraction methodology. For example, the biomass source may also contain additional fat soluble compounds, such as proteins or photosynthetic pigments (Balasubramanian et al., 2013), that can undermine the quality of the extracted bio-oil as a raw material for fuel processing. In this case the application of supercritical fluid extraction to photosynthetically grown microalgae will not provide a clean solution of lipids but rather a gooey, viscous, tarry substance that is most difficult to process. Depending upon their solubility in lipids the contaminants can be difficult to remove and doing so can add significant cost to the preprocessing of the extracted lipids prior to their conversion into fuel. The presence of these substances, then, has a great influence on the extraction process selected; the process selected must accommodate their initial inclusion in the bio-oil and must support the subsequent unit operations applied to purify the bio-oil.

9.1.3 Definition of Ionic Liquids and Their Emerging Role as Solvents

Room temperature ionic liquids (RTILs) are a new class of liquid solvents that possess negligible vapor pressure and whose macroscale properties of polarity and hydrophobicity can be manipulated through choice of cation and/or anion structure. Less discussed but no less important is their heterogeneous distribution of charge and hydrophobic/hydrophilic regions throughout the solvent space. These microscale distribution patterns can be modified through substitution of aliphatic groups attached to nitrogen groups within the cation's ring structure, as well as the choice of anion, actions that ultimately modify the solvent's macroscale properties. These features have led to their promotion as tunable and green solvents with increasing applications in liquid–liquid or liquid–gas extraction systems. Ionic liquids (ILs) have shown great potential for separations, including carbon dioxide capture and sequestration, and extraction and recovery in bioprocessing applications. ILs have also been shown to extract protein from aqueous two-phase systems (Pei et al., 2009), extract hormones from urine samples (He et al., 2006), esterify lipids (Lue et al., 2007), and be used as a green solvent for biodegradable copolymers (Zhua et al., 2008).

9.2 THE EXTRACTION PROCESS

9.2.1 Unit Operation versus Process

A process can be broken down into a series of integrated steps, called unit operations, each of which can be modeled using scientific principles. The general purpose of a process is to transform a material input to a desired final product while the general function of a given unit operation is to execute individual steps in that overall transformation. Typical unit operations include those that separate, crystallize, evaporate, filter, polymerize, isomerize, and cause other reactions. Although their performance is dependent upon the process into which they are integrated, the majority of literature presentations addressing unit operations that extract lipids evaluate them as standalone operations. Such analyses can be misleading since their performance is ultimately tied to their placement in a process that specifies rates, temperatures, and quantities of material inputs (to the unit operation), constraints on processing conditions (temperature, rates of flow through the system, heating or cooling transfers rates, etc.) applied, as well as input specifications of the next downstream unit operation. In terms of extracting lipids from oil-bearing biomass the purpose of the unit operation is to separate (ie, extract out) lipids from oil-bearing biomass. As will be described later, the decision as to the most appropriate unit operation is highly dependent upon the choice of upstream and downstream unit operations.

9.2.2 System Diagram

As discussed earlier, to select and design an effective extraction unit operation, its interrelationship with the entire process and, more specifically, the unit operations directly up and downstream should be considered. As an example,

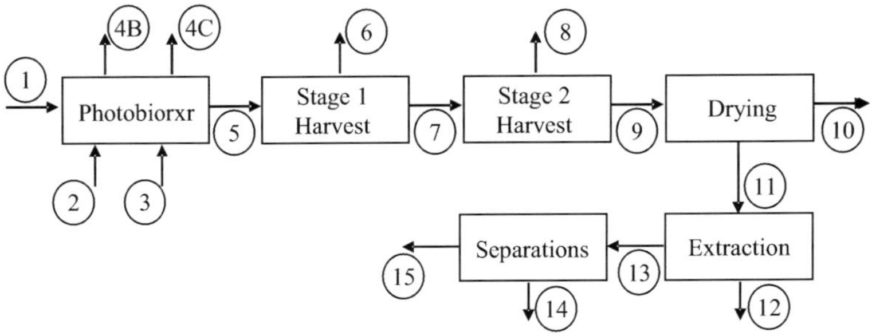

Stream #	Description	Stream #	Description
1	Inlet feed water	8	Water removed from stage 2
2	CO_2 or air	9	Stage 2 concentrated culture
3	Nutrients	10	Water removed from dryer
4a	Evaporation water loss	11	Dried culture (<1% water)
4b	Exiting gas stream	12	Solid biomass
5	Harvested culture	13	Extracted lipids
6	Water removed from stage 1	14	Polar lipids
7	Stage 1 concentrated culture	15	Nonpolar lipids

FIGURE 9.1 Microalgae to bio-oil process flow diagram.

Fig. 9.1 presents a representative process, in block diagram form, to extract bio-oil from bio-oil bearing biomass (Cooney et al., 2009). Typical biomass sources include microalgae or oil seeds with key characteristics being the degree of water present as well as the form of barriers separating the internal lipids from external solvent. Oil seeds are largely dehydrated while microalgae, depending upon the extent to which they have been dried, can possess over 90% water by weight. Water is a natural solvent whose molecules have a high degree of interaction through hydrogen bonding and as such the amount of water present can significantly impact the choice and/or efficiency of the extraction technology selected (Cooney et al., 2009). The water content of microalgae can be as high as 80–85% water (on a mass basis); even after centrifuging the harvested microalgae remains a paste possessing a suspended solid concentration of 15–20 wt%. Despite the fact that drying the microalgae further represents a major energy cost, the majority of lipid extraction technologies presented in the literature begin with biomass that has been completely dried (Cooney et al., 2009). Extraction unit operations that can operate in the presence of water have been proposed (Cooney et al., 2009) and from energy considerations (discussed elsewhere; (Cooney et al., 2009)), the application of bio-oil extraction methodologies that work on wet biomass is highly desired (Lidell, 2001).

Regardless of whether the extraction process is executed in the presence or absence of water, the extracted lipids (triglycerides and free fatty acids) must be recovered and converted into biofuels that meet the appropriate ASTM standards. As such the extraction technology needs also to be evaluated for the purity of the extracted lipids. For example, some biomass sources (such as photosynthetically grown microalgae) contain fat soluble pigments (eg, chlorophyll) that can not only interfere with the extraction process but also with downstream operations trying to convert the extracted lipids into a fuel. In the case of supercritical fluid extraction, for example, their interference on the extraction process is negligible but their presence produces a tarry lipid residue that is difficult to fractionate and convert into a usable biofuel. ILs, by contrast, offer the potential to selectively extract lipids although the pigments can, if charged, interfere at a molecular level with the action of the cation and anion molecules.

The extraction operation must also consider the strength, structure, and composition of the biomass barriers between the solvent and bio-oil, and how these will impact the efficiency and rate of lipid extraction. For example, the degree to which the biomass cell wall is damaged or degraded by the solvent can lead to altered properties in terms of the density or chemistry of the extracted biomass. This can lead to problems in the separation of the biomass from the extracted lipid (or solvent), particularly if the separation process is based upon differences in density. By contrast, extraction systems that are effective over a broad range of species and cell wall structures, such as those operated under more extreme conditions (ie, higher temperature and pressure associated with accelerated, subcritical water, or even supercritical fluid) can provide relatively robust solutions. Unfortunately, these processes are generally energy intensive, which works to minimize their effectiveness as the energy consumption can be high relative to the energy density of the extraction bio-oil. Finally, ILs have received attention due to the cost and environmental concerns associated with organic solvents. Despite promising reports, however, their application suffers the additional requirement of recovering the target compound from the RTIL (Sun et al., 2011)—a process made problematic by the negligible vapor pressure of RTILs.

9.3 THE EXTRACTION UNIT OPERATION

9.3.1 Overview

Solvent extraction is one of the most widely applied methods to recover lipids. Traditional chloroform-based lipid extraction is effective for the lab-scale exercises, but an alternative organic solvent, like hexane, which is easier to handle and more cost-effective, is needed for industrial scale activities (Halim et al., 2012). A challenge for solvent extraction methods is the physical algal cell envelope, which prevents direct contact between lipid and solvent. Cell disruption assists lipid extraction as it liberates microalgae lipids from the cellular matrix into the surrounding medium (Halim et al., 2012). As such, mechanical, chemical, and biological cell disruption techniques are widely studied as a means to overcome this problem (Lee et al., 2010; Prabakaran and Ravindran, 2011). Mechanical methods such as ultrasound, high pressure homogenization, and microwave pretreatment have been proven to be effective, but require high energy input (Halim et al., 2013). Alternative chemical and biological cell disruption techniques are attracting escalating attention. But despite the high cell-disruption performance, the potential to scale up these methods is still limited since they require continuous use of expensive chemicals and enzymes (Jin et al., 2012; Mendes-Pinto et al., 2001).

9.3.2 Organic Solvents

The simplest extraction system uses single homogeneous solvents to contact and solubilize the target lipid. This is based upon the concept that like dissolves like and a simple application would be the use of a bio-oil solvent to solubilize and extract lipids from bio-oil bearing biomass. In their 1996 report to the DOE, for example, Bennemann and Oswald (1996) suggested the use of hot oil to extract and recover carotenoids from *Dunaliella salina* (Kanel and Guelcher, 1999). The process exploits the observation that *Dunaliella* possesses a weak cell wall that easily breaks down to form a thick emulsion of water and oil when exposed to the shear forces present in a homogenizer (or pressure expansion valve). This water-based emulsion can be extracted with the hot solution of bio-oil, presumably a portion of which has already been collected and purified, with the entire mixture then passed through a three-phase centrifuge (Cooney et al., 2009). Although conceptually attractive, the lipid concentration (per unit volume of water being processed) is relatively small compared to the volume of extracting oil that would need to be used, making it difficult to achieve realistic recoveries.

In those cases where the cell barriers are not easily penetrated, the next level of complexity is the application of two solvents, one of which attaches the cell barrier and the other which dissolves the lipid. In general the solvents are chosen on the basis of their miscibility as well as their differentiation with respect to polarity. The more polar solvents will disrupt the hydrogen bonding and electrostatic forces between the lipids and proteins (Kates, 1986) while the more nonpolar solvents will favorably interact with the largely hydrophobic lipids. After the extraction, a third solvent (eg, water) that is immiscible with the lipid-bearing solvent is added to create a two-phase system wherein the lipid-bearing solvent can be separated and processed via evaporative distillation to recover the lipid. The most common application, credited to Bligh and Dyer (1959), uses methanol and chloroform. Other combinations have been proposed: hexane/isopropanol for tissue

(Hara and Radin, 1978), DMSO/petroleum ether for yeast (Park et al., 2007), hexane/ethanol for microalgae, and hexane/isopropanol for microalgae (Nagle and Lemke, 1990). The hexane system, in particular, has been promoted because the hexane and alcohol will readily separate into two separate phases when water is added, thereby improving downstream separations (Schafer, 1998).

There are a few important limitations, however, to these systems. The first is that the solubility (g/l) of the lipids in the lipid miscible cosolvent is often limited, which requires unreasonably large amounts of that solvent to be used. This emphasizes the point that the use of more polar solvents to improve the range of lipids extracted may also decrease the carrying capacity of the lipid-bearing cosolvent because, in general, solvents that extract polar lipids are less miscible with relatively high ratios of nonpolar lipids; lower miscibility leads to lower carrying capacity (Cooney et al., 2009). The second is that organic solvent systems are not always effective at penetrating cell barriers and thus require an initial step to physically disrupt the cell membrane to avoid the need of elevated temperature and pressures. Finally, energy is required to recover the extracted lipids from the less polar lipid-bearing solvent.

As discussed earlier the organic-based cosolvent systems are challenging to operate at commercial scale due to the toxicity of the organic solvents (ie, methanol and chloroform), and due to the low carrying capacity of the more nonpolar solvent (ie, the Bligh and Dyer method is only efficient on biomass samples containing less than 2% w/w lipids (Iverson et al., 2001)). In response, single solvent systems at elevated temperature and pressure have been proposed. These include those at moderate temperature and pressures (ie, accelerated solvent extraction (ASE) in the mid 1990s by Richter et al. (1996) and Romanik et al. (2007), and those at more extreme ranges such as subcritical water extraction (Ayala and Castro, 2001) and supercritical fluid extraction (Luque de Castro et al., 1994). The elevated temperature and pressure increase the rate of mass transfer and degree of single solvent access to all pores within the biomass matrix. The elevated pressures can also reduce the dielectric constant of an otherwise immiscible solvent (and by analogy the polarity) to better match that (ie, the polarity) of the lipids (Richter et al., 1996; Herrero et al., 2006); the miscibility of the lipid in the solvent is as important as its access to the lipids being extracted. In the case of ASE the solvents used are those normally proposed in typical distillation-based methods such as Soxhlet extraction. In supercritical fluid extraction carbon dioxide has been the most common. In addition to improving yields and dramatically reducing extraction time, these systems have been used to selectively extract polar compounds from lipid-rich samples and to fractionate lipids from biological samples (Cooney et al., 2009). Although effective, these systems are difficult to design and operate at a large scale and incur high energy loads to heat the system to sub- or critical temperatures. Large-scale systems will also require significant cooling systems to rapidly cool the product to room temperature in order to avoid product degradation. Such cooling systems are generally energy intensive.

9.3.3 Ionic Liquids

Since the introduction of the first air- and water-stable ILs, countless studies have explored the influence of cation and anion structure on the physicochemical properties of these unique solvents and on their utility in a wide variety of applications (Wilkes and Zaworotko, 1992). In addition to studies on density (Tokuda et al., 2005), water content (Huddleston et al., 2001), thermal (Ngo et al., 2000), lubricity (Weng et al., 2007), and conductivity (Hagiwara and Ito, 2000), ILs have been applied to extract protein from aqueous two-phase systems (Pei et al., 2009), hormones from urine samples (He et al., 2006), and increasingly lipids from bio-oil bearing biomass (Choi et al., 2014a,b). Not surprisingly they are considered by some to be preferable to water and organic solvent as catalysts and media for the purposes of synthesis or extraction (Marsh et al., 2004). The motivation to consider ILs as extraction solvents stems from how the nature of the cationic and anionic constituents, at a molecular scale, influence the macroscale physical properties of the solvent.

ILs, particularly those with Cl^- as the anion, are known competitors for hydrogen bonding. Li et al. (2013), for example, noted the essential role played by the short-range hydrogen-bonding interaction in the interfacial activity of ILs. As such, biomass with cell barriers that possess significant hydrogen bonding, such as microalgae, have become targets of research applying ILs to enhance lipid extraction due to their excellent properties for cellulosic biomass dissolution (Choi et al., 2014b). Choi et al. (2014a), for example, compared extraction efficiency of pure and mixtures of ILs against organic solvents. The results were mixed with some ILs (1-ethyl-3-methyl imidazolium acetate, 1-ethyl-3-methyl imidazolium diethylphosphate, 1-ethyl-3-methyl imidazolium tetrafluoroborate, and 1-ethyl-3-methyl imidazolium chloride) giving higher yields and others (1-ethyl-3-methyl imidazolium ethyl sulfate and 1-ethyl-3-methyl imidazolium thiocyanate) given lower yields than traditional organic solvents. Fujita et al. (2013) reported the successful dissolution of wet and saliferous marine microalgae in polar ILs under mild conditions and without heating. Of particular relevance was their reported success in completely dissolving the microalgae despite the presence of water (95 wt%).

9.3.4 Ionic Liquid Cosolvents

Despite the aforementioned benefits, the industrial use of ILs is limited by their relatively high cost. As such the use of ILs in the presence of additional solvents such as methanol to mediate the extraction of lipids has been explored. The goal has been to reduce overall solvent costs and to add additional function at the molecular level without losing the unique, tunable properties of pure IL solvents. One such example is the partnering of an IL with a polar covalent molecule (PCM) to form a cosolvent mixture (Young et al., 2009). In their groundbreaking work, Cooney and Young (2009) proposed the feasibility of lipid extraction processes using mixtures of [Emim] [$MeSO_4$] and polar organic solvents. Fundamental to their system was the self-partitioning of the extracted lipids to a separate immiscible phase at moderate temperatures and atmospheric pressure. It was proposed that although the cosolvent solution was hydrophilic, the presence of hydrophobic side groups attached to the nitrogen of the imidazolium cation created molecular regions of hydrophobic tunnels through which the otherwise immiscible lipids could diffuse, and that the driving force was a combination of the energy provided from mixing and a thermodynamic preference for ionic bonding between the cations and anions as opposed to hydrophobic bonding between the lipids and the side chains.

Since organic polar covalent molecules such as methanol, acetone, and such are relatively inexpensive compared to ILs, their use, depending on composition, could significantly reduce overall solvent costs and make these IL cosolvent systems more economically viable. IL/PCM mixtures have shown a promising ability to extract and autopartition lipids from biomass, with yields surpassing those achieved with pure ILs alone (Young et al., 2009; Cooney and Young, 2009; Young et al., 2011). The autopartitioning feature of the IL/PCM solvent requires little energy input to initiate (compared to traditional hexane solvent) and the separation is facilitated with gentle mixing under mild heating. These same IL/PCM mixtures have also solubilized amphiphilic molecules, such as phorbol esters, while leaving the protein behind with the native biomass (Severa et al., 2013a,b). In this context, IL/PCM cosolvent mixtures provide transformative separation power of multiple solutes/products from biomass and increase the potential for realizing true, large-scale bioprocessing facilities.

Following on the work of Young et al. (2009), Kim et al. (2012) reported on the efficiency of lipid extraction from *Chlorella vulgaris* using mixtures of ILs and methanol. Their results verified the successful dissolution of biomass and improved yields over those achieved using the Bligh and Dyer method. In another effort, Choi et al. explored the lipid extraction from *C. vulgaris* using molten-salt/IL mixtures (Choi et al., 2014b). Technically a class of ILs, molten salts, have been shown to dissolve, swell, and decompose cellulose (Fischer et al., 2003) and have thus gained attention as new, highly efficient solvents and media for use in cellulose modification (Heinze and Koschella, 2005). In their work Choi et al. reported improved lipid extraction from *C. vulgaris* through the synergistic effects of molten salt and ILs. In particular the recovery of the lipids into hexane was improved due to swelling from the ionic effects of molten salt and ILs on the hydrogen bonds (Lovejoy et al., 2013), which permitted increased hexane penetration into the microalgae cell walls.

In particular, after IL treatment, the cell walls maintained their shape and permitted the lipids along with some inner materials to pass through the cell walls without destroying them.

IL extraction cosolvents have also been proposed to facilitate chemical transformations of the extracted product. For example, Young et al. investigated the ability of their IL-alcohol cosolvent system (Young et al., 2009; Cooney and Young, 2009) to facilitate the acid catalyzed (HCl) direct transesterification of intercellular lipids to fatty acid methyl esters (FAMEs) (Young et al., 2011). With the exception of the *Chlorella* microalgae, the conversion from triglycerides to FAMEs was nearly complete (100%) for all biomass sources investigated, and the gravimetric yields of FAME product were found to be similar to those in the absence of acid catalyst.

It is important to note that the IL and polar covalent molecules may interact in such a way as to modify their overall solvent properties from what might be predicted from their individual weighted contributions. Data accumulated by the authors, for example, have revealed a nonlinear response in terms of lipid extraction yield plotted as a function of the IL to PCM ratio. In Fig. 9.2 extraction yields of bio-oil from two varieties of oil seeds as a function of the IL (1-ethyl-2-methyl imidazolium methyl sulfate (EMIN-MS)) to PCM (methanol, CH_3OH) ratio are presented. Although the data clearly shows that both solvents are required to achieve full extraction efficiency, the data also shows that the yields are likewise low, between 2 and 15 wt%, at low and high IL to PCM ratios. However, at ratios between these two extremes, the yields across all biomass sources plateaus at 40 wt%.

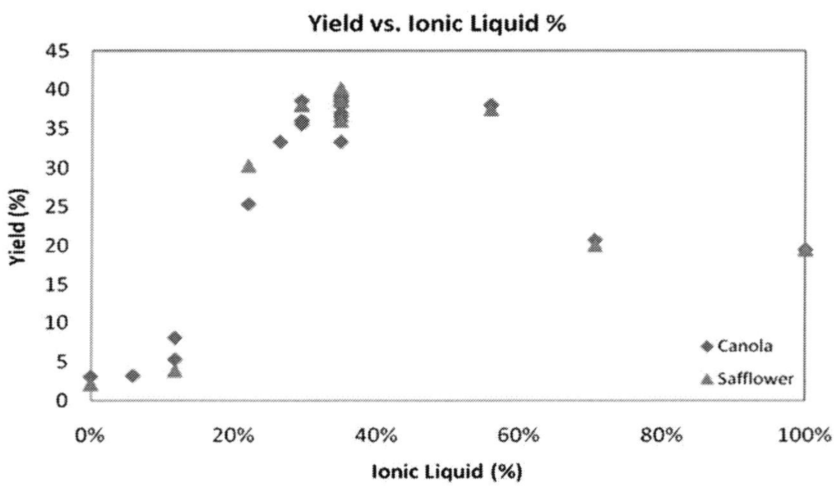

FIGURE 9.2 Bio-oil yield from jatropha, canola, and safflower oil seeds as a function of EMIN-MS to methanol concentration.

The authors explored this unexpected nonlinearity using steady-state polarization P_{SS} and light scattering data. Steady state fluorescent polarization (P_{SS}) is a method to yield information regarding the interaction of a fluorescent molecule with its immediate neighbors. In general its value is a measure of the degree to which the fluorescent molecule rotates around its axis. As fluorescent molecules become more fixed in space, as would occur if interactions with neighboring molecules is strong, the measured P_{SS} tends to zero. By contrast the P_{SS} tends to one for molecules that are fully free to rotate around their axis. As used in this work light scattering data measures the light (measured in photons) scattered at an angle of 90°. Micelles or other aggregates forming in solution can be detected at a molecular scale by comparing the photons scattered versus a baseline solution that has no aggregates.

The authors used these methods to test the hypothesis that the low extraction yields at low EMIN-MS to methanol ratios resulted from the aggregation of IL (EMIN-MS) micelles in the presence of high concentrations of methanol. Fig. 9.3A presents the P_{SS} of EMIN-MS as recorded in the presence of increasing concentrations of methanol. Methanol is also highly polar and has been reported to induce micelle formation in room temperature ILs (Hemmateenejad et al., 2011). As micelles form the rotation of the fluorescent cation molecules becomes more constrained due to increased molecular charge—charge interactions and the P_{SS} will decrease. Upon addition of methanol, however, the steady state polarization initially decreased suggesting that its initial addition supported greater rotational motility of the cation, likely reflecting either a dilution effect overriding any direct electrostatic interaction between the anion and cation, or a greater interaction between the methanol and anion (methyl sulfate), which would reduce the baseline cation—anion interaction. In either case, the cation was initially more free to rotate around its

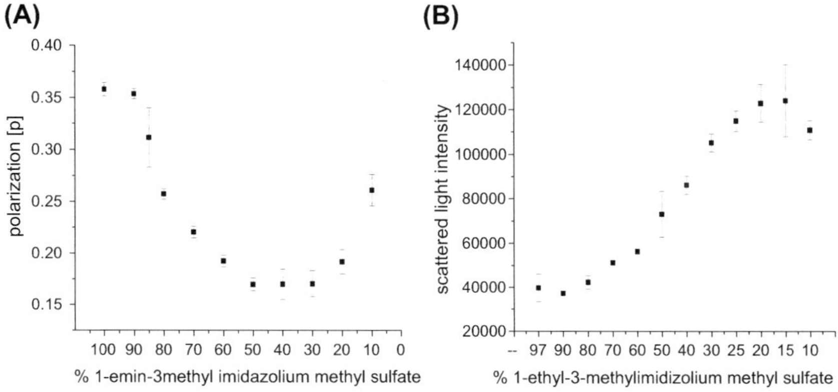

FIGURE 9.3 (A) Steady state polarization (excitation 340 nm, emission 370 nm) of EMIN methyl sulfate—methanol mixtures (EMIN-MS, 98% pure); (B) light scatter intensity EMIN methyl sulfate—methanol mixtures.

own axis. However, as the methanol reaches a critical concentration (~80% methanol, 20% RTIL) the P_{SS} began to increase, suggesting increased molecular interaction between the cation and solvent constituents that retarded its rotational motility. The literature suggests that imidazolium-based ILs will form micelles in aqueous solutions (Heintz et al., 2010; Łuczak et al., 2008, 2009; Zhang et al., 2008), a complex response that involves hydrogen bonding as well as coulombic and hydrophobic interactions; therefore, the authors propose this effect can be explained in terms of the anion and cation molecules approaching close proximity or aggregating together (Roth et al., 2012)—acts that would increase the electrostatic interaction between the two ions and thus retard the cation's rotational motility (Bandlamudi et al., 2014).

The light scattering data in Fig. 9.3B shows a similar trend, with light scatter increasing with increasing methanol concentration until approximately 20%, after which point the light scattering levels off. The increase in light scattering as a function of increasing methanol concentration reaffirms the formation of light scattering micelles. The slight drop off at the highest concentration is likely due to a transition in micelle size that either affects rapid phase separation (and thus less micelles to scatter light) or fewer but larger micelles that also reduce the effective light scatter. Combined, this data supports the theory of micelle formation at methanol concentrations above 80 wt%. They also further confirm the application of P_{SS} and LS as sensitive measurements for the initiation of micelle formation or precipitation in IL solvents that use a fluorescent cation (eg, those with an immidazolium ring).

9.4 THE RECOVERY UNIT OPERATION

9.4.1 Overview

Although extraction of lipids is the obvious first step, a necessary second step is to process the lipids into a biofuel. This requires recovering the lipids from the extraction solvent. When using conventional lipid soluble solvents such as hexane, for example, the lipids extract into the hexane solvent where they are miscible. To recover the lipids an additional unit operation must then be performed to separate the lipids from the hexane solvent. This is most often done by evaporative distillation wherein the lower boiling point of the hexane is exploited; as the hexane solvent is evaporated off the lipids concentrate into a recoverable bio-oil. Unfortunately, this process not only consumes extensive energy but also leaves behind any lipid soluble contaminants in the bio-oil (Cooney et al., 2009).

It would be more efficient to extract lipids with a solvent within which the lipids are insoluble such that they will naturally autopartition to their own separate phase. The most obvious choice is water but its efficacy is dependent upon cell barriers being efficiently disrupted or ruptured to facilitate rapid diffusion of the lipids from internal tissues into the water solvent. Rupture techniques include freeze-drying, sonication, homogenization, osmotic pressure, or the application of high temperature and pressure such as with subcritical water extraction (Cooney et al., 2009; Zhang et al., 2014). These are energy intensive and difficult to employ on a commercial scale. More, although conceptually attractive, extraction in the presence of water is also hampered by the tendency of water to form solvent shells around the lipids—a fact that hinders the rate at which the extracted lipids are able to aggregate into micelles large enough to rapidly autopartition into a separate immiscible phase. Water can also form troublesome emulsions with the extracted lipids in the presence of significant agitation and shear forces. For example, if the disruption step is applied to wet microalgae cell solutions, such as with multipass homogenizers, severe water–lipid emulsions can form (Samarasinghe et al., 2012). Techniques used to recover lipids from these systems range from addition of salts, countercurrent extraction with organic solvents, and even supercritical fluid extraction (Catchpole et al., 2008). These all require additional unit operations, however, that increase energy loads and costs.

9.4.2 Ionic Liquids in Recovery

As discussed earlier, the initial benefits of the IL and PCM solvents are their abilities to permeate and structurally modify biomass cell walls and tissues, so as to facilitate the release of internal biomass molecular constituents (lipids, esters, proteins, etc.) into the solvent medium. In addition to extraction an added feature to their system was the self-partitioning of the extracted lipids to a separate immiscible phase at moderate temperatures and atmospheric pressure. It was proposed that although the cosolvent solution was hydrophilic, the presence of hydrophobic side groups attached to the nitrogen of the imidazolium cation created regions of hydrophobic tunnels through which the otherwise immiscible lipids could diffuse, and that the driving force was a combination of the energy provided from mixing and a thermodynamic preference for ionic bonding between the cations and anions as opposed to hydrophobic bonding between the lipids and the side chains. The autopartitioning feature of the IL/PCM solvent required minimal energy input to initiate (compared to traditional hexane solvent) and the separation is facilitated with gentle mixing under mild heating. However, with the autopartitioning of lipids initially noted by Young et al. to a separate phase (Heinze and Koschella, 2005), came the potential of using ILs and their mixtures with polar covalent molecules to provide low-energy technologies supporting the recovery of the extracted lipids through their solubilization and/or aggregation (or phase separation), as a means of enhancing downstream purification of extracted analytes.

In pure water systems, lysed or ruptured microalgae cell solutions are not observed to rapidly produce a separate and immiscible bio-oil phase. Lipids from such systems are extracted by washes with organic solvents that extract and solubilize the lipids (Cooney, 2013). Similarly, large-scale extraction of bio-oil from oil seeds uses organic solvents such as hexane as opposed to water (Cooney, 2013). By contrast, IL cosolvent systems can promote the rapid autopartitioning of the extracted lipids to a separate immiscible phase. The authors have accumulated data that suggests once individual lipid molecules have dissolved into the IL-based cosolvent mixture the extracted lipids rapidly aggregate into micelles that rise to the surface where they form a separate immiscible phase. Fig. 9.4A presents light scattering data for jatropha oil seed in the presence of pure water. The data indicates that little to no oil has been extracted. Fig. 9.4B presents the same experiment but with the jatropha seed submerged in pure methanol. In this case after a short time period the photon count (light diffracted) immediately rose to a steady count of 400,000 where it remained, indicating micelles did not aggregate in sufficient size to coalesce and rise to the top. Instead an emulsion was formed and the light scattering remained constant.

This result confirms the accepted knowledge that methanol has some efficacy as a bio-oil extractant but also confirms the lack of rapid bio-oil separation that is observed with pure hydrophilic solvents such as water or methanol. By contrast, the same experiment using an IL-based cosolvent, as presented in Fig. 9.4C, shows discrete photon counts during the first 1200 s. The large spikes in the light scatter measurement are indicative of discrete particles (ie, oil micelles) passing through the light path. As the majority of spikes ceased after 1500 s, the surface oil extraction was further assumed to occur within the first 25 min and the longer term slow rise in the baseline (data not shown) was thought to represent the effect of color change (via pigments) to the solvent. This result is considered largely a thermodynamic effect, owing to the free energy changes of the lipid once in solution, and its propensity to aggregate or phase-separate, likely to avoid unfavorable solute–solvent interactions.

9.4.3 Recovery of Lipids from Ionic Liquids

In general the literature does not possess examples of pure ILs being used to extract lipids. Those that do exist span the use of a single IL (Kim et al., 2013) or a mixture (Choi et al., 2014a,b). If the IL used is hydrophilic the separation step can be efficient; as lipids possess long hydrophobic hydrocarbon chains they will aggregate rapidly in hydrophilic ILs and form a separate immiscible phase that surfaces or settles depending upon their density difference relative to the IL. That being said, the hydrophilic head of the lipid can facilitate a small degree of lipid solubilization within the IL as well as slightly emulsified interface. As such, to quantitate complete extraction yields researchers generally supplement the recovery of lipids with

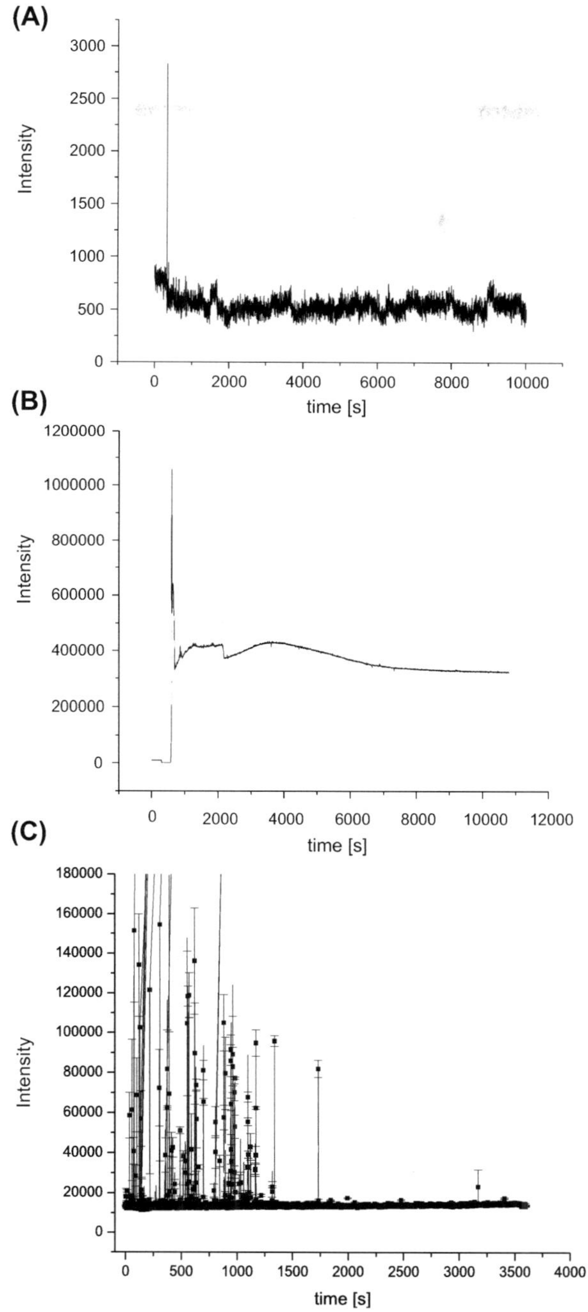

FIGURE 9.4 Intensity of light scatter as a function of time for a jatropha seed in the presence of distilled water (A) and in the presence of methanol (B). Intensity of light scatter from jatropha seed in RTIL cosolvent as a function of time for two separate experiments spanning 3600 s (C).

several rinses with an organic solvent such as hexane, which must then be distilled away to purify the lipids. In general approaches using pure hydrophobic ILs are not proposed owing to the difficulty in separating the extracted liquids. Molecular interactions between the hydrocarbon chains of the lipids and the associated hydrocarbon side chains linked of the imidazolium ring nitrogen groups form complex and strong hydrogen bonds that make their separation from the IL very difficult. The energy required for these separations disqualifies this approach for processes aimed at recovering the lipids as a fuel.

Unfortunately ILs are also not the most efficient solvents to penetrate through cell walls, contact the internal lipids, and catalyze their mass transfer out of the internal cell matrix and into the external solvent where they aggregate into micelles that form a separate immiscible phase. To enhance lipid extraction out of the cells and into an immiscible IL solvent researchers generally use an aggressive mechanical technique (such as ultrasonication (Kim et al., 2013)) or aggressive physical technique (such as high temperature heating (Choi et al., 2014a,b)) to lyse the cell walls and induce the lipid movement out of the cells and into the solvent.

9.5 MOLECULAR MODELING AND SIMULATION

9.5.1 Overview

The general activity of computer simulation is synergistic with experimental investigation. Within the larger activity of computer simulation is the specific branch of molecular simulation. Molecular simulations aim to simulate the actual configurations among and interactions between molecules, to such an extent that we can both compute relevant and accurate properties (thermodynamic and/or kinetic) and visualize molecular-level phenomena and mechanisms (through visualization software) not accessible by conventional experimental analytical laboratory techniques. Given that the main goals of a molecular simulation are to sample relevant configurations of molecules and calculate meaningful properties, we now turn our attention to the topics of sampling and properties/interactions.

9.5.1.1 Molecular Dynamics

One of the main ways to sample configurations within a molecular simulation is a technique known as molecular dynamics (MD). The MD method originated in 1959, and has been applied successfully to the study of condensed matter physics and biophysics (Alder and Wainwright, 1959). In MD, the classical physical behavior within a system is discovered through the application of Newton's laws of forces:

$$F = ma \quad [1]$$

where F is the force exerted on a molecule, m is the mass of the molecule, and a is the acceleration of the molecule. With knowledge of the instantaneous position and acceleration of each molecule, we can determine velocities and future atomic positions in a straightforward, deterministic manner. Computationally, a number of algorithms exist to solve (or integrate) these Newtonian equations of motion, including the well-known velocity Verlet and predictor–corrector algorithms. A comprehensive discussion of these various integration schemes is presented by Allen and Tildesley (1987).

MD simulations must be performed with a chosen ensemble, as dictated by statistical mechanics (McQuarrie, 2000). The most basic ensemble is that of constant number of molecules (N), constant volume (V), and constant energy (E)—the NVE (microcanonical) ensemble. While this ensemble is the most straightforward to simulate on a computer, it is also the least relatable to experimental conditions. Experimentally, we always deal with known or constant temperature (T) (instead of energy, E) and with either constant volume (V) or constant pressure (P) conditions. Therefore, more meaningful MD simulations are often conducted in the NVT (canonical) ensemble or the NPT (isothermal-isobaric) ensemble.

Maintaining a simulation environment of molecules at constant temperature requires heat transfer, just as a real system does. Therefore, constant temperature MD simulations require the configuration and implementation of a proper thermostat. Several thermostatting methods exist in the literature, including the Nose-Hoover, Andersen, and Berendsen versions (Andersen, 1980; Berendsen et al., 1984; Hoover, 1985; Nose, 1984). In a similar manner, constant pressure MD simulations require the incorporation of a barostat. Again, several options exist in the literature, with the Berendsen version being among the most popular (Berendsen et al., 1984).

MD simulations begin with the choice of ensemble (according to the property of interest), and the subsequent implementation of suitable integrator schemes, thermostats, and barostats. Ultimately, we are able to solve the time-dependent evolution of the system of molecules, to the point where we obtain either kinetic or thermodynamic (equilibrium) properties. However, a key component is the force field equation(s), which were referenced earlier in (Eq. [1]). These force fields contain all of the basic, fundamental information regarding how forces (or energies) between atoms vary with atomic position. Their development and quality is crucial to the success of any MD simulation. The construction and use of force fields are discussed in more detail in the next section.

9.5.1.2 Force Fields

Calculating macroscopic properties of condensed matter systems requires computing the energy of a collection of molecules. In one manner or another, thermophysical properties (density, heat capacity, solubility, viscosity, etc.) are related to the energy of a system of molecules. Therefore, a successful molecular simulation must be able to accurately compute the energy of a system of molecules, after it has generated a relevant sample configuration.

In considering the energy of a system of molecules, we must consider both intramolecular (bonded) as well as intermolecular (nonbonded) forms of energy. Intramolecular energies stem from phenomena such as bond-stretching, bond-angle bending, and torsion. Intermolecular energies arise from van der Waals and electrostatic forces between nonbonded pairs of atoms. An equation that successfully captures these (and other) intra- and intermolecular features is termed a force field, or an intermolecular potential (in reference to intermolecular potential energy). The relationship between the force field (equation for forces between atoms/molecules) and the intermolecular potential (equation for energies between atoms/molecules) is expressed by (Eq. [2]):

$$\frac{dE}{dr} = -F \qquad [2]$$

Several force fields (or families of force fields) have been proposed in the literature for a wide range of molecules/systems, including both hard and soft matter. For the family of soft matter (including organics, aqueous, ions, etc.), two of the most widely used force field types are CHARMM and OPLS (Brooks et al., 1983; Jorgenson et al., 1996). The equations used by CHARMM and OPLS for the various intramolecular portions of the force field are of these types:

$$E(r^N) = E_{\text{bond}} + E_{\text{angle}} + E_{\text{dihedral}} + E_{ab,\text{nonbonded}} \qquad [3]$$

$$E_{\text{bond}} = \sum_{\text{bonds}} K_r (r - r_{\text{eq}})^2 \qquad [4]$$

$$E_{\text{angle}} = \sum_{\text{angles}} k_\theta (\theta - \theta_{\text{eq}})^2 \qquad [5]$$

$$E_{\text{dihedral}} = \frac{V_1}{2}[1 + \cos(\phi)] + \frac{V_2}{2}[1 + \cos(2\phi)] + \frac{V_3}{2}[1 + \cos(3\phi)] + \frac{V_4}{2}[1 + \cos(4\phi)] \qquad [6]$$

Regardless of the specific functional form, each of the equation versions just listed (CHARMM or OPLS) are capable of reproducing real system behavior, with regard to intramolecular modes of motion/configuration within a molecule. Often, computational quantum chemistry is used to determine the appropriate values of the constants for these intramolecular portions of the force field.

For the nonbonded, intermolecular portion of the force field, both CHARMM and OPLS utilize the well-known Lennard-Jones potential for van der Waal interactions between molecules *a* and *b*:

$$E_{ab,LJ} = \sum_{i}^{\text{on } a} \sum_{j}^{\text{on } b} \left\{ 4\varepsilon_{ij} \left[\left(\frac{\sigma_{ij}}{r_{ij}} \right)^{12} - \left(\frac{\sigma_{ij}}{r_{ij}} \right)^{6} \right] \right\} \quad [7]$$

where σ is the collision diameter (and represents the crossover point between positive, repulsive energies and negative, attractive energies) and ε is the well depth (minimum energy) for a pair of atoms. It should be noted that the 6th-order portion of the *LJ* potential derives from the knowledge of London's dispersion forces, while the 12th-order portion exists to capture repulsion at short atomic separations, and is adopted for computational convenience only. It should be noted that the σ and ε values for the *LJ* potential are often determined by regression to pure component property data, such as vapor–liquid coexistence.

With regard to electrostatics, there exists more complexity in a molecular system. Atoms may possess instantaneous charges and dipole (and higher order) moments, but these can change due to polarization effects. While some force fields are constructed to incorporate multipoles and polarization, the most common approach is to assume point charges that are fixed in value, and do not change in response to the local molecular environment. This assumption allows the direct (and constant) application of Coulomb's law toward the fixed atomic (partial) charges on a molecule, for calculating the electrostatic force between atom pairs on molecules:

$$E_{ab,\text{Coulomb}} = \sum_{i}^{\text{on } a} \sum_{j}^{\text{on } b} \frac{q_i q_j e^2}{r_{ij}} \quad [8]$$

As with the intramolecular parameters, these partial charges q_i are often computed with computational quantum chemistry, and mapped according to standard practices (Breneman and Wiberg, 1990).

9.5.1.3 Radial Distribution Functions

To provide a framework for assessing local fluid structure in solution, radial distribution functions are generated from MD simulations. Radial distribution functions (*rdfs*; also known as pair correlation functions), often noted as $g(r)$, are quantities determined by counting the number of atom pairs within various volume shells around a central atom. In this manner, $g_{ij}(r)$ gives the probability of finding an atom of type *j* a distance *r* from another central atom of type *i*. In the full formula development (not presented here; the interested reader is referred to the book by Allen and Tildesley (1987)), this probability is computed relative to the ideal gas distribution of atoms in three-dimensional space, which has a $g(r)$ value of unity. Therefore, increases in values of

$g_{ij}(r)$ indicate there is a greater probability of finding atom type j near atom type i, and provide a convenient means of discerning increases (or decreases) in aggregation (or nearest neighboring atoms) in molecular systems. *rdf* peaks ($g(r)$ values) will be used in this chapter as one means for describing the extent of aggregation in IL/PCM mixtures.

9.5.1.4 Equilibrium Volumetric and Thermal Properties

With regard to properties of molecular systems, we are most often interested in equilibrium properties, such as densities, enthalpies, heat capacities, viscosities, and so on. According to the ergodic hypothesis within statistical mechanics, the true ensemble average of a property (such as energy) is equal to the time-average of that property, provided the system has been equilibrated (and was averaged only over equilibrated configurations) and sampled long enough in time. For an MD simulation, this requires confirming that the system has been equilibrated, and then running for a sufficient length of time after equilibration (often known as production). A number of metrics exist for confirming equilibration, such as tracking energy profiles (total, kinetic, and potential) and velocity autocorrelation functions over time (Allen and Tildesley, 1987).

The most relevant equilibrium properties include both volumetric and thermal properties. Volumetric properties include equation of state (pressure-volume-temperature-composition) and phase equilibrium behavior. Thermal properties include internal energies, enthalpies, and heat capacities. For fluid-phase mixtures, such as that comprised of ILs and PCMs, some of the most relevant quantities are volumes and heats of mixing. Volumes of mixing are best determined through NPT (constant pressure) MD simulations, whereas heats of mixing are best determined via NVT (constant volume) MD simulations, holding the volume fixed at the ensemble-averaged volume value predicted by a preceding NPT simulation.

With regard to mixture thermodynamics properties, volumes and heats of mixing are calculated from standard, classical thermodynamic relationships:

$$\Delta h_{\mathrm{mix}} = h_{\mathrm{mix}} - x_1 h_1 - x_2 h_2 \qquad [9]$$

$$\Delta v_{\mathrm{mix}} = v_{\mathrm{mix}} - x_1 v_1 - x_2 v_2 \qquad [10]$$

where x_1 and x_2 are the mole fractions of the two components in the mixture (in this case, IL and PCM), h_i and v_i are the molar enthalpies and volumes of the pure components, and h_{mix} and v_{mix} are the molar enthalpy and volume of the mixture.

9.5.1.5 Free Energy Calculations

Thermodynamically, the most important quantity is free energy. Free energy dictates stability, and subsequently, controls phenomena in solution such as solubilization and aggregation. As such, the accurate determination of free energies by molecular simulation is an important area of computational research. Over the years, a number of methods have been developed and applied for determining free energies, including umbrella sampling, thermodynamic integration, perturbation methods, and steered molecular dynamics. Excellent reviews of these methods have been provided by Frenkel and Smit (2002) and Park et al. (2003).

For our work in IL/PCM mixtures, we have chosen to explore IL ion (cation/anion) aggregation in solution using the umbrella sampling method (Torrie and Valleau, 1977). In umbrella sampling, two atoms/molecules/particles are constrained at fixed reaction coordinates (fixed interparticle distance, for the case of ion aggregation) using a fictitious restraining force. Oftentimes a harmonic, spring-like restraining force is used, and modeled energetically as:

$$E_{\text{harmonic}} = \frac{1}{2}k(x - x_0)^2 \qquad [11]$$

where the equilibrium spring distance x_0 is the chosen interparticle separation and the spring constant k reflects the strength of this restraining potential. During the course of the MD simulation, the constrained/tethered entities (ions, in this case) will experience their natural interactions (solute/solute, solute/solvent) and fluctuations, but these are modulated by the imposed restraining force. These restrained fluctuations about the reaction coordinate (chosen equilibrium spring length) produce umbrella curves at each chosen reaction coordinate. In effect, a well-tuned restraining force (tuned via the choice of the spring constant) serves as a probe of the natural solute–solute (ion–ion) aggregation tendency (relative to the effect of the solvent to separate the ion pair). Proper tuning of the restraining force spring constant is achieved by inspecting histograms at all sampled reaction coordinates, and ensuring sufficient overlap between all neighboring reaction coordinate points. The cumulative umbrella sampling MD data is then processed using the weighted histogram analysis method to compute potentials of mean force (PMF) between ions in solution, where the PMF in the NVT ensemble is equivalent to the Helmholtz free energy (Ferrenberg and Swendsen, 1989).

9.5.2 Molecular Simulation of Ionic Liquid/Cosolvent Mixtures

While the effectiveness and potential of IL/PCM mixtures has been demonstrated experimentally, the true potential of IL cosolvent systems can only be achieved through an intimate knowledge of the molecular scale phenomena dominating the forces between molecules. To date, there is little molecular-level information describing IL/PCM environments, which are heterogeneous collections of molecular constituents (IL cation, IL anion, alcohols, acids, hydrocarbons, etc.) that possess significantly different size, structure, interactions, and chemistry. This section is devoted to the presentation and discussion of molecular simulation data regarding the IL/PCM cosolvent mixtures under investigation for lipid extraction from biomass. Specifically, results on cosolvent structure, mixture thermodynamics, and free energy of ion aggregation/association in solution are presented and discussed, including in relation to experimental data previously presented.

9.5.2.1 Structure

While the structure of condensed matter is intimately linked to its properties, we will nonetheless present and discuss the structure of IL/PCM mixtures here first, and then make connections to properties in subsequent subsections. As mentioned previously, we make use of radial distribution functions (*rdfs*) in presenting our MD simulation results for structure. In our *rdf* plots, we chose to focus on the structure (state of aggregation) between IL cation and IL anion. Among these ions, the most attractive portions are those atoms that possess the positive and negative point charges of largest magnitude, respectively. For this reason, we have generated *rdfs* between the C_2 carbon atom of the imidazolium ring (according to the published molecular model (Kelkar et al., 2008)) and the sulfur of the methyl sulfate anion, for the case of [emim][$MeSO_4$]/methanol mixtures. The force fields for emim cation, $MeSO_4$ anion, and methanol solvent are available in the literature (Jorgensen et al., 1996; Kelkar et al., 2008; Lopes et al., 2008).

Although the experimental data strongly suggests the formation of EMIN-MS micelles at high concentrations of methanol, even greater confidence can be realized if simulation predictions of the extent of aggregation (or number of nearest neighboring atoms) between the cation and anion is likewise predicted to increase at high methanol concentrations. Typical radial distribution data between cation and anion generated from molecular simulation are shown in Fig. 9.5. These *rdf* peak heights indicate the local structure involving the C_2 carbon atoms of the cation imidazolium ring, as they preferentially interact with the sulfur of the anion. As discussed earlier, it has been reported that ILs tend to form aggregates in the presence of water (Hemmateenejad et al., 2011; Heintz et al., 2010; Łuczak et al., 2009). Since methanol is highly polar (similar to water) we expected to see more IL aggregation in solution with methanol. Fig. 9.5 presents cation/anion *rdfs* for solution of pure EMIN-MS, 30 mol% EMIN-MS with 70 mol% methanol, and 10 mol% EMIN-MS with 90 mol% methanol. In the case of the pure EMIN-MS, mild aggregation is observed. By contrast, a significant increase in aggregation is observed in the presence of high methanol concentration (90 mol%). Specifically, the first (neighbor) peaks for the $C_2{-}S$ atom pairs increased by a factor of three to four relative to that in pure EMIN-MS alone. These simulated increases in *rdf* peak heights correspond to increases in ion aggregation (more nearest neighbors) and are also in good agreement with polarization data obtained for a similar EMIN-MS-methanol mixture (see Fig. 9.3A). The

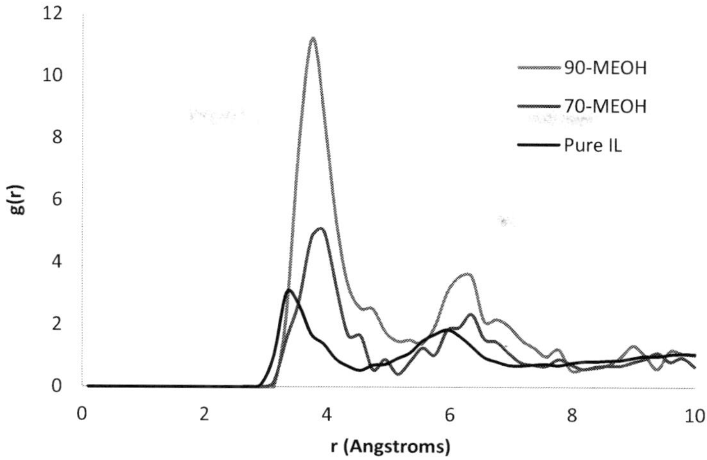

FIGURE 9.5 Variation of IL ion aggregation with methanol concentration. The black, red, and blue lines correspond to the radial distribution functions for the C_2—S atom (cation—anion) pair for concentrations of 0 mol%, 70 mol%, and 90 mol% methanol, respectively.

increased polarization for the EMIN-MS-methanol mixture at high methanol concentrations observed in Fig. 9.3A is believed to be a result of increased resistance to rotation of the cation around its own axis due to stronger attractive electrostatic interactions. Experimentally, this would suggest a larger extent of anion—cation pairing.

Simulation results presented in Fig. 9.5 also help explain experimental light scattering data in Fig. 9.3B. As discussed earlier, the light scattering data reports the amount of light (measured in photons) scattered at 90°degrees as a function of cosolvent composition. As the methanol concentration increases the amount of light scattered likewise increases until reaching a maximum at around 80% methanol, after which the amount of light scattered leveled off. From the *rdf* plots in Fig. 9.5, we observe that the *rdf* peak heights also grow in size as methanol concentration increases, predicting a continuous increase in IL ion aggregation with increasing methanol concentration—in good agreement with the light scattering data. Fig. 9.5 also shows how IL ion aggregation continues to increase as methanol concentration increases above 80%. As shown in Fig. 9.2, the extraction yields decrease rapidly at the limit of high methanol concentration. The simulations suggest that the rapid drop off in extraction yields corresponds to extreme aggregation of the IL molecules, wherein the IL ions are restricted to a very small volume in solution, effectively reducing the solvent (as it pertains to the biomass) properties to pure methanol. This extreme extent of IL ion aggregation and the realization of essentially pure methanol solvent behavior for $X_{MeOH} \geq 0.9$ is consistent with one other IL—methanol system reported in the literature (Bandlamudi et al., 2014).

9.5.2.2 *Mixture Thermodynamics*

The thermodynamic properties of IL/PCM mixtures are also of interest, for their potential use in biomass processing applications. As mentioned before, relevant properties are volumes and heats of mixing, which are readily computed from molecular simulation. Volumes of mixing for [emim][$MeSO_4$] with several PCMs are shown in Figs. 9.6A–F. As each of the figures shows, the volumes of mixing are relatively small in magnitude, regardless of composition (% PCM) or PCM type. As a whole, these IL/PCM mixtures can reasonably be described as ideal, volumetrically.

The thermal behavior of four of these same [emim][$MeSO_4$]/PCM mixtures (acetic acid, acetone, DMSO, and methanol) is shown in Figs. 9.7A–D. As the figures indicate, these mixtures are mildly exothermic. This finding is in agreement with observations of mixtures of ILs with water (Kelkar et al., 2008). The exothermic nature of the [emim][$MeSO_4$]/PCM curves is likely due to the solvation of the PCM molecules by IL ions. In addition to the

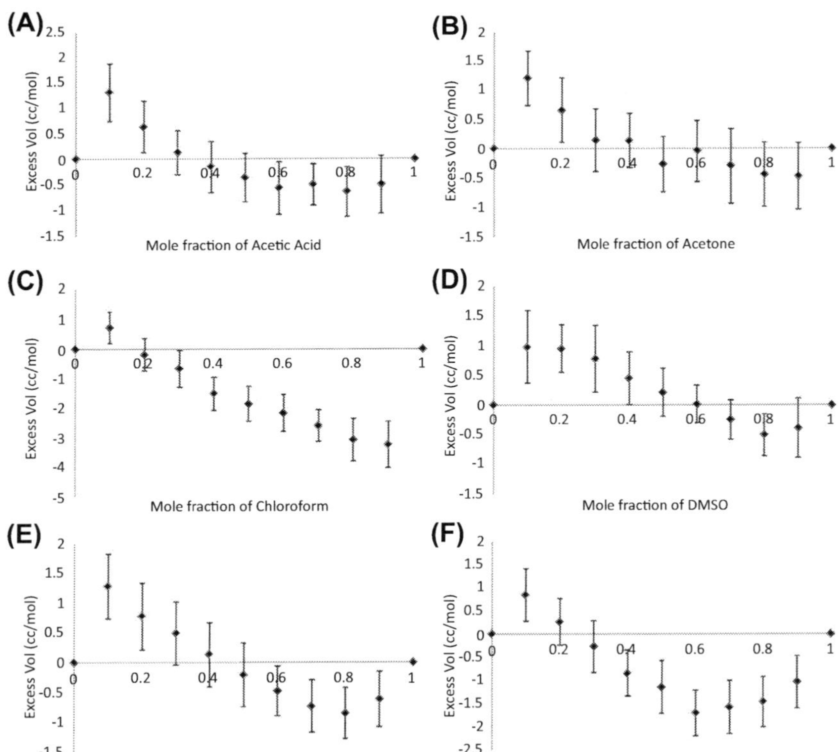

FIGURE 9.6 Excess volumes for (A) IL-acetic acid, (B) IL-acetone, (C) IL-chloroform, (D) IL-DMSO, (E) IL-methanol, and (F) IL-propanol mixtures.

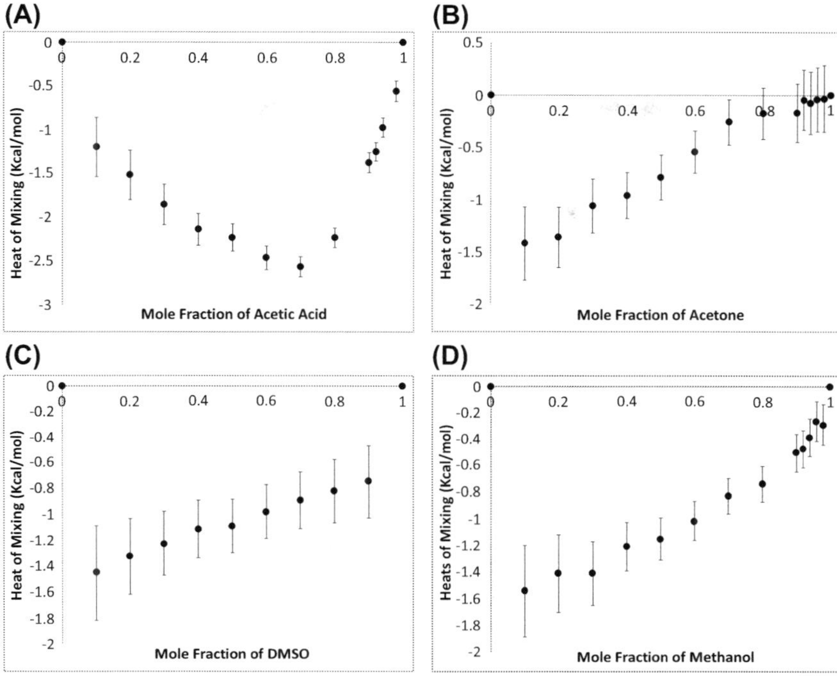

FIGURE 9.7 Heats of mixing for (A) IL-acetone, (B) IL-chloroform, (C) IL-DMSO, (D) IL-Methanol, respectively.

mixture thermodynamic properties, it is interesting to analyze changes in the local structure of these IL/PCM mixtures as a function of composition. Specifically, it is interesting to note any changes in hydrogen bonding (native to compounds such as acetic acid) within these IL/PCM mixtures. For this analysis, we present and inspect *rdfs* between the carbonyl oxygen atom and the alcohol hydrogen atom of acetic acid, the main participants in the formation of hydrogen bonds within acetic acid. As Fig. 9.8 shows, as IL is added to pure acetic acid, there is a decrease in the extent of hydrogen bonding, as represented by the *rdf* peak near 2.0 Å, up to 50 mol% acetic acid. This is expected, as the IL ions would likely disrupt the native hydrogen bonding network that exists within acetic acid. However, as the mixture is further diluted to 10 mol% acetic acid, the extent of hydrogen bonding increases (green curve in Fig. 9.8). This may indicate a level of clustering between acetic acid molecules in this IL rich region.

9.5.2.3 Free Energy of Ion Aggregation

As discussed, the calculation of PMFs, the equivalent of the Helmholtz free energy in the canonical ensemble, is a more exact way of confirming IL ion

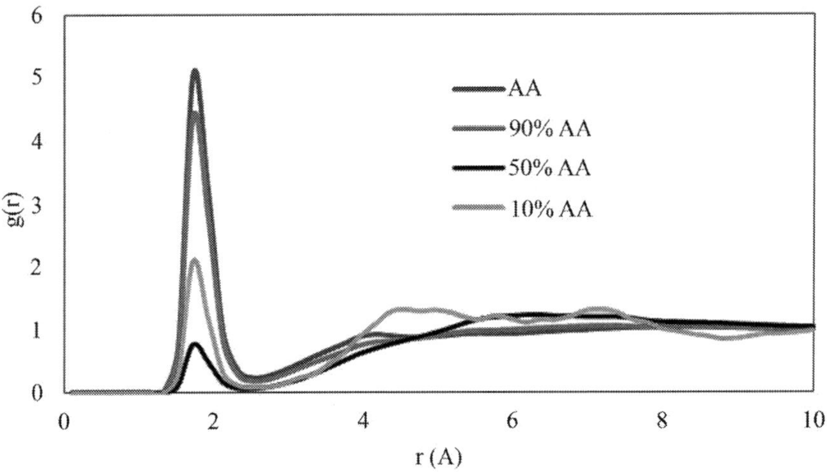

FIGURE 9.8 Hydrogen bonding in IL-acetic acid (AA) mixtures as a composition of mole fraction AA.

aggregation. Fig. 9.9 shows cation/anion PMFs for 3 M compositions of [emim][$MeSO_4$] with methanol—10% MeOH, 50% MeOH, and 90% MeOH. Consistent with the experimental polarization and light scattering measurements presented earlier (Figs. 9.3 and 9.4), these PMF curves suggest a deeper free energy well for cation/anion aggregation as methanol concentration is increased. Thermodynamically, this confirms that IL ion aggregation is more favorable at higher methanol concentrations.

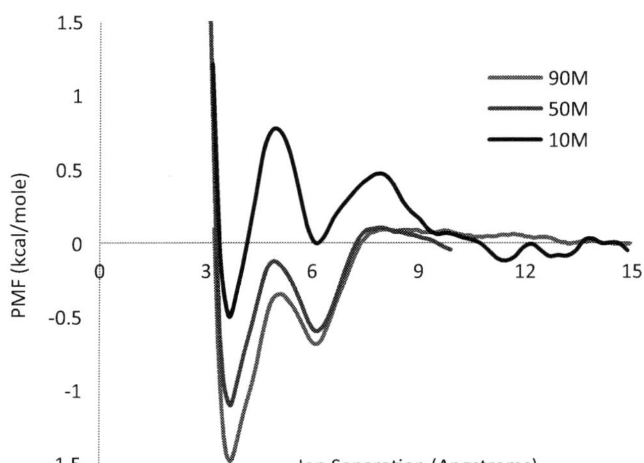

FIGURE 9.9 Potential of mean force (PMF; equivalent of Helmholtz free energy) of cation/anion aggregation/association in [emim][MeSO4]/methanol mixtures.

TABLE 9.1 Correlation of Lipid Aggregation and Yield with Varying Methanol Concentration

Weight% Methanol	*rdf* Peak Height	Extracted Yield (%)
10	4	22
35	21	43
85	9	18

9.5.2.4 Lipid Aggregation

Experimental light scattering results (Fig. 9.4C) suggest the formation of lipid micelles/aggregates in IL/PCM mixtures, as the mechanism for lipid autopartitioning and separation. This result is considered largely a thermodynamic effect, owing to the free energy changes of the lipid once in solution, and its propensity to aggregate or phase-separate, likely to avoid unfavorable solute–solvent interactions. This phenomena has also been probed, using MD simulations to predict the extent of aggregation between lipid molecules in RTIL-based cosolvents. The force field for the model lipid triolein was taken from the work of Sum et al. (2003). Table 9.1 shows *rdf* peak heights (intermolecular) for triolein–triolein as a function of methanol composition. Going from low to high weight percent methanol, we note a maximum in the *rdf* peak, suggesting that the extent of lipid aggregation goes through a maximum. This maximum in aggregation coincides with the experimentally observed maximum in extracted yield at intermediate methanol concentration (see Fig. 9.2), as shown in Table 9.1.

This suggests that the large extent of lipid extraction is directly tied to the ability of the IL—methanol cosolvent system to facilitate lipid aggregation. Fig. 9.10 shows a molecular simulation snapshot of a lipid aggregate formed in the IL–methanol mixture (red atoms: lipid molecules; green atoms: methanol molecules; and white atoms: IL cation and anion; all lipids in one aggregate due to periodic boundary conditions). This indicates a high level of aggregation in agreement with the light scattering data presented in Fig. 9.4C, and further explains the experimentally observed autopartitioning behavior (thermodynamically). In the highly polar IL–methanol cosolvent mixture, it is thermodynamically favorable for the lipids to form aggregates (to minimize their free energy, especially at the molecular surfaces exposed to the more polar IL–methanol molecules), and to potentially form their own separate stable phase. This is consistent with the experimental observation that the lipid phase is easily extracted off the top of the separation vessel.

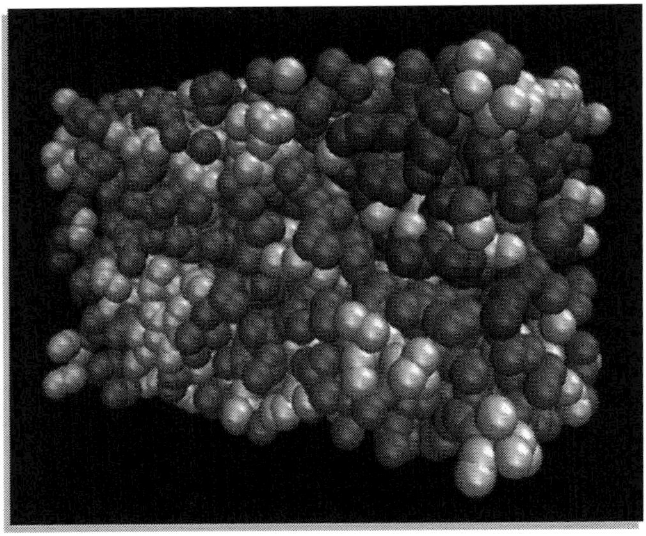

FIGURE 9.10 Visualization of lipid aggregate formed in [EMIN] [$MeSO_4$]-methanol mixture.

9.6 FUTURE TRENDS

9.6.1 Application of Ionic Liquids

The future role of ILs in extraction of lipids from biomass will focus on several key actions of solvents that ILs could potentially improve: the degradation of cell wall biomass to enable solvent penetration, solvent contact with lipids within the cellular matrix, the ability of the solvent to facilitate mass transfer of the lipid outside the cell, and finally recovery of the lipid. Research will focus on the use of ILs applied in mixtures (Young et al., 2009; Choi et al, 2014a,b; Kim et al., 2012; Lab, 2008) and/or under extreme physical conditions (Kim et al., 2013) to improve extraction yields. Young et al. demonstrated the application of polar covalent molecules mixed with ILs to significantly improve extraction yields owing to the cell wall penetration of the polar covalent molecule. Choi et al. investigated the role of ILs mixed with other ILs or molten salts to improve extraction yields. Molten inorganic hydrates have gained attention as new, highly efficient solvents and media for use in cellulose modification and as such they promise improved solvent penetration of the cell wall. Kim et al. applied ultrasonication and Young et al. considered increased temperature as useful tools for strengthening mass transfer of liquid–liquid heterogeneous systems. Their application, however, will require energy efficiency if they are to be used.

There is also the issue of solvent loss. ILs are charged molecules and can be expected to bond with elements of cell debris. ILs are expensive solvents and make-up costs can pose significant impediments on a commercial scale.

Research focused on IL solvent recovery from delipified biomass will be important. The ability of ILs to support extraction of additional value added products will add additional value to lipid extraction. Godwin et al., for example, has demonstrated the capacity of IL cosolvent systems to concomitantly extract phorbol esters from jatropha oil seeds in addition to lipids (Severa et al., 2013a,b). The recovery of phorbol esters opens up pathways to both produce a natural pesticide and a delipified biomass with value as an animal feed. These and other pathways to extract multiple products in addition to lipids will be useful. Finally, the action of ILs on biomass also offers future potential opportunities for research. In another report Godwin et al. demonstrated the ability to recover fermentable sugars from delipified bio-oil bearing biomass. In the process of extracting the lipids the IL cosolvent system also degraded the cell wall matrix to the point that mixtures of hydrolysis enzymes could be used to break down the biomass to simple sugars that were then recovered and used in subsequent fermentation of the yeast *Rhodospridium toruloides* to produce additional lipids (Severa et al., 2013a,b). Along with the coextraction of value added products like phorbol esters this approach also potentially adds value to any process dedicating the use of ILs (or their mixtures) to the extraction of lipids. In summary, future trends of IL research in lipid extraction and recovery will likely focus on ILs used in combination with other ILs or polar covalent molecules to enhance cell wall degradation, solvent penetration, and lipid aggregation. Also, these systems will be further investigated for their ability to corecover additional value added products.

9.6.2 Solvent Design

The results presented in section Molecular Modeling and Simulation demonstrate the ability of molecular simulation (particularly molecular dynamics simulations) to compute the structure and properties of IL-based solvents for lipid extraction from biomass. Specific properties computed included volumes and heats of mixing and free energies of cation/anion association/aggregation in solution. The structural information produced from these molecular simulations included radial distribution functions, which confirmed such solution phase spatial information as cation/anion and lipid aggregation and the persistence/disappearance of hydrogen bonds within IL/PCM mixtures. Many of these results were obtained for the base case of [emim][$MeSO_4$] in methanol, which is the main cosolvent system that has been studied experimentally. The agreement and synergy between molecular simulation and experimentation for this base case yields confidence in using molecular modeling and simulation as a driver for future solvent design.

Solvent design by molecular modeling and simulation can proceed by considering specific IL functionalities (ie, stronger anion) and then exploring their effect on computed macroscopic solvent/mixture properties. Since this new functionality involves specific atoms in specific arrangements with specific interactions, we would next obtain the necessary force field information regarding the atoms of the functionality under consideration. In some cases, the force field information may already be present in the literature (ie, switching PCM cosolvent). In other cases, force field information will have be computed using computational quantum chemistry (CQC). A representative sample of bond-stretched, angle-bend, and torsion configurations can be examined and their energies computed, using ab initio or density functional theory methods (depending on the molecular complexity, required accuracy, and available computational resources). These energies can be used to parameterize the intramolecular portion of the force field, likely in one of the standard, successful forms proposed and used in the literature (Brooks et al., 1983; Jorgensen et al., 1996). With regard to nonbonded electrostatic interactions, CQC calculations can also be performed (typically energy minimizations) and the resulting electron densities (from the numerical solution of the electronic wave function within Schroedinger's equation) mapped to point charges on the respective atoms (Breneman and Wiberg, 1990). Given the highly polar, electrostatic nature of ILs, this step is crucial for accurate property determination. In some instances, a more precise electrostatic treatment (beyond simple point charges) may be required. This is especially true for transport properties, such as viscosity and diffusivity. In such instances, the parameterization and use of a polarizable force field may be required (Borodin, 2009). Polarizable force fields take into account the fact that molecules (or ions) may experience induced dipoles, resulting from the surrounding electronic environment inherent to the other atoms in solution. For the nonbonded van der Waals interactions, these values (often expressed in the form of a Lennard-Jones type potential), can either be obtained by CQC calculations or by fitting to pure component experimental data (such as densities, viscosities, heat capacities, phase equilibrium, etc.).

The force fields generated by the aforementioned procedure allow the exploration of the effect of basic intermolecular forces on macroscopic solvent properties. These force fields can be used in subsequent molecular simulations, either MD, Monte Carlo (MC), or both. In general, MD simulations are useful for computing single-phase properties, thermodynamic or kinetic/transport. In contrast, MC methods can be used only for thermodynamic properties, but are more amenable to solubility and phase equilibrium studies. For instances where we want to compute solubilities of target solutes in IL-based solvent systems, techniques such as Widom insertions and ensembles such as the grand canonical and Gibbs are more naturally implemented in MC algorithms (vs MD) (Frenkel and Smit, 2002).

While molecular simulation is perhaps the most common computational atomistic approach for liquid phase studies, we should not leave out the other possibilities afforded by CQC. Ab initio and density functional theory calculations provide additional avenues for simulation-based solvent design, beyond their usage for developing force fields for molecular simulation (as described earlier). One example is the use of CQC to calculate FTIR spectra, as a means of probing local (bonded atom) shifts in solvent microstructure. Vibrational spectra predicted by CQC can be compared readily to experimental FTIR measurements, and provide an additional, local means of predicting solvent structure. These local structural predictions complement the radial distribution functions (or structure factors) generated from molecular simulation, which provide a better picture or longer-range (nonbonded) solvent microstructure. A second example is the use of continuum solvent models within CQC calculations to predict solvation and mixture thermodynamics and solubility (Klamt et al., 2010). These continuum solvent approaches largely rely on a dielectric description of the bulk solvent environment. In the case of large, macromolecular solutes (such as lipids) dissolved in dense media (such as ILs), these continuum solvent model approaches to determining solubility are much preferred over conventional Monte Carlo molecular simulation methods, which suffer from low particle (Widom) insertion acceptance rates.

Regardless of whether molecular simulation or computational quantum chemistry or both are used, the possibilities and philosophy for solvent design are the same. We can make changes to the solvent environment on an atomistic level (based on chemical intuition, analogy, experimental input, etc.) and then determine macroscopic structure and properties from first principles. In addition, it should be noted that the cost for this type of computational solvent design and exploration pales in comparison to a complimentary experimental approach. Moreover, after the fact we will possess a powerful molecular-level basis (or perspective) for affecting changes in desired solvent properties. From this continued process and the cumulative knowledge gained, it will be possible to develop molecular-level guidelines and rules to drive solvent development and process optimization for any targeted separation, including lipid extraction from biomass.

REFERENCES

Adarme-Vega, T.C., Thomas-Hall, K.R., Schenk, P.M., 2014. Towards sustainable sources for omega-3 fatty acids production. Curr. Opin. Biotechnol. 26, 14–18.

Adeoti, I.A., Hawboldt, K., 2014. A review of lipid extraction from fish processing by-product for use as a biofuel. Biomass Bioenerg. 63, 330–340.

Alder, B.J., Wainwright, T.E., 1959. Studies in molecular dynamics. I. General Method. J. Chem. Phys. 31, 459–466.

Allen, M.P., Tildesley, D.J., 1987. Computer Simulation of Liquids. Oxford University Press.

Andersen, H.C., 1980. Molecular dynamics at constant pressure and/or temperature. J. Chem. Phys. 72, 2384–2393.

Ayala, R.S., Castro, L., 2001. Continuous subcritical water extraction as a useful tool for isolation of edible essential oils. Food Chem. 75, 109–113.

Balasubramanian, R.K., Doan, T.T.Y., Obbard, J.P., 2013. Factors affecting cellular lipid extraction from marine microalgae. Chem. Eng. J. 215–216, 929–936.

Bandlamudi, S., Benjamin, K.M., Cooney, M., Martin, G., 2015. Molecular simulation and experimental characterization of ionic liquid based co-solvent extraction solvents. Ind. Eng. Chem. Res. (submitted).

Bennemann, J., Oswald, W.J., 1996. Systems and Economic Analysis of Microalgal Ponds for Conversion of CO_2 to Biomass. Department of Energy, p. 201.

Berendsen, H.J.C., Postma, J.P.M., van Gunsteren, W.F., DiNola, A., Haak, J.R., 1984. Molecular-Dynamics with coupling to an external bath. J. Chem. Phys. 81, 3684−3690.

Bligh, E.G., Dyer, W.J., 1959. A rapid method for total lipid extraction and purification. Can. J. Biochem. Phys. 37, 911−917.

Borodin, O., 2009. Polarizable force field development and molecular dynamics simulations of ionic liquids. J. Phys. Chem. B 113, 11463−11478.

Breneman, C.M., Wiberg, K.B., 1990. Determining atom-centered monopoles from molecular electrostatic potentials. The need for high sampling density in formamide conformational analysis. J. Comput. Chem. 11, 361−373.

Brooks, B.R., Bruccoleri, R.E., Olafson, B.D., States, D.J., Swaminathan, S., Karplus, M., 1983. CHARMM: a program for macromolecular energy, minimization, and dynamics calculations. J. Comput. Chem. 4, 187−217.

Cartens, M., Moina Grima, E., Robels Medina, A., Gimenez Gimenez, A., Ibanez Gonzalez, J., 1996. Eicosapentaenoic acid (20:5n-3) from the marine microalgae *Phaeodactylum tricornutum*. J. Am. Oil Chem. Soc. 73, 1025−1031.

Catchpole, O.J., Tallon, S.J., Grey, J.B., Fletcher, K., Fletcher, A.J., 2008. Extraction of lipids from a specialist dairy stream. J. Supercrit. Fluids 45, 314−321.

Choi, S.-A., Oh, Y.-K., Jeong, M.-J., Kim, S.W., Lee, J.-S., Park, J.-Y., 2014. Effects of ionic liquid mixtures on lipid extraction from *Chlorella vulgaris*. Renew. Energ. 65, 169−174.

Choi, S.-A., Le, J.-S., Oh, Y.-K., Jeong, M.-J., Kim, S.W., Park, J.-Y., 2014b. Lipid extraction from *Chlorella vulgaris* by molten-salt/ionic-liquid mixtures. Algal Res. 3, 44−48.

Cooney, M.J., 2013. Case Studies of Separation in Biorefineries−extraction of Algae Oil from Microalgae. In: Ramaswamy, S., Huang, H.-J., Ramarao, B.V. (Eds.). John Wiley & Sons.

Cooney, M.J., Young, G., Nagle, N., 2009. Extraction of bio-oils from microalgae: a review. Sep. Purif. Rev. 38, 291−325.

Cooney, M.J., Young, G.L., Pate, R., 2010. Bio-oil from photosynthetic microalgae: case study. Bioresour. Technol. 102, 166−177.

Cooney, M.J., Young, G.Y., 2009. Methods and compositions for extraction and transesterification of biomass components. In: U.S.P.T. Office (Ed.), Patent Application Full Text and Image Database. University of Hawaii, United States.

FAO, 2010. The State of World Fisheries and Aquaculture. FAO, Rome, Italy.

Ferrenberg, A.M., Swendsen, R.H., 1989. Optimized Monte Carlo data analysis. Phys. Rev. Lett. 63, 1195−1198.

Fischer, S., Leipner, H., Thummler, K., Brendler, E., Peter, J., 2003. Inorganic molten salts as solvents for cellulose. Cellulose 10, 227−236.

Frenkel, D., Smit, B., 2002. Understanding Molecular Simulation: from Algorithms to Applications. Academic Press, San Diego.

Fujita, K., Kobayashi, D., Nakamura, N., Ohno, H., 2013. Direct dissolution of wet and saliferous marine microalgae by polar ionic liquids without heating. Enzyme Microb. Technol. 52, 199−202.

Gallagher, B.J., 2011. The economics of producing biodiesel from algae. Renew. Energ. 36, 158−162.

Global omega 3 ingredients market (EPA/DHA), 2013. Industry Analysis, Market Size, Share, Growth and Forecast. http://www.prweb.com/releases/2013/9/prweb11097689.htm.

Godwin Severa, K.G., Cooney, M.J., 2013. Co-recovery of bio-oil and fermentable sugars from oil-bearing biomass. Int. J. Chem. Eng., Article ID 617274. http://dx.doi.org/10.1155/2013/617274.

Hagiwara, R., Ito, Y., 2000. Room temperature ionic liquids of alkylimidazolium cations and fluoroanions. J. Fluorine Chem. 105, 221–227.

Halim, R., Danquah, M.K., Webley, P.A., 2012. Extraction of oil from microalgae for biodiesel production: a review. Biotechnol. Adv. 30, 709–732.

Halim, R., Rupasinghe, T.W.T., Tull, D.L., Webley, P.A., 2013. Mechanical cell disruption for lipid extraction from microalgal biomass. Bioresour. Technol. 140, 53–63.

Hara, A., Radin, N.S., 1978. Lipid extraction of tissues with a low-toxicity solvent. Anal. Biochem. 90, 420–426.

He, C., Li, S., Liu, H.-J., Li, K., Liu, F., 2006. Extraction of testosterone and epitestosterone in human urin using aqueous two-phase systems of ionic liquids and salt. J. Chromatogr. 1082, 143–149.

Heintz, A., Lehmann, J.K., Kozlova, S.A., Balantseva, E.V., Bazyleva, A.B., Ondo, D., 2010. Micelle formation of alkylimidazolium ionic liquids in water and in ethylammonium nitrate ionic liquid: a calorimetric study. Fluid Phase Equilib. 294 (1–2), 187–196.

Heinze, T., Koschella, A., 2005. Solvents applied in the field of cellulose chemistry—A mini review. Polímeros 15, 84–90.

Hemmateenejad, B., Safavi, A., Dorostkar, S., 2011. Aggregation of imidazolium based ionic liquids in binary methanol-water solvents: a linear solvation free energy relationship study. J. Mol. Liq. 160, 35–39.

Herrero, M., Cifuentes, A., Ibanez, E., 2006. Sub- and supercritical fluid extraction of functional ingrediants from different natural sources: plants, food-by-products, algae and microalgae. A review. Food Chem. 98, 136–148.

Hoover, W.G., 1985. Canonical dynamics: equilibrium phase-space distributions. Phys. Rev. A: At. Mol. Opt. Phys. 31, 1695–1697.

Huddleston, J.G., Visser, A.E., Reichert, W.M., Willauer, H.D., Broker, B.A., Rogers, R.D., 2001. Characterization and comparison of hydrophilic and hydrophobic room-temperature ionic liquids incorporating the imidazolium cation. Green Chem. 3, 156–164.

Hutchings, J.A., Reynolds, J.D., 2004. Marine fish population collapses: consequences for recovery and extinction risk. Bioscience 54, 297–309.

Iverson, S.J., Lang, S.L.C., Cooper, M.H., 2001. Comparison of the bligh and dyer and folch methods for total lipid determination in a broad range of marine tissue. Lipids 36 (11), 1283–1287.

Jin, G., Yang, F., Hu, C., Shen, H., Zhao, Z.K., 2012. Enzyme-assisted extraction of lipids directly from the culture of the oleaginous yeast *Rhodosporidium toruloides*. Bioresour. Technol. 111, 378–382.

Jorgensen, W.L., Maxwell, D.S., Tirado-Rives, J., 1996. Development and testing of the OPLS all-atom force field on conformational energetics and properties of organic liquids. J. Am. Chem. Soc. 118, 11225–11236.

Kanel, J.S., Guelcher, S.A., 1999. Method of Rupturing Algae Cells. Eastman Chemical Company.

Kates, M., 1986. Techniques of lipidology: isolation, analysis, and identification of lipids. In: Bruden, R.H., van Knippenberg, P.H. (Eds.), Laboratory Techniques: Biochemistry and Molecular Biology. Elsevier, New York, pp. 100–111.

Kelkar, M.S., Shi, W., Maginn, E.J., 2008. Determining the accuracy of classical force fields for ionic liquids: atomistic simulation of the thermodynamic and transport properties of 1-ethyl-3-methylimidazolium ethylsulfate ([emim][EtSO4]) and its mixtures with water. Ind. Eng. Chem. Res. 47, 9115–9126.

Kim, Y.-H., Choi, Y.-K., Park, J., Lee, S., Yang, Y.-H., Kim, H.J., Park, T.-J., Kim, Y.H., Lee, S.H., 2012. Ionic liquid-mediated extraction of lipids from algal biomass. Bioresour. Technol. 109, 312–315.

Kim, Y.-H., Park, S., Kim, M.H., Choi, Y.-K., Yang, Y.-H., Kim, H.J., Kim, H., Kim, H.-S., Song, K.-G., Lee, S.H., 2013. Ultrasound-assisted extraction of lipids from *Chlorella vulgaris* using [Bmim][MeSO4]. Biomass Bioenerg. 56, 99–103.

Klamt, A., Eckert, F., Arlt, W., 2010. COSMO-RS: an alternative to simulation for calculating thermodynamic properties of liquid mixtures. Ann. Rev. Chem. Biomol. Eng. 101–122.

Lab, T.N.F., 2008. Analytical Report CN1144.

Lee, J.Y., Yoo, C., Jun, S.Y., Ahn, C.Y., Oh, H.M., 2010. Comparison of several methods for effective lipid extraction from microalgae. Bioresour. Technol. 101, S75–S77.

Li, H., Imai, Y., Takiue, T., Matsubara, H., Aratono, M., 2013. Effect and mixing of counter anions at the surface of aqueous solution of imidazolium-based ionic liquids. Colloids Surf. A 427, 26–32.

Liddell, J.M., 2001. Extraction of Triglycerides from Micro-Organisms. Zeneca Limited.

Lopes, J.N.C., Padua, A.A.H., Shimizu, K., 2008. Molecular force field for ionic liquids IV: trialkylimidazolium and alkoxycarbonyl-imidazolium cations; alkylsulfonate and alkylsulfate anions. J. Phys. Chem. B 112, 5039–5046.

Lopez, R., Higgins, S., Pagaling, E., Yan, T., Cooney, M.J., 2014. High rate anaerobic digestion of wastewater separated from grease trap. J. Renew. Energ. 62, 232–242.

Lovejoy, K.S., Davis, L.E., McClellan, L.M., Lillo, A.M., Welsh, J.D., Schmidt, E.N., Sanders, C.K., Lou, A.J., Fox, D.T., Koppisch, A.T., Sesto, R.E.D., 2013. Evaluation of ionic liquids on phototrophic microbes and their use in biofuel extraction and isolation. J. Appl. Phycol. 25, 973–981.

Lue, M., Guo, Z., Xu, X., 2007. Lipid processing in ionic liquids. Lipid Technol. 19, 204–207.

Luque de Castro, M.D., Valcárcel, M., Tena, M.T., 1994. Analytical Supercritical Fluid Extraction. Springer Verlag, Heidelberg.

Łuczak, J., Hupka, J., Thöming, J., Jungnickel, C., 2008. Self-organization of imidazolium ionic liquids in aqueous solution. Colliod Surf. A 329 (3), 125–133.

Łuczak, J., Jungnickel, C., Joskowska, M., Thöming, J., Hupka, J., 2009. Thermodynamics of micellization of imidazolium ionic liquids in aqueous solutions. J. Colloid Int. Sci. 336 (1), 111–116.

Marsh, K.N., Boxall, J.A., Lichtenthaler, R., 2004. Room temperature ionic liquids and their mixtures—a review. Fluid Phase Equilib. 219, 93–98.

McQuarrie, D.A., 2000. Statistical Mechanics. University Science Books, Sausalito, CA.

Mendes-Pinto, M.M., Raposo, M.F.J., Bowen, J., Young, A.J., Morais, R., 2001. Evaluation of different cell disruption processes on encysted cells of *Haematococcus pluvialis*: effects on astaxanthin recovery and implications for bio-availability. J. Appl. Phycol. 13, 19–24.

Nagle, N., Lemke, P., 1990. Production of methyl ester fuel from microalgae. Appl. Biochem. Biotechnol. 24/25, 355–361.

Ngo, H.L., LeCompte, K., Hargens, L., McEwen, A.B., 2000. Thermal properties of imidazolium ionic liquids. Thermochim. Acta 357 (8), 97–102.

Nose, S., 1984. A unified formulation of the constant temperature molecular-dynamics methods. J. Chem. Phys. 81, 511–519.

Park, P.K., Kima, E.Y., Chub, K.H., 2007. Chemical disruption of yeast cells for the isolation of carotenoid pigments. Sep. Purif. Technol. 53 (2), 148–152.

Park, S., Khalili-Araghi, F., Tajkhorshid, E., Schulten, K., 2003. Free energy calculation from steered molecular dynamics simulations using Jarzynski's equality. J. Chem. Phys. 119, 3559–3566.

Pei, Y., Wang, J., Wu, K., Xuan, X., Lu, X., 2009. Ionic liquid-based aqueous two-phase extraction of selected proteins. Sep. Purif. Technol. 64, 288–295.

Prabakaran, P., Ravindran, A.D., 2011. A comparative study on effective cell disruption methods for lipid extraction from microalgae. Lett. Microbiol. 53, 150–154.

Richter, B.E., Jones, B.A., Ezzell, J.L., Porter, N.L., Avdalovic, N., Pohl, C., 1996. Accelerated solvent extraction: a technique for sample preparation. Anal. Chem. 68, 1033–1039.

Romanik, G., Gilgenast, E., Przyjanzy, A., Kaminski, M., 2007. Techniques of preparing plant material for chromotographic separation and analysis. J. Biochem. Biophys. Methods 70, 253–261.

Roth, C., Appelhagen, A., Jobst, N., Ludwig, R., 2012. Microheterogeneities in ionic-liquid-methanol solutions studied by FTIR spectroscopy, DFT calculations and molecular dynamics simulations. Chem. Phys. Chem. 13, 1708–1717.

Samarasinghe, N., Fernando, S., Lacey, R., Faulkner, W.B., 2012. Algal cell rupture using high pressure homogenization as a prelude to oil extraction. Renew. Energ. 48, 300–308.

Schafer, K., 1998. Accelerated solvent extraction of lipids for determining the fatty acid composition of biological material. Anal. Chim. Acta 358, 69–77.

Severa, G., Kumar, G., Cooney, M.J., 2013a. Corecovery of bio-oil and fermentable sugars from oil-bearing biomass. Int. J. Chem. Eng., Article ID 617274. http://dx.doi.org/10.1155/2013/617274.

Severa, G., Kumar, G., Troung, M., Young, G., Cooney, M.J., 2013b. Ionic liquid co-solvent assisted extraction of phorbol esters from jatropha biomass. Sep. Purif. Technol. Submitted for publication.

Shackleton, S., Shackelton, C., Wynberg, R., Sullivan, C., Leakey, R., Mander, M., McHardy, T., Den Adel, S., Botelle, A., Du Plessis, P., Lombard, C., Laird, S., Cunningham, T., O'Regan, D., 2006. Livelihood Trade-offs in the Commercialization of Multiple-Use NTFPs: Lessons from Marula (Sclerocarya birrea subsp. caffra) in Southern Africa. Non-timber Forest Products: Integrating Ecology, Management and Policy. ATREE Press, India.

Sum, A.K., Biddy, M.J., de Pablo, J.J., Tupy, M.J., 2003. Predictive molecular model for the thermodynamic and transport properties of triacylgylcerols. J. Phys. Chem. B 107, 14443–14451.

Sun, Y., Li, W., Wang, J., 2011. Ionic liquid based ultrasonic assisted extraction of isoflavones from *Iris tectorum* Maxim and subsequently separation and purification by high-speed counter-current chromatography. J. Chromatogr. B 879 (13–14), 975–980.

Tokuda, H., Hayamizu, K., Ishii, K., Susan, M., Watanabe, M., 2005. Physicochemical properties and structures of room temperature ionic liquids. 2. Variations of alkyl chain length in imidazolium cation. J. Phys. Chem. B 109, 6103–6110.

Torrie, G.M., Valleau, J.P., 1977. Nonphysical sampling distributions in Monte Carlo free-energy estimation: umbrella sampling. J. Comput. Phys. 23, 187–199.

U.S. Energy Outlook, 2009. Table A11. Liquid Fuels: Biodiesel. U.S. Energy Information Administration. http://www.eia.

Vermaak, I., Kamatou, G.P.P., Komane-Mofokeng, B., Viljoen, A.M., Beckett, K., 2011. African seed oils of commercial importance—Cosmetic applications. S. Afr. J. Bot. 77, 920–933.

Weng, L.J., Liu, X.Q., Liang, Y.M., Xue, Q.J., 2007. Effect of tetraalkylphosphonium based ionic liquids as lubricants on the tribological performance of a steel on steel system. Tribol. Lett. 26, 11–17.

Wilkes, J.S., Zaworotko, M.J., 1992. Air and water stable 1-ethyl-3-methylimidazolium-based ionic liquids. J. Chem. Soc. Chem. Comm. 965–967.

Young, G., Nippen, F., Titterbrandt, S., Cooney, M.J., 2009. Extraction of biomass using an ionic liquid co-solvent system. Sep. Purif. Technol. 72, 118–121.

Young, G., Nippen, F., Titterbrandt, S., Cooney, M.J., 2011. Direct transesterification of biomass using in ionic liquid co-solvent system. Biofuels 2 (3), 261–266.

Zhang, H., Li, K., Liang, H., Wang, J., 2008. Spectroscopic studies of the aggregation of imidazolium-based ionic liquids. Colliod Surf. A 329, 75–81.

Zhang, X., Yan, S., Tyagi, R.D., Drogui, P., Surampalli, R.Y., 2014. Ultrasonication assisted lipid extraction from oleaginous microorganisms. Bioresour. Technol. 158, 253–261.

Zhua, J., Wang, W., Wang, X., Li, B., Wang, Y., 2008. Green synthesis of a novel biodegradable copolymer base on cellulose and poly(p-dioxanone) in ionic liquid. Carbohydr. Polym. 76, 138–144.

Zimba, N., Wren, S., Stucki, A., 2005. Three major tree nut oils of southern central Africa: their uses and future as commercial base oils. Int. J. Aromather. 15, 177–182.

Chapter 10

Ionic Liquids in the Synthesis of Antioxidant Targeted Compounds

Mia Falkeborg
Aarhus University, Department of Engineering, Aarhus, Denmark; Danish Technological Institute, Center for Food Technology, Aarhus, Denmark

Claire C. Berton-Carabin
Wageningen University, Food Process Engineering Group, Wageningen, The Netherlands

Ling-Zhi Cheong
Wilmar Global Research and Development Center, Shanghai, China; Department of Food Science, School of Marine Science, Ningbo University, China

10.1 INTRODUCTION

Oxidation of polyunsaturated lipids is a major cause of degradation of the sensory and nutritional quality of food products. The oxidation reactions lead to formation of volatile compounds generally associated with unpleasant flavors, which damages the sensory quality of foods (Grosch, 1982). Lipid oxidation is also responsible for a loss of macro- and micronutrients, and leads to the formation of potentially toxic compounds (Ursini et al., 1998; Kanazawa et al., 2002; Riemersma, 2002; Eder and Ringseis, 2010; Serini et al., 2011). The general mechanisms and reaction pathways of lipid oxidation are now well established, and have been extensively reviewed (Frankel, 1985, 2005; Gardner, 1989; Schaich, 2005; Cheng and Li, 2007). The reaction is generally described as following three main steps: initiation, propagation, and termination (Fig. 10.1).

Initiation consists of the abstraction of a hydrogen atom (H) in an allylic position relative to a fatty acid double bond in an unsaturated fatty acid (LH). An alkyl radical ($L^{\bullet}$) is formed (Reaction [i]). The direct reaction between LH and triplet oxygen ($^3O_2$) cannot occur spontaneously because of the very high activation energy induced by the spin barrier between lipids and $^3O_2$ (Chan, 1987; Belitz et al., 2004). This spin barrier can, however, be

Ionic Liquids in Lipid Processing and Analysis. http://dx.doi.org/10.1016/B978-1-63067-047-4.00010-6

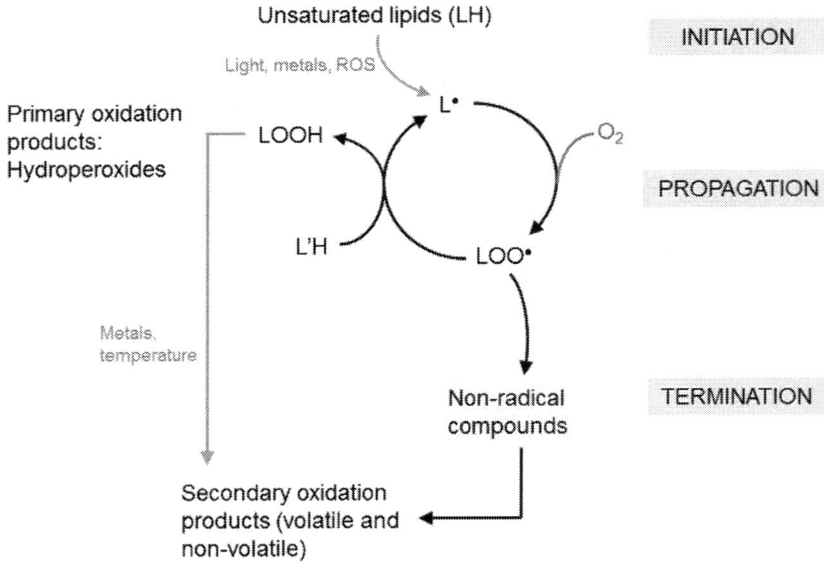

FIGURE 10.1 General scheme of lipid oxidation. *ROS*, reactive oxygen species.

overcome under certain environmental conditions such as heat, ionizing radiations, light radiations, and/or in the presence of chemical initiators such as transition metals. The chemical initiators can produce lipid radicals (Reaction [i]) or convert $^3O_2$ into reactive oxygen species, such as the hydroxyl radical ($OH^{\bullet}$).

$$LH \rightarrow H + L^{\bullet} \quad [i]$$

Different initiation pathways may be distinguished: autoxidation, enzymatic oxidation, and singlet oxygen ($^1O_2$) oxidation. The latter mechanism is referred to as photoinitiation and it occurs when lipids are exposed to light in the presence of a photosensitizer. All three pathways are characterized by specific mechanisms of formation of the first radical, but the later stages of the reactions are, overall, similar.

During the propagation step, alkyl radicals rapidly react with $^3O_2$, forming peroxyl radicals ($LOO^{\bullet}$, Reaction [ii]).

$$L^{\bullet} + O_2 \rightarrow LOO^{\bullet} \quad [ii]$$

Peroxyl radicals are unstable, and can abstract hydrogen atoms from other unsaturated fatty acids ($L'H$), leading to the formation of increasing amounts of hydroperoxides (LOOH) and alkyl radicals (Reaction [iii]).

$$LOO^{\bullet} + L'H \rightarrow LOOH + L'^{\bullet} \quad [iii]$$

The newly formed alkyl radicals can react with $^3O_2$, hence propagating the chain reaction. Eventually, radical molecules may recombine, forming stable, nonradical compounds (Reaction [iv]). This is referred to as the termination stage.

$$R^{\bullet} + R'^{\bullet} \rightarrow \text{non-radicals products} \quad [iv]$$

Concomitantly, hydroperoxides can undergo oxidation, reduction, or decomposition (Reactions [v]–[viii]), which, as for the termination reactions, leads to the formation of secondary lipid oxidation products.

Metal-catalyzed oxidation or reduction of hydroperoxides (LOOH):

$$LOOH + M^{2+} \rightarrow LO^{\bullet} + OH^{-} + M^{3+} \quad [v]$$

$$LOOH + M^{3+} \rightarrow LOO^{\bullet} + H^{+} + M^{2+} \quad [vi]$$

Thermal decomposition of hydroperoxides:

$$LOOH \rightarrow LO^{\bullet} + OH \quad [vii]$$

$$2LOOH \rightarrow LO^{\bullet} + H_2O + LOO^{\bullet} \quad [viii]$$

The decomposition of hydroperoxides starts with the formation of an alkoxyl radical ($LO^{\bullet}$), after which carbon–carbon cleavage on either side of the alkoxyl group can occur. A broad range of compounds can be produced throughout the reaction, depending, for instance, on the type of hydroperoxide initially involved, on the location of the homolytic cleavage, and on the postcleavage reactions that may develop. The compounds that are formed include aldehydes, alcohols, and ketones, of which many are volatile compounds. These compounds are largely responsible for the flavor defects that characterize oxidized lipids.

The most common way to counteract lipid oxidation in food or other biobased matrices is to use antioxidants (Lea and Ward, 1959; Uri, 1961; Sherwin, 1976; Porter, 1980; Frankel, 1996; Laguerre et al., 2007; Jacobsen et al., 2008; Shahidi and Zhong, 2010). This term relates to a broad range of molecules (Fig. 10.2), which can intervene at different steps of the lipid oxidation reaction. Accordingly, a distinction can be made between antioxidants that prevent or delay the initiation step, referred to as preventive antioxidants, and antioxidants that interrupt or limit the propagation of the radical chain reaction after the first lipid radicals have been formed, namely, chain-breaking antioxidants.

Preventive antioxidants act by suppressing the prooxidant effect of lipid oxidation catalysts or initiators (Laguerre et al., 2007; Shahidi and Zhong, 2010). Hence, this category of antioxidants includes metal chelators that form strong and thermodynamically stable complexes with metal ions, for example, ethylenediaminetetraacetic acid, citric acid, or polyphosphates. The efficiency of metal chelators depends on the microstructure and composition (eg, pH) of the lipid-based systems (Jacobsen et al., 2008), and on the chelator-to-metal ion molar ratio (Samokyszyn et al., 1990).

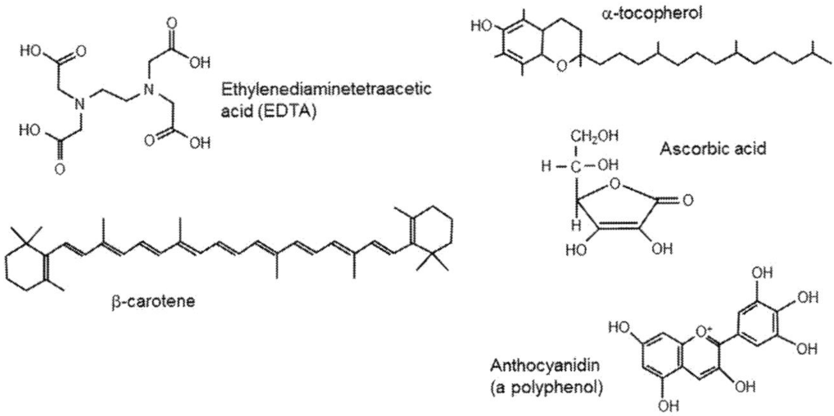

FIGURE 10.2 Molecular structures of examples of preventive antioxidants (left) and chain-breaking antioxidants (right).

Singlet oxygen quenchers also belong to the category of preventive antioxidants, and their mechanism of action is based on the deactivation of singlet oxygen into ground state triplet oxygen (Laguerre et al., 2007; Shahidi and Zhong, 2010). Carotenoids, such as β-carotene, are the most efficient singlet oxygen quenchers (Reaction [ix]). The absorption of energy from $^1O_2$ brings the quencher to an excited state (*), after which the energy can be dissipated in the form of heat and the quencher is accordingly regenerated (Reaction [x]).

$$^1O_2 + \beta\text{-carotene} \rightarrow {}^3O_2 + \beta\text{-carotene}^* \qquad [ix]$$

$$\beta\text{-carotene}^* \rightarrow \beta\text{-carotene} + \text{heat} \qquad [x]$$

Chain-breaking antioxidants (AH) act by donating a hydrogen atom to a radical molecule (for example, to a peroxyl radical), which, if present in sufficient amounts, competes with the labile lipid substrate (LH) and interrupts the propagation of the chain radical reactions (Reaction [xi]).

$$AH + LOO^{\bullet} \rightarrow A^{\bullet} + LOOH \qquad [xi]$$

An antioxidant radical ($A^{\bullet}$) is thereby generated, in which the unpaired electron is delocalized, for example, around conjugated double bonds, forming stable resonance structures. This pathway implies that chain-breaking antioxidants are progressively consumed over time, unless they are regenerated, for example, by ascorbic acid (Tappel, 1968). The main natural chain-breaking antioxidants in foods and biobased matrices are tocopherols (Lea and Ward, 1959; Frankel, 1996; Laguerre et al., 2007), and notably α-tocopherol (also referred to as vitamin E). Other common chain-breaking antioxidants include tocotrienols, flavonoids, and phenolic acids.

10.2 EXTRACTION OF NATURAL ANTIOXIDANTS WITH IONIC LIQUIDS

Antioxidants from natural sources are gaining increasing popularity due to consumers' preferences for natural antioxidants (Botterweck et al., 2000). Various methods have been employed for extraction of antioxidants from natural materials, such as plants. Some of the conventional extraction methods include Soxhlet extraction, solid–liquid extraction and liquid–liquid extraction (LLE). These methods involve high solvent consumption and prolonged exposure to thermal treatment, resulting in degradation of the labile antioxidants (Ballard, 2008). Thus, alternative extraction procedures have been developed, including microwave-assisted extraction (MAE), ultrasound-assisted extraction (UAE), accelerated solvent extraction, supercritical fluid extraction, solid phase microextraction (SPME), liquid phase microextraction (LPME), and negative pressure cavitation-assisted extraction (NPCAE). Compared to the conventional extraction methods, these methods reduce extraction time and solvent consumption, increase sample throughput, and improve product recovery.

10.2.1 Ionic Liquids as Extraction Solvents

Ionic liquids (ILs) have been used as extraction solvents to improve recovery of antioxidants from natural sources (Sun and Armstrong, 2010; Ho et al., 2011). ILs are organic salts consisting of an organic cation paired with an organic or inorganic anion. They are widely recognized as green solvents due to their low volatility and nonflammable nature. Extraction of antioxidants from plants using ILs as extraction solvents is promising, as it can alleviate environmental pollution and improve the selectivity and extraction yields compared to conventional organic solvents. ILs have low combustibility, excellent thermal stability, high viscosity, and wide liquid regions; thus, they are favorable for solvating a range of polar and nonpolar compounds (Huddleston et al., 2001; Yang and Dionysiou, 2004). ILs can interact with different analytes through anion exchange, hydrogen bonding, $\pi-\pi$ or hydrophobic interactions, deeming it suitable for use as an extraction solvent.

Thermal stability is an important consideration in selecting extraction solvent, particularly in MAE, LPME, and SPME. Thermal stability of IL depends on its cation and anion. For cations, imidazolium-based cations exhibit higher thermal stability compared to tetraalkylammonium cations. In terms of anions, organic anions are generally reported to be more thermally stable than inorganic anions (Ngo et al., 2000). It has been reported that thermal stability of imidazolium-based ILs increases in the following order: $[Cl]^-$, $[Br]^-$, $[I]^- < [BF_4]^- < [TfO]^- < [Tf_2N]^- < [PF_6]^-$ (Ngo et al., 2000; Awad et al., 2004; Tang et al., 2012).

The viscosity of ILs also affects their efficiency as an extraction solvent. The viscosity of ILs is dependent on intermolecular interactions, such as electrostatic interactions, van der Waals forces, and hydrogen bonding. Tang et al. (2012) reported that the viscosity of IL decreases with increasing asymmetry of anion. For example, the viscosity of imidazolium-based ILs can be increased by increasing the alkyl substituent chain length or branching, which results in higher van der Waals interactions between the ions (Tang et al., 2012). Table 10.1 shows a list of ILs used in different types of antioxidant extractions.

10.2.2 Liquid–Liquid Extraction

Some ILs are immiscible with water and organic solvent and form biphasic systems with such solvents, which allows them to be used in LLEs. IL-mediated LLE has been employed to extract antioxidants, including tocopherols (Yang et al., 2009), gallic, vanillin, and syringic acids (Almeida et al., 2014), and ferulic and caffeic acids (Yan-Ying et al., 2007).

Tocopherols are able to act as free radical scavengers; thus, they play an important role in human health due to their antioxidative capacity. Tocopherols are present in four different homologs, namely α-, β-, γ-, and δ-tocopherol. There is a strong need to purify and isolate α-tocopherol from the other forms of tocopherol, since it has been reported to possess the highest biological activity (Chen and Payne, 2001). However, purification and isolation of individual tocopherol homologs is challenging due to their high structural similarity. The available methods, including various chromatographic technologies, can only be applied on a laboratory scale due to low capacity and large consumption of absorbents and solvents (Chen and Payne, 2001; Pyka et al., 2001). ILs were found to be effective in selectively separating tocopherol homologs. A selectivity of δ-tocopherol to α-tocopherol of up to 21.3 was achieved using $[bmim]^+[Cl]^-$ as extractant diluted by methanol. The proposed separation mechanism was based on the hydrogen bonding interaction between IL anions and the hydroxyl group on the tocopherols (Yang et al., 2009).

Ferulic and caffeic acids are widely used in medicine, food, and cosmetic industries due to their antiinflammatory, anticancer, and antioxidant properties. Two ILs, namely $[bmim]^+[PF_6]^-$ and $[hmim]^+[PF_6]^-$, were found to be effective extracting agents of ferulic and caffeic acids (Yan-Ying et al., 2007).

Vanillic, gallic, and syringic acids are phenolic compounds present naturally in wood, barks, fruits, and vegetables. These compounds are garnering much research interest due to their antioxidant and radical scavenging properties (Lu et al., 2007). A novel aqueous two-phase system composed of polyethylene glycol and sodium sulfate (Na_2SO_4), using ILs as additives (at 5%w/w) was found to result in high extraction efficiencies of up to 90%. Hydrophilic and aromatic ILs, particularly $[bmim]^+[Cl]^-$, enhances the partition coefficient and extraction efficiencies compared to hydrophobic and

TABLE 10.1 List of Ionic Liquids Used in Different Types of Antioxidant Extractions

Ionic Liquid	Structure	Extraction Methods	Antioxidants	References
$[emim]^+[Br]^-$	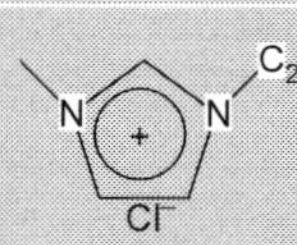	NPCAE	Flavonoids	Duan et al. (2013)
		MAE	Phenolic alkaloids	Lu et al. (2008)
$[emim]^+[BF_4]^-$	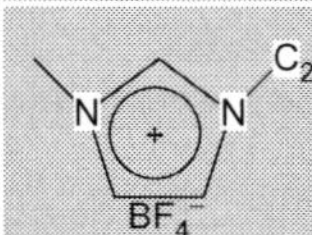	NPCAE	Flavonoids	Duan et al. (2013)
		MAE	Phenolic alkaloids	Lu et al. (2008)
		UAE-MISPE	Astaxanthin	Bi et al. (2010)
$[emim]^+[Cl]^-$	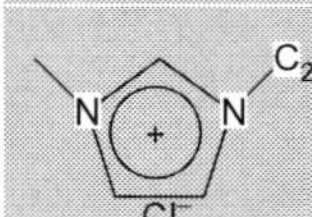	NPCAE	Flavonoids	Duan et al. (2013)
		MAE	Phenolic alkaloids	Lu et al. (2008)
$[emim]^+[HSO_4]^-$	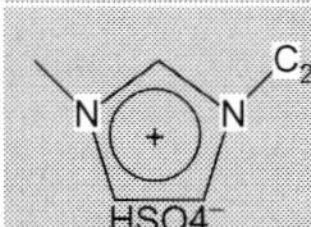	NPCAE	Flavonoids	Duan et al. (2013)
$[bmim]^+[BF_4]^-$	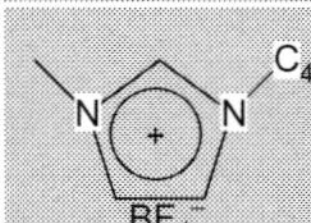	UAE	Phenolic compounds	Han et al. (2011)
		MAE	Phenolic alkaloids	Lu et al. (2008)
		UAE	Piperine	Cao et al. (2009)
		LLE	Tocopherol	Yang et al. (2009)
$[bmim]^+[Br]^-$	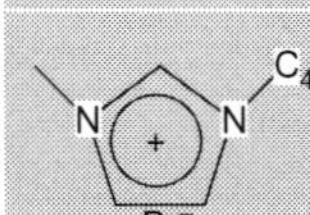	MAE	Polyphenolics	Du et al. (2009)
		UAE-MISPE	Astaxanthin	Bi et al. (2010)
		NPCAE	Flavonoids	Duan et al. (2013)

Continued

TABLE 10.1 List of Ionic Liquids Used in Different Types of Antioxidant Extractions—cont'd

Ionic Liquid	Structure	Extraction Methods	Antioxidants	References
$[bmim]^+[Cl]^-$	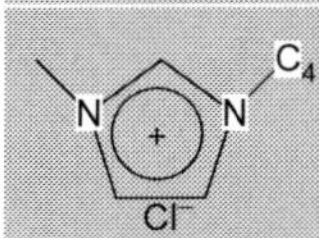	MAE	Phenolic alkaloids	Lu et al. (2008)
		LLE	Tocopherol	Yang et al. (2009)
		UAE	Piperine	Cao et al. (2009)
		UAE-MISPE	Astaxanthin	Bi et al. (2010)
$[bmim]^+[CH_3CO_2]^-$	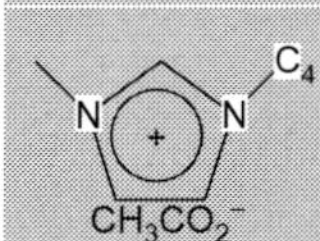	LLE	Gallic acid	Almeida et al. (2014)
			Vanillic acid	
			Syringic acid	
$[bmim]^+[H_2PO_4]^-$	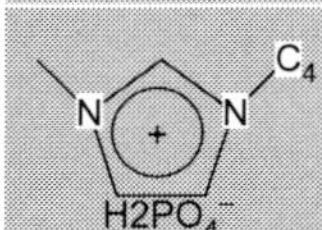	UAE-MISPE	Astaxanthin	Bi et al. (2010)
		LLE	Gallic acid	Almeida et al. (2014)
			Vanillic acid	
			Syringic acid	
$[bmim]^+[N(CN)_2]^-$	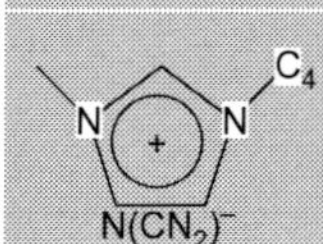	NPCAE	Flavonoids	Duan et al. (2013)
$[bmim]^+[SCN]^-$	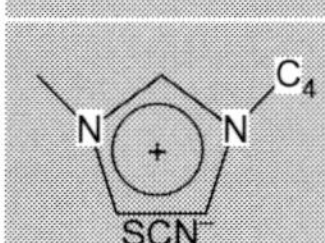	UAE	Piperine	Cao et al. (2009)
$[bmim]^+[TOS]^-$	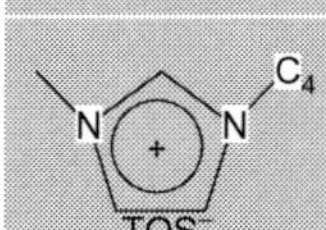	LLE	Gallic acid	Almeida et al. (2014)
			Vanillic acid	
			Syringic acid	

TABLE 10.1 List of Ionic Liquids Used in Different Types of Antioxidant Extractions—cont'd

Ionic Liquid	Structure	Extraction Methods	Antioxidants	References
$[bmim]^+[PF_6]^-$	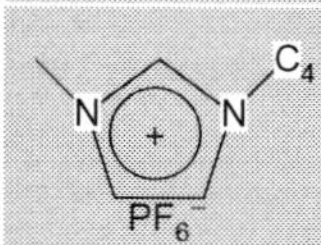	LLE	Ferulic acid	Yan-Ying et al. (2007)
		LLE	Caffeic acid	Almeida et al. (2014)
			Gallic acid	
			Vanillic acid	
			Syringic acid	
$[hmim]^+ [Br]^-$	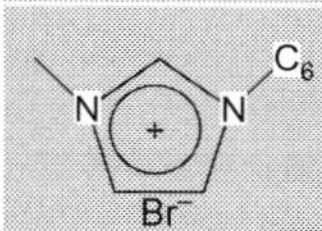	LLE	Gallic acid	Almeida et al. (2014)
			Vanillic acid	
			Syringic acid	
$[hmim]^+ [BF_4]^-$	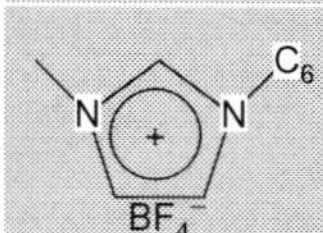	NPCAE	Flavonoids	Duan et al. (2013)
		MAE	Phenolic alkaloids	Lu et al. (2008)
		LLE	Gallic acid	Almeida et al. (2014)
			Vanillic acid	
			Syringic acid	
$[hmim]^+ [PF_6]^-$	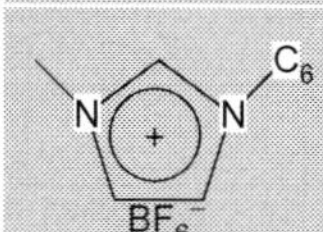	LLE	Ferulic acid	Yan-Ying et al. (2007)
			Caffeic acid	
$[omim]^+ [BF_4]^-$	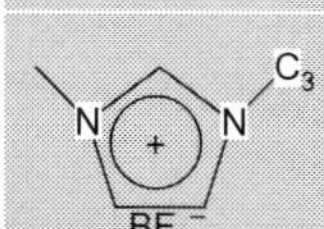	UAE	Piperine	Cao et al. (2009)
		UAE-MISPE	Astaxanthin	Bi et al. (2010)
		MAE	Phenolic alkaloids	Lu et al. (2008)

nonaromatic ILs, possibly due to the increase in hydrogen-bonding ability between the IL and antioxidant (Almeida et al., 2014). Since only small amounts of ILs were used to increase the extraction efficiency, this method presents an interesting possibility for large-scale extractions.

10.2.3 Ultrasound-Assisted Extraction

A rapid and effective IL-mediated UAE has been developed for extraction of phenolic compounds from natural sources. *Laminaria japonica* is a type of brown algae that contains large amounts of phenolic compounds. These phenolic compounds possess antioxidant, antimicrobial, and antiviral activities (Kim et al., 2006; Jeong et al., 2009). Conventional methods for extraction of such phenolic compounds include heat-reflux extraction and Soxhlet extraction. Hydrophilic ILs with longer alkyl chains, such as $[bmim]^+[BF_4]^-$, were found to be more effective in extracting the phenolic compounds. Compared with conventional extraction methods, this approach was able to obtain up to 88.3% of phenolic compounds (Han et al., 2011).

Piperine, which is the main constituent of white pepper, exhibits a variety of pharmacological activities including antifungal, antidiarrheal, and anti-inflammatory activities. Heat-reflux extraction is the most widely used traditional technique for extraction of piperine (Navickiene et al., 2000). IL-mediated UAE was employed to extract piperine by Cao et al. (2009). A series of ILs, differing in composition of alkyl chain and anion, were evaluated for their extraction efficiency. Anions and cations both affected the extraction efficiency. Due to the hydrophobicity of piperine, hydrophobic interactions between piperine and the water phase containing IL were the main driving force for extraction. Hydrophobic ILs with short alkyl chains were found to be most effective. Compared with the conventional heat-reflux extraction and regular UAE, IL-mediated UAE showed significant improvement in extraction efficiency (from 1.950% to 3.577%) (Cao et al., 2009).

10.2.4 Microwave-Assisted Extraction

As MAE has been developed, the method is proposed to be a suitable alternative for extraction of organic compounds directly from solid matrixes. Compared with traditional and other modern extraction techniques, it is rapid, effective, and exhibits a promising alternative in sample pretreatment. MAE has been employed in extraction of phenolic alkaloids in embryo of the seed of *Nelumbo nucifera*, which were frequently used as remedy in the treatment of tissue inflammation, nervous disorders, and cardiovascular diseases (Ling et al., 2005; Lu et al., 2008). Hydrophilic anions ($[Cl]^-$, $[Br]^-$, and $[BF_4]^-$) were found to be more effective than hydrophobic ($[PF_6]^-$), possibly due to increased water miscibility. Unlike extraction of most hydrophilic compounds, increase in cation alkyl chain length did not show significant improvement in extraction efficiencies. This approach showed an increased extraction efficiency of up to 20–50% compared to the conventional heat-reflux extraction (Lu et al., 2008).

Psidium guajava Linn. leaves and *Smilax china* tubers contain polyphenolic compounds including gallic acid, ellagic acid, quercetin, and *trans*-resveratrol, which have antioxidant, antiinflammatory, antimicrobial, and antitumor effects. They can be extracted with volatile organic solvents and microwave-assisted heating extraction at temperatures up to 100°C for 20 min without degradation (Du et al., 2011). Extraction efficiency of polyphenolic compounds with ILs is reported to be anion-dependent. For $[bmin]^+$ based ILs with $[Br]^-$, $[Cl]^-$, $[BF_4]^-$, $[SO_4]^{2-}$, $[N(CN)_2]^-$, and $[H_2PO_4]^-$ anions, the extraction yields indicated that $[Br]^-$ and $[H_2PO_4]^-$ were more efficient in the MAE of polyphenolic compounds from *P. guajava* leaves; and $[Br]^-$ and $[BF_4]^-$ were more efficient in the MAE of *trans*-resveratrol and quercetin from *S. china* tubers (Du et al., 2009).

10.2.5 Negative Pressure Cavitation-Assisted Extraction

In negative pressure cavitation-assisted extraction, cavitation is generated by negative pressure, and nitrogen is continuously introduced into a liquid–solid system to increase the turbulence, collision, and mass transfer between the extraction solvent and matrix. Presence of nitrogen throughout the extraction process can reduce oxidation of thermosensitive components (Kong et al., 2011; Duan et al., 2013). IL-mediated negative pressure cavitation-assisted extraction showed higher extraction yields and shorter extraction time in comparison to conventional heat-reflux extraction. The extraction yields of genistin, genistein, and apigenin were 0.477 ± 0.013, 0.480 ± 0.014, and 0.271 ± 0.021 mg/g, respectively (Duan et al., 2013).

10.3 INCENTIVES OF ANTIOXIDANTS MODIFICATION

10.3.1 Lipid Oxidation in Bulk versus Multiphase Systems

The efficiency of antioxidants at protecting unsaturated lipids against oxidation not only depends on their chemical structure and concentration, but also largely on the physical microstructure of the matrix in which they are contained. In fact, many food products or other biobased matrices do not contain lipids as a bulk, structurally homogeneous phase, but rather as colloidal dispersions (eg, emulsions). Lipid oxidation generally occurs earlier and faster when the oil is part of a multiphase matrix (for instance, emulsions), as compared to bulk oil (Frankel et al., 2002; Lethuaut et al., 2002). Such a favored oxidation is attributed to several causes. The interfacial area between the oil and other phases can promote the contacts between oxidizable lipids and prooxidant compounds, such as metal ions dissolved in the aqueous phase. In addition, the emulsification process itself may enhance oxidative reactions due to incorporation of oxygen, and increase in temperature (Berton-Carabin et al., 2014).

10.3.2 Food Emulsions

Food systems, as well as other biobased matrices, can often be described as structurally heterogeneous systems constituted of several immiscible phases (air, aqueous phase, lipid phase, insoluble material, etc.). Systems in which oil is dispersed in an aqueous-based matrix in the form of droplets are called oil-in-water (O/W) emulsions (eg, mayonnaise, dairy drinks, cosmetic creams, etc.). Such systems are characterized by the presence of distinct compartments, or phases, between which molecules can partition (Fig. 10.3). The physical organization of O/W emulsions has already been extensively reviewed in many articles and books, and we can cite, among others, the work of Dickinson (1992), McClements (2005), and Leal-Calderon et al. (2007).

In O/W emulsions, oil droplets form the dispersed phase and are surrounded by an aqueous continuous phase. The narrow region (typically, a few nanometers) between the dispersed and continuous phase is the interface. The volume represented by the interface is often only a very low fraction of the total emulsion volume, except when the droplet size becomes extremely small, for example in microemulsions (McClements and Decker, 2000).

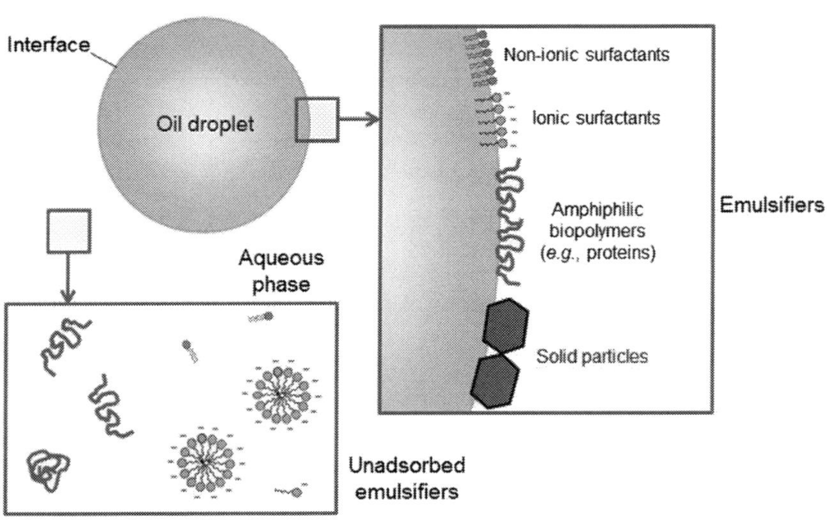

FIGURE 10.3 Schematic representation (not to scale) of the compartments and possible components of an oil-in-water (O/W) emulsion.

The surface represented by the interface directly depends on the oil phase volume fraction, and on the droplet size: The smaller the droplets, the higher their number and the total interfacial surface. A broad range of droplet sizes, from less than 0.1 μm to 100 μm, or even more, can usually be encountered in food emulsions. The interfacial area per mass unit of dispersed phase, or specific surface area (A, m^2/g oil), can be calculated according to (Eq. [1]), where r is the average oil droplet radius (μm) and ρ is the density of the oil (g/mL).

$$A = \frac{3}{r * \rho} \qquad [1]$$

This specific surface area often reaches several m^2 per g of oil, showing that a large surface between both immiscible phases exists in emulsions. Because of the molecular incompatibility of both liquids, this surface area tends to decrease the free energy of the system (ΔG), which is why emulsions are thermodynamically unstable ((Eq. [2]), where γ is the interfacial tension between both liquids (N/m), and ΔA is the total interfacial area (m^2)).

$$\Delta G = \gamma \, \Delta A \qquad [2]$$

Hence, emulsion droplets have to be physically stabilized to avoid rapid, spontaneous separation of the phases. A first category of stabilizers is thickening agents, which increase the viscosity of the emulsion's continuous phase and thereby slow down the creaming of oil droplets. A second category of emulsion stabilizers is emulsifiers, which are amphiphillic molecules. Because of their dual affinity for both polar and nonpolar media, such molecules can adsorb at the oil—water interface in emulsions, which decreases the interfacial tension and lowers the free energy of the system (Eq. [2]). Importantly, in practice, only a fraction of the emulsifier molecules used in conventional O/W emulsion formulations actually adsorbs at the interface; the residual fraction (excess emulsifiers) remains unadsorbed in the aqueous phase in the form of monomers, or aggregated, colloidal structures (Fig. 10.3). This unadsorbed emulsifier fraction can be considered as another phase, or pseudophase, of the emulsion system. It can tremendously affect the location and reactivity of molecular species, and hence the physicochemical properties of emulsions (Berton-Carabin et al., 2014).

Different types of emulsifiers can be used to stabilize O/W emulsions (Hasenhuettl and Hartel, 2008; Leal-Calderon and Schmitt, 2008). In food applications, two main categories of emulsifiers are generally described: low molecular weight emulsifiers (LMWEs, also referred to as surfactants), and amphiphillic biopolymers. The molecular structure of LMWEs comprises a hydrophilic head (which can be nonionic yet polar, or fully charged) and a hydrophobic tail, which generally is constituted by one or more acyl chains. Unadsorbed LMWEs may be present as monomers in the aqueous phase, and can form micelles above a certain concentration, called critical micelle concentration. Food-grade amphiphillic polymers are mostly represented by proteins. Most of the work on protein-stabilized emulsions has been conducted with dairy proteins (caseins or whey proteins) (Dickinson, 1994, 1998) even though other kinds of proteins, such as soy proteins, also have emulsifying properties (Kinsella, 1979).

In addition to the aforementioned conventional emulsifiers, oil–water interfaces in emulsions can also be stabilized by solid particles that have dual affinity for both phases, and can thereby strongly adsorb (Fig. 10.3). Such systems are referred to as Pickering emulsions (Pickering, 1907), and are generally characterized by a good resistance of the emulsion droplets to coalescence, due to the high desorption energy of the particles (Binks, 2002).

10.3.3 Partitioning of Antioxidants in Emulsions and the Polar Paradox Theory

The efficiency of antioxidants in food or biobased emulsions depends on several factors, one of the most important being their localization within the system. An early theory related to this matter is the paradigm of the polar paradox (Porter, 1980; Porter et al., 1989; Frankel et al., 1994). It states that polar, more hydrophilic antioxidants (for example, ascorbic acid) are more effective in nonpolar media (such as bulk oils) as compared to nonpolar, more hydrophobic antioxidants (for example, α-tocopherol). Conversely, in multiphase systems such as O/W emulsions, hydrophobic antioxidants tend to be more active than hydrophilic antioxidants. Such an apparently paradoxical effect would be because, in bulk oil, hydrophilic antioxidants locate preferentially at the air–oil interface, that is, where oxidizable lipid molecules get into contact with atmospheric oxygen, whereas hydrophobic antioxidants remain buried in the bulk phase. Conversely, in O/W emulsions, hydrophilic antioxidants are located in the aqueous phase, whereas hydrophobic antioxidants are located in oil droplets, where they can effectively counteract the propagation of lipid oxidation, or in close vicinity to the interface, where lipid oxidation is initiated (Fig. 10.4).

Despite the conceptual logic behind the polar paradox theory, many experimental results have shown deviations or even contradictions from this model, suggesting that factors other than the strict oil–water partitioning of the antioxidants affect their activity in O/W emulsions (Shahidi and Zhong, 2011). In particular, it has been repeatedly shown that the antioxidant activity of lipophilized antioxidants (ie, hydrophilic antioxidants onto which an alkyl chain is esterified) does not increase linearly with the alkyl chain length. Instead, in many cases, optimal antioxidant activity has been achieved for a medium alkyl chain length, while further increasing the alkyl chain length has led to a drastic decrease of the antioxidant activity (Fig. 10.5). This phenomenon has been referred to as the cutoff effect, and

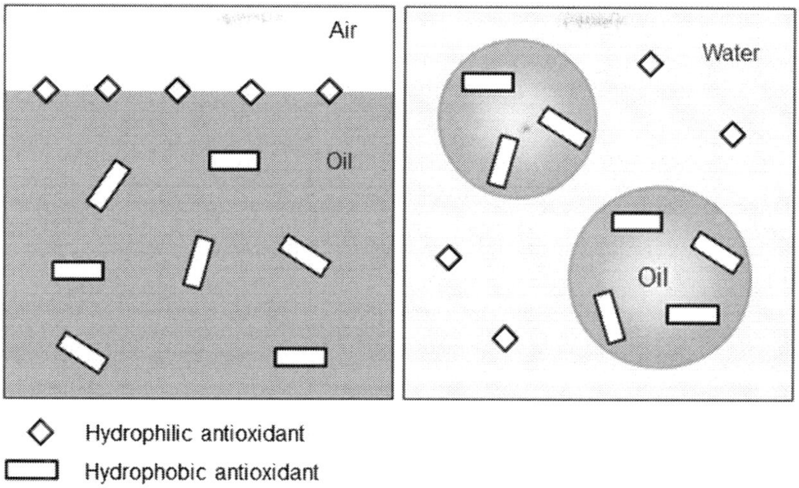

FIGURE 10.4 Schematic representation of the distribution of hydrophilic and hydrophobic antioxidants in bulk oil (left) or in O/W emulsion (right): the basis for the polar paradox theory.

has been substantially documented in the literature over the past few years (Laguerre et al., 2009, 2010, 2013; Shahidi and Zhong, 2011; Liu et al., 2014).

Several hypotheses have been proposed to explain the optimum found for medium alkyl chain, and the cutoff effect. First, increasing the alkyl chain length to a too-high extent could reduce the molecular mobility and promote self-aggregation of antioxidant molecules under the form of colloidal structures, which would decrease their mobility and diffusion rate, and thereby their activity. Second, the optimal activity observed with medium alkyl chain lengths would correspond to a favored location of lipophilized antioxidants at the oil—water interface, since the balance between the hydrophilic and hydrophobic moieties would make them optimally amphiphillic (Shahidi and Zhong, 2011; Laguerre et al., 2013). Recent work of the groups of C. Bravo-Diaz and L. Romsted has given experimental evidence in favor of this second hypothesis (Barreiro et al., 2013). In this work, the antioxidant activity of gallic acid and its alkyl derivatives (3—12 carbons) was investigated in emulsion systems. The location of antioxidant molecules was concomitantly determined using the pseudophase kinetic model (Gunaseelan et al., 2006), which confirmed that the higher the antioxidant fraction located at the oil—water interface, the greater the antioxidant efficiency.

The location of antioxidant molecules within emulsion systems, and thereby their ability to counteract lipid oxidation, can be affected by other factors, mostly related to the composition and microstructure of the system. Surfactant micelles in the aqueous phase of O/W emulsions can solubilize nonpolar antioxidants in their hydrophobic core, or form comicelles with amphiphillic

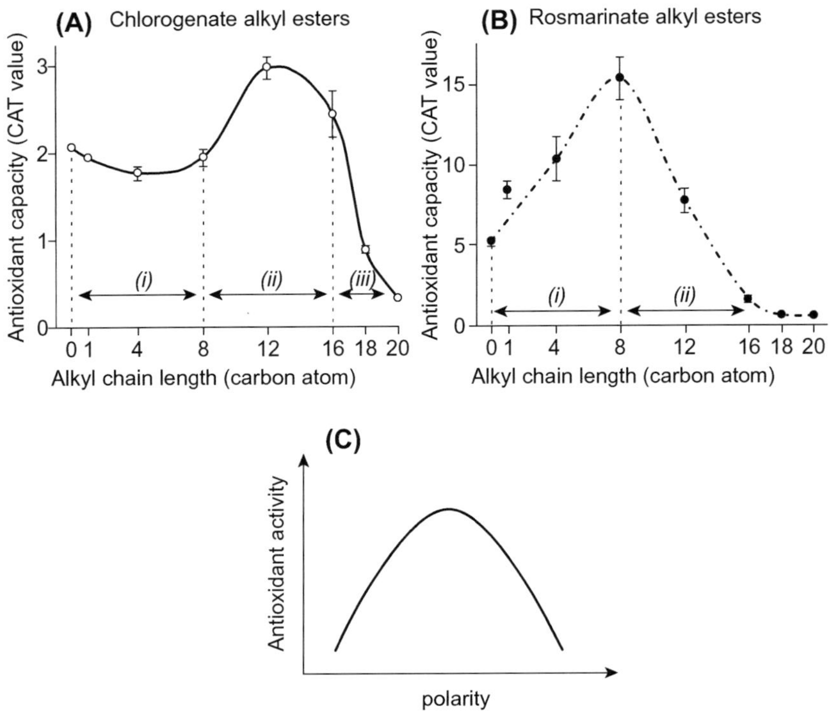

FIGURE 10.5 (A, B) Antioxidant activity of lipophilized antioxidants as a function of the alkyl chain length. *(Reprinted with permission from Laguerre, M., Giraldo, L.J.L., Lecomte, J., Figueroa-Espinoza, M.C., Barea, B., Weiss, J., Decker, E.A., Villeneuve, P., 2009. Chain length affects antioxidant properties of chlorogenate esters in emulsion: the cutoff theory behind the polar paradox. J. Agric. Food Chem. 57, 11335–11342. Copyright (2009) American Chemical Society; and from Laguerre, M., Giraldo, L.J.L., Lecomte, J., Figueroa-Espinoza, M.C., Barea, B., Weiss, J., Decker, E.A., Villeneuve, P., 2010. Relationship between hydrophobicity and antioxidant ability of phenolipids in emulsion: A parabolic effect of the chain length of rosmarinate esters. J. Agric. Food Chem. 58, 2869–2876. Copyright (2010) American Chemical Society.)* (C) Generalization of the cutoff effect. *(Reprinted with permission from Shahidi, F., Zhong, Y., 2011. Revisiting the polar paradox theory: a critical overview. J. Agric. Food Chem. 59, 3499–3504. Copyright (2011) American Chemical Society.)*

antioxidants (Richards et al., 2002; Panya et al., 2012). Interestingly, contradicting effects on such a solubilization/comicellization on the antioxidant activity has been found, depending on the hydrophobicity of the antioxidant. Antioxidants may also chemically interact with adsorbed emulsifiers (Almajano et al., 2007; Conde et al., 2011; von Staszewski et al., 2014), which can enhance their accumulation at the oil–water interface. However, such chemical interactions may also negatively affect their antioxidant efficiency, possibly by decreasing the hydrogen-donating activity of chain-breaking antioxidants (Pekkarinen et al., 1999).

10.4 MODIFICATION OF ANTIOXIDANTS IN IONIC LIQUIDS

In addition to their application as sorbent for extraction of antioxidant from natural sources, ILs are also applied as reaction media for the lipophilization of natural antioxidant. Lipophilization of antioxidants improve the oil solubility and tailor the antioxidants to specific applications, according to the polar paradox and cutoff theories. Numerous examples exist on the acylation of hydrophilic antioxidants in ILs, as summarized in Table 10.2. The structures of selected antioxidants modified in ILs are presented in Table 10.3.

The use of ILs as reaction media for the lipophilization of antioxidants leads to a more environmentally friendly process compared to organic solvents, mainly due to their low volatility and flammability, which significantly reduces health and safety risks (Alevizou and Voutsas, 2013). Their low toxicity has gained special attention in the lipophilization of antioxidants, as these compounds are usually produced for use in food products. Toxicological tests of a series of ILs with imidazolium cations and various anions showed that most of the ILs were less toxic than organic solvents by several orders of magnitude (Ranke et al., 2004).

A central challenge in designing a suitable reaction system for the lipophilization of antioxidants is to find a reaction medium in which the hydrophobic antioxidant and hydrophilic acylating agent both are soluble, while also providing a suitable environment for the catalyst. Generally, comparable or higher yields, faster reaction rates, and shorter reaction times are observed when antioxidants are acylated in ILs compared to conventional solvents. Katsoura et al. (2006) acylated flavonoid glycosides (naringin and rutin) with vinyl esters of varying chain length, using lipase catalysts, in the ILs $[bmim]^+[BF_4]^-$ or $[bmim]^+[PF_6]^-$. They found that the initial reaction rates were 2–4 times higher compared to those reported in conventional organic solvents. Later, the same research group (Katsoura et al., 2007) acylated esculin, salicin, helicin, and naringin with vinyl butyrate in the same ILs, and compared the yields to those obtained in acetone. For esculin and helicin, the yield after 72 h of reaction was higher in ILs, whereas the yield of salicin and naringin was comparable and lower in ILs compared to acetone, respectively.

Ha et al. (2012) synthesized caffeic acid phenethyl ester enzymatically from phenethyl alcohol and caffeic acid in a range of ILs. Caffeic acid phenethyl ester is an active flavonoid, which has been shown to have good antioxidant properties, and is an interesting compound for food applications. Of the ILs tested in their study (listed in Table 10.2), the highest conversion (92% of caffeic acid after 48 h) was obtained in $[emim]^+[Tf_2N]^-$ at 70°C, using Novozym 435 (immobilized lipase from *Candida antartica*) as catalyst, and a 30:1 ratio of substrates, phenethyl alcohol and caffeic acid. Compared to isooctane, the use of all ILs except $[omin]^+[PF_6]^-$ led to higher substrate conversions.

TABLE 10.2 List of Antioxidant-Lipophilizations in Ionic Liquids

Antioxidant	Acylating Agent	Catalysts	Ionic Liquids	References
Naringin	Vinyl esters with varying carbon chain length (C_2, C_3, C_4, C_{10}, C_{12}, and C_{18})	Novozym 435	$[bmim]^+[BF_4]^-$	Katsoura et al. (2006)
Rutin	Oleic acid	Lipozyme TLIM	$[bmim]^+[PF_6]^-$	
		Lipozyme RMIM		
		Lipase from *C. rugosa*		
Naringin	Vinyl butyrate	Novozym 435	$[bmim]^+[BF_4]^-$	Katsoura et al. (2007)
Helicin			$[bmim]^+[PF_6]^-$	
Salicin				
Esculin				
Caffeic acid	Phenethyl alcohol	Novozym 435	$[emim]^+[Tf_2N]^-$	Ha et al. (2012)
		Lipozyme RMIM	$[bmim]^+[Tf_2N]^-$	
		Lipozyme TLIM	$[hmim]^+[Tf_2N]^-$	
			$[omim]^+[Tf_2N]^-$	
			$[emim]^+[Tf_2N]^-$	
			$[bmim]^+[PF_6]^-$	
			$[hmim]^+[PF_6]^-$	
			$[omim]^+[PF_6]^-$	
			$[emim]^+[BF_4]^-$	
			$[emim]^+[TfO]^-$	
Salicylic acid	Alkyl halides	$NaHCO_3$	$[bmim]^+[Br]^-$	Ambika et al. (2008)
	Dialkyl sulfates	NaOH	$[bmim]^+[BF_4]^-$	
			$[bmim]^+[PF_6]^-$	
			$[bmim]^+[Br]^-$	
			$[bmim]^+[PF_6]^-$	
			$[bmim]^+[BF_4]^-$	
			$[bmim]^+[Br]^-$	
			$[bmim]^+[PF_6]^-$	
			$[bmim]^+[BF_4]^-$	

TABLE 10.2 List of Antioxidant-Lipophilizations in Ionic Liquids—cont'd

Antioxidant	Acylating Agent	Catalysts	Ionic Liquids	References
Caffeic acid	1-propanol	Novozym 435	$[hmim]^+[HSO_4]^-$	Pang et al. (2013)
		Lipozyme RMIM	$[omim]^+[BF_4]^-$	
		Lipozyme TLIM	$[emim]^+[TfO]^-$	
			$[bmim]^+[Tf_2N]^-$	
			$[bmim]^+[BF_4]^-$	
			$[bmim]^+[PF_6]^-$	
			$[bmim]^+[TfO]^-$	
			$[nmim]^+[HSO_4]^-$	
			$[hmim]^+[Cl]^-$	
Methyl caffeate	Cyclohexylethanol	Novozym 435	$[bmim]^+[PF_6]^-$	Kurata et al. (2010)
	Cyclohexylpropanol	Lipozyme RMIM	$[bmim]^+[BF_4]^-$	
		Lipozyme TLIM	$[bmim]^+[TfO]^-$	
		B. cepacia lipase	$[bmim]^+[Tf_2N]^-$	
			$[PMPra]^+[Tf_2N]^-$	
			$[MPPro]^+[Tf_2N]^-$	
			$[MPPip]^+[Tf_2N]^-$	
			$[TMPA]^+[Tf_2N]^-$	

Ambika et al. (2008) reported the chemoselective synthesis of phenolic esters from the corresponding phenolic acids with dialkyl sulfates or alkyl halides in the presence of sodium hydrogen carbonate ($NaHCO_3$) in imidazolium-based ILs. They compared the reaction to that in organic solvents (dimethyl formamide, dimethyl sulfoxide, and tetrahydrofuran). They found that the use of ILs led to higher reaction yields (88−94%) and shorter reaction times (20−40 min), compared to the corresponding reactions in organic solvents (35−50% yield and 180−240 min, respectively).

10.4.1 Factors Affecting Reactions in Ionic Liquids Media

Several factors influence the acylation reaction of hydrophilic antioxidants in ILs. The nature of the IL directly influences the reaction due to the effects on substrate solubility, reaction medium viscosity, and enzymatic stability. Additionally, factors such as nature of the catalyst, reaction temperature, and water content also influence the reaction rate and yield of lipophilized antioxidants in ILs.

10.4.1.1 Influence of Anions on Substrate Solubility

Alevizou and Voutsas (2013) measured the solubility of the antioxidative compounds *p*-coumaric acid and caffeic acid in six alkyl-substituted imidazolium-based ILs with $[PF_6]^-$, $[BF_4]^-$, $[TfO]^-$, and $[Tf_2N]^-$ anions; and in two organic solvents, *t*-pentanol and ethyl acetate. They showed that ILs based on the hydrophilic anions ($[BF_4]^-$ and $[TfO]^-$) were better solvents for these antioxidants compared to ILs based on the hydrophobic anions ($[PF_6]^-$ and $[Tf_2N]^-$). In terms of solubility of the antioxidant substrates, the anion ranking was as follows: $[BF_4]^- > [TfO]^- > [PF_6]^- > [Tf_2N]^-$, for the same cationic moiety ($[bmin]^+$). This ranking is generally in agreement with the relative polarity, hydrophobicity, and hydrogen bonding capacity of the ILs. Hence, the hydrophilic ILs interacted more strongly with the hydrophilic antioxidants, and specific interactions such as hydrogen bonding were likely to occur.

Katsoura et al. (2006) reported that the solubility of naringin and rutin was up to four times higher in $[bmin]^+[BF_4]^-$ at 60°C compared to conventional organic solvents such as acetone or *tert*-butanol. Katsoura et al. (2007) further found that $[bmim]^+[BF_4]^-$ could dissolve significantly more esculin, salicin, helicin, and naringin at 50°C compared to $[bmim]^+[PF_6]^-$, and that $[bmim]^+[BF_4]^-$ and $[bmim]^+[PF_6]^-$ both dissolved more of the antioxidants compared to acetone (except from naringin, which showed lower solubility in $[bmim]^+[BF_4]^-$ compared to acetone). They further observed that the higher solubility of the antioxidant compounds in ILs corresponded to a higher reaction rate.

10.4.1.2 Influence of Cations on Substrate Solubility

In their study of solubility of *p*-coumaric acid and caffeic acid in ILs and organic solvents, Alevizou and Voutsas (2013) found a correlation between the IL cation and the solubility of the antioxidants. When comparing ILs with the same anion ($[BF_4]^-$ or $[PF_6]^-$), Alevizou et al. showed that, as the alkyl chain length increases from butyl to octyl, the solubility of antioxidants increased in case of $[BF_4]^-$ ILs, but decreased in case of $[PF_6]^-$ ILs. In their work with enzymatic synthesis of caffeic acid phenethyl ester, Ha et al. (2012) showed that, as the chain length of the alkyl substitution on the imidazolium-based cation increased from ethyl to octyl, the solubility of caffeic acid decreased for $[PF_6]^-$ and $[Tf_2N]^-$ ILs. In case of $[Tf_2N]^-$ based ILs, the solubility of caffeic acid increased remarkably when the imidazole was substituted with two methyl groups, in addition to an ethyl alkyl chain, or with a benzyl and one methyl. For the latter, Ha et al. argues that the similarity in structure between the caffeic acid and the benzyl substitution was the reason for the observed higher solubility of caffeic acid in this IL.

10.4.1.3 Influence of Viscosity and Reaction Temperature

As with all reactions, temperature plays an important role in the lipophilization of antioxidants in ILs. By increasing temperature, the solubility of the substrates generally increases, and especially important in the case of ILs, the viscosity decreases, which decreases mass transfer limitations (Pang et al., 2013). The viscosity of an IL-based reaction medium is also dependent on the nature of the IL. The viscosity generally increases with an increase in the length of the alkyl group on the cation, as well as with increasing symmetry of the anion (Tang et al., 2012). In addition to influencing mass transfer, viscosity may also affect the enzyme conformation, which can influence its activity (Sureshkumar and Lee, 2009). When enzymatic catalysts are employed, thermal inactivation of the catalysts usually sets the upper limit to the reaction temperature. This point was shown by Ha et al. (2012) in their synthesis of caffeic acid phenethyl ester. They found that both the caffeic acid solubility and enzymatic activity increased when temperature increased from 50 to 70°C, which led to an increased conversion. Although the substrate solubility continued to increase with further increases in temperature, optimum conversion was observed at 70°C, probably due to thermal inactivation of the enzyme at higher temperatures.

10.4.1.4 Influence of Catalysts

Lipophilizations of antioxidants in ILs can be catalyzed by chemical and enzymatic catalysts. ILs serve only as the solvent and do not catalyze the esterification of antioxidants, as was shown by Ha et al. (2012). Generally, the use of enzymatic catalysts over chemical catalysts is considered advantageous due to the high specificity and mild reaction conditions generally seen in enzymatic reactions. Examples of chemically catalyzed modification of antioxidants in ILs exist (Ambika et al., 2008), yet the majority of the reported lipophilizations of antioxidants in ILs are enzymatically catalyzed. Enzyme activity in ILs is generally dependent on the nature of both the cation and the anion. The degree of hydrophobicity of ILs is of special importance, as hydrophilic ILs are believed to be capable of stripping off the essential water layer of the enzyme, or directly interacting with the lipase and inactivating it. This was shown by Ha et al. (2012), who found that no enzymatic

TABLE 10.3 Structures of Selected Antioxidant Compounds Modified in Ionic Liquids

Antioxidant	Modified Antioxidant	References
Naringin 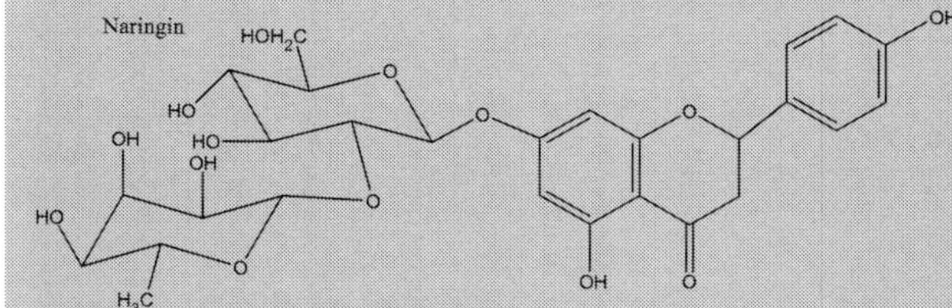	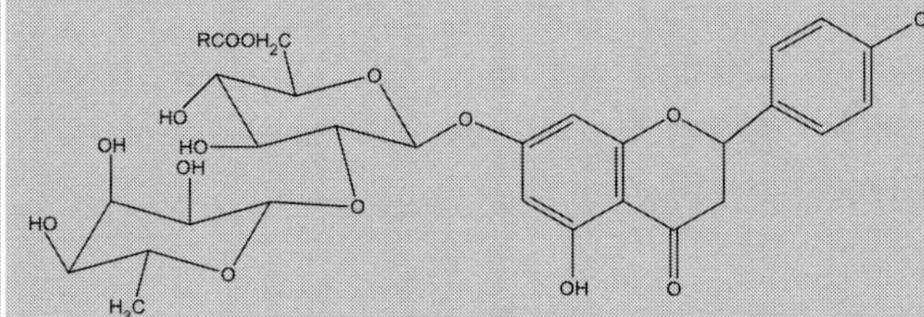	Katsoura et al. (2006)
Helicin 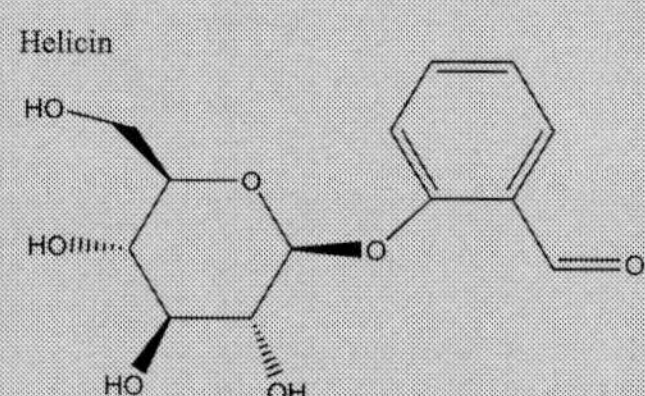	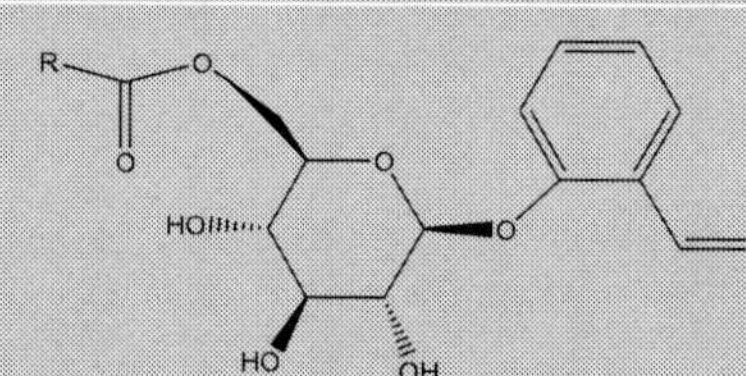	Katsoura et al. (2007)

Salicin Salicin 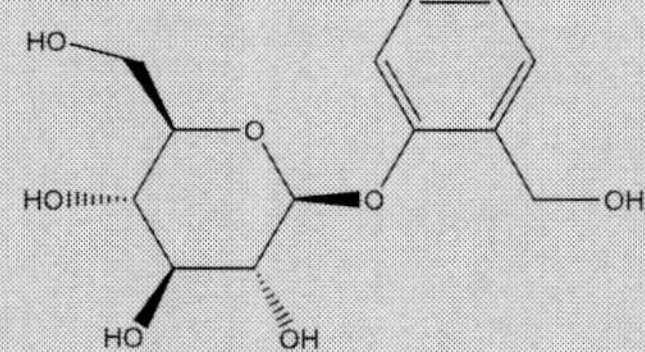	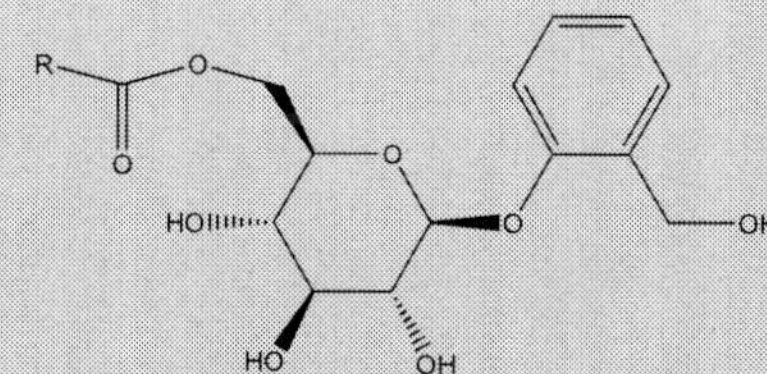	Katsoura et al. (2007)
Caffeic acid Caffeic acid 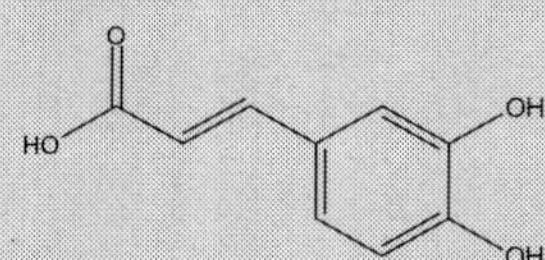	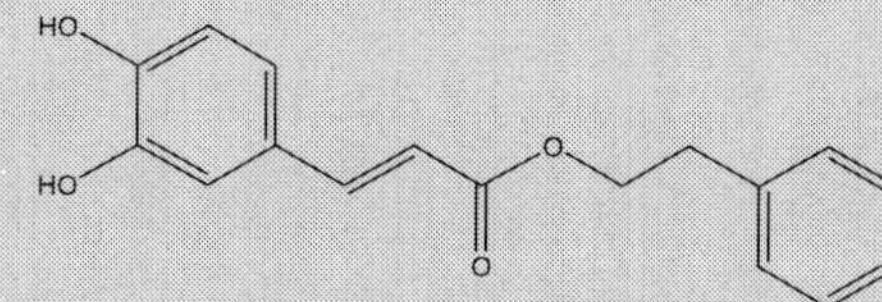	Ha et al. (2012)

reaction between caffeic acid and phenethyl ester took place in hydrophilic ILs (preliminary data).

In their studies on acylation of flavonoids, Katsoura et al. (2006) found that only immobilized lipases (Novozym 435, Lipozyme TLIM (immobilized lipase from *Thermomyces lanuginosus*), and Lipozyme RMIM (immobilized lipase from *Mucor miehei*)) were effective catalysts in the IL media ($[bmim]^+[BF_4]^-$ and $[bmim]^+[PF_6]^-$). The nonimmobilized lipase from *Candida rugosa* did not catalyze the reaction in $[bmim]^+[BF_4]^-$ and only led to $<5\%$ conversion in $[bmim]^+[PF_6]^-$. Ha et al. (2012) found that only Novozym 435, and not Lipozyme RMIM nor Lipozyme TLIM, catalyzed the reaction between caffeic acid and phenethyl ester in $[emim]^+[Tf_2N]^-$. Kurata et al. (2010) reported similar results for the enzymatic transesterification of methyl caffeate with 2-cyclohexylethanol in $[bmin]^+[Tf_2N]^-$. Of the four commercial lipases tested, Novozym 435 had the highest activity, while Lipozyme TLIM and lipase from *Burkholderia cepacia* had very low activity. They found, however, that Lipozyme RMIM catalyzed the reaction relatively well. Kurata et al. further tested the influence of various ILs on the performance of Novozym 435. They found that the enzyme activity in ILs was mainly anion dependent; the enzyme was active in $[bmin]^+[Tf_2N]^-$ and $[bmim]^+[PF_6]^-$, but inactive in $[bmim]^+[BF_4]^-$ and $[bmim]^+[TfO]^-$. Slightly different observations were obtained by Pang et al. (2013) in their study of the transesterification of methyl caffeate and 1-propanol using Novozym 435. They reported that considerably higher yields of propyl caffeate were obtained in $[bmim]^+[TfO]^-$, $[bmim]^+[PF_6]^-$, and $[bmin]^+[Tf_2N]^-$ compared to the other ILs (listed in Table 10.2).

10.4.1.5 Influence of Water Content

For all esterification reactions, the water content plays a critical role. The water content of ILs varies with type and manufacturer. Alevizou and Voutsas (2013) examined the influence of water content of six ILs on the solubility of *p*-coumaric acid and caffeic acid, and found that drying did not change the solubility of the antioxidants. However, Katsoura et al. (2007) found that, as the amount of water in the reaction system increased the reaction rate and the yield of the antioxidant (salicin, helicin, esculin, and naringin) vinyl butyrate esters decreased, when the reaction took place in $[bmim]^+[BF_4]^-$ or $[bmim]^+[PF_6]^-$. Generally, for enzymatic esterification reactions, the amount of water influences the hydrolysis rate of the system, and limits the transport of the hydrophobic substrate from the surrounding solvent through the hydration layer to the enzyme. In the special case of ILs, the influence of water has also been attributed to the dissociation of ILs into their individual ions, which then interacts with the substrate or product molecules, or the catalyst (Zhao, 2005).

10.4.2 Reusability of Ionic Liquids

A clear advantage of the use of ILs in the synthesis of antioxidant-targeted compounds is the easy product purification and the possibility of reusing the IL in subsequent reactions. Ambika et al. (2008) synthesized methyl salicylate from salicylic acid and dimethyl sulfate in presence of $NaHCO_3$ in $[bmin]^+[PF_6]^-$. They found that the product esters could be easily separated from the reaction mixture by simple extraction with ethyl acetate. Furthermore, they showed that the IL could be reused in subsequent reactions; the reaction yield was 94%, 92%, 87%, and 87% in four cycles of IL reuse, respectively (Ambika et al., 2008). Successful reuse of ILs has also been reported by others (Katsoura et al., 2007; Kurata et al., 2010).

10.5 ANTIOXIDANT SYNTHESIS IN IONIC LIQUIDS

The main area in which ILs has been applied for synthesis of antioxidant-targeted compounds is in the acylation of natural antioxidants, which has been the topic of the previous sections. However, other types of antioxidants are being modified, or synthesized, in ILs, too. Luo et al. (2013) prepared Maillard reaction products from chitosan and xylan in ILs. They found that this chitosan-xylan conjugate had a high antioxidant activity making it a promising natural antioxidant agent. The challenge associated with different solubility of chitosan and xylan (chitosan is only soluble in specific organic acid, while xylan, as most hemicelluloses, only is soluble in alkaline solution) was overcome by the use of IL ($[bmin]^+[Cl]^-$). The Maillard reaction products that were successfully formed from chitosan and xylan in ILs were found to have increased antioxidant activity compared to their parent polymers.

10.6 SUMMARY AND FUTURE PERSPECTIVES

The interest in using antioxidants from natural sources is increasing, and ILs have attracted considerable attention as solvents in extraction and separation of antioxidants from natural materials. Although only a few studies have examined the extraction and separation efficiency of antioxidant-targeted compounds using ILs, it has been successfully applied for separation and extraction of a wide range of analytes. ILs offer high specificities and can be used to separate compounds with a high degree of similarities. Nevertheless, wide use of ILs as extraction solvents may present challenges, as it presents problems in terms of purification of the extracted compounds from the IL phase. Solvents are still required to extract the antioxidant-targeted compounds from the IL phase.

The application of antioxidants in multiphase systems is increasingly understood, and it is now evident that controlled modification of natural antioxidants is necessary to provide optimal protection against oxidation in multiphase systems. This prompts the need for environmentally friendly modification systems, in which natural antioxidants can be modified without the use of toxic organic solvents. The low toxicity of ILs makes them promising alternatives to organic solvents, and with the increasing research in the area, careful control of the acylation reactions has become possible. The high cost of ILs remain, however, one of the main challenges in the widespread use of ILs as reaction media. Promising results have indicated that ILs can be successfully reused, with only minor reductions in conversion yields after repeated cycles. This can significantly lower the cost of the reaction system, and the use of ILs in the synthesis and modification of antioxidant-targeted compounds is believed to gain increasing potential in the future.

REFERENCES

Alevizou, E.I., Voutsas, E.C., 2013. Solubilities of p-coumaric and caffeic acid in ionic liquids and organic solvents. J. Chem. Thermodyn. 62, 69–78.

Almajano, M.P., Delgado, M.E., Gordon, M.H., 2007. Albumin causes a synergistic increase in the antioxidant activity of green tea catechins in oil-in-water emulsions. Food Chem. 102, 1375–1382.

Almeida, M.R., Passos, H., Pereira, M.M., Lima, A.S., Coutinho, J.A.P., Freire, M.G., 2014. Ionic liquids as additives to enhance the extraction of antioxidants in aqueous two-phase systems. Sep. Purif. Technol. 128, 1–10.

Ambika, Singh, P.P., Chauhan, S.M.S., 2008. Chemoselective esterification of phenolic acids in the presence of sodium bicarbonate in ionic liquids. Synth. Commun. 38.

Awad, W.H., Gilman, J.W., Nyden, M., Harris, R.H., Sutto, T.E., Callahan, J., Trulove, P.C., DeLong, H.C., Fox, D.M., 2004. Thermal degradation studies of alkyl-imidazolium salts and their application in nanocomposites. Thermochim. Acta 409, 3–11.

Ballard, T., 2008. Optimizing the Extraction of Phenolic Antioxidant Compounds from Peanut Skins (Ph.D. thesis). Virginia Polytechnic Institute and State University.

Barreiro, S.L., Bravo-Diaz, C., Paiva-Martins, F., Romsted, L.S., 2013. Maxima in antioxidant distributions and efficiencies with increasing hydrophobicity of gallic acid and its alkyl esters. The pseudophase model interpretation of the "cutoff effect." J. Agric. Food Chem. 61, 6533–6543.

Belitz, H.D., Grosch, W., Schieberle, P., 2004. Lipids. In: Belitz, H.D., Grosch, W., Schieberle, P. (Eds.), Food Chemistry, third revised edition. Springer-Verlag, Berlin, pp. 157–244.

Berton-Carabin, C.C., Ropers, M.-H., Genot, C., 2014. Lipid oxidation in oil-in-water emulsions: involvement of the interfacial layer. Comp. Rev. Food Sci. Food Saf. 13, 945–977.

Bi, W., Tian, M., Zhou, J., Row, K.H., 2010. Task-specific ionic liquid-assisted extraction and separation of astaxanthin from shrimp waste. J. Chromatogr. B. Anal. Technol. Biomed. Life Sci. 878, 2243–2248.

Binks, B.P., 2002. Particles as Surfactants—Similarities and differences. Curr. Opin. Coll. Int. Sci. 21–41.

Botterweck, A.A.M., Verhagen, H., Goldbohm, R.A., Kleinjans, J., van den Brandt, P.A., 2000. Intake of butylated hydroxyanisole and butylated hydroxytoluene and stomach cancer risk: results from analyses in the Netherlands cohort study. Food Chem. Toxicol. 38, 599–605.

Cao, X.J., Ye, X.M., Lu, Y.B., Yu, Y., Mo, W.M., 2009. Ionic liquid-based ultrasonic-assisted extraction of piperine from white pepper. Anal. Chim. 640, 47–51.

Chan, H.W.S., 1987. The mechanisms of autoxidation. In: Chan, H.W.S. (Ed.), Autoxidation of Unsaturated Lipids. Academic Press, London, pp. 141–206.

Chen, T.H., Payne, G.F., 2001. Separation of alpha- and delta-tocopherols due to an attenuation of hydrogen bonding. Ind. Eng. Chem. Res. 40, 3413–3417.

Cheng, Z., Li, Y., 2007. What is responsible for the initiating chemistry of iron-mediated lipid peroxidation: an update. Chem. Rev. 107, 748–766.

Conde, E., Gordon, M.H., Moure, A., Dominguez, H., 2011. Effects of caffeic acid and bovine serum albumin in reducing the rate of development of rancidity in oil-in-water and water-in-oil emulsions. Food Chem. 129, 1652–1659.

Dickinson, E., 1992. An Introduction to Food Colloids. Oxford University Press, New York, USA.

Dickinson, E., 1994. Protein-stabilized emulsions. J. Food Eng. 59–74.

Dickinson, E., 1998. Proteins at interfaces and in emulsions. Stability, rheology and interactions. J. Chem. Soc. Faraday Trans. 1657–1669.

Du, F.Y., Xiao, X.H., Li, G.K., 2011. Ionic liquid aqueous solvent-based microwave-assisted hydrolysis for the extraction and HPLC determination of myricetin and quercetin from *Myrica rubra* leaves. Biomed. Chromatogr. 25, 472–478.

Du, F.Y., Xiao, X.H., Luo, X.J., Li, G.K., 2009. Application of ionic liquids in the microwave-assisted extraction of polyphenolic compounds from medicinal plants. Talanta 78, 1177–1184.

Duan, M.H., Luo, M., Zhao, C.J., Wang, W., Zu, Y.G., Zhang, D.Y., Yao, X.H., Fu, Y.J., 2013. Ionic liquid-based negative pressure cavitation-assisted extraction of three main flavonoids from the pigeonpea roots and its pilot-scale application. Sep. Purif. Technol. 107, 26–36.

Eder, K., Ringseis, R., 2010. Health aspects of oxidized dietary fats. In: Decker, E.A., Elias, R.J., McClements, D.J. (Eds.), Oxidation in Foods and Beverages and Antioxidant Applications, vol. 1. Woodhead Publishing Ltd, Sawston, Cambridge, UK, pp. 142–180.

Frankel, E.N., 1985. Chemistry of autoxidation: mechanism, products and flavor significance. In: Min, D.B., Smouse, T. (Eds.), Flavor Chemistry of Fats and Oils. American Oil Chemists Society, Urbana, IL, USA, pp. 1–37.

Frankel, E.N., 1996. Antioxidants in lipid foods and their impact on food quality. Food Chem. 57, 51–55.

Frankel, E.N., 2005. Lipid Oxidation. The Oily Press Ltd, Bridgwater, England.

Frankel, E.N., Huang, S.W., Kanner, J., German, J.B., 1994. Interfacial phenomena in the evaluation of antioxidants—bulk oils vs emulsions. J. Agric. Food Chem. 42, 1054–1059.

Frankel, E.N., Satue-Gracia, T., Meyer, A.S., German, J.B., 2002. Oxidative stability of fish and algae oils containing long-chain polyunsaturated fatty acids in bulk and in oil-in-water emulsions. J. Agric. Food Chem. 50, 2094–2099.

Gardner, H.W., 1989. Oxygen radical chemistry of polyunsaturated fatty acids. Free Radic. Biol. Med. 7, 65–86.

Grosch, W., 1982. Lipid degradation products and flavour. In: Morton, I.D., McLeod, A.D. (Eds.), Food Flavours. Elsevier, Amsterdam, pp. 325–398.

Gunaseelan, K., Romsted, L.S., Gallego, M.J.P., Gonzalez-Romero, E., Bravo-Diaz, C., 2006. Determining alpha-tocopherol distributions between the oil, water, and interfacial regions of macroemulsions: novel applications of electroanalytical chemistry and the pseudophase kinetic model. Adv. Coll. Int. Sci. 123, 303–311.

Ha, S.H., Anh, T.V., Lee, S.H., Koo, Y.M., 2012. Effect of ionic liquids on enzymatic synthesis of caffeic acid phenethyl ester. Bioprocess Biosyst. Eng. 35, 235–240.

Han, D., Zhu, T., Row, K.H., 2011. Ultrasonic extraction of phenolic compounds from *Laminaria japonica* Aresch using ionic liquid as extraction solvent. Bull. Korean Chem. Soc. 32, 2212–2216.

Hasenhuettl, G.L., Hartel, R.W., 2008. Food Emulsifiers and Their Applications. Springer Science + Business Media, New York, USA.

Ho, T.D., Canestraro, A.J., Anderson, J.L., 2011. Ionic liquids in solid-phase microextraction: a review. Anal. Chim. Acta 695, 18–43.

Huddleston, J.G., Visser, A.E., Reichert, W.M., Willauer, H.D., Broker, G.A., Rogers, R.D., 2001. Characterization and comparison of hydrophilic and hydrophobic room temperature ionic liquids incorporating the imidazolium cation. Green Chem. 3, 156–164.

Jacobsen, C., Let, M.B., Nielsen, N.S., Meyer, A.S., 2008. Antioxidant strategies for preventing oxidative flavour deterioration of foods enriched with n-3 polyunsaturated lipids: a comparative evaluation. Trends Food Sci. Technol. 19, 76–93.

Jeong, J.B., Hong, S.C., Jeong, H.J., 2009. 3,4-dihydroxybenzaldehyde purified from the barley seeds (*Hordeum vulgare*) inhibits oxidative DNA damage and apoptosis via its antioxidant activity. Phytomedicine 16, 85–94.

Kanazawa, A., Sawa, T., Akaike, T., Maeda, H., 2002. Dietary lipid peroxidation products and DNA damage in colon carcinogenesis. Eur. J. Lipid Sci. Technol. 104, 439–447.

Katsoura, M.H., Polydera, A.C., Katapodis, P., Kolisis, F.N., Stamatis, H., 2007. Effect of different reaction parameters on the lipase-catalyzed selective acylation of polyhydroxylated natural compounds in ionic liquids. Process Biochem. 42, 1326–1334.

Katsoura, M.H., Polydera, A.C., Tsironis, L., Tselepis, A.D., Stamatis, H., 2006. Use of ionic liquids as media for the biocatalytic preparation of flavonoid derivatives with antioxidant potency. J. Biotechnol. 123, 491–503.

Kim, K.H., Kim, Y.W., Kim, H.B., Lee, B.J., Lee, D.S., 2006. Anti-apoptotic activity of laminarin polysaccharides and their enzymatically hydrolyzed oligosaccharides from *Laminaria japonica*. Biotechnol. Lett. 28, 439–446.

Kinsella, J.E., 1979. Functional-properties of soy proteins. J. Am. Oil Chem. Soc. 56, 242–258.

Kong, Y., Wei, Z.F., Fu, Y.J., Gu, C.B., Zhao, C.J., Yao, X.H., Efferth, T., 2011. Negative-pressure cavitation extraction of cajaninstilbene acid and pinostrobin from pigeon pea [*Cajanus cajan* (l.) Millsp.] leaves and evaluation of antioxidant activity. Food Chem. 128, 596–605.

Kurata, A., Kitamura, Y., Irie, S., Takemoto, S., Akai, Y., Hirota, Y., Fujita, T., Iwai, K., Furusawa, M., Kishimoto, N., 2010. Enzymatic synthesis of caffeic acid phenethyl ester analogues in ionic liquid. J. Biotechnol. 148, 133–138.

Laguerre, M., Bayrasy, C., Lecomte, J., Chabi, B., Decker, E.A., Wrutniak-Cabello, C., Cabello, G., Villeneuve, P., 2013. How to boost antioxidants by lipophilization? Biochimie 95, 20–26.

Laguerre, M., Giraldo, L.J.L., Lecomte, J., Figueroa-Espinoza, M.C., Barea, B., Weiss, J., Decker, E.A., Villeneuve, P., 2009. Chain length affects antioxidant properties of chlorogenate esters in emulsion: the cutoff theory behind the polar paradox. J. Agric. Food Chem. 57, 11335–11342.

Laguerre, M., Giraldo, L.J.L., Lecomte, J., Figueroa-Espinoza, M.C., Barea, B., Weiss, J., Decker, E.A., Villeneuve, P., 2010. Relationship between hydrophobicity and antioxidant ability of "phenolipids" in emulsion: a parabolic effect of the chain length of rosmarinate esters. J. Agric. Food Chem. 58, 2869–2876.

Laguerre, M., Lecomte, J., Villeneuve, P., 2007. Evaluation of the ability of antioxidants to counteract lipid oxidation: existing methods, new trends and challenges. Prog. Lipid Res. 46, 244–282.

Lea, C.H., Ward, R.J., 1959. Relative antioxidant activities of the seven tocopherols. J. Sci. Food Agric. 10, 537–548.

Leal-Calderon, F., Schmitt, V., 2008. Solid-stabilized emulsions. Curr. Opin. Colloid Interface Sci. 13, 217–227.

Leal-Calderon, F., Schmitt, V., Bibette, J., 2007. Emulsion Science—Basic Principles. Springer Science + Business Media, New York, USA.

Lethuaut, L., Metro, F., Genot, C., 2002. Effect of droplet size on lipid oxidation rates of oil-in-water emulsions stabilized by protein. J. Am. Oil. Chem. Soc. 79, 425–430.

Ling, Z.Q., Xie, B.J., Yang, E.L., 2005. Isolation, characterization, and determination of antioxidative activity of oligomeric procyanidins from the seedpod of *Nelumbo nucifera* Gaertn. J. Agric. Food Chem. 53, 2441–2445.

Liu, L.Y., Jin, C., Zhang, Y., 2014. Lipophilic phenolic compounds (Lipo-PCS): emerging antioxidants applied in lipid systems. RSC Adv. 4, 2879–2891.

Lu, J.J., Wei, Y., Yuan, Q.P., 2007. Preparative separation of gallic acid from chinese traditional medicine by high-speed counter-current chromatography and followed by preparative liquid chromatography. Sep. Purif. Technol. 55, 40–43.

Lu, Y.B., Ma, W.Y., Hu, R.L., Dai, X.J., Pan, Y.J., 2008. Ionic liquid-based microwave-assisted extraction of phenolic alkaloids from the medicinal plant *Nelumbo nucifera* Gaertn. J. Chrom. A 1208, 42–46.

Luo, Y.Q., Ling, Y.Z., Wang, X.Y., Han, Y., Zeng, X.J., Sun, R.C., 2013. Maillard reaction products from chitosan-xylan ionic liquid solution. Carbohydr. Polym. 98, 835–841.

McClements, D.J., 2005. Food Emulsions: Principles, Practices and Techniques. CRC Press, Boca Raton.

McClements, D.J., Decker, E.A., 2000. Lipid oxidation in oil-in-water emulsions: impact of molecular environment on chemical reactions in heterogeneous food systems. J. Food Sci. 65, 1270–1282.

Navickiene, H.M.D., Alecio, A.C., Kato, M.J., Bolzani, V.D., Young, M.C.M., Cavalheiro, A.J., Furlan, M., 2000. Antifungal amides from *Piper hispidum* and *Piper tuberculatum*. Phytochemistry 55, 621–626.

Ngo, H.L., LeCompte, K., Hargens, L., McEwen, A.B., 2000. Thermal properties of imidazolium ionic liquids. Thermochim. Acta 357, 97–102.

Pang, N., Gu, S.S., Jun, W.A., Hong-Sheng, C.A., Fang-Qin, W.A., Xi, L.A., Xing-Yu, Z.A., Wu, F.A., 2013. A novel chemoenzymatic synthesis of propyl caffeate using lipase-catalyzed transesterification in ionic liquid. Bioresour. Technol. 139, 337–342.

Panya, A., Laguerre, M., Bayrasy, C., Lecomte, J., Villeneuve, P., McClements, D.J., Decker, E.A., 2012. An investigation of the versatile antioxidant mechanisms of action of rosmarinate alkyl esters in oil-in-water emulsions. J. Agric. Food Chem. 60, 2692–2700.

Pekkarinen, S.S., Stockmann, H., Schwarz, K., Heinonen, I.M., Hopia, A.I., 1999. Antioxidant activity and partitioning of phenolic acids in bulk and emulsified methyl linoleate. J. Agric. Food Chem. 47, 3036–3043.

Pickering, S.U., 1907. Emulsions. J. Chem. Soc. 91, 2001–2021.

Porter, W.L., 1980. Recent trends in food applications of antioxidants. In: Simic, M.G., Karel, M. (Eds.), Autoxidation in Food and Biological Systems. Plenum Press, New York.

Porter, W.L., Black, E.D., Drolet, A.M., 1989. Use of polyamide oxidative fluorescent test on lipid emulsions: contrast in relative effectiveness of antioxidants in bulk versus dispersed systems. J. Agric. Food Chem. 37, 615–624.

Pyka, A., Sliwoik, J., 2001. Chromatographic separation of tocopherols. J. Chromatogr. A 935, 71–76.

Ranke, J., Molter, K., Stock, F., Bottin-Weber, U., Poczobutt, J., Hoffmann, J., Ondruschka, B., Filser, J., Jastorff, B., 2004. Biological effects of imidazolium ionic liquids with varying chain lengths in acute *Vibrio fischeri* and wst-1 cell viability assays. Ecotoxicol. Environ. Saf. 58, 396–404.

Richards, M.P., Chaiyasit, W., McClements, D.J., Decker, E.A., 2002. Ability of surfactant micelles to alter the partitioning of phenolic antioxidants in oil-in-water emulsions. J. Agric. Food Chem. 50, 1254–1259.

Riemersma, R.A., 2002. Analysis and possible significance of oxidised lipids in food. Eur. J. Lipid Sci. Technol. 104, 419–420.

Samokyszyn, V.M., Miller, D.M., Reif, D.W., Aust, S.D., 1990. Iron-catalyzed lipid peroxidation. In: Vigo-Pelfrey, C. (Ed.), Membrane Lipid Oxidation. CRC Press, Boca Raton.

Schaich, K.M., 2005. Lipid oxidation: theoretical aspects. In: Shahidi, F. (Ed.), Bailey's Industrial Oil and Fat Products. John Wiley & Sons, Inc.

Serini, S., Fasano, E., Piccioni, E., Cittadini, A.R.M., Calviello, G., 2011. Dietary n-3 polyunsaturated fatty acids and the paradox of their health benefits and potential harmful effects. Chem. Res. Toxicol. 24, 2093–2105.

Shahidi, F., Zhong, Y., 2010. Lipid oxidation and improving the oxidative stability. Chem. Soc. Rev. 39, 4067–4079.

Shahidi, F., Zhong, Y., 2011. Revisiting the polar paradox theory: a critical overview. J. Agric. Food Chem. 59, 3499–3504.

Sherwin, E.R., 1976. Antioxidants for vegetable oils. J. Am. Oil Chem. Soc. 53, 430–436.

von Staszewski, M., Ruiz-Henestrosa, V.M.P., Pilosof, A.M.R., 2014. Green tea polyphenols-beta-lactoglobulin nanocomplexes: interfacial behavior, emulsification and oxidation stability of fish oil. Food Hydrocolloids 35, 505–511.

Sun, P., Armstrong, D.W., 2010. Ionic liquids in analytical chemistry. Anal. Chim. Acta 661, 1–16.

Sureshkumar, M., Lee, C.K., 2009. Biocatalytic reactions in hydrophobic ionic liquids. J. Mol. Catal. B-Enzym. 60, 1–12.

Tang, B., Bi, W., Tian, M., Row, K.H., 2012. Application of ionic liquid for extraction and separation of bioactive compounds from plants. J. Chromatogr. B., Analyt. Technol. Biomed. Life Sci. 904, 1–21.

Tappel, A.L., 1968. Will antioxidant nutrients slow aging processes? Geriatrics 23, 97–105.

Uri, N., 1961. Mechanism of antioxidation. In: Lundberg, W. (Ed.), Autoxidation and Antioxidants. Wiley, New York, USA, pp. 133–169.

Ursini, F., Zamburlini, A., Cazzolato, G., Maiorino, M., Bon, G.B., Sevanian, A., 1998. Postprandial plasma lipid hydroperoxides: a possible link between diet and atherosclerosis. Free Radic. Biol. Med. 25, 250–252.

Yan-Ying, Y., Wei, Z., Shu-Wen, C., 2007. Extraction of ferulic acid and caffeic acid with ionic liquids. Chin. J. Anal. Chem. 35, 1726–1730.

Yang, Q.L., Dionysiou, D.D., 2004. Photolytic degradation of chlorinated phenols in room temperature ionic liquids. J. Photochem. Photobiol. A: Chem. 165, 229–240.

Yang, Q.W., Xing, H.B., Cao, Y.F., Su, B.G., Yang, Y.W., Ren, Q.L., 2009. Selective separation of tocopherol homologues by liquid–liquid extraction using ionic liquids. Ind. Eng. Chem. Res. 48, 6417–6422.

Zhao, H., 2005. Effect of ions and other compatible solutes on enzyme activity, and its implication for biocatalysis using ionic liquids. J. Mol. Catal. B-Enzym. 37, 16–25.

Chapter 11

Ionic Liquids in the Synthesis of Sugar/Carbohydrate and Lipid Conjugates

Worawan Panpipat, Manat Chaijan
Walailak University, School of Agricultural Technology, Department of Agro-Industry, Nakhon Si Thammarat, Thailand

11.1 INTRODUCTION

Nowadays, the interest for synthesis of sugar/carbohydrate and lipid conjugates is increasing. These compounds are widely used in food, cosmetic, detergent, and pharmaceutical industries (Yan et al., 1999). Amphiphilic polysaccharides can function as polymeric nonionic surfactants showing emulsifying, gelling, and film-forming properties combined with partial water solubility or permeability. Their functional properties can be fine-tuned by adjusting the ratio between the hydrophilic (oligo- or polysaccharide), and hydrophobic (lipophilic) parts (amount and chain length of the alkyl (fatty acid) residue). Derivatives with low degree of substitution (DS) exhibit insufficient amphiphilic character, whereas esters with too high DS are water insoluble. Due to their unique properties, amphiphilic sugar/carbohydrate lipid conjugates find multiple applications including as structural components in food products due to their impact on texture of food (Chang and Shaw, 2009). They can also be used in topical formulations to increase solubility of insoluble or poorly soluble drugs, where they replace synthetic polymers such as poloxamer (a copolymer of polyoxoethylene—polyoxopropylene), poly(vinyl ethers), or poly(metacrylic acids). Moreover, fatty acid esters of polysaccharides can be applied for delivery and controlled release of agrochemicals (fungicides and insecticides). Due to their lipophilic character, they enhance adhesion of the formulation to the hydrophobic plant leaves.

Esters of carbohydrates are commercially manufactured by chemical esterification with acid chlorides, in organic solvents (eg, dimethyl sulfoxide (DMSO)) using organic and inorganic catalysts (pyridine, tionyl chloride, K_2CO_3, celite) that leave residual traces in the final products. Generally,

Ionic Liquids in Lipid Processing and Analysis. http://dx.doi.org/10.1016/B978-1-63067-047-4.00011-8

esterification reactions require high temperatures which, in combination with the catalysts used, might induce partial degradation of the polysaccharide chains and discoloration. Currently, solvent-free chemical processes for esterification of starch are under development. Nevertheless, this approach does not eliminate the problem of chain degradation and high volume of wastes produced in the downstream processing step.

Chemical esterification is nonselective; both primary and secondary hydroxyl groups are substituted, making this process suitable only for production of carbohydrate esters with high DS. This is a severe limitation when products with low DS are targeted, as in the case of amphiphilic polysaccharides, since chemical esterification results in complex mixtures of mono-, di-, and triesters with random distribution of the ester groups both on the carbohydrate monomer and the polymer backbone. Enzymatic processes offer an attractive alternative route for synthesis of oligo- and polysaccharide esters. Selective processes catalyzed by enzymes may be performed under mild conditions of temperature and pressure, thereby avoiding polymer degradation. Application of enzymes for modification of polysaccharides will bring the advantage of high specificity and regioselectivity of the reaction, which produces products with controlled structure and functionality. Formation of ester linkages is thermodynamically favored in low water content media; hence, alternative enzymatic procedures should employ solvent-free organic solvents or organic solvent mixtures as reaction medium. In addition, reaction media like ionic liquids (ILs) and supercritical fluids, in particular ILs, can be used for enzyme catalyzed reactions.

A growing interest in applying ILs as polysaccharide solvent has been observed. ILs are organic salts, which are liquid at ambient temperatures. ILs can dissolve various substances including biomass (eg, cellulose, lignin, and directly wood as well as other renewable polymers such as starch and chitin) (Wilpiszewska and Spychaj, 2011). ILs are composed of ions and they differ considerably from traditional solvents. The unique properties of ILs are high polarity, nonvolatility, nonflammability, solvent miscibility, and excellent chemical and thermal stability (Yang and Pan, 2005). ILs can be designed for specific reaction such as to increase the solubility of substrate, and modify the selectivity of enzyme and rate of reaction. Their special characteristics made them superior to conventional organic solvents. ILs are an ideal solvent due to their ability to dissolve a wide range of different substances including polar and nonpolar organic, inorganic, and polymeric compounds. Despite their high polarity, most ILs are hydrophobic and can dissolve up to 1% of water, and the presence of water may affect the physical properties of the ILs (Seddon et al., 2000). As a consequence, ILs hold potential as green solvents for chemical and/or enzymatic synthesis of novel substances including sugar/carbohydrate and lipid conjugates.

ILs have been found capable of dissolving cellulose and carbohydrates to a great extent and research is going on to develop ILs that are less hazardous for enzymes (Zhao et al., 2008). When compared to conventional organic solvents, enzymes in ILs presented enhanced activity, stability, and selectivity. In comparison with normal organic solvents, ILs are able to dissolve a wide range of substrates; especially those highly polar ones (Yang and Pan, 2005).

This chapter gives a survey on the latest most representative developments and progress concerning ILs as a media for synthesis of sugar/carbohydrate and lipid conjugates. It also highlights their emerging production strategies and also their applications.

11.2 POTENTIAL CARBOHYDRATE FATTY ACID ESTERS: AVAILABILITY AND PROPERTIES

Carbohydrate fatty acid esters are nonionic biosurfactants, which can be synthesized from esterification of mono-, di-, oligo-, and polysaccharides by chemical/enzymatic catalysis. These compounds are increasingly used as important commodity chemicals, such as low caloric sweeteners, flavorings, fat replacer and biosurfactants/emulsifiers in foods; pharmaceutical, biomedical, detergent, and cosmetic industries; and oral-care products because of their amphiphilic nature, nontoxicity, and biodegradability (Chang and Shaw, 2009; Ferrer et al., 2005; Marshall and Bullermann, 1994; Nakamura, 1997; Okabe et al., 1999). In general, sugar fatty acid esters are used as emulsifiers with tasteless and odorless characteristics. Since the types that constitute fatty acid and sugar moieties can be easily changed, their properties widely vary. For example, each sugar ester has a different critical micelle concentration and hydrophilic–lipophilic balance (Kobayashi, 2011). Certain carbohydrate esters have shown their therapeutic potential with antitumor activity, plant growth inhibition, insecticidal and miticidal properties, antibiotic activity, and antifriction property (Chang and Shaw, 2009; Ferrer et al., 2005; Puterka et al., 2003; Peterson et al., 1997; Singh et al., 2014).

In food applications, sucrose esters of fatty acids are mostly used as food additives. The safety of sucrose esters of fatty acids has been recognized by the European Commission Scientific Committee for Food and the Joint FAO/WHO Expert Committee on Food Analysis (JECFA). Sucrose esters of fatty acids consist of a mixture of essentially mono-, di-, and triesters of sucrose with food fatty acids. The fatty acids form esters with the hydroxyl groups in the sucrose molecule and in principle one sucrose molecule could accommodate a total of eight fatty acid molecules. Depending on the number of esterified fatty acids on a sucrose molecule, commercial sucrose ester products are classified in five groups including sucroglycerides (mono- to triesters 40–60%; contain residual mono-, di-, and triglycerides), sucrose esters of fatty acids (mono- to triesters $\geq$ 80%), sucrose oligoesters type I (tetra- to octaesters 20–80%, hepta- and octaesters $\leq$ 50%, octaesters $\leq$ 20%), sucrose oligoesters type II (mono- to octaesters 20–80%, hepta- and octaesters $\leq$ 20%, octaesters $\leq$ 10%), and olestra (hexa-, hepta-, and octaesters $\geq$ 97%) (EFSA, 2010). Sucrose esters of fatty acids are nonionic surfactants consisting of sucrose as a hydrophilic group and fatty acids as lipophilic groups. The lipophilic character increases with an increasing degree of esterification and the nature of the fatty acid in the ester group. Beside their surface activities, sucrose esters of fatty acids can be used to improve the water vapor barrier property of fish skin gelatin-based film (Jongjaroenrak et al., 2006).

11.3 PRODUCTION STRATEGY OF CARBOHYDRATE FATTY ACID ESTERS

Synthesis of carbohydrate esters can be achieved by both enzymatic and chemical routes. The synthesis using biocatalyst might obtain high yields of specific isomers from cheap renewable material (Somashekar and Divakar, 2007; Plou et al., 2002; Arcos et al., 1998; Degn et al., 1999) in comparison to the poor selectivity of chemical synthesis method (Kennedy et al., 2006). Their current chemical synthesis is usually base-catalyzed at high temperatures, has a poor selectivity, and gives rise to colored side-products (Polat and Linhardt, 2001). Chemical routes require more energy than enzymatic routes (Gumel et al., 2011). Some of the by-products from chemical pathways are allergenic and possibly carcinogenic (Cauglia and Canepa, 2008). In addition, Chang and Shaw (2009) pointed out that the particularly challenging target for regioselective modifications of carbohydrate molecules was because their multiple hydroxyl groups resulted in lower conversions. Thus, application of lipases/peptidases as biocatalyst for carbohydrate fatty acid esters production has gained more attention as the preferred option. The yield of enzyme-catalyzed synthesis of particular carbohydrate ester is greatly dependent on the different enzymes and reaction conditions used. Solvents and solubilizing agents such as ILs can improve the yields of sugar esters. However, the cost of the final enzymatic product remains a hurdle compared to the cheaper alternative of using chemical catalysis (Chang and Shaw, 2009).

Enzymes including lipases, esterases, and proteases from different sources can be used as biocatalysts for regioselective esterification/transesterification of sugars and peracylated sugars in different solvent systems (Shaw and Klibanov, 1987). Several sugars have been acylated at a high conversion rate with high regioselectivity catalyzed by lipases from microbial origin such as *Candida antarctica*, *Candida rugosa*, *Candida cylindracea*, *Rhizomucor miehei*, *Mucor miehei*, *Bacillus substilis*, and *Bacillus licheniformis*, and from porcine pancrease (Sharma and Chattopadhyay, 1993; Sharma et al., 2001; Chang and Shaw, 2009; Shu et al., 2010; Treichel et al., 2010). Catalytic activity and specificity of hydrolases in organic solvents correlate well with solvent hydrophobicity (log P), which resulted in highest activities and stability at high log P value (Carrea et al., 1995). However, in case of carbohydrates, the hydrophobic solvent is not always a good choice because hydrophilic carbohydrates are used as acyl acceptors in hydrolase-catalyzed direct- or transesterification (Degn and Zimmermann, 2001). A strongly hydrophobic solvent is not suitable for lipase mediated syntheses because hydrophilic sugars such as glucose tend to be poorly soluble in such solvents and this can adversely affect the product yield of the reaction (Gumel et al., 2011). The polarity and ionization capacity of a nonaqueous reaction medium may be favorably modified by blending two or more solvents (Akoh and Mutua, 1994; Jia et al., 2010).

Lipase-catalyzed reaction is greatly useful to obtain sugar esters with chemically defined structures. The lipase-catalyzed method theoretically can introduce a fatty acid into a specific position of the sugar (Kobayashi, 2011; Gumel et al., 2011). Sugar fatty acid esters are produced by an esterification reaction between a sugar ($C_n(H_2O)_n$) and a fatty acid (RCO_2H) (Gumel et al., 2011) as shown:

$$C_n(H_2O)_n + RCO_2H \leftrightarrow C_n(H_2O)_{n-1}(OCOR) + H_2O$$

In synthesis of sugar esters, lipases are almost always used in an immobilized form, typically attached to porous beads of a plastic material (Gumel et al., 2011). Increasing concentration of lipase in the reaction mixture is known to enhance the rate of the initial reaction, but too high a concentration of the immobilized enzyme makes mixing of the reaction slurry difficult. This reduces the reaction rate by adversely affecting mass transfer (Sabeder et al., 2006).

The synthesis of sugar esters in nonaqueous media has been performed either in an organic solvent with or without a polar adjuvant or in an IL, or by using a hydrophobic sugar derivative (Kobayashi, 2011). The lipase-catalyzed synthesis of sugar esters by dehydrative condensation or transesterification can be relatively easily applied for food processing in comparison with the general organic ones from the viewpoint of safety because one of the organic ones is performed by adding an acyl chloride into the mixture of a sugar, *N*, *N*-dimethylformamide (DMF), and pyridine (Park et al., 1996; Sin et al., 1998). The consumers would accept the products of sugar esters produced by lipase-catalyzed reaction as a natural product (Kobayashi, 2011).

Industrial sugar esters are currently being manufactured by chemical syntheses. However, the current methods generally produce several isomers with different degrees of acylation and acylated positions. Many trials have been achieved to produce sugar esters with chemically defined structures (Kobayashi, 2011). In the case of sucrose esters of fatty acids, according to the Commission Directive 96/77/EC, sucrose esters of fatty acids are either prepared from sucrose and the methyl and ethyl esters of food fatty acids or by extraction from sucroglycerides. No organic solvent other than dimethylsulphoxide, DMF, ethyl acetate, propane-2-ol, 2-methyl-1-propanol, propylene glycol, and methyl ethyl ketone may be used for their preparation. The new method of manufacturing sucrose esters of fatty acids is different from existing production methods. The starting fatty acid esters used are vinyl esters rather than traditional methyl and ethyl esters of fatty acids. The initial starting material also contains a stabilizer, *p*-methoxyphenol. This change in manufacturing method results in trace levels of other residual chemicals; for example, vinyl fatty acid residues, acetaldehyde (formed from the vinyl portion of the vinyl fatty acid esters), and *p*-methoxyphenol, not found in sucrose esters of fatty acids manufactured by the traditional methods. The reaction sequence involves the initial reaction of sucrose with fatty acid esters followed by a series of purification steps to remove reaction solvents and fatty acid vinyl ester starting materials. The solvent is removed by distillation and the residue mixed with water. The mixture is sprayed into supercritical CO_2 to remove solvent residues (EFSA, 2010).

11.4 ROLE OF IONIC LIQUIDS IN ENZYMATIC SYNTHESIS OF CARBOHYDRATE FATTY ACID ESTERS

Enzymes generally do not dissolve in ILs but remain suspended as a powder or immobilized by a solid support, like when working in organic solvents (Yang and Pan, 2005). Previous studies showed that a variety of enzymes such as lipase, alcohol dehydrogenase, thermolysin, α-chymotrypsin, esterase, subtilisin, β-galactosidase, peroxidase, laccase, and formate dehydrogenase are able to exhibit catalytic activities in ILs, which are generally comparable with or higher than those found in conventional organic solvents (Yang and Pan, 2005).

When a sugar ester is synthesized by a lipase in a general organic solvent, it is necessary to use an adjuvant to dissolve the sugars in some cases. In addition, organic solvents usually have a high volatility and are also harmful. To tackle such problems, lipase-catalyzed reaction in an IL has attracted attention. IL is an asymmetric organic salt and is in liquid state below 100°C or close to room temperature. It is composed of ions instead of the neutral molecules of solvents such as acetone. It also has unique characteristics such as being suitable for green chemistry with almost zero vapor pressure, or nonvolatile, it is easy to change the combination of the cation and anion species of the IL, and is a good choice of an IL with high stability of enzyme and the high solubility of sugars (Diego et al., 2005; Kaar et al., 2003; Lozano et al., 2004; Zhao et al., 2008; Kobaysahi, 2011). ILs dissolve sugars and other complex carbohydrates well (Fort et al., 2006a,b; Zakrzewska et al., 2010). For example, the solubility of sugars such as glucose, sucrose, and lactose in an IL may exceed 100 g/L (Ganske and Bornscheuer, 2005; Lee et al., 2008b). The solubility of certain sugars and other carbohydrates in ILs is presented in Table 11.1. Park and Kazlauskas (2001) reported that additives such as aqueous sodium carbonate can be used to further modify the suitability of ILs for lipase catalyzed reaction. Washing with aqueous sodium carbonate yields ionic solvents that are suitable for reactions in ILs because it may add small amounts of a buffer or water to the IL.

TABLE 11.1 Solubility of Sugar/Carbohydrate in Ionic Liquids

Sugar/Carbohydrate	Ionic Liquid	Solubility	Temperature (°C)	References
Glucose	*tert*-Butyl alcohol	0.3 g/L	25	Liu et al. (2005)
	[BMIm][BF_4]	<0.5 g/L	25	Liu et al. (2005)
	[BMIm][PF_6]	<0.5 g/L	25	Liu et al. (2005)
	[M*O*MMIm][Tf_2N]	0.5 g/L	25	Liu et al. (2005)
	[M*O*MMIm][Bf_4]	4.4 g/L	25	Liu et al. (2005)
	[M*O*MMIm][TfO]	4.3 g/L	25	Liu et al. (2005)
	[M*O*EMIm][Tf_2N]	0.5 g/L	25	Liu et al. (2005)
	[M*O*EMIm][Pf_6]	2.5 g/L	25	Liu et al. (2005)
	[M*O*EMIm][BF_4]	2.8 g/L	25	Liu et al. (2005)
	[M*O*EMIm][TfO]	3.2 g/L	25	Liu et al. (2005)
	[E*O*EMIm][Tf_2N]	0.5 g/L	25	Liu et al. (2005)
	[E*O*EMIm][PF_6]	0.7 g/L	25	Liu et al. (2005)
	[E*O*EMIm][BF_4]	2.8 g/L	25	Liu et al. (2005)
	[M*O*MMIm][dca]	66 g/L	25	Liu et al. (2005)
	[M*O*EMIm][dca]	91 g/L	25	Liu et al. (2005)
	[E*O*EMIm][dca]	70 g/L	25	Liu et al. (2005)
	[BMIm][dca]	145 g/L	25	Liu et al. (2005)

Continued

TABLE 11.1 Solubility of Sugar/Carbohydrate in Ionic Liquids—cont'd

Sugar/Carbohydrate	Ionic Liquid	Solubility	Temperature (°C)	References
Fructose	[bmim][Cl]	56 wt%	110	Egorov et al. (2007)
	[bmim][BF_4]	15.9 g/L	60	Lee et al. (2008c)
	[bmim][CF_3SO_3]	87.5 g/L	60	Lee et al. (2008c)
	[cmim][BF_4]	25.7 g/L	60	Lee et al. (2008c)
	[cmim][CF_3SO_3]	123.9 g/L	60	Lee et al. (2008c)
Xylose	[bmim][PF_6]	0.25 M	95–97	Moens and Khan (2002)
	[hmim][Cl]	50 g/L	22	Spear et al. (2002)
	[omim][Cl]	0.25 M	25	Moens and Khan (2002)
Mannose	[bmim][PF_6]	0.25 M	103–105	Moens and Khan (2002)
	[omim][Cl]	0.25 M	25	Moens and Khan (2002)
Arabinose	[bmim][PF_6]	0.25 M	110–112	Moens and Khan (2002)
	[omim][Cl]	0.25 M	25	Moens and Khan (2002)
Sucrose	MOMMIm][dca]	249 g/L	25	Liu et al. (2005)
	MOMMIm][dca]	352 g/L	60	Liu et al. (2005)
	[MOEMIm][dca]	220 g/L	25	Liu et al. (2005)
	[EOEMIm][dca]	50 g/L	25	Liu et al. (2005)

	[EOEMIm][dca]	240 g/L	60	Liu et al. (2005)
	[BMIm][dca]	195 g/L	25	Liu et al. (2005)
	[BMIm][dca]	282 g/L	60	Liu et al. (2005)
Lactose	[BMIm][dca]	51 g/L	25	Liu et al. (2005)
	[BMIm][dca]	225 g/L	75	Liu et al. (2005)
Dextrin	[amim][HCOO]	3 wt%	30	Fukaya et al. (2006)
	[amim][HCOO]	25 wt%	45	Fukaya et al. (2006)
α-Cyclodextrin	[bmim][BF_4]	<1 wt%	–	Armstrong et al. (1999)
	[bmim][Cl]	30 wt%	–	Armstrong et al. (1999)
	[bmim][PF_6]	<1 wt%	–	Armstrong et al. (1999)
	[MeOEtmim][Br]	350 g/L	–	Kimizuka and Nakashima (2001)
	[MeOMemim][Br]	350 g/L	–	Kimizuka and Nakashima (2001)
β-Cyclodextrin	[bmim][BF_4]	<1 wt%	–	Armstrong et al. (1999)
	[bmim][Br]	25 wt%	25	Liu et al. (2009)
	[bmim][Cl]	21	–	Armstrong et al. (1999)
	[bmim][$N(CN)_2$]	750 g/L	75	Liu et al. (2005)
	[bmim][PF_6]	<1 wt%	–	Armstrong et al. (1999)
γ-Cyclodextrin	[bmim][BF_4]	<1 wt%	–	Armstrong et al. (1999)
	[bmim][Cl]	30 wt%	–	Armstrong et al. (1999)
	[bmim][PF_6]	<1 wt%	–	Armstrong et al. (1999)

Continued

TABLE 11.1 Solubility of Sugar/Carbohydrate in Ionic Liquids—cont'd

Sugar/Carbohydrate	Ionic Liquid	Solubility	Temperature (°C)	References
Cellulose	[amim][Cl]	8–14.5 wt%	80	Zhang et al. (2005)
	[ammim][Br]	4–12 wt%	80	Barthel and Heinze (2006)
	[amim][HCOO]	10–20 wt%	60–85	Fukaya et al. (2006)
	[bmim][Br]	<5 wt%	90	Zavrel et al. (2009)
	[bmim][Cl]	10 wt%	100	Swatloski et al. (2002)
	[bmim][I]	5 wt%	90	Zavrel et al. (2009)
	[bmim][HCOO]	8 wt%	110	Zhao et al. (2008)
	[bmim][$N(CN_2)$]	1 wt%	110	Zhao et al. (2008)
	[bmim][$N(CF_3SO_2)_2$]	<0.5 wt%	110	Zhao et al. (2008)
	[bmpy][Cl]	12–39 wt%	105	Heinze et al. (2005)
	[emim][Cl]	5 wt%	90	Zavrel et al. (2009)
	[emim][CH_3COO]	15 wt%	110	Zhao et al. (2008)
	[emim][$(CH_3O)_2PO_2$]	10 wt%	65	Fukaya et al. (2008)
	[hmim][Cl]	<5 wt%	90	Zavrel et al. (2009)
	[$H(OEt)_2$-mim][CH_3COO]	5 wt%	110	Zhao et al. (2008)
	[$H(OEt)_2$-mim][Cl]	1 wt%	110	Zhao et al. (2008)
	[mmim][$(CH_3O)_2PO_2$]	5 wt%	90	Zavrel et al. (2009)

	[MeOMemim[Br]	10 g/L	–	Kimizuka and Nakashima (2001)
	[Me(OEt)$_2$eim[Cl]	2 wt%	110	Zhao et al. (2008)
	[Me(OEt)$_2$eim[CH$_3$COO]	12 wt%	110	Zhao et al. (2008)
	[omim][CH$_3$COO]	<1 wt%	110	Zhao et al. (2008)
	[P66614][N(CN)$_2$]	<0.5 wt%	110	Zhao et al. (2008)
	[pmpy][Cl]	5 wt%	105	Zavrel et al. (2009)
	[tpbm][HCOO]	6 wt%	110	Zhao et al. (2008)
	[Bu$_4$N][HCOO]	1.5 wt%	110	Zhao et al. (2008)
	[Me(OEt)$_2$-Et$_3$N][CH$_3$COO]	10 wt%	110	Zhao et al. (2008)
	[MeMe(EtOH)NH][CH$_3$COO]	<0.5 wt%	110	Zhao et al. (2008)
	[Amm110][CH$_3$COO]	0.5 wt%	110	Zhao et al. (2008)
Agarose	[MeOEtmim][Br]	20 g/L	–	Kimizuka and Nakashima (2001)
	[MeOMemim][Br]	10 g/L	–	Kimizuka and Nakashima (2001)
Amylopectin	[bmim][Cl]	5 wt%	70	Fort et al. (2006a,b)
Amylose	[amim][HCOO]	4 wt%	30	Fukaya et al. (2006)
	[amim][HCOO]	19 wt%	60	Fukaya et al. (2006)
	[bmim][N(CN)$_2$]	4 g/L	25	Liu et al. (2005)
	[MeOEtmim][Br]	30 g/L	–	Kimizuka and Nakashima (2001)
	[MeOMemim][Br]	30 g/L	–	Kimizuka and Nakashima (2001)

Continued

TABLE 11.1 Solubility of Sugar/Carbohydrate in Ionic Liquids—cont'd

Sugar/Carbohydrate	Ionic Liquid	Solubility	Temperature (°C)	References
Chitin	[amim][Br]	10 wt%	100	Yamazaki et al. (2009)
	[bmim][CH_3COO]	3–7 wt%	110	Wu et al. (2008)
	[bmim][Cl]	10 wt%	110	Xie et al. (2006)
Chitosan	[amim][Cl]	8 wt%	110	Wu et al. (2008)
	[bmim][CH_3COO]	12 wt%	110	Wu et al. (2008)
	[bmim][Cl]	10 wt%	110	Xie et al. (2006)
Inulin	[amim][HCOO]	2 wt%	30	Fukaya et al. (2006)
	[amim][HCOO]	25%	55	Fukaya et al. (2006)
Pectin	[amim][HCOO]	1.5 wt%	65	Fukaya et al. (2006)
	[amim][HCOO]	2.5 wt%	80	Fukaya et al. (2006)
Starch	[amim][Cl]	15 wt%	80	Xu et al. (2008)
	[amim][Cl]	20 wt%	100	Xu et al. (2008)
	[bmim][Cl]	10 wt%	80	Biswas et al. (2006)
	[amim][$N(CN)_2$]	10 wt%	90	Biswas et al. (2006)
Xylan	[amim][HCOO]	1.5 wt%	45	Fukaya et al. (2006)
	[amim][HCOO]	21 wt%	95	Fukaya et al. (2006)

According to these characteristics, ILs are becoming increasingly used as reaction media in lipase mediated synthesis of fatty acid sugar esters (Chang and Shaw, 2009; Chen et al., 2007; Fan and Qian, 2010; Ganske and Bornscheuer, 2005; Park and Kazlauskas, 2001; Lee et al., 2008a,b; Moniruzzaman et al., 2010). For instance, Kim et al. (2003) studied the acetylation of the monoprotected glycosides. Biswas et al. (2006) and Zhao et al. (2008) completed the acylation of starch and glucose/cellulose, respectively. Ganske and Bornscheuer (2005) optimized the lipase-catalyzed glucose fatty acid ester synthesis. Lee et al. (2008a) succeeded in the synthesis of sugar ester with the aid of ultrasound. The enzymatic synthesis of some carbohydrate fatty acid esters in ILs is presented in Table 11.2.

Polar organic solvents such as methanol inactivate lipases, whereas ILs with similar polarities did not exhibit such characteristics (Park and Kazlauskas, 2001). However, in some cases, ILs affected the enzyme activity depending on the type of ILs. Lee et al. (2008b) investigated the stability of *C. antarctica* lipase B in different ILs and their mixtures. Data showed that the enzyme activity was stable in [Bmim][Tf2N] but the activity decreased rapidly in [Bmim][TfO]. However, the stability of lipase in a 1:1 (v/v) mixture of [Bmim][Tf2N] and [Bmim][TfO] was similar to its stability in pure [Bmim][Tf2N]. High conversion (86%) of fatty acid to ester was obtained when a supersaturated solution of glucose in [Bmim][TfO] was used. However, the conversion dropped to 61% when the enzyme was reused for a second time (Lee et al., 2008b).

Park and Kazlauskas (2001) were the first to introduce lipase catalyzed acylation reaction of unmodified sugar ILs. The regioselective acylation of both mono- and disaccharides (eg, glucose and maltose, respectively) with vinyl acetate, catalyzed by Novozym 435, can produce a high conversion and a high regioselectivity for monoacylation at the 6-position. Among the eight ILs tested, [MOEMIm][BF_4] was the best, yielding a maximum conversion of 99% and an excellent monoacylation rate of 93% within a reaction period of 36 h. This result was probably due to the ability of the solvent to dissolve glucose, which was responsible for high reaction rate and regioselectivity in the ILs. The reaction rate and regioselectivity obtained in ILs were significantly enhanced relative to those obtained in organic solvents (Kim et al., 2003).

Ganske and Bornscheuer (2005) found that glucose fatty acid esters can be enzymatically synthesized from glucose and fatty acid vinyl esters through transesterification in ILs. *Candida antarctica* lipase B (CAL B) modified with polyethylene glycol was able to catalyze the transesterification of glucose with lauric and myristic acid vinyl esters in pure [BMIm][BF_4] and [BMIm][PF_6], with a conversion of 30% and 35%, respectively. The conversion was markedly improved (up to 90%) when the reaction was performed with an immobilized CAL B in a bisolvent (biphasic) system composed of an IL [BMIm][BF_4] or [BMIm][PF_6] (60%) and *t*-BuOH (40%). Although direct esterification reaction using free lipase did not proceed in pure ILs, a high

TABLE 11.2 The Enzymatic Synthesis of Some Carbohydrate Fatty Acid Esters in Ionic Liquids

Ionic Liquid	Enzyme	Sugar(s)/ Carbohydrate(s)	Fatty Acid(s)	Product	Yield (%)	References
[Bmim][Cl] and [Bmim][BF_4]	*C. rugosa* lipase	Starch	Lauric acid	Starch laurate	–	Lu et al. (2013)
[Bmim]Ac and [Bmim][BF_4]	*C. rugosa* lipase	Starch	Palmitic acid	Starch palmitate	–	Lu et al. (2012)
[Bmim][TfO]	Novozym 435	Glucose	Vinyl laurate	6-*O*-glucose dodecanoate	86	Lee et al. (2008b)
[Bmim][PF_6]: *tert*-butanol (3:2, v/v)	CAL B	Glucose	Palmitic acid	6-*O*-glucose hexadecanoate	45	Ganske and Bornscheuer (2005)
$[Bmim]^+PF_6^-$, $[Moemim]^+PF_6^-$	*C. rugosa* lipase	Monoprotected glycosides (methyl-6-*O*-trityl-glucosides)	Vinyl acetate	2-*O*-acetyl-glycosides (major products) and 3-*O*-acetyl-glycosides (minor products)	20–95	Kim et al. (2003)
[Bmim][BF_4] and [Bmim][PF_6]	CAL B	Salicin, helicin, esculin, and naringin	Vinyl butyrate	Salicin-6″-*O*-butyrate, helicin-6″-*O*-butyrate, esculin-6″-*O*-butyrate, and naringin-6″-*O*-butyrate,	55–92	Katsoura et al. (2007)
[Bmpyrr][TfO]	Novozym 435	D-mannose	Vinyl myristate	Mannosyl myristate	72	Galonde et al. (2013)
[Bmim][BF_4]:DMSO (20:1, v/v)	Lipozyme RM IM (*R. miehei* lipase)	D-galactose	Oleic acid	Galactose oleate	87	Abdulmalek et al. (2012)
[emim][$MeSO_3$]	Novozym 435	D-maltose	Linoleic acid	6-and 6′-*O*-linoleyl-α-D-maltose	82	Fischer et al. (2013)

conversion of 64% was obtained by esterification between glucose and palmitic acid in the binary solvent system. Without doubt, addition of the cosolvent reduces the viscosity of the IL-containing medium and hence improves the mass transfer in the reaction system; this may be the major reason for the initial increase in the enzyme reactivity. However, too much *t*-BuOH may lead to a reduction in the glucose solubility, thus resulting in a lower conversion yield.

By employing the water-mediated supersaturation method, Lee et al. (2008c) have succeeded to produce a glucose solution with high concentration for synthesizing glucose fatty acid esters in ILs using transesterification and esterification reactions. [BMIm][TfO] was the best solvent with the highest conversion. The initial rates and conversions were notably increased when the glucose solution was prepared by using the water-mediated supersaturation method. Therefore, this study has clearly indicated that the concentration of the homogeneously dissolved glucose in the IL is critical to the synthesis of the sugar esters. Thereafter, to optimize the esterification process, the reaction was performed in an IL mixture of [BMIm][TfO]/[OMIm][Tf_2N] for the purpose of achieving maximal productivity. The selection of these two ILs was based on the fact that the lipase (Novozym 435) exhibited the highest activity but a rather low stability in the hydrophilic [BMIm][TfO], whereas the hydrophobic [OMIm][Tf2N] offered a higher enzyme stability, although the enzyme activity in this IL was much lower than in [BMIm][TfO]. As a result of compromise, the mixture of these two ILs with a 1:1 volume ratio was selected as the best reaction medium for the lipase-catalyzed synthesis of sugar esters. By using this new solvent system, a series of glucose fatty acid esters has been synthesized through direct esterification with different fatty acids. As expected, use of a supersaturated glucose solution offered a higher conversion relative to the reaction with a saturated glucose solution. Interestingly, the conversion was higher for the glucose ester with a shorter fatty acid chain, and this may be related to the difference in the solubility of the fatty acid in the ILs.

Lee et al. (2008a) have introduced ultrasound irradiation into the lipase-catalyzed sugar ester synthesis. Data showed that in both esterification and transesterification reactions, this new technique enabled the conversions to be greatly enhanced without significantly affecting enzyme stability. Abdulmalek et al. (2012) have developed another strategy to significantly improve sugar ester synthesis through direct esterification. In this case, DMSO is a good solvent for dissolving sugars and it was introduced into the IL system as a cosolvent as well. After examining different reaction parameters (type of solvent and enzyme, amount of enzyme, reaction time, temperature, stirring rate, and molar ratio of the two substrates), the optimal reaction conditions for the synthesis of galactose oleate was obtained. The esterification between galactose and oleic acid was performed at 60°C with a stirring rate of 300 rpm in a [BMIm][BF4]/DMSO (20:1 v/v) bisolvent system containing the two substrates in the molar ratio of 1:3, with addition of 2% (w/w) Lipozyme RM IM as the catalyst. Maximum conversion (over 90%) was achieved after reaction for only 4 h, a much faster process as compared to others that usually require a few days (Lee et al., 2008c; Ganske and Bornscheuer, 2005).

The feasibility of ILs as novel solvents for biosynthesis of sugar fatty acid esters has been reported by Mai et al. (2014), who successfully synthesized glucose laurate using lipase in ILs. Data showed that under optimal reaction conditions (66.86°C, vinyl laurate/glucose molar ratio of 7.63, enzyme load of 73.77 g/L), an experimental conversion yield of 96.4% was obtained. Interestingly, the enzyme and IL could be recycled and reused effectively for up to 10 cycles.

Besides sugars, other polysaccharides such as glucomannan, cellulose, and starch have been used to conjugate with fatty acids in ILs via enzymatic route (Chen et al., 2006; Zhao et al., 2008; Lu et al., 2012, 2013).

The lipase catalyzed acylation of konjac glucomannan, a water soluble polysaccharide, with vinyl acetate in pure ILs and IL/t-BuOH bisolvent systems has been carried out by Chen et al. (2006). Novozym 435 lipase-mediated acylation reactions proceeded in the ILs ([CnMIm][BF_4] and [C4MIm][PF_6]) with a higher substitution degree and a higher thermostability, as compared to the results obtained in *t*-BuOH. The degrees of substitution and the production yields were further increased by addition of *t*-BuOH as the cosolvent.

In the case of enzymatic synthesis of cellulose fatty acid esters, ILs can function as both reaction medium and solubilizing agent. For example, Avicel cellulose was dissolved in [BMIm][Cl] to make the structure more accessible. Thereafter, the IL was removed and the remaining cellulose was incubated with immobilized esterase from hog liver or with cutinase from *Fusarium solani*. FT-IR analysis revealed the formation of ester linkages using vinyl propionate, vinyl laurate, or vinyl stearate as acyl donor. Lipase from *C. antarctica*, *C. cylindracea*, and *Aspergillus niger* were not able to esterify the pretreated cellulose (Gremos et al., 2011). The esterification of cellulose with methyl methacrylate in [Me(Oet) 3-Et3N][OAc], [Me(OEt)2-Et-Im][OAc], and [Me(OEt)7-Et-Im][OAc] has been reported using lipase from *C. antarctica* (Zhao et al., 2008).

In addition, Lu et al. (2012) reported the synthesis of starch palmitate in mixed [bmim][Ac] and [bmim][BF_4] catalyzed by *C. rugosa* lipase with the DS value ranging from 0.053 to 0.144. This report was found in synthesis of starch laurate in [bmim][Cl] and [bmim][BF_4] using two-step enzymatic processes (*C. rugosa* lipase) with the maximum DS value of 0.171 (Lu et al., 2013). They also studied the recyclable and reusable possibilities of ILs and found that the ILs can be effectively recycled and reused after each modification cycle.

11.5 ROLE OF IONIC LIQUID IN CHEMICAL SYNTHESIS OF CARBOHYDRATE FATTY ACID ESTERS

Synthesis of carbohydrate fatty acid esters in ILs with chemical route has been reported for several compounds, particularly cellulose fatty acid esters (Singh et al., 2014; Huang et al., 2011a,b) and starch esters (Biswas et al., 2006; Wang and Xie, 2010; Xie et al., 2011; Lehmann and Volkert, 2011; Gao et al., 2012). Some ILs (containing Cl^- anions) are able to destroy the semicrystalline structure of native starch granules and disrupt hydrogen bonding between hydroxyl groups of polysaccharide (Wilpiszewska and Spychaj, 2011) leading to increase in their solubility, which is important for the synthesis of carbohydrate esters. Chemical modification of polysaccharides (eg, esterification reactions of mostly 1–8 wt% polysaccharide dissolved in ILs) are achieved using five or more molar equivalents of carboxylic acid anhydride or carboxylic acid chlorides per anhydroglucose unit resulting in highly substituted polysaccharide esters (Liu et al., 2007; Heinze et al., 2008).

Cellulose and starch are abundant natural polymers and interest in manufacturing value-added products from these polysaccharides is increasing. The use of ILs for dispersing cellulose and starch gained increasing attention in recent years. Cellulose fatty acid esters with different fatty acids such as lauric, myristic, palmitic, and oleic acids having the same degree of substitution (DS ~ 2) have been prepared by chemical methods in IL, $[C_4mim]^+Cl^-$, in order to study their antifriction properties (Singh et al., 2014). Results showed that the cellulose esters exhibited antifriction character in the order as myristic > palmitic > laurate > stearate > oleate. Homogeneous synthesis of cellulose stearates with different degrees of substitution in IL $[C_4mim]^+Cl^-$ was reported by Huang et al. (2011a). Data showed that cellulose stearates with DS values of 2.15, 2.19, and 2.57 exhibited good solubility in nonpolar organic solvents and their melting temperatures ranged from 38.66 to 56.63°C. Cellulose stearate with DS value of 2.57 has better thermal stability than cellulose. In addition, a new way for high-value-added utilization of sugarcane bagasse, an agro-industrial residue, was achieved by converting such waste to cellulose acetate butyrate and cellulose acetate propionate in an [AmimCl] IL system (Huang et al., 2011b). Those cellulose acetates are widely used in the paint industry for top grade cars and furniture, as well as printing ink (Huang et al., 2011b).

Additionally, biomass from poplar wood meal, sugarcane bagasse, switchgrass, and ball-milled bamboo can be also used as an alternative raw material for production of carbohydrate esters in ILs (Wang et al., 2012; Wen et al., 2011; Huang et al., 2011b; Yuan et al., 2010; Li et al., 2010).

Many papers regarding preparation of esterified starches using ILs have been reported periodically (Table 11.3). Using IL as reaction medium has advantages over other solvents, such as DMSO, which exhibit high toxicity and low stability during the modification process as well as difficulties with solvent recovery (Xie and Shao, 2009). Obtained values of DS of ester derivatives prepared from acidic anhydrides usually exceeded 2. It has been observed that regardless of starch origin, susceptibility of hydroxyl groups to esterification in IL decreased in order of carbon atoms of glycosidic unit: C6 > C2 > C3 (Lehmann and Volkert, 2009). However, Shogren and Biswas (2010) observed that in the case reaction between starch (maltodextrin) and

TABLE 11.3 Starch Ester Synthesis in Ionic Liquids

Compound	Ionic Liquid	Reagent	Degree of Substitution	References
Starch acetate	[bmim][Cl]	Acetic anhydride[a]	0.3–2.6	Biswas et al. (2006)
	[mim][Cl]	Acetic anhydride[a]	2.28	Lehmann et al. (2010)
	[bmim]dca	Vinyl acetate[a]	0.84–1.10	Shogren and Biswas (2010)
Starch propionate	[bmim][Cl]	Propionic anhydride[a]	2.1–3.0	Lehmann and Volkert (2009)
	[EtIM]Cl	Propionic anhydride[a]	0.63	Lehmann et al. (2010)
Starch hexanoate	[bmim][Cl]	Hexanoic anhydride[a]	1.92	Lehmann and Volkert (2009)
Starch laurate	[AcIM]Cl	Lauroyl chloride[a]	0.1	Lehmann et al. (2010)
	[bmim][Cl]	Methyl laurate[a]	0.06–0.38	Xie and Wang (2011)
	[bmim][Cl] and [bmim][BF_4]	Lauric acid[b]	0.171	Lu et al. (2013)
Starch palmitate	[bmim][Ac] and [bmim][BF_4] mixture	Methyl palmitate[b]	0.053–0.144	Lu et al. (2012)
Starch stearate	[bmim][Cl]	Methyl stearate[a]	0.03–0.29	Xie and Wang (2011)
	[bmim]dca	Stearic acid[a]	0.03–0.72	Biswas et al. (2009)
	[bmim]dca	Vinyl stearate[a]	0.01–0.64	Biswas et al. (2009)

[a]*Chemical synthesis.*
[b]*Enzymatic synthesis.*

vinyl acetate, the substitution position depends on the type of IL medium used. In [bmim] dicyanamide (dca) acetylation at C2 was preferred, while in [bmim] acetate was more uniform. Biswas et al. (2009) used IL [bmim][dca] as a solvent and catalyst for direct esterification of starch (maltodextrin) with stearic acid or vinyl stearate. Reaction occurred even at mild temperature (40°C). At higher temperature (75°C), starch stearates with high degree of substitution (DS = 2.4) might be obtained (Shogren et al., 2010). Xie and Wang (2011) have performed starch esterification with fatty acid methyl esters in [bmim][Cl] using base catalyst; the highest DS values were close to 0.4.

Biswas et al. (2006), Wang and Xie (2010), and Xie et al. (2011) investigated synthesis of starch ester in IL using pyridine as a catalyst. However, the reaction system reported caused environmental pollution because of the toxicity and volatility of pyridine. Biswas et al. (2006) found starch to be soluble at 80°C in ILs such as [BMIMCl] and [BMIMdca] at concentration up to 10% (w/w). Higher concentrations of starch in these ILs resulted in viscous solution. Solution of starch in [BMIMCl] can be acylated with anhydrides in the presence of pyridine to give acetyl starch with various degrees of substitution. Without pyridine, the acylation reaction did not proceed.

Recently, synthesis of starch ester in ILs without catalyst has been proposed by Gao et al. (2012). In this case, [BMIM]Cl and [EMIM]Ac were used to disperse corn starch for synthesis of fatty acid starch esters without catalysts (Gao et al., 2012). Results showed corn starch was completely dissolved in ILs at 100°C for 60 min. [EMIM]Ac was a better solvent for starch than [BMIM]Cl. The esterification of corn starch with lauric, palmitic, and stearic acids using [BMIM]Cl, [EMIM]Ac, and the mixture of [BMIM]Cl/DMSO, respectively, were successfully accomplished in a homogeneous system without catalysts. The DS value of starch laurate was the highest, up to 0.053, 0.068, and 0.10 when [BMIM]Cl, [EMIM]Ac, and the mixture of [BMIM]Cl/DMSO were the reaction media, respectively. The reaction condition is lauric acid/anhydroglucose unit molar ratio 1:5, temperature 120°C, at a reaction time of 3 h.

11.6 REMARKS AND FUTURE ASPECTS

Carbohydrate fatty acid esters can be synthesized from esterification of sugars or polysaccharides with fatty acids by chemical and enzymatic reactions. These compounds are increasingly used as active biosurfactants in several applications. While almost all the commercial carbohydrate esters available in the market are manufactured by chemical syntheses, there has been a growing interest in developing enzymatic approaches to synthesize them. Comparatively, the enzymatic method is more advantageous in terms of mild reaction conditions, simple operational procedures, high productivity, excellent regioselectivity, and easy product separation.

Recently, ILs are being increasingly used as reaction media as well as solubilizing agents in both chemical and enzymatic synthesis of carbohydrate fatty acid esters, particularly in lipase-mediated synthesis reaction. ILs are good solvents for dissolving carbohydrates leading to increase in their solubility, which is important for synthesis of carbohydrate esters using both chemical and enzymatic catalysts. The special characteristics made ILs superior to conventional organic solvents, which allow designing an IL for specific reaction conditions, such as to increase the solubility of substrate, to modify the selectivity of enzyme, and to modify the rate of reaction. In comparison with conventional nonaqueous solvents, ILs are substantially better for use in lipase-mediated syntheses of carbohydrate fatty acid esters. Both sugars and polysaccharides have been intensively used as substrates for conjugation with fatty acids. Sugar fatty acid esters synthesized in ILs have gained attention for having multipurposes such as being used as food additives, in cosmetics, for pharmaceutical and medical applications, and in other chemical industries. Polysaccharides, mainly starch and cellulose, from different sources, particularly agro-industrial by-products or renewable resources, have been chemically/enzymatically esterified with fatty acids in ILs to produce functional polysaccharide fatty acid esters. The synthesized polysaccharide fatty acid esters can be used in a whole range of products and applications such as novel biosurfactants, antifriction agents, and paint. However, to make the carbohydrate fatty acid ester synthesis in ILs possible on an industrial scale more research is needed in the field.

REFERENCES

Abdulmalek, E., Saupi, H.S.M., Tejo, B.A., Barsi, M., Salleh, A.B., Rahman, R.N.Z.R.A., Rahman, M.B.A., 2012. Improved enzymatic galactose oleate ester synthesis in ionic liquids. J. Mol. Catal. B. Enzym. 76, 37–43.

Akoh, C.C., Mutua, L.M., 1994. Synthesis of alkyl glucoside fatty acid esters: effect of reaction parameters and the incorporation of n-3 polyunsaturated fatty acids. Enzyme Microb. Technol. 16, 115–119.

Arcos, J.M., Bernabé, M., Otero, C., 1998. Quantitative enzymatic production of 6-O-acylglucose esters. Biotechnol. Bioeng. 57, 505–509.

Armstrong, D.W., He, L., Liu, Y.S., 1999. Examination of ionic liquids and their interaction with molecules, when used as stationary phases in gas chromatography. Anal. Chem. 71, 3873–3876.

Barthel, S., Heinze, T., 2006. Acylation and carbanilation of cellulose in ionic liquids. Green Chem. 8, 301–306.

Biswas, A., Shogren, R.L., Willett, J.L., 2009. Ionic liquid as a solvent and catalyst for acylation of maltodextrin. Ind. Crop Prod. 30, 172–175.

Biswas, A., Shogren, R.L., Stevenson, D.G., Willett, J.L., Bhowmik, P.K., 2006. Ionic liquids as solvents for biopolymers: acylation of starch and zein protein. Carbohydr. Polym. 66, 546–550.

Carrea, G., Ottolina, G., Riva, S., 1995. Role of solvent in the control of enzyme selectivity in organic media. Trends Biotechnol. 13, 63–70.

Cauglia, F., Canepa, P., 2008. The enzymic synthesis of glucosylmyristate as a reaction model for general considerations on 'sugar esters' production. Biores. Technol. 99, 4065–4072.

Chang, S.W., Shaw, J.F., 2009. Biocatalysis for the production of carbohydrate esters. New Biotechnol. 26, 109–116.

Chen, Z., Zong, M., Gu, Z., 2007. Enzymatic synthesis of sugar esters in ionic liquids. Chin. J. Org. Chem. 27, 1448–1452.

Chen, Z.G., Zong, M.H., Li, G.J., 2006. Lipase-catalyzed acylation of konjac glucomannan in ionic liquids. J. Chem. Technol. Biotechnol. 81, 1225–1231.

Degn, P., Zimmermann, W., 2001. Optimization of carbohydrate fatty acid ester synthesis in organic media by a lipase from *Candida antarctica*. Biotechnol. Bioeng. 74, 483–491.

Degn, P., Pedersen, L.H., Duus, J.O., Zimmermann, W., 1999. Lipase-catalyzed synthesis of glucose fatty acid esters in tert-butanol. Biotechnol. Lett. 21, 275–280.

Diego, T.D., Lozano, P., Gmouh, S., Vaultier, M., Iborra, J.L., 2005. Understanding structure-stability relationships of *Candida antarctica* lipase B in ionic liquids. Biomacromolecules 6, 1457–1464.

Egorov, V.M., Smirnova, S.V., Formanovsky, A.A., Pletnev, I.V., Zolotov, Y.A., 2007. Dissolution of cellulose in ionic liquids as a way to obtain test materials for metal-ion detection. Anal. Bioanal. Chem. 387, 2263–2269.

European Food Safety Authority (EFSA), 2010. Scientific opinion on the safety of sucrose esters of fatty acids prepared from vinyl esters of fatty acids and on the extension of use of sucrose esters of fatty acids in flavourings. Eur. Food Saf. Autority J. 8, 1–36.

Fan, Y., Qian, J., 2010. Lipase catalysis in ionic liquids/supercritical carbon dioxide and its applications. J. Mol. Catal. B. Enzym. 66, 1–7.

Ferrer, M., Soliveri, J., Plou, F., Lopez-Cortes, N., Reyes-Duarte, D., Christensen, M., Copa-Patino, J.L., Ballesteros, A., 2005. Synthesis of sugar esters in solvent mixtures by lipases from *Thermomyces lanuginosus* and *Candida antarctica* B, and their antimicrobial properties. Enzym. Microb. Tech. 36, 391–398.

Fischer, F., Happe, M., Emery, J., Fornage, A., Schutz, R., 2013. Enzymatic synthesis of 6- and 6′-O-Linoleyl-α-D-Maltose: from solvent-free to binary ionic liquid reaction media. J. Mol. Catal. B. Enzym. 90, 98–106.

Fort, D.A., Remsing, R.C., Swatloski, R.P., Moyna, P., Monya, G., Rogers, R.D., 2006a. Can ionic liquids dissolve wood? Processing and analysis of lignocellulosic materials with 1-n-butyl-3-methylmidazolium chloride. Green Chem. 9, 63–69.

Fort, D.A., Swatloski, R.P., Moyna, P., Rogers, R.D., Moyna, G., 2006b. Use of ionic liquids in the study of fruit ripening by high-resolution ^{13}C NMR spectroscopy: "Green" solvents meet green bananas. Chem. Commun. 714–716.

Fukaya, Y., Hayashi, K., Wada, M., Ohno, H., 2008. Cellulose dissolution with polar ionic liquids under mild conditions: required factors for anions. Green Chem. 10, 44–46.

Fukaya, Y., Sugimoto, A., Ohno, H., 2006. Superior solubility of polysaccharides in low viscosity, polar, and halogen-free 1,3-dialkylimidazolium formates. Biomacromolecules 7, 3295–3297.

Galonde, N., Brostaux, Y., Richard, G., Nott, K., Jerome, C., Fauconnier, M.L., 2013. Use of response surface methodology for the optimization of the lipase-catalyzed synthesis of mannosyl myristate in pure ionic liquid. Process. Biochem. 48, 1914–1920.

Ganske, F., Bornscheuer, U.T., 2005. Optimization of lipase-catalyzed glucose fatty acid ester synthesis in a two-phase system containing ionic liquids and t-BuOH. J. Mol. Catal. B. Enzym. 36, 40–42.

Gao, J., Luo, Z.G., Luo, F.X., 2012. Ionic liquids as solvents for dissolution of corn starch and homogeneous synthesis of fatty-acid starch esters without catalysts. Carbohydr. Polym. 89, 1215–1221.

Gremos, S., Zarafeta, D., Kekos, D., Kolisis, F., 2011. Direct enzymatic acylation of cellulose pretreated in BMIMCl ionic liquid. Biores. Technol. 102, 1378–1382.

Gumel, A.M., Annuar, M.S.M., Heidelberg, T., Chisti, Y., 2011. Lipase mediated synthesis of sugar fatty acid esters. Process. Biochem. 46, 2079–2090.

Heinze, T., Dorn, S., Schobitz, M., Liebert, T., Kohler, S., Meister, F., 2008. Interactions of ionic liquids with polysaccharides—2: cellulose macromolecular. Macromol. Symp. 262, 8–22.

Heinze, T., Schwikal, K., Barthe, S., 2005. Ionic liquids as reaction medium in cellulose functionalization. Macromol. Biosci. 5, 520–525.

Huang, K., Xia, J., Li, M., Lian, J., Yang, X., Lin, G., 2011a. Homogeneous synthesis of cellulose sterates with different degrees of substitution in ionic liquid 1-butyl-3-methylimidazolium chloride. Carbohydr. Polym. 83, 1631–1635.

Huang, K., Wang, B., Cao, Y., Li, H., Wang, J., Lin, W., Mu, C., Liao, D., 2011b. Homogeneous preparation of cellulose acetate propionate (CAP) and cellulose acetate butyrate (CAB) from sugarcane bagasse cellulose in ionic liquid. J. Agric. Food Chem. 59, 5376–5381.

Jia, C., Zhao, J., Feng, B., Zhang, X., Xia, W.A., 2010. A simple approach for the selective enzymatic synthesis of dilauroyl maltose in organic media. J. Mol. Catal. B. Enzym. 62, 265–269.

Jongjaroenrak, A., Benjakul, S., Visessanguan, W., Tanaka, M., 2006. Fatty acids and their sucrose esters affect the properties of fish skin gelatin-based film. Eur. Food Res. Technol. 222, 650–657.

Kaar, J.L., Jesionowski, A.M., Berberich, J.A., Moulton, R., Russell, A.J., 2003. Impact of ionic liquid physical properties on lipase activity and stability. J. Am. Chem. Soc. 125, 4125–4131.

Katsoura, M.H., Polydera, A.C., Katapodis, P., Kolisis, F.N., Stamatis, H., 2007. Effect of different reaction parameters on the lipase-catalyzed selective acylation of polyhydroxylated natural compounds in ionic liquids. Process Biochem. 42, 1326–1334.

Kennedy, J.F., Harish Kumar, H., Panesar, P.S., Marwaha, S.S., Goyal, R., Parmar, A., Kaur, S., 2006. Enzyme-catalyzed regioselective synthesis of sugar esters and related compounds. J. Chem. Technol. Biotechnol. 81, 866–876.

Kim, M.J., Choi, M.Y., Lee, J.K., Ahn, Y., 2003. Enzymatic selective acylation of glycosides in ionic liquids: significantly enhanced reactivity and regioselectivity. J. Mol. Catal. B. Enzym. 26, 115–118.

Kimizuka, N., Nakashima, T., 2001. Spontaneous self-assembly of glycolipid bilayer membranes in sugar-philic ionic liquids and formation of ionogels. Langmuir 17, 6759–6761.

Kobayashi, T., 2011. Lipase-catalyzed synthesis of sugar esters in non-aqueous Media. Biotechnol. Lett. 33, 1911–1919.

Lee, S.H., Nguyen, H.M., Koo, Y.M., Ha, S.H., 2008a. Ultrasound-enhanced lipase activity in the synthesis of sugar ester using ionic liquids. Process Biochem. 43, 1009–1012.

Lee, S.H., Ha, S.H., Hiep, N.M., Chang, W.J., Koo, Y.M., 2008b. Lipase-catalyzed synthesis of glucose fatty acid ester using ionic liquids mixtures. J. Biotechnol. 133, 486–489.

Lee, S.H., Dang, D.T., Ha, S.H., Chang, W.J., Koo, Y.M., 2008c. Lipase-catalyzed synthesis of fatty acid sugar ester using extremely supersaturated sugar solution in ionic liquids. Biotechnol. Bioeng. 99, 1–8.

Lehmann, A., Volkert, B., 2009. Investigations on esterification reactions of starches in 1-n-Buthyl-3-methylimidazolium chloride and resulting substituent distribution. J. App. Polym. Sci. 114, 369–376.

Lehmann, A., Volkert, B., 2011. Preparing esters from high-amylose starch using ionic liquids as catalysts. Carbohydr. Polym. 83, 1529–1533.

Lehmann, A., Volkert, B., Hassan-Nejad, M., Greco, T., Fink, H.P., 2010. Synthesis of thermoplastic starch mixed esters catalyzed by the in situ generation of imidazolium salts. Green Chem. 12, 2164–2171.

Li, C., Knierim, B., Manisseri, C., Arora, R., Scheller, H.V., Auer, M., Vogel, K.P., Simmons, B.A., Singh, S., 2010. Comparison of dilute acid and ionic liquid pretreatment of switchgrass: biomass recalcitrance, delignification and enzymatic saccharification. Biores. Technol. 101, 4900–4906.

Liu, C.F., Sun, R.C., Zhang, R.P., Ren, J.L., 2007. Preparation of sugarcane bagasse cellulosic phthalate using an ionic liquid as reaction medium. Carbohydr. Polym. 68, 17–25.

Liu, Q., Janssen, M.H.A., van Rantwijk, F., Sheldon, R.A., 2005. Room-temperature ionic liquids that dissolve carbohydrates in high concentrations. Green Chem. 7, 39–42.

Liu, Z.T., Shen, L.H., Liu, Z.W., Lu, J., 2009. Acetylation of β-cyclodextrin in ionic liquid green solvent. J. Mater. Sci. 44, 1813–1820.

Lozano, P., Diego, T., Gmouh, S., Vaultier, M., Iborra, J.L., 2004. Criteria to design green enzymatic processes in ionic liquid/supercritical carbon dioxide systems. Biotechnol. Progr. 20, 661–669.

Lu, X., Lou, Z., Yu, S., Fu, X., 2012. Lipase-catalyzed synthesis of starch palmitate in mixed ionic liquids. J. Agric. Food Chem. 60, 9273–9279.

Lu, X., Luo, Z., Fu, X., Xiao, Z., 2013. Two-step method of enzymatic synthesis of starch laurate in ionic liquids. J. Agric. Food Chem. 61, 9882–9891.

Mai, N.L., Ahn, K., Bae, S.W., Shin, D.W., Morya, V.K., Koo, Y.M., 2014. Ionic liquids as novel solvents for the synthesis of sugar fatty acid ester. Biotechnol. J. [Online early access]. http://dx.doi.org/10.1002/biot.201400099 (accessed 01/11/14).

Mashall, D.L., Bullermann, L.B., 1994. Antimicrobial properties of sucrose fatty acid esters. In: Akoh, C.C., Swanson, B.G. (Eds.), Carbohydrate Polymers as Fat Substitutes. Marcel Dekker, New York, pp. 9–35.

Moens, L., Khan, N., 2002. Room-temperature ionic liquids as new solvents for carbohydrate chemistry: a new tool for the processing of biomass feedstocks? In: Rogers, R.D., Seddon, K.R. (Eds.), Ionic Liquids: Industrial Applications to Green Chemistry. American Chemical Society, Washington, DC, pp. 360–372.

Moniruzzaman, M., Nakashima, K., Kamiya, N., Goto, M., 2010. Recent advances of enzymatic reactions in ionic liquids. Biochem. Eng. J. 48, 295–314.

Nakamura, S., 1997. Using sucrose esters as food emulsifiers. Oleochem 8, 866–874.

Okabe, S., Suganuma, M., Tada, Y., Ochiai, Y., Sueoka, E., Kohya, H., Shibata, A., Takahashi, M., Mizutani, M., Matsuzaki, T., Fujiki, H., 1999. Disaccharide esters screened for inhibition of tumor necrosis factor-alpha release are new anticancer agents. Cancer Res. 90, 669–676.

Park, O.J., Park, H.G., Yang, J.W., 1996. Enzymatic trans-esterification of monosaccharides and amino acid esters in organic solvents. Biotechnol. Lett. 18, 473–478.

Park, S., Kazlauskas, R.J., 2001. Improved preparation and use of room-temperature ionic liquids in lipase-catalyzed enantio-and regioselective acylations. J. Org. Chem. 66, 8395–8401.

Peterson, J.K., Harrison, H.F., Chortyk, O.T., 1997. Effects of various sucrose esters on weed seed germination and crop growth: structure-activity and dose-response relationship. J. Agric. Food Chem. 45, 4833–4837.

Plou, F.J., Cruces, M.A., Ferrer, M., Fuentes, G., Pastor, E., Bernabe, M., 2002. Enzymatic acylation of di-and trisaccharides with fatty acids: choosing the appropriate enzyme, support and solvent. J. Biotechnol. 96, 55–66.

Polat, T., Linhardt, R.J., 2001. Synthesis and applications of sucrose-based esters. J. Surf. Deterg. 4, 415–421.

Puterka, G.J., Farone, W., Palmer, T., Barrington, A., 2003. Structure-function relationships affecting the insecticidal and miticidal activity of sugar esters. J. Econ. Entomol. 96, 636–644.

Sabeder, S., Habulin, M., Knez, Z., 2006. Lipase-catalyzed synthesis of fatty acid fructose esters. J. Food Eng. 77, 880–886.

Seddon, K.R., Stark, A., Torres, M.J., 2000. Influence of chloride, water, and organic solvents on the physical properties of ionic liquids. Pure Appl. Chem. 72, 2275–2287.

Sharma, A., Chattopadhyay, S., 1993. Lipase catalyzed acetylation of carbohydrates. Biotechnol. Lett. 15, 1145–1146.

Sharma, R., Chisti, Y., Banerjee, U.C., 2001. Production, purification, characterization, and applications of lipases. Biotechnol. Adv. 19, 627–662.

Shaw, J.F., Klibanov, A.M., 1987. Preparation of various glucose esters via lipase-catalyzed hydrolysis of glucose pentaacetate. Biotechnol. Bioeng. 29, 648–651.

Shogren, R.L., Biswas, A., 2010. Acetylation of starch with vinyl acetate in imidazolium ionic liquids and characterization of acetate distribution. Carbohydr. Polym. 81, 149–151.

Shogren, R.L., Biswas, A., Willett, J.L., 2010. Preparation and physical properties of maltodextrin stearates of low to high degree of substitution. Starch-Staerke 62, 333–340.

Shu, Z.Y., Jiang, H., Lin, R.F., Jiang, Y.M., Lin, L., Huang, J.Z., 2010. Technical methods to improve yield, activity and stability in the development of microbial lipases. J. Mol. Catal. B Enzym. 62, 1–8.

Sin, Y.M., Cho, K.W., Lee, H.T., 1998. Synthesis of fructose esters by *Pseudomonas* sp. lipase in anhydrous pyridine. Biotechnol. Lett. 20, 91–94.

Singh, R.K., Gupta, P., Sharma, O.P., Ray, S.S., 2014. Homogeneous synthesis of cellulose fatty acid esters in ionic liquid (1-butyl-3-methylimidazolium chloride) and study of their comparative antifriction property. J. Ind. Eng. Chem. [Online early access]. http://dx.doi.org/10.1016/j/jiec/2014.09.031 (accessed 01/11/14).

Somashekar, B.R., Divakar, S., 2007. Lipase catalyzed synthesis of l-alanyl esters of carbohydrates. Enzym. Microb. Technol. 40, 299–309.

Spear, S.K., Visser, A.E., Rogers, R.D., 2002. Ionic liquids: green solvents for carbohydrate studies. In: Proceedings of the Sugar Processing Research Institute (SPRI) 2002 Conference on Sugar Processing Research, New Orleans, LA, March 14–15, 2002, pp. 336–340.

Swatloski, R.P., Spear, S.K., Holbrey, J.D., Rogers, R.D., 2002. Dissolution of cellulose with ionic liquids. J. Am. Chem. Soc. 124, 4974–4975.

Treichel, H., de Oliveira, D., Mazutti, M.A., Di Luccio, M., Oliveira, J.V., 2010. A review on microbial lipases production. Food Bioproc. Technol. 3, 182–196.

Wang, Y.B., Xie, W.L., 2010. Synthesis of cationic starch with a high degree of substitution in an ionic liquid. Carbohydr. Polym. 80, 1172–1177.

Wang, H.T., Yuan, T.Q., Meng, L.J., She, D., Geng, Z.C., Sun, R.C., 2012. Structural and thermal characterization of lauroylated hemicelluloses synthesized in an ionic liquid. Polym. Degrad. Stab. 97, 2323–2336.

Wen, J., Sun, Y.C., Meng, L.Y., Yuan, T.Q., Xu, F., Sun, R.C., 2011. Homogeneous lauroylation of ball-milled bamboo in ionic liquid for bio-based composites production part I: modification and characterization. Ind. Crop Prod. 34, 1491–1501.

Wilpiszewska, K., Spychaj, T., 2011. Ionic liquids: media for starch dissolution, plasticization and modification. Carbohydr. Polym. 86, 424–428.

Wu, Y., Sasaki, T., Irie, S., Sakurai, K., 2008. A novel biomass-ionic liquid platform for the utilization of native chitin. Polymer 49, 2321–2327.

Xie, H., Zhang, S., Li, S., 2006. Chitin and chitosan dissolved in ionic liquids as reversible sorbents of CO_2. Green Chem. 8, 630–633.

Xie, W., Wang, Y., 2011. Synthesis of high fatty acid starch esters with 1-Butyl-3-Methylimidazolium chloride as a reaction medium. Starch-Stärke 63, 190–197.

Xie, W.L., Shao, L., Liu, Y.W., 2011. Synthesis of starch esters in ionic liquids. J. Appl. Polym. Sci. 116, 218–224.

Xie, W., Shao, L., 2009. Phosphorylation of corn starch in an ionic liquid. Starch-Stärke 61, 702–708.

Xu, Q., Kennedy, J.F., Liu, L., 2008. An ionic liquid as reaction media in the ring opening graft polymerization of ε-caprolactone onto starch granules. Carbohydr. Polym. 72, 113–121.

Yamazaki, S., Takegawa, A., Kaneko, Y., Kadokawa, J.I., Yamagata, M., Ishikawa, M., 2009. An acidic cellulose–chitin hybrid gel as novel electrolyte for an electric double layer capacitor. Electrochem. Commun. 11, 68–70.

Yan, Y., Bornscheuer, U.T., Cao, L., Schmid, R.D., 1999. Lipase catalyzed solid-phase synthesis of sugar fatty acid esters. Enzym. Microb. Technol. 25, 725–728.

Yang, Z., Pan, W., 2005. Ionic liquids: green solvents for nonaqueous biocatalysis. Enzym. Microb. Technol. 37, 19–28.

Yuan, T.Q., Sun, S.N., Xu, F., Sun, R.C., 2010. Homogeneous esterification of poplar wood in an ionic liquid under mild conditions: characterization and properties. J. Agric. Food Chem. 58, 11302–11310.

Zakrzewska, M.E., Bogel-Lukasik, E., Bogel-Lukasik, R., 2010. Solubility of carbohydrate in ionic liquids. Energy Fuels 24, 737–745.

Zavrel, M., Bross, D., Funke, M., Büchs, J., Spiess, A.C., 2009. High-throughput screening for ionic liquids dissolving (ligno-) cellulose. Biores. Technol. 100, 2580–2587.

Zhang, H., Wu, J., Zhang, J., He, J., 2005. 1-Allyl-3-methylimidazolium chloride room temperature ionic liquid: a new and powerful nonderivatizing solvent for cellulose. Macromolecules 38, 8272–8277.

Zhao, H., Baker, G.A., Song, Z., Olubajo, O., Crittlea, T., Peters, D., 2008. Designing enzyme-compatible ionic liquids that can dissolve carbohydrates. Green Chem. 10, 696–705.

Chapter 12

Ionic Liquids in the Production of Biodiesel and Other Oleochemicals

Bethala Lakshmi Anu Prabhavathi Devi, Tangadanchu Vijai Kumar Reddy
CSIR—Indian Institute of Chemical Technology, Center for Lipid Research, Hyderabad, India

Mohd Firdaus Mohd Yusoff
Aarhus University, Department of Engineering, Aarhus, Denmark; Universiti Kebangsaan Malaysia, UKM, Fakulti Sains dan Teknologi, Pusat Pengajian Sains Kimia dan Teknologi Makanan, Bangi, Selangor, Malaysia

Ionic liquids (ILs) have been shown to be promising green solvents with several advantages compared to traditional organic solvents (Wilkes et al., 2008). ILs are composed of ions and are liquids at room temperature. Interest in these compounds stems from their near-zero vapor pressure (Earle et al., 2006) and their widely tunable properties (polarity, hydrophobicity, and solvent miscibility behavior) through appropriate modification of the cation and anion. Most notably, their nonvolatility, high thermal and chemical stability, and extraordinary solvent properties make them important solvents for the chemical industry (Plechkova and Seddon, 2008). ILs have been explored in organic synthesis, catalysis, biocatalysis, extraction and dissolution processes, nanomaterials synthesis, polymerization reactions, and electrochemistry. The oleochemical industry is one of the most important industries as a promising alternate to the petrochemical industry. Recently, ILs have gained a lot of attention as alternative solvents and/or catalysts in chemical or enzymatic (biocatalytic) production of oleochemicals like biodiesel, biolubricants, surfactants, and so on.

12.1 PRODUCTION OF BIODIESEL

12.1.1 What Is Biodiesel?

The major components of vegetable oils and animal fats are triacylglycerides (TAG), which are esters of fatty acids with glycerol (1,2,3-propanetriol; also

Ionic Liquids in Lipid Processing and Analysis. http://dx.doi.org/10.1016/B978-1-63067-047-4.00012-X

known as glycerine). TAGs of vegetable oils and animal fats typically contain several different fatty acids attached to one glycerol backbone. This feature contributes to the physical characteristics of the oil or fat. The fatty acid profile becomes the most important parameter that influences the properties of the vegetable oils and animal fats. Biodiesel is obtained when vegetable oil/animal fat is subjected to a chemical reaction termed transesterification (Fig. 12.1) in the presence of a catalyst (usually base) with an alcohol (methanol or ethanol).

Based on Fig. 12.1, transesterification reaction requires a molar ratio of alcohol to oil of 3:1, however in normal practice a higher amount of alcohol is used (6:1) in order to push the reaction toward higher yield. Vegetable oils or animal fats are transesterified with methanol to fatty acid methyl esters mainly because kinematic viscosity of these esters is almost similar to petro diesel. High viscosity of untreated oils and fats will cause operational problems in normal diesel engines such as deposits on its interior parts. Vegetable oils and animal fats also demonstrate ignition quality, which is expressed by an octane number that is similar to diesel fuel. Biodiesel standards such as ASTM D6751 (United States) and EN 14,214 (Europe) contain several specifications, which often relate to production or storage matters, to ensure that the product can be used in diesel engines. Common feed stocks for biodiesel production are soybean, cottonseed, palm, peanut, rapeseed, safflower, sunflower, coconut, tallow, lard, and also waste cooking oil. Choices of feedstock are based on accessibility of the raw material, pricing, and local temperature where biodiesel is produced. However, due to the issue of food versus fuel, which ignites a discussion between global science academia and economical figures, there is concern of finding other resources that are cheaper without jeopardizing human food source.

Nonedible vegetable oils are not suitable for human food because of some toxic compounds present in the oils. Problems of food versus fuel, environmental, and economic issues related to edible vegetables can be overcome if this nonedible resource is used as a feedstock for biodiesel production (Atabani et al., 2013). A collection of some potential nonedible sources and their oil content for the biodiesel industry is shown in Table 12.1.

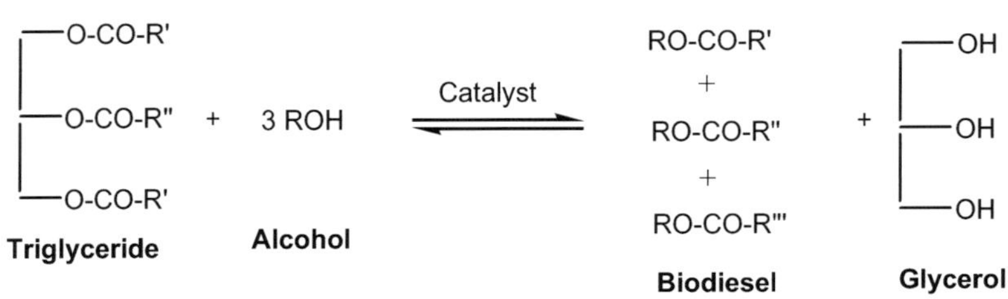

FIGURE 12.1 Transesterification of triglycerides to biodiesel. R′, R″, and R‴ are mixtures of various fatty acid chains.

TABLE 12.1 Plants that Produce Nonedible Oil for Potential Biodiesel Feedstock

Common Name	Scientific Name	Oil Content (wt%)		Nativity
		Seed	Kernel	
Jarak pagar, Tampate, Galumbang	*Jatropha curcas* L.	20–60	40–60	Indonesia, Thailand, Malaysia, Philippines, India, Pakistan, Nepal
Karanja	*Pongamis pinnata* L.	25–50	30–50	Native Western Ghats in India, Northern Australia, Eastern Asia
Castor	*Ricinus communis*	45–50	–	Cuba, Brazil, China, India, Italy, France
Linseed	*Linum usitatissimum*	35–45	–	Canada, Argentina, Eastern Mediterranean to India
Rubber seed	*Hevea brasiliensis*	40–60	40–50	Nigeria, India, Brazil, Southeast Asia, West Africa

Source: Atabani et al. (2013).

Some of the advantages of using nonedible oil are:

- No competition with existing agricultural resources
- Eliminates competition for human consumption as food
- More efficient and more environmentally friendly than normal oil crops (due to the fact that they grow naturally and organically without pesticides and chemical fertilizers)
- Able to produce side products such as biomass that can be used in other chemical processes or as burning material to generate heat and power
- More resilient to crop disease and pests

Biodiesel has several advantages in comparison to petrodiesel and is considered competitive in most technical aspects such as:

1. Biodegradability
2. Feedstocks are renewable domestic resources unlike the dependence of conventional diesel fuel to ever-decreasing petroleum stock
3. Higher flash points meaning safer handling and storage
4. Reduction of most air pollutants that often exist due to petrodiesel combustion (with the exception of nitrogen oxides, NO_x)
5. Excellent lubricity; ever since low-sulfur petrodiesel fuels (with reduced lubricity) were introduced to the market, there is a significant need to add biodiesel into the formulation at low levels (1–2%) to restore lubricity

12.1.2 Biodiesel Production with Conventional Catalyst

The general synthetic route for biodiesel production is the transesterification of vegetable oils or animal fats with alcohol (normally methanol). This reaction can be catalyzed by acids, alkaline metal hydroxides, alkoxides, and nonionic bases such as amines and amidines (Schuchardt et al., 1998). These methods however, suffer from several major drawbacks such as:

1. Acid/base processes are often related to corrosion and emulsification problems.
2. Acid-catalyzed reactions proceed slower than the base-catalyzed processes (Formo, 1954; Freedman et al., 1984). On the other hand, the base-catalyzed process may cause unnecessary saponification of the free fatty acids that are present in the feed. Thus, base catalyst requires high quality feedstock with minimal acid value.
3. Oils and fats are not completely soluble in alcohols, resulting in barriers for the triglyceride conversions.
4. The process is usually energy intensive, requires alkaline wastewater treatment, and there is interference of free fatty acids and water (Freedman et al., 1984).
5. A significantly large excess of alcohol is required to push the equilibrium to ester formation and to achieve good separation of biodiesel from by-products formed during reaction (glycerol). This will create a recycling problem; that is, the investment and operating cost required to recover the alcohol used.

In order to overcome these drawbacks, several innovations in biodiesel production have been aggressively developed such as the use of IL-catalyzed transesterification (Wu et al., 2007), conducting the alcoholysis in supercritical methanol (He et al., 2007), use of the lipase-catalyzed transesterification (Fukuda et al., 2008; Akoh et al., 2007), and development of heterogeneous catalysts (Alexandre et al., 2008). Generally, there are two accepted methods to purify biodiesel, namely wet and dry washing. The wet washing method is widely used to remove excess contaminants and by-product chemicals from the biodiesel. Since both glycerin and methanol are highly soluble in water, water washing is very effective in removing both contaminants. This technique still remains the most common method of purification. The wet washing technique can also remove any residual sodium salts and soaps. Since additional water was incorporated into the process, several disadvantages emerged, including increased operating cost and production time in terms of special waste treatment. Significant product losses were also observed in the water phase. Besides that, formation of emulsion was also observed, especially if the feedstock is low quality oil such as waste cooking oils or other feeds that contain high FFA content due to soap formation (Berriosa and Skelton, 2008).

Another technique for product separation is dry washing. Ion exchange resin or magnesium silicate powder was used to neutralize the impurities present in the biodiesel. Currently, both the wet and dry washing methods are being used in industrial plants. Other possible methods that can be used to purify biodiesel include membrane reactors and the addition of lime and phosphoric acid (Chiu et al., 2006; Yori et al., 2007). Nonetheless, there were numerous problems reported, especially on the costs and complications faced in operating the process at industrial scale.

12.1.3 Role of Ionic Liquid in Biodiesel Production

For the past few years, ILs have been accepted as a suitable green catalyst in industrial chemical processes. This molten salt substance can stay at liquid state below 100°C, hence it is known as room temperature IL (RTIL) (Welton, 1999). IL was dubbed a green catalyst due to its relatively low volatility, ability to stay in liquid form in a wide temperature range, and also less toxicity compared to other organic solvents (Shen et al., 2008; Gui et al., 2004). ILs can selectively dissolve various chemical compounds (organic, inorganic, and organometallic materials) because of polarity, which can be adjusted based on the specific purposes. By tuning their anion−cation combinations, they can be either miscible or immiscible with water and organic solvents. This unique feature provides great advantages compared to conventional solvents.

12.1.3.1 Ionic Liquid as Catalyst for Biodiesel Production

IL is considered as a prospective catalyst for biodiesel production due to its excellent properties; it is recyclable, mild, and less hazardous in handling, easily separable, suitable for continuous reaction, and produces less effluents (Earle and Seddon, 2000; Xie et al., 2008). In biodiesel production, product separation and purification is the most crucial step in order to get the highest yield. IL catalysts are able to reduce the number of reactions and purification step, hence the production process consumes less energy and is more economically efficient. Besides that, ILs are able to overcome some drawbacks associated with product separation and catalyst recycling in reactions where in most cases it was impossible to achieve 100% reaction conversion with 100% selectivity due to the thermodynamic limitations and the competition from parallel reactions (Zhao et al., 2002). The choice of using IL as an excellent substitute for traditional solvents or catalysts in organic synthesis grows along with the prospect of developing environmentally safe green catalysts. Academics from around the world have reported their studies, which demonstrate successful applications of ILs as catalysts in hydrogenation, isomerization, cleavation of carbon and ether bonds (eg, catalytic cracking of poly alkanes), and C−C coupling reactions with excellent yields and selectivity (Zhao et al., 2002). Brønsted acid IL is a typical pathway for biodiesel since it possess several advantages because the manufacturer/researcher can combine and design solid and mineral acids to replace hazardous conventional acid catalyst such as sulfuric acid (Fang et al., 2007). Brønsted acid is widely used in the esterification process, resulting in high yield of products. In 2002, Cole and coworkers produced aliphatic and aromatic esters using Brønsted acid IL. However their research provides little information on preparation of fatty acid alkyl esters. Nonetheless, their work sparked academia's interest in using ILs in the production of biodiesel (Cole et al., 2002).

12.1.3.1.1 Ionic Liquid With Inorganic Support Material as Catalyst

Immobilizing catalysts on a suitable support material has been proven to be a good solution in terms of product separation and purification. Study on alcoholysis of soybean oil using two multiphase systems has been reported (Frederique et al., 2005). Using acidic resin as a support material, the two systems were developed from *Sn*(3-hydroxy-2-methyl-4-pyrone)$_2$(H_2O_2) by dissolving it in 1-butyl-3-methyl-imidazolium hexafluorophosphate ([BMIM][PF_6]). However, recycling of the IL of the multiphase system failed to achieve more than 1% yield. To improve the system, additional research was conducted to combine the advantages of IL and heterogeneous catalysis by immobilizing the same Sn complex in 1-n-butyl-3-methyl-imidazolium tetrachloroindate ($BMIM^+ \cdot InCl_4^-$) and comparing their performance with different catalysts. Tests on other catalysts ($LaCl_3$, $CeCl_3$, $ZrCl_4$, $CuCl_2$, $MgCl_2$, $ZnCl_2$, $BaCl_2$, $NiCl_2$, $PtCl_4$, $PtCl_2$, $TiCl_3$, $SnCl_2$, $CdCl_2$, $InCl_3$, Al_2O_3, GeO_2, $CuBr_2$, CuI, $Na_2O_5(H_2O)$, Na_2SiF_6, BF_3, and OEt_2) showed very low conversion of vegetable oil to biodiesel. Replacing all of these catalysts with tin complex improved the conversion rate to 84% after 4 h of reaction using 1 wt% catalyst. However, recycling of the catalyst showed a poor conversion, thus making it economically inefficient for continuous reaction. These two works conclude that biphasic systems are expensive due to the number of chemicals used for synthesizing the IL and for the preparation of complex mixtures. Brønsted acidic IL can also be supported on iron complex. A study on efficiency of Brønsted acid on Fe-SBA-15 as a catalyst for esterification of oleic acid with short-chain alcohol was conducted. Synergistic effect of Lewis site of Fe and Brønsted acid site of IL improves the efficiency by producing biodiesel at 87.7% under reaction temperature of 363K. Molar ratio of alcohol to fatty acid is most optimum at 6:1 with the amount of catalyst used at 5 wt% for 3 h of reaction (Zhang et al., 2012).

Another novel approach was conducted using tungstophosphoric anion-based IL. The complex was incorporated on the surface of silica and the resultant complex was used to convert oleic acid into methyl oleate. The yield, however, is quite low (48.49%). Besides producing biodiesel from pure fatty acid, IL has also been proven to work on low-quality oils and fats. Raw jatropha oil, which contains high FFA (acid value 13.8 mg KOH/g), is transesterified with alcohol using IL as catalyst. [BMIm][CH_3SO_3] was coupled with $FeCl_3$ and the complex was reacted with low-quality jatropha oil. The yield increases from 12% to 93% when only IL is used as catalyst at elevated temperature (Fang et al., 2007). When $FeCl_3$ was added, the yield increases even more, up to 99.7% (reaction temperature 120°C). These features of high yield at low temperature were established due to metal ions having Lewis acid sites, and more of the sites could be provided by trivalent metallic ions than those of bivalent ones. Furthermore, the complex is easily separated and has excellent reusability for the next reaction cycle.

12.1.3.1.2 Ionic Liquid Without Support Material as Catalyst

Some IL can also be used as catalyst solely without support material. Refined cottonseed oil was transesterified using several Brønsted acid ILs of various cation side coupled with butyl and propyl sulfonic acid group as its side chains and hydrogen sulfate as the anion. It was reported that 1-(4-sulfonic acid) butylpyridinium hydrogen sulfate gives the best catalytic performance (Wu et al., 2007). After 5 h of reaction, 92% conversion was achieved with optimum condition reaction at 12:1 oil to IL molar ratio, at 170°C and the catalyst concentration of 5.7 wt%. Other than that, Ghiachi et al. (2011) conducted a repeated transesterification of jatropha oil using $[BSPy]CF_3SO_3$ and found biodiesel production in high yields even after seven cycles. The anion in the IL structure plays an important role in the acidic nature of IL and the cation side chain may also influence the accessibility of acid active sides. ILs containing benzyl side chains such as imidazolium-based ILs have different efficiency on catalytic conversion based on the size of alkyl side chains. For example (Fig. 12.2), between $BMIMHSO_4$, $BIMHSO_4$, and

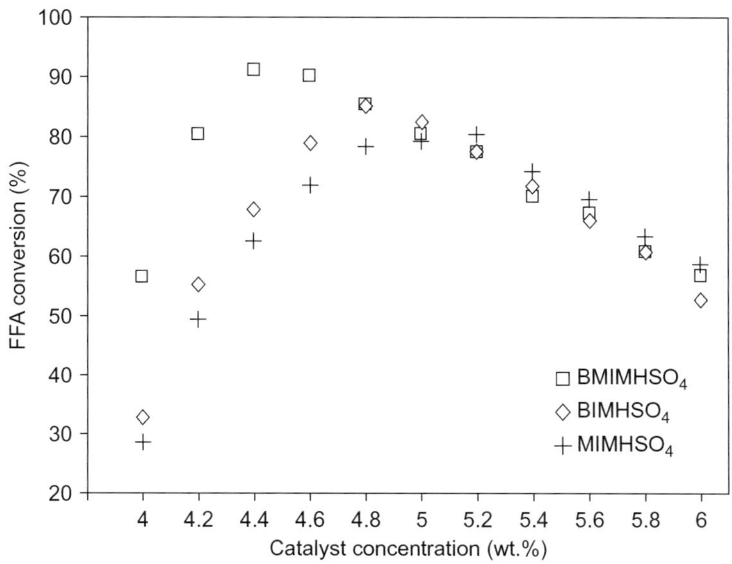

FIGURE 12.2 Comparison of catalytic performance of different ionic liquids based on alkyl side chain.

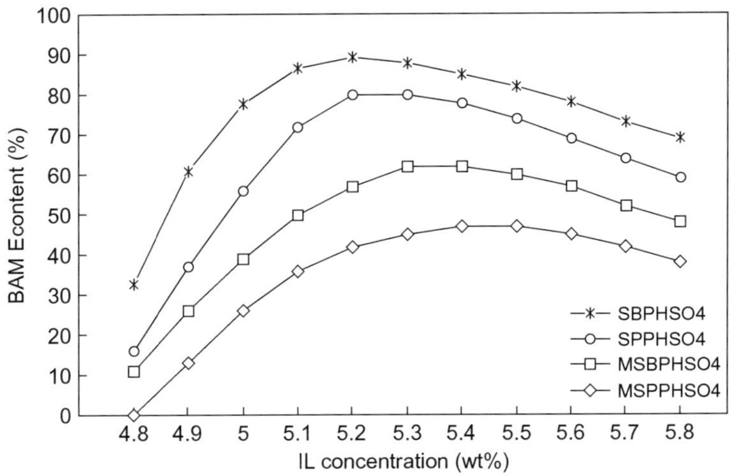

FIGURE 12.3 Effect of ionic liquid types and concentrations on the production bitter apple methyl esters (BAME). Molar ratio alcohol to bitter apple oil (15:1), with temperature optimized at 170°C agitated for 6 h at 800 rpm.

$MIMHSO_4$, the first compound shows a better efficiency than other compounds because of its longer alkyl side chain, which provides steric hindrance on cation moiety (Elsheikh et al., 2011).

Besides the imidazolium compound, pyrazolium-based ILs also show the same results. The extension of a sulfonate side chain from propyl to butyl that linked to N_2 in pyrazole increases the catalytic activity. One of the possible reasons is the longer carbon chain in Brønsted acidic ILs of the same nitrogen groups; that is, pyrazolium, which has higher tendency to dissociate into H^+, giving rise to stronger acidity (Wu et al., 2007). A similar trend (Fig. 12.3) was also observed during the transesterification of bitter apple oil to its methyl ester (Elsheikh, 2014). Replacing the methyl side chain in $MSBPHSO_4$ and $MSPPHSO_4$ with proton to obtain $SBPHSO_4$ and $SPPHSO_4$, respectively results in an increase in the activity and acidity of the catalyst.

12.1.3.2 Ionic Liquid as Solvents for Biodiesel Production

ILs are an ideal medium for separation of biodiesel and removal of glycerol by-product. Nevertheless, more study is needed to investigate whether accumulated glycerol affects recovery of IL. It is still unknown, or more accurately, the investigations are rather vague, whether ILs possess any potential toxicity. Few preliminary reports on toxicological properties of IL were available. IL as solvent for various chemical reactions had recently been the center of interest among academia. A special type of IL, known as deep eutectic solvents (DES) was developed. It consists of a mixture of two compounds that has the lowest melting point. Two components that can be used in chemical reaction usually consist of organic halide salts such choline

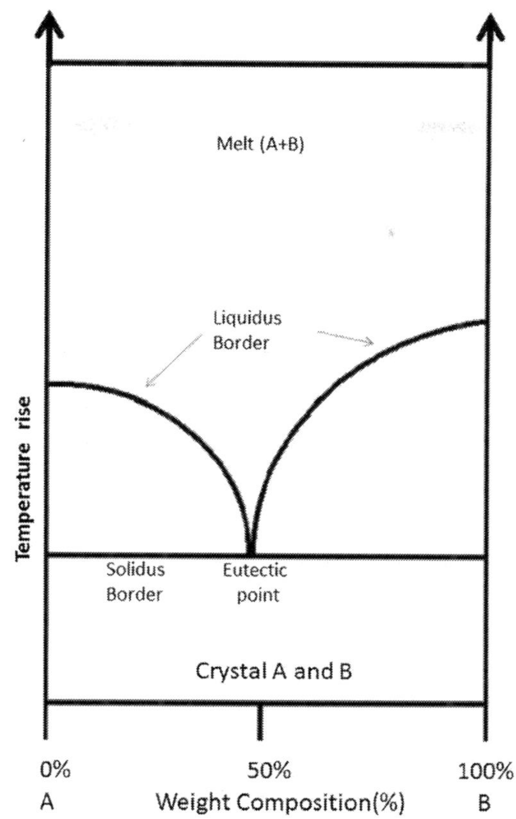

FIGURE 12.4 Deep eutectic solvent phase scheme.

chloride with an organic compound that acts as a hydrogen bond donor (HBD) capable of forming H bond with halide ions. Examples of HBD are amines, amides, carboxylic acids, and alcohols (Liu et al., 2008). Interaction between the two components, whereby hydrogen-bonding interactions between an anion and an HBD, are more energetically favored relative to the lattice energies of the pure constituents causing depression of freezing point (Fig. 12.4). As a result, DES stays at liquid state. Zhao and Baker believed that a combination of conventional ILs with DES can be applied in biodiesel production. DES has several advantages over normal ILs. They are easier to be prepared in pure state and do not react with water. They are also biodegradable and less toxic (Zhao and Baker, 2013). Several works have been published in recent years, such as the one that uses phosphonium-based DES (*p*-DES) for esterification of low grade crude palm oil. Besides that, ammonium-based DES was also being used with the same feedstock (Hayyan et al., 2013a,b). The hydrogen bond donor used was *p*-toluenesulfonic acid, which was combined with *N*,*N*-diethylene ethanol ammonium chloride as a novel recyclable catalyst.

12.1.3.2.1 Ionic Liquid as a Solvent for Enzymatic Transesterification

Vegetable oils and animal fats have poor miscibility with methanol. In addition, methanol also inactivates a biocatalyst. To overcome this problem, stepwise addition of methanol was introduced during reaction. Other acyl acceptors like ethyl acetate and methyl acetate, which have less poisoning effects, are also used. In addition, conventional organic solvents are also found to be toxic to lipase, which will lead to poor reactivity and stability. Enzymes are also deactivated by glycerol via adsorption onto the surface of the enzyme (Rios et al., 2011). Several groups of chemists had proposed an enzymatic transesterification of vegetable oils in IL to produce biodiesel. Water-immiscible ILs, such as 1-ethyl-3-methylimidizolium bis(trifluoromethylsulfonyl) imide and butyltrimethyl ammonium bis(trifluoromethylsulfonyl) imide, were used for butyl butyrate synthesis from vinyl butyrate catalyzed by *Candida antarctica* lipase B (Diego et al., 2005). The result of the test shows that synthetic activity and stability of the enzyme in these ILs were enhanced as compared to those in hexane. A schematic representation of biocatalytic transesterification in IL is shown in Fig. 12.5.

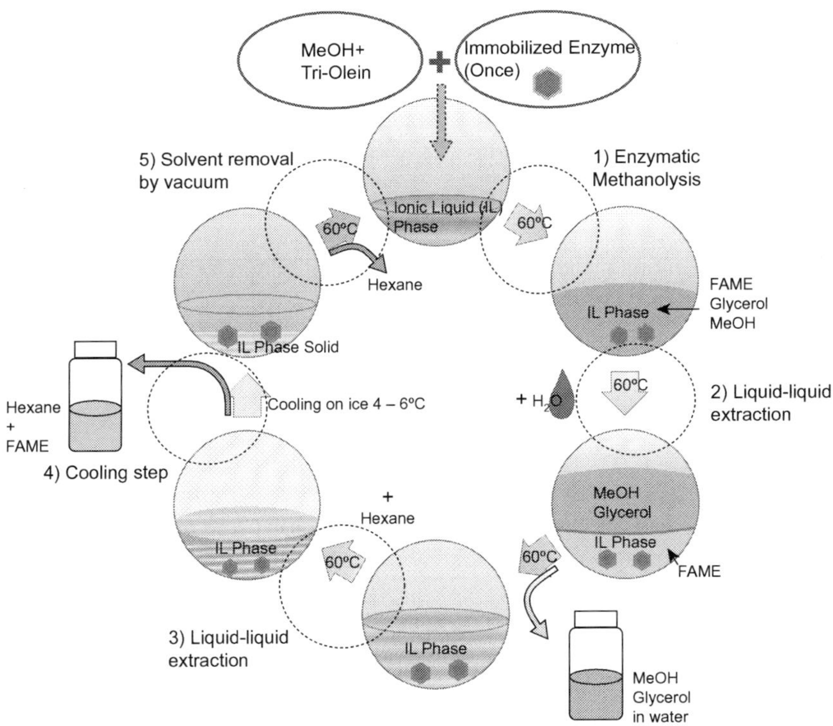

FIGURE 12.5 Enzymatic biodiesel processes in ionic liquid.

The reaction was conducted in a biphasic environment where a certain amount of water is required. Using shorter chain IL such as 1,3-dialkylimidazolium cation ([BMIM]PF_6 or [BMIM]NTf_2) combined with lipase, 96.3% yield were produced from soybean oil (Gamba et al., 2008). Normally, addition of salt hydrate will diminish the water requirement of lipase-catalyzed reaction, particularly with [BMIM]PF_6.. The salt hydrate may influence specific ion effects on the enzyme activities in an IL system. Addition of water into a biphasic system will create a third phase where the water phase will help to dissolve by-products and hence, help recover IL. However after six cycles of reaction, a decrease of lipase activity was observed (Ruzich and Bassi, 2010).

Longer chain imidazolium-based ILs when used in a homogeneous single phase system for lipase catalyzed biodiesel production helps to protect enzymes from direct contact with methanol (Diego et al., 2005). These long chain and lipophilic ILs create a water-free environment for oil trans-esterification which, at the end of reaction, creates a third phase system that facilitates biodiesel extraction. On the other hand, hydrophilic IL often causes enzyme dissolution and denaturation (Zhao et al., 2011). A study has been reported demonstrating the effect of the molecular structures of the ILs on enzymatic production of biodiesel using an oil/IL biphasic system (Devi et al., 2011). Among different ILs evaluated, Ammoeng 102 containing tetraaminum cation with C-18 acyl and oligoethyleneglycol units was found to be more capable of forming an oil/IL biphasic reaction system, thus resulting in 98% biodiesel yield. In addition, the oil/Ammoeng 102 system also exhibited some unique properties, such as much less MG and DG in the final product, only small amounts of IL needed to host an efficient reaction, a protective effect for enzymes against inactivation by methanol, and automatic phase separation for biodiesel recovery. Acidic ILs give a good catalytic activity for biodiesel production. Utilization of IL gives a better separation of products and side-products such as glycerol. Moreover, its reactivity and stability after several reaction cycles shows a promising and economical solution for achieving a much greener and environmentally friendly production of biodiesel. IL also shows potential as a solvent or cosolvent in biodiesel production. A hydrophobic system is preferable for enzymatic reaction since it avoids enzyme denaturation. These characteristics are of paramount importance to explore the industrial interests of the system.

12.2 PREPARATION OF BIOLUBRICANTS

12.2.1 What Is a Biolubricant?

The basic functions of a lubricant are friction and wear reduction, heat removal, and contaminant suspension. Lubricant oil consists of 90–95% base stocks, 5–10% chemical additives, and other components. Lubricant additives are important constituents for enhancing the inherent properties of the base oil and also adding new characteristics to improve their performance. There are a wide range of lubricant base oils including mineral oils, synthetic oils, rerefined oils, and plant oils. Among these, mineral oils are the most commonly used. They consist predominantly of hydrocarbons but also contain some sulfur and nitrogen compounds with traces of a number of metals. Mineral oils contribute approximately 85% of the base stock for the production of lubricants, which are in great demand and have long served in applications such as automobile engines and hydraulic machinery. They are not ecofriendly or biodegradable, and they can pollute the environment during use or at the time of disposal. The ultimate consequence of the prolonged use of such lubricants might be detrimental to human beings, animals, plants, and aquatic life (Betton, 2010; Bartz, 1998, 2006). Due to the damaging potential of conventional lubricants, the US Environmental Protection Agency and other government agencies have introduced stringent regulations on the use, containment, and disposal of lubricants. These regulations have altered the market for lubricants. Now, new ecofriendly lubricants from natural resources need to be developed to address the environmental, health, and economic issues, as well as the need for high performance (Nagendramma and Kaul, 2012). To solve this problem, lubricants should be prepared from plant oil derivatives. Plant-based oils show great potential and are a highly attractive candidate to replace conventional mineral oils for the use in lubricant production, In addition to being structurally similar to the long-chained hydrocarbons, plant-based oils are renewable, nontoxic, and economically and environmentally friendly (Gryglewicz et al., 2003; Rudnick and Erhan, 2006).

Biolubricants can be produced by esterification of plant oil containing oleic acid or other long-chain fatty acids (12–20 carbon atoms) with long-chain alcohol (5–12 carbon atoms) in the presence of an acid catalyst. Biodegradability is one of the most important properties of a lubricant; a good lubricant should be highly biodegradable when released into the environment. Petroleum-based lubricants are poorly biodegradable while the lubricants from alternative resources are highly biodegradable and will not cause any adverse effect to the environment (Table 12.2). Lubricants from alternative resources

TABLE 12.2 The Biodegradability of Petroleum-Based Lubricants and Lubricants from Alternative Resources

Entry	Lubricant	Biodegradability (%)
1	Petroleum-based lubricants	20–30
2	Lubricants from alternative resources	90–100

are biodegradable in the range of 90–100%, which means that these lubricants quickly degrade as biological organisms. Thus the adverse effects of lubricants to the environment are reduced to a greater extent, which is the demand of today's world.

Biolubricants have several advantages compared to petroleum-based lubricants and are considered competitive in most technical aspects such as:

1. Biodegradability
2. Very low or no pollution-causing effect on the environment
3. Zero toxicity, so no skin diseases to the operator in case of leak or accident
4. Similar features like viscosity, lubricity, viscosity index

Controlling environmental pollution caused by petroleum-based lubricants is of prime importance, because these lubricants have an adverse effect on the ecological system and on all the living beings. Table 12.3 shows the comparison of petroleum-based lubricants with lubricants from alternative resources (Kumar and Suhane, 2013).

It is clear from the Table 12.3 that petroleum-based lubricants are the main cause of environmental pollution, while the lubricants from alternative

TABLE 12.3 Comparison of Petroleum-Based Lubricants with Lubricants from Alternative Resources

Environmental Pollution Control			
Entry		Petroleum-based Lubricants	Lubricants from Alternative Resources
1	Biodegradable	No	Yes
2	Toxic	Yes	No
3	Cause of global warming	Yes	No
4	Harmful by-products of emission	Yes	No
5	Highly toxic to humans and animals	Yes	No
6	Pollutes environment	Yes	No
7	Economic gain to Indian farmers	No	Yes
8	Dangerous to handle and store	Yes	No
9	Domestic production	No	Yes
10	Alternative resource	No	Yes

resources are best to reduce environmental pollution and also to reduce the harmful effects on human beings and other living beings. Alternative resources like nonedible vegetable oils can be produced domestically while petroleum oil cannot be produced domestically, and is also on the verge of depletion so its price is also increasing. Petroleum-based lubricants are dangerous to handle and store in case of spill, leakage, and such due to their poor biodegradability and high toxicity, whereas this is not the case with the lubricants from alternative lubricants.

12.2.2 Biolubricant Preparation with Conventional Catalyst

Biolubricants are generally prepared by chemical modification of plant-based oils such as transesterification, epoxidation, and esterification reactions. The esterification reaction between fatty acid and polyol is an industrially important tool for obtaining desired polyhydric esters. Traditionally, homogeneous acids and alkali were used as esterification catalysts. However, alkali catalysts may cause saponification and inorganic acids are responsible for the dark color of the reaction products. More importantly, they are not renewable and may result in a difficult separation process, which causes environmental pollution. Heterogeneous catalysts, such as metallic or metal oxide, can also serve as an alternative choice for the preparation of biolubricants. Although these catalysts can be recycled easily, they have other shortcomings such as deactivation from coking and high cost, and some of them still have the tendency to cause saponification (Bondioli, 2004).

Biocatalysts such as lipases have attracted much attention because they are ecofriendly. Modern enzymology has achieved improvements in the development and application of lipase as a catalyst. New immobilization techniques make it possible to use enzymes in industrial processes in a similar way to the classical catalysis for heterogeneous reactions. For example, esters produced from the reaction of long-chain fatty acids with long-chain alcohols (12–20 carbon atoms) also have important applications as plasticizers and lubricants (Dossat et al., 2002; Jaidi et al., 2002). Compared with conventional chemical synthesis from alcohols and fatty acids using mineral acids as a catalyst, the use of enzymes such as lipases to produce these high value added fatty acid esters in solvent-free media may offer many significant advantages (Yadav and Lathi, 2003). These include the use of any hydrophobic substrate, higher selectivity, milder processing conditions, and the ease of product isolation and enzyme reuse. However, long reaction time, deactivation from polyols, and higher amounts of lipases required compared with other catalysts limited the broad application of enzymes (Kiriliauskaite et al., 2011; Linko et al., 1997). Therefore, in order to overcome these drawbacks, several innovations in biolubricant production have been developed such as the use of IL-catalyzed reactions (Li et al., 2012) and the use of the lipase-catalyzed reactions in ILs (Banyai et al., 2011).

12.2.3 Role of Ionic Liquid in the Preparation of Biolubricants

12.2.3.1 Ionic Liquid as Catalyst

ILs have been shown as promising catalysts for both esterification and transesterification owing to their negligible volatility, excellent thermal stability, and the variety of structures available. Yan et al. (2011) reported preparation of biolubricants by transesterification of vegetable oil with trimethylpropane (1−3:1 wt/wt) in the presence of acidic IL (0.5−1.5%) at 120−180°C for 4−8 h. Recently, ILs containing SO_3H have received much attention as catalysts in the production of biolubricants because they are flexible, noncorrosive, and immiscible with many organic solvents. Li et al. (2012) prepared trimethylpropane triesters of oleic acid as biodegradable lubricant by the esterification of TMP with oleic acid (Fig. 12.6) employing a multi-SO_3H-functionalized strong Brønsted acidic IL as catalyst.

Wang et al. (2013) prepared a microemulsion type biolubricant formulation with vegetable oil as the continuous phase, room-temperature IL 1-butyl-3-methyl-imidazolium tetrafluoroborate as the polar phase, TritonX-100 as the surfactant, and 1-butanol as the cosurfactant. A pseudo-ternary phase diagram, dynamic light scattering data, UV−Vis spectra, and kinematic viscosity measurements were employed to characterize and analyze the phase behavior and microstructure of the vegetable oil-based IL microemulsions. The tribological properties of the designed microemulsions were evaluated using a four-ball tester which confirmed formation of an IL microemulsion containing vegetable oil. The designed petroleum-free microemulsions exhibit excellent viscosity-temperature characteristics and friction−reduction properties, indicating the tremendous potential of IL microemulsions as renewable biolubricant basestocks.

12.2.3.2 Ionic Liquid as Solvent for Enzymatic Esterification

Many enzymes, especially lipases, showed higher activity and greater selectivity in ILs than in organic solvents (Jain et al., 2005). The stability of *C. antarctica* lipase B (CAL-B) in IL was found to be higher than in organic solvents (Romero et al., 2005). The main advantages of IL and enzyme systems are milder, lower reaction temperature compared with conventional chemical synthesis; it needs less energy investment, can be better-controlled,

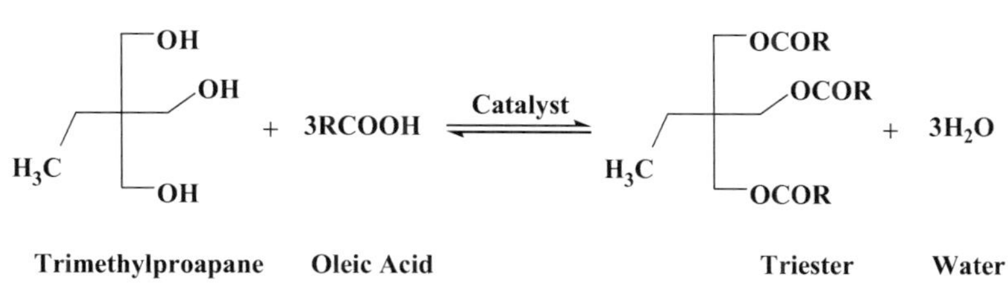

FIGURE 12.6 Esterification of TMP and OA over $[PyBS]_3PW_{12}O_{40}$ catalyst.

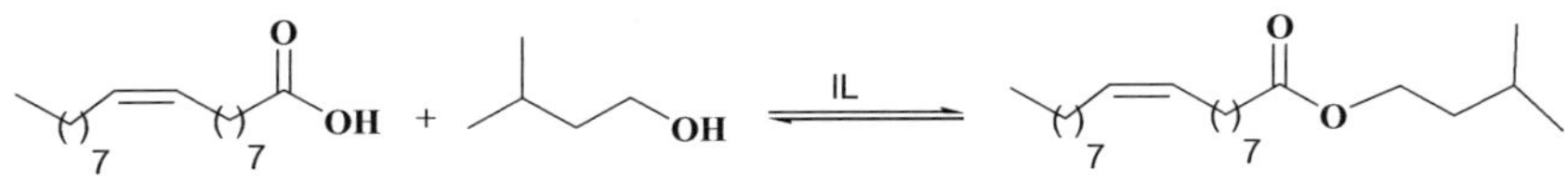

FIGURE 12.7 Preparation of isoamyl-oleate.

and attended with less side-reaction; furthermore, the product is more readily separable. ILs are not only green solvents, they can have a catalytic effect. In order to investigate separately the enzyme stabilization effect, two of the ILs (Cyphos-105 and Cyphos-110) as a mixture were used for the preparation of isoamyl-oleate (Fig. 12.7) biolubricant (Banyai et al., 2011). From this study it was found that the phosphonium-type ILs namely, Cyphos-201 and Cyphos-104, greatly catalyze the esterification process without the presence of enzymes. They have also reported that an ester-type biolubricant could be prepared from isoamyl-alcohol and oleic acid by a lipase enzyme in an IL two-phase system with higher conversion in shorter time when compared to a solvent-free system.

12.2.3.3 Ionic Liquids as Lubricant Additives

12.2.3.3.1 Ionic Liquid in Water

One remarkable characteristic of ILs is their organization in solution, with polar and nonpolar domains and their miscibility with polar and nonpolar solvents (Dupont, 2007; Consorti et al., 2005). ILs have been used as lubricating additives both in water and in lubricating oil. Ye et al. studied the interactions of the IL lubricants with water and reported that the addition of a small amount of water ($\leq$5 wt%) to the ILs improve the antiwear behavior for some metal–metal contacts (Ye et al., 2001). When used as additives in water (Phillips and Zabinski, 2004; Omotowa et al., 2004), ILs reduced the running-in period (the initial period of high friction) of ceramic–ceramic sliding contacts. The proposed mechanism was the initial smoothening of the surface by mechanical wear, the formation of BFx and PFx films on the surface, and the creation of an electric double layer of IL molecules, which increases the local viscosity near the surface and the load carrying ability and lowers friction and wear.

12.2.3.3.2 Ionic Liquids as Mineral Oil Additive

1-*n*-Alkyl-3-methylimidazolium ILs with different alkyl side chains on the cations and different fluorine-containing anions have been studied as mineral oil additives (Jimenez et al., 2006). The more effective neat IL lubricants were those containing anions [BF_4^-] and cations with longer alkyl chains: octyl-([$OMIM^+$]) better than hexyl-([$HMIM^+$]) and better than ethyl-([$EMIM^+$]). In contrast, when used as additives, the more effective wear reducing ILs are those with shorter alkyl chains: [$EMIM^+$] better than [$HMIM^+$], better than [$OMIM^+$]. For the same cation ([$EMIM^+$]) and different anions, the wear rate of aluminum decreases in the order: $MeC_6H_4SO_3^- > CF_3SO_3^- > BF^{4-}$.

12.2.3.3.3 ILs as Synthetic Oil Additive

1-*n*-Alkyl-3-methylimidazolium ILs, with different side chain lengths and the fluorine-containing anions hexafluorophosphate [PF_6^-], tetrafluoroborate [BF_4^-] and triflate [$CF_3SO_3^-$] have also been studied as 1 wt% additives of the synthetic ester propylene glycol dioleate in steel-aluminum contacts at 25 and 100°C (Jimenez and Bermudez, 2008). At 100°C, all additives reduce both friction and wear with respect to the base oil. Friction coefficients for IL additives are similar or lower than for neat ILs, while wear rates for 1 wt% ILs can be several orders of magnitude lower than those for neat ILs. In this case the contact conditions for optimum lubrication were the same for all additives independently of composition. Dicationic bis(imidazolium) ILs with the same long side-chain substituted cation and different anions were evaluated as additives in polyethylene glycol at room temperature (Yao et al., 2009). Results showed that they could effectively reduce the friction and wear of steel—steel sliding pairs compared with the base oil without additives. The dicationic IL with [$(CF_3SO_2)_2N^-$] anion showed better antiwear properties for an optimum concentration of 3 wt%. IL additives not only reduce the friction and wear with respect to base oils, but they show better performance than neat ILs. The excellent tribological properties of IL additives are attributed to the formation of physically adsorbed films, formation of tribochemical products during friction without tribocorrosion, and good miscibility with the base oil.

12.3 PREPARATION OF POLYMERS AND PLASTICIZERS

Vegetable oils represent an interesting renewable source for the production of useful chemicals and new materials (Biermann et al., 2000). Polymers derived from renewable resources are gaining great interest due in part to the rarefaction of fossil raw materials (Raquez et al., 2010). The chemistry based on vegetable oil feedstock fell into disuse during the last decades with the emergence of synthetic polymers. Beside polysaccharides, polypeptides, and so on, lipids are a class of versatile materials with complex compositions that can be transformed by chemical processes and open the way to new materials (Guner et al., 2006) such as polyurethanes (Boutevin et al., 1990, 1991) and polyamides (Rybak and Meier, 2007). Vegetable oils containing unsaturated fatty acids possess double bonds, which are the reactive sites for coatings and paints; they need to be functionalized by the introduction of epoxy, hydroxyl, or carboxyl groups in order to be used for preparation of polymeric materials. Epoxy-based resins are widely used as adhesives and as matrices in composite materials because of their good physical and chemical properties. Epoxidized vegetable oils prepared from the most unsaturated oils such as soybean oil or linseed oil can be used for such purpose.

12.3.1 Preparation of Polymers and Plasticizers with Conventional Catalyst

Much emphasis has been placed on the utilization of renewable resources as an alternative to petrochemical-based feedstocks. In this context, vegetable oils are undoubtedly promising candidates. Not only they are widespread, inexpensive, ecocompatible, renewable, and nonnoxious, but, above all, several multifunctionalized molecules can be obtained by their chemical modification. For example, the unsaturated olefinic bonds in vegetable oils can be epoxidized to give epoxy oils, which have been already used widely as PVC-plasticizers and stabilizers, as reactive diluents for paints, and as additives in lubricants. Traditionally, epoxidized vegetable oils are commercially produced via the Prileshajev epoxidation process or the conventional process by the peroxy acids (Petrovic et al., 2002). Meyer et al. (2008) had reported that vegetable oils could be epoxidized by reacting carbon−carbon double bonds of unsaturated fatty acids with per acids. Dinda et al. worked on the epoxidation kinetics of cottonseed oil using the aqueous H_2O_2 catalyzed by liquid inorganic acids, hydrochloric acid, sulfuric acid, nitric acid, and phosphoric acid. It was found that the degree of conversion to oxirane in cottonseed oil with H_2SO_4 or H_3PO_4 as catalyst were higher than those with HCl and HNO_3. The greatest oxirane conversion ($\sim$70.4%) could be achieved in the shortest period of time (after 6 h) using H_2SO_4 as catalyst compared to those using H_3PO_4 (Dinda et al., 2008). Cai et al. (2008) reported in situ epoxidation of soybean oil, sunflower oil, and corn oil by peroxyacetic acid catalyzed with sulfuric acid. Goud et al. (2006a,b) described the studies on epoxidation of mahua and karanja oil by H_2O_2. Identical to the trend reported by Dinda et al. (2008), the increment of the processing temperature, stirring speed, H_2O_2 concentration, as well as mole portion of acetic acid to ethylenic unsaturation up to their critical level would enhance the degree of oxirane conversion to their optimum level. In terms of the concentration of catalysts, it was again found that H_2SO_4 was the most powerful catalyst in boosting up the epoxidation of vegetable oils compared to those with HNO_3. All the processes described earlier have several drawbacks that require improvement: (1) selectivity to epoxidized products is relatively low due to acid-catalyzed oxirane ring opening side reactions; (2) the separation of acidic by-products, whose presence may be detrimental for further applications, is not easy; and (3) the handling of highly concentrated hydrogen peroxide and strong acids is dangerous and causes corrosion problems. For all these reasons there is considerable interest in finding an alternative method for the epoxidation of these substrates.

With the intention of eliminating the practice of using strong acid as a catalyst, which would lead to side reactions to catalyze peroxy acid formation during the epoxidation process, and to make the process more environmentally friendly (Rosana et al., 2009), Klass reported on the epoxidation of plant oils such as rapeseed, sunflower, soybean, and linseed using an immobilized lipase catalyzed perhydrolysis process (Klass and Warwel, 1999). From an industrial point of view, the most interesting progress in the enzymatic epoxidation of plant oils is the discovery that the lipase enzyme is able to catalyze the epoxidation of alkene analog, including fatty acid derivatives. Lipase (triacylglycerol acylhydrolase E3.1.1.3) is a versatile hydrolase that uses triglycerides (ie, oils and fats) as its natural substrates. The chemoenzymatic epoxidation mediated by lipase was discovered by Novozymes scientists (Bjorkling et al., 1990, 1992; Kirk et al., 1994). According to Biermann et al., Novozym 435 is easy to separate from the reaction media, and it is highly reusable with little loss of enzymatic activity. Additionally, Novozym 435 demonstrated a turnover rate of more than 200,000 mol of products per mole of catalyst (Biermann et al., 2000). Enzymatic epoxidation can be carried out in a solvent (Klaas and Warwel, 1996; Hilker et al., 2001) or in a solvent-free system (Orellana et al., 2005a,b; Hagstrom et al., 2010). A major problem for lipase-mediated epoxidation is the inactivation of the enzyme by H_2O_2 (Tornvall et al., 2009, 2007; Yadav and Devi, 2002). As an alternative, Orellana-Coca et al. (2007) reported that the stepwise addition of the enzyme may also prevent enzymatic inactivation by H_2O_2. Although the side reactions during enzyme-mediated epoxidation are minimal due to the mild reaction conditions, two major side reactions, hydrolysis and dihydroxylation, are still present during the epoxidation of glycerides. To overcome these problems, there is considerable interest in finding an alternative method for the production of biobased polymers and plasticizers by using ILs.

12.3.2 Role of Ionic Liquids in the Preparation of Polymers and Plasticizers

12.3.2.1 Ionic Liquid as Solvent

The asymmetric epoxidation of olefins catalyzed by (salen)Mn^{III} (Song and Roh, 2000) using sodium hypochlorite (NaOCl) as an oxidant in 1-butyl-3-methylimidazolium hexaflurophosphate ([BMIm][PF_6]) can also be implemented with methyltrioxorhenrium as the catalyst in 1-ethyl-3-methylimidazolium tetrafluroborate ([EMIm][BF_4]) using UHP (Owens and Abu-Omar, 2000). Both methods demonstrated that homogeneous catalytic oxidation in ILs provides a cleaner chemical transformation pathway and easier product separation from the reactants compared to epoxidation in organic solvents. A recent study demonstrated that the epoxidation of unsaturated fatty acid methyl esters in the presence of a SO_3H-functional Brønsted acidic IL as a catalyst (Cai and Wang, 2001) resulted in a faster conversion rate compared to the use of sulfuric acid. Additionally, this study demonstrated that the IL catalyst could be recycled up to four times. Cai et al. (2009) investigated the catalytic epoxidation of unsaturated fatty acid in ILs using oxo-diperoxo molybdenum (VI) complex [$MoO(O_2)_2 \cdot 2QOH$

(QOH = 8-quinilinol)] as catalyst and $NaHCO_3$ as cocatalyst. It was found that the addition of IL as a solvent such as [bmim][BF_4], [hydemim][BF_4] and [bmim][PF_6] could enhance the efficiency of the conversion of unsaturated fatty acid during the epoxidation process up to 92%. Also, a higher degree of epoxy conversion would be obtained using IL containing systems compared to those systems with organic solvents.

12.3.2.2 Ionic Liquid as Support

Hydrophobic ILs, namely 1-n-butyl-3-methylimidazolium bis(trifluoromethylsulfonyl)imide [BMIm][NTf_2], 1-n-butyl-3-methylimidazolium hexafluriphosphate [BMIm][PF_6], and hydrophilic IL 1-n-butyl-3-methylimidazolium tetrafluoroborate [BMIm][BF_4] have been reported as efficient supports for the enzyme-catalyzed epoxidation of methyl oleate (biodiesel) (Silva et al., 2011). Nine species of lipases, including *Aspergillus niger* (Amano A) and *C. antarctica* B, were studied. This study revealed that the hydrophobic ILs served as a better media for the *A. niger* lipase catalyzed epoxidization of methyl oleates, which yielded 89% of epoxides after 1 h with a high degree of selectivity. Contrarily, hydrophilic ILs are prone to deprive an enzyme of water and thus, decrease the enzymatic activity (Cai and Wang, 2001; Moreira and Nascimento, 2007; Chiappe et al., 2004; Persson and Bornscheuer, 2003). An additional advantage that was demonstrated by the use of ILs as media was the efficient recovery and reusability of both the enzyme and ILs.

12.4 PREPARATION OF OTHER OLEOCHEMICALS

The sources of oils and fats are various vegetable and animal raw materials (eg, tallow, lard) with the vegetable raw materials soybean, palm, rapeseed, and sunflower oil being the most important ones regarding the amounts involved (Hill, 2000). Of the approximately 101 million tonnes of fats and oils that were produced worldwide in 1998, by far the largest share was used in human foodstuffs. For oleochemistry, 14 million tonnes were available. In recent years, the amounts produced have continuously increased by approximately 3% per year. It is predicted that this trend will continue in the medium and long terms. The composition of the fatty acids contained in the oil (fatty acid spectrum) determines the further use of the oils. Vegetable oils and fats are important constituents of human and animal foodstuffs. Certain grades are industrially used and, together with carbohydrates and proteins, are important renewable resources compared to fossil and mineral raw materials, whose occurrence is finite. In concepts for new products, the price, performance, and product safety criteria are equally important and have a correspondingly high importance right at the start of product development. To ensure a high degree of product safety for consumers and the environment, renewable resources have often been shown to have advantages when compared with petrochemical raw materials and can therefore be regarded as being the ideal raw material basis. Results from oleochemistry show that the use of vegetable fats and oils allows the development of competitive, powerful products, which are both consumer-friendly and environment-friendly. Oils and fats are triglycerides which typically consist of glycerin and saturated and unsaturated fatty acids. There are a few exceptions from this rule, such as castor oil, a glycerol triester of 12-hydroxyoleic acid (ricinoleic acid). From a chemical point of view, triglycerides offer two reactive sites, the double bond in the unsaturated fatty acid chain and the acid group of the fatty acid chain. With regard to product development based on triglycerides, the majority of derivatization reactions is carried out at the carboxylic group (>90%) whereas oleochemical reactions involving the alkyl chain or double bond represent less than 10%. For most of the further uses oils and fats must be split into the so-called oleochemical base materials: fatty acid methyl esters, fatty acids, glycerol, and as hydrogenation products of the fatty acid methyl esters, fatty alcohols.

12.4.1 Role of Ionic Liquids in the Preparation of Other Oleochemicals

Recently, ILs have found increasing attention as new reaction media, especially for homogeneously catalyzed syntheses in liquid–liquid biphasic systems (Sheldon, 2001; Wasserscheid and Keim, 2000). Their physical and chemical properties are very attractive and can be tuned by a proper choice of the organic cation as well as of the anion (Olivier, 1999). Mostly, quaternary nitrogen or phosphorous compounds, such as tetraalkylammonium, tetraalkylphosphonium, *N*-alkylpyridinium, and di- or trialkylimidazolium compounds, are used as cations. Depending on the choice of the anion, ILs may work as catalyst and solvent, especially in the case of Lewis-acidic ILs like choloroaluminates or cholostannates, or as solvent for homogeneous organometallic catalysts. 1,3-Dialkylimidazolium IL has been used for the Diels–Alder reaction of oleochemicals (Behr et al., 2003). Oleic acid and its methyl esters are important oleochemicals used for the development of novel substances. In addition to food, oleic acid and its derivatives are used in many commercial applications, for example, detergents and soaps, lubricants, cosmetics, and emulsifying agents. Due to the wide variety of applications, Biswas et al., developed a straightforward and ecofriendly reaction to functionalize methyl oleate with a model aryl amine moiety. In this study, they used 1-methylimidazolium (MIM) IL as a recyclable catalyst for the selective synthesis of oleate-aniline adduct (Fig. 12.8) without the formation of fatty amide in a solvent-free system (Biswas et al., 2009).

Aliphatic γ-lactones are industrially important oleochemicals having applications as flavor and food additives (Hislop et al., 2004; Schultt et al., 2007). They are also useful synthons for the development of novel

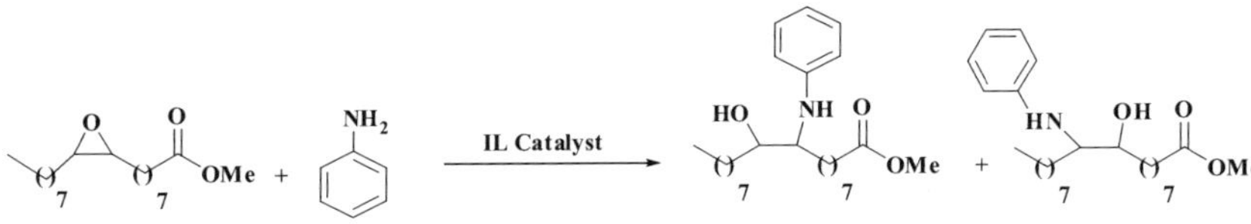

FIGURE 12.8 Methylimidazolium ionic liquid catalyzed synthesis of oleate-aniline adduct.

biodegradable detergents and other organic intermediates (Cermak and Isbell, 2000). The most sustainable method for the synthesis of aliphatic γ-lactone is from unsaturated fatty acid, derived from renewable fats and oils. Fatty acid lactones are usually prepared following a tandem isomerization–lactonization pathway. Shivaraju et al. reported the tandem isomerization–lactonization of unsaturated fatty acids to their corresponding γ-lactones in the presence of a Lewis acidic IL, choline chloride.2ZnCl$_2$ (Fig. 12.9). The IL initially catalyzes the stepwise migration of the double bond along the carbon chain toward the carboxyl group at the Δ4 position, which subsequently undergoes lactonization resulting in the formation of γ-lactones. This one-step process allows the formation of γ-lactone in good yield with little or no formation of δ-lactones (Shivaraju et al., 2012). This IL plays the dual role of solvent as well as catalyst.

Very recently, novel basic ILs, namely alkylguanidinium and alkylamidinium hydroxides, as new basic ILs were reported for the first time by Nowicki et al. (2014) from cyclic guanidine (TBD) and amidines (DBN and DBU) (Fig. 12.10) as new catalysts for the transesterification of oleochemicals. The method of preparation of [bmim]OH described in the literature was adopted for the synthesis of alkylguanidine and alkylamidine bromides using cyclohexane as the solvent (Fig. 12.11). These ILs exhibited good thermal stability and catalytic activity for the transesterification of fatty acid methyl

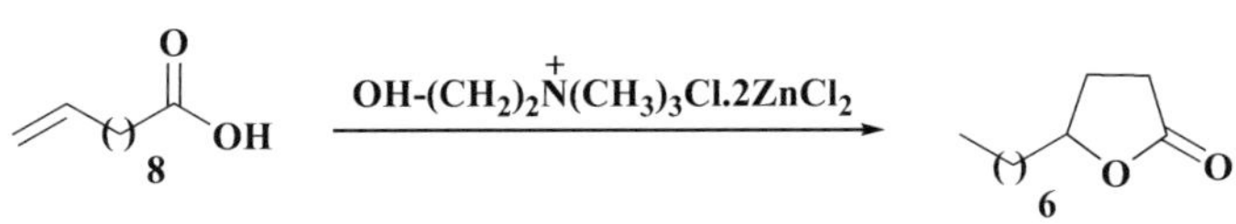

FIGURE 12.9 Tandem isomerization–lactonization of UDA to undecalactone by choline chloride·2ZnCl$_2$.

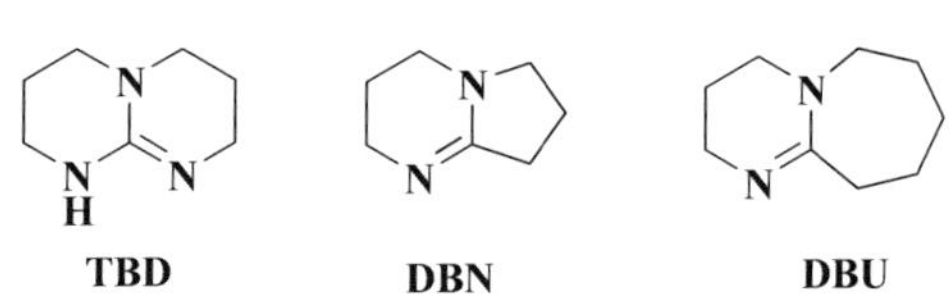

FIGURE 12.10 General structures of cyclic guanidines and amidines.

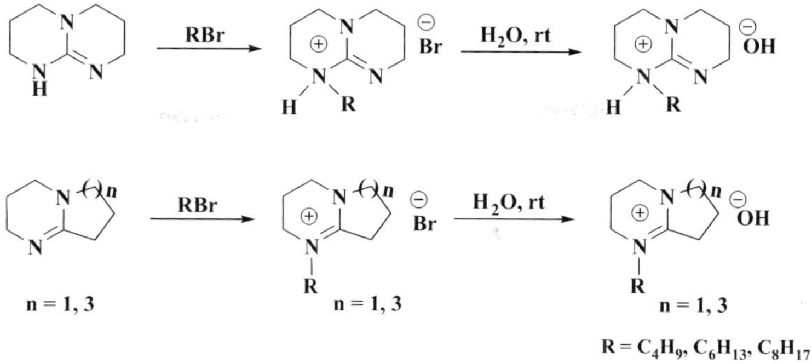

FIGURE 12.11 Synthesis of new ionic liquids from cyclic guanidine and amidines.

esters with trimethylpropane, neopentyl glycol and 2-ethylhexyl alcohol to corresponding esters. The highest conversion of FAME (83%) was obtained for TMP ester of oleic acid with [TBD-Bu]OH IL. Among the ILs derived from cyclic guanidine and amidine, only the ILs obtained from TBD were found to be good catalysts for transesterification of FAME and polyols. In comparison with [bmim]OH, the catalytic activity of TBD IL was comparable in the synthesis of TMP ester and better in the synthesis of 2-ethylhexyl esters.

12.5 CONCLUSION

Over the years, ILs have gained significant attention in the oleochemical industry as a medium and catalyst. The use of ILs in the chemoenzymatic preparation of biodiesel, biolubricants, and other oleochemicals have great potential due to the ease of product isolation from the reaction, the possibility of reusing the catalyst, and the low risk of environmental contamination. These characteristics are of paramount importance to explore the industrial interests of the system. However, there are additional challenges that need to be explored, such as reducing the cost of ILs and increasing their recyclability. ILs have higher cost compared to traditional catalysts and hence, studies that focus on the reduction of the production costs of these ILs and their reuse would increase the feasibility of using this system.

REFERENCES

Akoh, C.C., Chang, S.W., Lee, G.C., Shaw, J.F., 2007. Enzymatic approach to biodiesel production. J. Agric. Food Chem. 55, 8995–9005.

Alexandre, A.M.L., Luciane, F.O., Brenno, A.D.N., Jairton, D., 2008. Ionic liquid supported acid/base-catalyzed production of biodiesel. ChemSusChem. 1, 759–762.

Atabani, A.E., Silitonga, A.S., Ong, H.C., Mahlia, T.M.I., Masjuki, H.H., Irfan Anjum Badruddin, Fayaz, H., 2013. Non-edible vegetable oils: a critical evaluation of oil extraction, fatty acid compositions, biodiesel production, characteristics, engine performance and emissions production. Renew. Sustainable Energy Rev. 18, 211–245.

Banyai, T., Bako, K.B., Nemestothy, N., Gubicza, L., 2011. Biolubricant production in ionic liquids by enzymatic esterification. Hungarian J. Ind. Chem. 39, 395–399.

Bartz, W.J., 2006. Ecotribology: environmentally acceptable tribological practices. Tribol. Int. 39, 728–733.

Bartz, W.J., 1998. Lubricants and the environment. Tribol. Int. 31, 35–47.

Behr, A., Naendrup, F., Nave, S., 2003. Use of ionic liquids for the Diels-Alder reaction of oleochemicals. Eng. Life Sci. 3, 325–327.

Berriosa, M., Skelton, R.L., 2008. Comparison of purification methods for biodiesel. Chem. Eng. J. 144, 459–465.

Betton, C.I., 2010. Lubricants and their environmental impact. In: Mortier, R.M., Fox, M.F., Orszulik, S.T. (Eds.), Chemistry and Technology of Lubricants. Springer, Heidelberg, Germany, pp. 435–457.

Biermann, U., Friedt, W., Lang, S., Luhs, W., Machmüller, G., Metzger, J.O., Klaas, M.R., Schäfer, H.J., Schneider, M.P., 2000. New syntheses with oils and fats as renewable raw materials for the chemical Industry. Angew. Chem. Int. Ed. Engl. 39, 2206–2224.

Biswas, A., Sharma, B.K., Doll, K.M., Erhan, S.Z., Willett, J.L., Cheng, H.N., 2009. Synthesis of an amine-oleate derivative using an ionic liquid catalyst. J. Agric. Food Chem. 57, 8136–8141.

Bjorkling, F., Frykman, H., Godtfredsen, S.E., Kirk, O., 1992. Lipase catalyzed synthesis of peroxycarboxylic acids and lipase mediated oxidations. Tetrahedron 48, 4587–4592.

Bjorkling, F., Godtfredsen, S.E., Kirk, O., 1990. Lipase-mediated formation of peroxycarboxylic acids used in catalytic epoxidation of alkenes. J. Chem. Soc. 19, 1301–1303.

Bondioli, P., 2004. The preparation of fatty acid esters by means of catalytic reactions. Top. Catal. 27, 77–82.

Boutevin, B., Chaib, M., Robin, J.J., 1990. Synthese et application de diols isomolecularies aux resins polyurethanes. Macromol. Chem. Phys. 191, 737–747.

Boutevin, B., Chaib, M., Robin, J.J., 1991. Synthesis of monodispers macromolecular diisocyanate. Polym. Bull. 26, 177–179.

Cai, C., Dai, H., Chen, R., Su, C., Xu, X., Zhang, S., Yang, L.T., 2008. Studies on the kinetics of in situ epoxidation of vegetable oils. Eur. J. Lipid Sci. Technol. 110, 341–346.

Cai, S., Wang, L., 2001. Epoxidation of unsaturated fatty acid methyl esters in the presence of SO_3H-functional Bronsted acidic ionic liquid as catalyst. Chin. J. Chem. Eng. 19, 57–63.

Cai, S.F., Wang, L.S., Fan, C.L., 2009. Catalytic epoxidation of a technical mixture of methyl oleate and methyl linoleate in ionic liquids using $Moo(O_2)_2.2QOH$ (QOH = 8-Quinilinol) as catalyst and $NaHCO_3$ as co-catalyst. Molecules 14, 2935–2946.

Cermak, S.C., Isbell, T.A., 2000. Synthesis of δ–stearolactone from oleic acid. J. Am. Oil Chem. Soc. 77, 243–248.

Chiappe, C., Leandri, E., Lucchesi, S., Pieraccini, D., Hammock, B.D., Morisseau, C., 2004. Biocatalysis in ionic liquids: the stereoconvergent hydrolysis of trans-β-methylstyrene oxide catalyzed by soluble epoxide hydrolase. J. Mol. Catal. B: Enzym. 27, 243–248.

Chiu, C.W., Dasari, M.A., Sutterlin, W.R., Suppes, G.J., 2006. Removal of residual catalyst from simulated biodiesel's crude glycerol for glycerol hydrogenolysis to propylene glycol. Ind. Eng. Chem. Res. 45, 791–795.

Cole, A.C., Jensen, J.L., Ntai, I., Tran, K.L.T., Weaver, K.J., Forbes, D.C., Davis, J.H., 2002. Novel Brønsted acidic ionic liquids and their use as dual solvent-catalysts. J. Am. Chem. Soc. 124, 5962–5963.

Consorti, C.S., Suarez, P.A.Z., de Souza, R.A., Farrar, D.H., Lough, A.J., Loh, W., da Silva, L.H.M., Dupont, J., 2005. Identification of 1,3-dialkylimidazolium salt supramolecular aggregates in solution. J. Phys. Chem. B 109, 4341–4349.

Devi, B.L.A.P., Guo, Z., Xu, X., 2011. Characterization of ionic liquid-based biocatalytic two-phase reaction system for production of biodiesel. AIChE J. 57, 1628–1637.

Diego, T.D., Lozano, P., Gmouh, S., Vaultier, M., Iborra, J.L., 2005. Understanding structure-stability relationships of *Candida antarctica* lipase B in ionic liquids. Biomacromolecules 6, 1457–1464.

Dinda, A., Patwardhan, A.V., Goud, V.V., Pradhan, N.C., 2008. Epoxidation of cottonseed oil by aqueous hydrogen peroxide catalyzed by liquid inorganic acids. Bioresour. Technol. 99, 3737–3744.

Dossat, A., Combes, D., Marty, A., 2002. Efficient lipase catalysed production of a lubricant and surfactant formulation using a continuous solvent-free process. J. Biotechnol. 97, 117–124.

Dupont, J., 2007. On the solid, liquid and solution structural organization of imidazolium ionic liquids. J. Braz. Chem. Soc. 15, 341–350.

Earle, M.J., Seddon, K.R., 2000. Ionic liquids. Green solvents for the future. Pure Appl. Chem. 72, 1391–1398.

Earle, M.J., Esperanca, J.M.S.S., Gilea, M.A., Canongia Lopes, J.N., Rebelo, L.P.N., Magee, J.W., Seddon, K.R., Widegren, J.A., 2006. The distillation and volatility of ionic liquids. Nature 439, 831–834.

Elsheikh, Y.A., Man, Z., Bustam, M.A., Suzana, Y., Wilfred, C.D., 2011. Brønsted imidazolium ionic liquids: synthesis and comparison of their catalytic activities as pre-catalyst for biodiesel production through two stage process. Energy Convers. Manage 52, 804–809.

Elsheikh, Y.A., 2014. Optimization of novel pyrazolium ionic liquid catalysts for trans-esterification of bitter apple oil. Process. Saf. Environ. Prot. 92, 828–834.

Fang, D., Gong, K., Shi, Q., Liu, Z., 2007. A green procedure for the protection of carbonyls catalyzed by novel task-specific room-temperature ionic liquid. Catal. Commun. 8, 1463–1466.

Formo, M.W., 1954. Ester reactions of fatty materials. J. Am. Oil Chem. Soc. 31, 548–559.

Frederique, R.A., Melquizedeque, B.A., Caio, C.S.M., Luiz, F.Z., Paulo, A.Z.S., 2005. New multi-phase catalytic systems based on tin compounds active for vegetable oil transesterificaton reaction. J. Mol. Catal. A: Chem. 227, 263–267.

Freedman, B., Pryde, E.H., Mounts, T.L., 1984. Variables affecting the yields of fatty esters from transesterified vegetable oils. J. Am. Oil Chem. Soc. 61, 1638–1643.

Fukuda, H., Hama, S., Tamalampudi, S., Noda, H., 2008. Whole-cell biocatalysts for biodiesel fuel production. Trends Biotechnol. 26, 668–673.

Gamba, M., Lapis, A.A.M., Dupont, J., 2008. Supported ionic liquid enzymatic catalysis for the production of biodiesel. Adv. Synth. Catal. 350, 160–164.

Ghiaci, M., Aghabarari, B., Habibollahi, S., Gil, A., 2011. Highly efficient Brønsted acidic ionic liquid-based catalysts for biodiesel synthesis from vegetable oils. Bioresour. Technol. 102, 1200–1204.

Goud, V.V., Patwardhan, A.V., Pradhan, N.C., 2006a. Studies on the epoxidation of mahua oil (*Madhumica indica*) by hydrogen peroxide. Bioresour. Technol. 97, 1365–1371.

Goud, V.V., Pradhan, N.C., Patwardhan, A.V., 2006b. Epoxidation of karanja (*Pongamia glabra*) oil by H_2O_2. J. Am. Oil Chem. Soc. 83, 635–640.

Gryglewicz, S., Piechocki, W., Gryglewicz, G., 2003. Preparation of polyol esters based on vegetable and animal fats. Bioresour. Technol. 87, 35–39.

Gui, J., Cong, X., Liu, D., Zhang, X., Hu, Z., Sun, Z., 2004. Novel Brønsted acidic ionic liquid as efficient and reusable catalyst system for esterification. Catal. Commun. 5, 473–477.

Guner, F.S., Yagci, Y., Erciyes, A.T., 2006. Polymers from triglyceride oils. Prog. Polym. Sci. 31, 633–670.

Hagstrom, A.E.V., Tornall, U., Nordblad, M., Hatti-Kaul, R., Woodley, J.M., 2010. Chemo-enzymatic epoxidation process options for improving biocatalytic productivity. Biotechnol. Progr 27, 67–76.

Hayyan, A., Hashim, M.A., Hayyan, M., Mjalli, F.S., Hayyan, M., Nashef, I.M.A., 2013a. A novel ammonium based eutectic solvent for the treatment of free fatty acid and synthesis of biodiesel fuel. Ind. Crops Prod. 46, 392–398.

Hayyan, A., Hashim, M.A., Mjalli, F.S., Hayyan, M., Nashef, I.M.A., 2013b. A novel phosphonium-based deep eutectic catalyst for biodiesel production from industrial low grade crude palm oil. Chem. Eng. Sci. 92, 81–88.

He, H., Wang, T., Zhu, S., 2007. Continuous production of biodiesel fuel from vegetable oil using supercritical methanol process. Fuel 86, 442–447.

Hilker, I., Bothe, D., Pruss, J., Warnecke, H.J., 2001. Chemo-enzymatic epoxidation of unsaturated plant oils. Chem. Eng. Sci. 56, 427–432.

Hill, K., 2000. Fats and oils as oleochemical raw materials. Pure Appl. Chem. 72, 1255–1264.

Hislop, J.A., Hunt, M.B., Fielder, S., Rowan, D.D., 2004. Synthesis of deuterated γ-lactones for use in stable isotope dilution assays. J. Agric. Food Chem. 52, 7075–7083.

Jaidi, A., Gainer, J.L., Carta, G., Mrani, A., Kadiri, T., Belarbi, Y., Mir, A., 2002. Esterification of fatty acids using nylon-immobilized lipase in *n*-hexane: kinetic parameters and chain-length effects. J. Biotechnol. 93, 209–216.

Jain, N., Kumar, A., Chauhan, S., Chauhan, S.M.S., 2005. Chemical and biochemical transformations in ionic liquids. Tetrahedron 61, 1015–1060.

Jimenez, A.E., Bermudez, M.D., 2008. Imidazolium ionic liquids as additives of the synthetic ester propylene glycol dioleate in aluminium-steel lubrication. Wear 265, 787–798.

Jimenez, A.E., Bermudez, M.D., Carrion, F.J., Martinez-Nicolas, G., 2006. Room temperature ionic liquids as lubricant additives in steel-aluminium Contacts: Influence of sliding velocity, normal load and temperature. Wear 261, 347–359.

Kiriliauskaite, V., Bendikiene, V., Juodka, B., 2011. Synthesis of trimethylolpropane esters of oleic acid by Lipoprime 50T. J. Ind. Microbiol. Biotechnol. 38, 1561–1566.

Kirk, O., Christensen, M.W., Damhus, T., Godtfredsen, S.E., 1994. Enzyme catalyzed degradation and formation of peroxycarboxylic acids. Biocatalysis 11, 65–77.

Klaas, M.R., Warwel, S., 1996. Chemoenzymatic epoxidation of unsaturated fatty acid esters and plant oils. J. Am. Oil Chem. Soc. 73, 1453–1457.

Klass, M.R., Warwel, S., 1999. Complete and partial epoxidation of plant oils by lipase-catalyzed perhydrolysis. Ind. Crops Prod. 9, 125–132.

Kumar Jain, A., Suhane, A., 2013. Biotechnology: a way to control environmental pollution by alternative lubricants. Res. Biotechnol. 4, 38–42.

Li, R.J., Chen, L., Yan, Z.C., 2012. Synthesis of trimethylpropane esters of oleic acid using a multi-SO_3H-functionalized ionic liquid as an efficient catalyst. J. Am. Oil Chem. Soc. 89, 705–711.

Liu, L., Kong, Y., Xu, H., Li, J.P., Dong, J.X., Lin, Z., 2008. Ionothermal synthesis of a three-dimensional zinc phosphate with DFT topology using unstable deep-eutectic solvent as template-delivery agent. Microporous Mesoporous Mater. 115, 624–628.

Linko, Y.Y., Tervakangas, T., Lamsa, M., Linko, P., 1997. Production of trimethylolpropane esters of rapeseed oil fatty acids by immobilized lipase. Biotechnol. Tech. 11, 889–892.

Meyer, P.P., Techaphattana, N., Manundawee, S., Sangkeaw, S., Junlakan, W., Tongurai, C., 2008. Epoxidation of soybean oil and jatropha oil. Thammasat Int. J. Sc. Tech. 13, 1–5.

Moreira, M.A., Nascimento, M.G., 2007. Chemo-enzymatic epoxidation of (+)-3-carene. Catal. Commun. 8, 2043–2047.

Nagendramma, P., Kaul, S., 2012. Development of ecofriendly/biodegradable lubricants: an overview. Renew. Sustainable Energy Rev. 16, 764–774.

Nowicki, J., Muszynski, M., Gryglewicz, S., 2014. Novel basic ionic liquids from cyclic guanidines and amidines-new catalysts for transesterification of oleochemicals. J. Chem. Technol. Biotechnol. 89, 48–55.

Olivier, H., 1999. Recent developments in the use of non-aqueous ionic liquids for two-phase catalysis. J. Mol. Catal. A: Chem. 146, 285–289.

Omotowa, B.A., Phillips, B.S., Zabinski, J.S., Shreeve, J.M., 2004. Phosphazene-based ionic liquids: synthesis, temperature-dependent viscosity, and effect as additives in water lubrication of silicon nitride ceramics. Inorg. Chem. 43, 5466–5471.

Orellana, C.C., Billakanti, J.M., Mattiasson, B., Hatti-Kaul, R., 2007. Lipase mediated simultaneous esterification and epoxidation of oleic acid for the production of alkylepoxystearates. J. Mol. Catal. B: Enzym. 44, 133–137.

Orellana, C.C., Camocho, S., Adlercreutz, D., Mattiasson, B., Hatti-Kaul, R., 2005a. Chemo-enzymatic epoxidation of linoleic acid: parameter influencing the reaction. Eur. J. Lipid Sci. Technol. 107, 864–870.

Orellana, C.C., Tornvall, U., Adlercreutz, D., Mattiasson, B., Hatti-Kaul, R., 2005b. Chemo-enzymatic epoxidation of oleic acid and methyl oleate in solvent-free medium. Biocatal. Biotransform. 23, 431–437.

Owens, G.S., Abu-Omar, M.M., 2000. Methyltrioxorhenium-catalyzed epoxidations in ionic liquids. Chem. Commun. 13, 1165–1166.

Persson, M., Bornscheuer, U.T., 2003. Increased stability of an esterase from *Bacillus stearothermophilus* in ionic liquids as compared to organic solvents. J. Mol. Catal. B: Enzym. 22, 21–27.

Petrovic, Z.S., Zlatanic, A., Lava, C.C., Sinadinovic, F.S., 2002. Epoxidation of soybean oil in toluene with peroxoacetic and peroxoformic acids-kinetics and side reactions. Eur. J. Lipid Sci. Technol. 104, 293–299.

Phillips, B.S., Zabinski, J.S., 2004. Ionic liquid lubrication effects on ceramics in a water environment. Tribol. Lett. 17, 533–541.

Plechkova, N.V., Seddon, K.R., 2008. Applications of ionic liquids in the chemical industry. Chem. Soc. Rev. 37, 123–150.

Raquez, J.M., Deleglise, M., Lacrampe, M.F., Krawczak, P., 2010. Thermosetting (bio) materials derived from renewable resources: a critical review. Prog. Polym. Sci. 35, 487–509.

Rios, A.P., Fernández, F.J.H., Gómez, D., Rubio, M., Víllora, G., 2011. Biocatalytic transesterification of sunflower and waste cooking oils in ionic liquid media. Process. Biochem. 46, 1475–1480.

Romero, M.D., Calvo, L., Alba, C., Danesfar, A., Ghaziaskar, H.S., 2005. Enzymatic synthesis of isoamyl acetate with immobilized *Candida antarctica* lipase in *n*-hexane. Enzym. Microb. Technol 37, 42–48.

Rosana, C.S.S., Luciano, R.S.L., Thiago, B.B., Maria, G.N., Marta, R.S.N., 2009. Chemo-enzymatic epoxidation of Sunflower oil methyl esters. J. Braz. Chem. Soc. 20, 1473–1477.

Rudnick, L.R., Erhan, S.Z., 2006. Natural oils as lubricants. In: Rudnick, L.R. (Ed.), Synthetics, Mineral Oils and Bio-Based Lubricants. CRC Taylor and Francis, Boca Raton, pp. 353–360.

Ruzich, N.I., Bassi, A.S., 2010. Investigation of enzymatic biodiesel production using ionic liquid as a co-solvent. Can. J. Chem. Eng. 88, 277–282.

Rybak, A., Meier, M.A.R., 2007. Cross-metathesis of fatty acid derivatives with methyl acrylate: renewable raw materials for the chemical Industry. Green Chem. 9, 1356–1361.

Schuchardt, U., Sercheli, R., Vargas, R.M., 1998. Transesterification of vegetable oils: a review. J. Braz. Chem. Soc. 9, 199–210.

Schultt, B., Moran, N., Schieberle, P., Holmann, T., 2007. Sensory-directed identification of creaminess-enhancing volatiles and semivolatiles in full-fat cream. J. Agric. Food Chem. 55, 9634–9645.

Sheldon, R., 2001. Catalytic reactions in ionic liquids. Chem. Commun. 23, 2399–2407.

Shen, J., Wang, H., Liu, H., Sun, Y., Liu, Z., 2008. Brønsted acidic ionic liquids as dual catalyst and solvent for environmentally friendly synthesis of chalcone. J. Mol. Catal. A: Chem. 280, 24–28.

Shivaraju, A., Phani Kumar, P., Prasad, R.B.N., Kanjilal, S., 2012. Tandem isomerization-lactonization of olefinic fatty acids using the Lewis acidic ionic liquid, choline chloride·$2ZnCl_2$. Tetrahedron Lett. 53, 3471–3473.

Silva, W.S.D., Lapis, A.A.M., Suarez, P.A.Z., Neo, B.A.D., 2011. Enzyme-mediated epoxidation of methyl oleate supported by imidazolium-base ionic liquids. J. Mol. Catal. B: Enzym. 68, 98–103.

Song, C.E., Roh, E.J., 2000. Practical method to recycle chiral (salen)-Mn epoxidation catalyst by using an ionic liquid. Chem. Commun. 10, 837–838.

Tornvall, U., Borjesson, P., Tufvesson, L.M., Hatti-Kaul, R., 2009. Biocatalytic production of fatty epoxides from rapeseed and tall oil derivatives: process and environmental evaluation. Ind. Biotechnol. 5, 184–192.

Tornvall, U., Orellana, C.C., Hatti-Kaul, R., Adlercreutz, D., 2007. Stability of immobilized *Candida antarctica* lipase B during chemo-enzymatic epoxidation of fatty acids. Enzym. Microb. Technol. 40, 447–451.

Wang, A., Chen, L., Jiang, D., Yan, Z.C., 2013. Vegetable oil-based ionic liquid microemulsions and their potential as alternative renewable biolubricants basestocks. Ind. Crops Prod. 51, 425–429.

Wasserscheid, P., Keim, W., 2000. Ionic liquids-new solutions for transition metal catalysis. Angew. Chem. Int. Ed. 39, 3772–3789.

Welton, T., 1999. Room temperature ionic liquids. Solvents for synthesis and catalysis. Chem. Rev. 99, 2071–2083.

Wilkes, J.S., Wasserscheid, P., Welton, T., 2008. Introduction. In: Wasserscheid, P., Welton, T. (Eds.), Ionic Liquids in Synthesis. Wieley-VCH, Weinheim, Germany, pp. 1–6.

Wu, Q., Chen, H., Han, M., Wang, D., Wang, J., 2007. Transesterification of cottonseed oil catalyzed by Brønsted acidic ionic liquids. Ind. Eng. Chem. Res. 46, 7955–7960.

Xie, C., Li, H., Li, L., Yu, S., Liu, F., 2008. Synthesis of plasticizer ester using acid-functionalized ionic liquid as catalyst. J. Hazard. Mater. 151, 847–850.

Yadav, G.D., Devi, K.M., 2002. Enzymatic synthesis of perlauric acid using Novozym 435. Biochem. Eng. J. 10, 93–101.

Yadav, G.D., Lathi, P.S., 2003. Kinetics and mechanism of synthesis of butyl isobutyrate over immobilized lipases. Biochem. Eng. J. 16, 245–252.

Yan, Z., Chen, L., Wang, H., Li, K., 2011. Biological lubricant prepared with ionic liquid as catalyst and preparation method. Eur. Pat. Appl. CN20101286234 20100916.

Yao, M., Liang, Y., Xia, Y., Zhou, F., 2009. Bisimidazolium ionic liquids as the high-performance antiwear additives in poly(ethylene glycol) for steel-steel contacts. ACS Appl. Mater. Interfaces 1, 467–471.

Ye, C.F., Liu, W.M., Chen, Y.X., Yu, L.G., 2001. Room-temperature ionic liquids: a novel versatile lubricant. Chem. Commun. 21, 2244–2245.

Yori, J.C., D'Ippolito, S.A., Pieck, C.L., Vera, C.R., 2007. Deglycerolization of biodiesel streams by adsorption over silica beds. Energy Fuels 21, 347–353.

Zhang, L., Cui, Y., Zhang, C., Wang, L., Wan, H., Guan, G., 2012. Biodiesel production by esterification of oleic acid over Brønsted acidic ionic liquid supported onto Fe-incorporated SBA-15. Ind. Eng. Chem. Res. 51, 16590–16596.

Zhao, D., Wu, M., Kou, Y., Min, E., 2002. Ionic liquids: applications in catalysis. Catal. Today 74, 157–189.

Zhao, H., Baker, G.A., 2013. Ionic liquids and deep eutectic solvents for biodiesel synthesis: a review. J. Chem. Technol. Biotechnol. 88, 3–12.

Zhao, H., Baker, G.A., Holmes, S., 2011. New eutectic ionic liquids for lipase activation and enzymatic preparation of biodiesel. Org. Biomol. Chem. 9, 1908–1916.

Chapter 13

Toxicity of Ionic Liquids: Past, Present, and Future

Diego O. Hartmann, Cristina Silva Pereira
Universidade Nova de Lisboa, Instituto de Tecnologia Química e Biológica António Xavier, Oeiras, Portugal

Ionic liquids (ILs) have been described as molten salts that have, by definition, a melting point below 100°C (Stark and Seddon, 2007). They are entirely ionic in nature and usually consist of a large asymmetrical organic cation associated with a polyatomic anion that may be either organic or inorganic. In the design of ILs, the asymmetry of the cation or anion is preferred over symmetrical species, since the former do not tend to pack so easily, therefore rendering salts with a lower melting point (Stark and Seddon, 2007). Structural modifications can be made either to the anion, to the cation, or in substituents on these two species. Therefore, millions of formulations become possible and, by altering their cationic or anionic components, their properties can be tailored (Plechkova and Seddon, 2007; Stark and Seddon, 2007). Their physical and chemical properties (eg, melting point, viscosity, solubility, or hydrophobicity and chemical polarity or hydrogen bonding ability, respectively) can be tuned to best suit a specific process. The tunability of such properties rendered ILs an excellent choice as alternative solvents in many chemical processes. They already found applications in many reactions (Plechkova and Seddon, 2008), as solvents (Holbrey and Seddon, 1999; Plechkova and Seddon, 2008) and catalysts (Cole et al., 2002), and show great potential in electrochemistry (Armand et al., 2009; Sakaebe et al., 2007).

ILs' physical and chemical properties, together with their negligible vapor pressure and bulk nonflammability, suggested them as potential green solvents (Deetlefs and Seddon, 2006, 2010; Dong et al., 2007). Their negligible volatility distinguishes them from traditional organic solvents, reducing risk of atmospheric pollution. However, the potential release of IL vapors should be considered when used at elevated temperatures (Ranke et al., 2007a,b). Nonetheless, soil and water contamination, either by accidental spills or waste disposal, constitutes a serious hazard. The ecotoxicological assessment of this class of compounds should be considered a priority, and has been taken into

Ionic Liquids in Lipid Processing and Analysis. http://dx.doi.org/10.1016/B978-1-63067-047-4.00013-1

account for the conscious design of novel ILs (Petkovic et al., 2011; Petkovic and Silva Pereira, 2012).

13.1 PAST: EXTENSIVE STUDIES ON IONIC LIQUID TOXICITY

Interest in IL toxicity began in the early 2000s. Pernak et al. (2001a,b) developed new cationic surfactants with benzimidazolium or pyridinium cores and evaluated their antimicrobial properties (minimal inhibitory and lethal concentrations against bacterial and yeast strains relevant for human health). The evaluation of the antimicrobial activity of 1-alkoxymethyl-(3-nicotionylaminomethyl)benzimidazolium, 1-alkoxymethyl-3-(1-benzimidazolmethylamino)-pyridinium, 1-alkoxymethyl-3-[1-(benzotriazol-1-yl)methylamino]pyridinium chlorides, and *N,N'*-bis[3-(1-alkoxymethyl)pyridinium chloride]methylenediamine revealed that their toxicity increased with the elongation of the alkoxy chain, which varied between 2 and 12 carbon atoms. This constituted one of the first systematic studies on the toxicity of ILs, and served afterward as a basis for evaluating the toxicity of many newly synthesized imidazolium- and pyridinium-based compounds (Fig. 13.1), including 1-alkoxymethylcarbamoylpyridinium chlorides (Pernak et al., 2001a,b), 1-alkyl-3-methylimidazolium chlorides, and bromides; 1-alkyl-3-hydroxyethyl-2-methylimidazolium chlorides (Demberelnyamba et al., 2004), 1-alkoxymethyl-3-methylimidazolium chlorides, tetrafluoroborates and hexafluorophosphates (Pernak et al., 2003), 1,3-dialkoxymethylimidazolium chlorides (Pernak et al., 2004a,b,c), 1-alkyl- and 1-alkoxy-methylimidazolium lactates (Pernak et al., 2004a,b,c), 1-alkoxymethyl-3-hydroxypyridinium, and 1-alkoxymethyl-3-dimethylaminopyridinium chlorides (Pernak and Branicka, 2003). The data revealed a clear trend toward a stronger toxic effect of ILs with the increase in the length of the alkyl or the alkoxy side chain. Further studies revealed that the incorporation of an oxygen atom in the side chain significantly reduced the toxicity of the ILs (Morrissey et al., 2009). Symmetrical chains in positions R^1 and R^3 of the imidazolium ring, however, resulted in lower toxicity, probably due to steric effects with the cell surface (Pernak et al., 2004a,b,c). The effect of anions on the observed toxicities seemed to be secondary to the effect of the cations, but their broad diversity did not allow a conclusive analysis.

These initial observations inspired an escalating number of studies on toxicity of several ILs, expanding the research for other biological models, particularly model eukaryotic organisms. For example, the toxicity of ILs with 1-alkyl-3-methylimidazolium as cation were analyzed against leukemia and glioma rat cell lines (Ranke et al., 2004), human HeLa cell lines (Stepnowski et al., 2004), and the worm *Caenorhabditis elegans* (Swatloski et al., 2004). Additionally, the toxicity of pyridinium-based ILs has been analyzed using freshwater snail *Physa acuta* (Bernot et al., 2005), the zebra mussel *Dreissena polymorpha* (Costello et al., 2009), and green algae *Pseudokirchneriella subcapitata* (Pham et al., 2008), focusing on the effect of the alkyl side chain

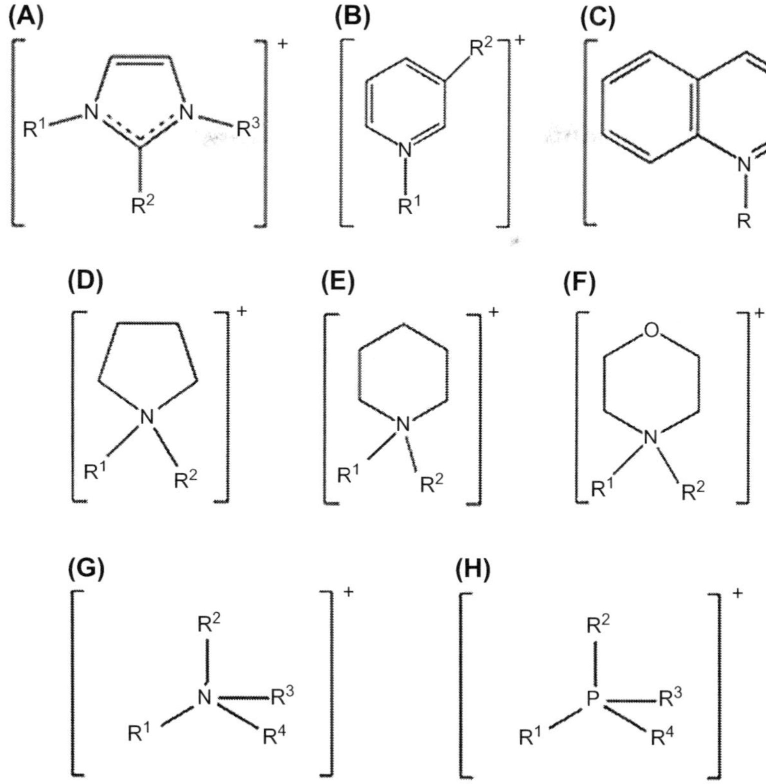

FIGURE 13.1 Common IL cation cores and some of their possible structural modifications: (A) imidazolium, (B) pyridinium, (C) quinilinium, (D) pyrrolidinium, (E) piperidinium, (F) morpholinium, (G) quaternary ammonium, and (H) quaternary phosphonium cations.

of the pyridinium cation. The effects of imidazolium- and pyridinium-based ILs in the bioluminescence of the bacterium *Vibrio fischeri* were also analyzed (Docherty and Kulpa, 2005). Bacterial bioluminescence is directly linked to cellular respiration, so decrease in luminescence is indicative of toxicity.

Other cationic head groups were also evaluated on their toxic effects, such as the aromatic quinolinium-based cations (Fig. 13.1). Despite rarely being studied, these ILs showed great antimicrobial properties and cytotoxicity potential that increased with elongation of substituted alkyl chain (Busetti et al., 2010; Ranke et al., 2007a,b). The effects of ILs with nitrogen-containing alicyclic cations, namely pyrrolidinium, piperidinium, and morpholinium (Fig. 13.1), were also investigated; for example, their aquatic toxicity against the zebrafish *Danio rerio* (Pretti et al., 2006) and their cytotoxicity against mammal cell lines (Kumar et al., 2009; Stolte et al., 2007a,b). It was observed that increasing numbers of carbon atoms in an alicyclic ring generally increases toxicity; for example, the piperidinium cation (six-member rings) was more toxic than the pyrrolidinium cation (five-member rings) (Kumar et al., 2009). The morpholinium ILs, due to incorporation of an oxygen atom in the ring, were the least toxic of these alicyclic cations. Furthermore, the nonaromatic head groups appeared to be generally less toxic than their aromatic equivalents (Kumar et al., 2009; Stolte et al., 2007a,b). Their lower toxicity, relative to the corresponding aromatic rings, was generally evident in the bioassays with models such as *V. fischeri*, the algae *Scenedesmus vacuolatus*, and the aquatic plant *Lemna minor* (Stolte et al., 2007a,b). As observed for imidazolium- and pyridinium-based ILs, the length of the alkyl chains in the cation played a major role in toxicity of piperidinium- and pyrrolidinium-based ILs (Kumar et al., 2009; Ranke et al., 2007a,b). Anions have been considered to have an unpredictable effect, largely reflecting the lack of systematization in their ecotoxicological assessment.

Quaternary ammonium salts (Fig. 13.1) have been largely studied and explored in numerous applications, such as disinfectants, surfactants, antistatic agents, and catalysts (Jones, 2001). Their properties depend on chain length and functional groups of anion (Ying, 2006). Pernak and coauthors were the first to consider their antimicrobial activity against relevant clinical strains, which was, as expected, governed by length of the alkoxy and alkyl side chains (Pernak and Chwała, 2003; Pernak and Feder-Kubis, 2005). The toxicity of ammonium ILs against other biological models was also tested, including bioluminescent bacterium *V. fischeri* (Couling et al., 2006), fresh water crustacean *Daphnia magna*, and algae *P. subcapitata* (Pretti et al., 2009; Wells and Coombe, 2006), zebrafish (Pretti et al., 2009), and cell lines (Ranke et al., 2007a,b). The same toxicity trends previously described for other cationic head groups were observed; the highest toxicity was detected for those with longest alkyl chains and could be well correlated with the cation lipophilicity.

Some of the most interesting groups of quaternary ammonium-based ILs are those containing the cholinium cation (2-hydroxyethyl-trimethylammonium). Its combination with benign anions constituted a major advance in the design of biocompatible ILs (Abbott et al., 2001, 2002, 2003, 2004; Abbott and Davies, 2000; Fukaya et al., 2007; Pernak et al., 2007; Renshaw, 1909). The low toxicity of many cholinium ILs has been demonstrated, including those carrying as anions; for example, saccharinate and acesulfamate (Nockemann et al., 2007), dimethylphosphate (Dipeolu et al., 2009), phosphate-based anions (Weaver et al., 2010), lactates (Petkovic et al., 2009), and alkanoates (Petkovic et al., 2010). The last were analyzed by our group for the first time against filamentous fungi of the genus *Penicillum*, and included a range of linear alkanoate anions and two structural isomers (Petkovic et al., 2010). Their toxicity increased with the elongation of the anion, with the branched isomers less toxic than corresponding linear ones with an equal number of carbon atoms. These cholinium alkanoates also displayed high biodegradability potential and great solvent ability (Garcia et al., 2010). These findings have been strengthened by other groups, which also evaluated their cytotoxicity (Klein et al., 2013).

The research focus on toxicity of ILs is moving toward other cationic groups aside imidazolium- and pyridinium-based ones, such as the quaternary phosphonium cations (Fig. 13.1). These ILs are currently produced in tonne quantities, particularly by Cytec Industries Inc. (Bradaric et al., 2003), further highlighting the need for their ecotoxicological assessment, which is still seldom investigated. Some halides (eg, with the tetrabutylphosphonium cation) showed levels of toxicity against the freshwater snail *P. acuta* that were comparable to imidazolium-based ILs carrying the same chain length (Bernot et al., 2005). The antimicrobial properties of a series of alkyltrihexylphosphonium halides demonstrated a significant role of the structure of the cation in their toxicity, since the antimicrobial activity decreased for the longest alkyl chains (8–14 carbon atoms) (Cieniecka-Rosłonkiewicz et al., 2005). The apparently high toxicity of tetraalkylphosphonium halides against *V. fischeri*, and *D. magna* (Couling et al., 2006; Wells and Coombe, 2006), and *P. subcapitata* (Cho et al., 2008; Wells and Coombe, 2006) was also demonstrated, as well as their cytotoxicity against human and rat cell lines (Ranke et al., 2007a,b, Wang et al., 2007). The lack of systematization in these studies, however, did not allow a conclusive rationalization, a general problem when evaluating the toxicity of the highly diverse groups of ILs (Petkovic et al., 2011; Petkovic and Silva Pereira, 2012).

These initial studies were essential to establish the foundations for evaluating the ecotoxicity of ILs, particularly for understanding how the cationic or anionic head groups and their structural modifications impact toxicity. Despite the massive amount of significant data produced so far, evidence of their specific mechanisms of toxicity is still largely lacking. In-depth knowledge of their toxicity is critical for developing safer IL based-processes and may provide unforeseen opportunities in IL frontier research.

13.2 PRESENT: UNDERSTANDING THE MOLECULAR BASIS OF IONIC LIQUID TOXICITY

The hydrophobic interactions between ILs and biological membranes have been proposed as a main mechanism of toxicity of ILs, largely supported by the observation that their toxic effects increase with the elongation of the side chains. This nonspecific toxicity, called baseline toxicity or narcosis (Mackay et al., 2009), appears to be correlated with 1-octanol/water partition coefficients. This is suggestive of an increase in either the cation or anion lipophilicity with elongation of the side chains (Stolte et al., 2007a,b), leading to greater probability of interaction with biological membranes. Liposomes (ie, phospholipid vesicles) were used to assess the effects of some ILs in cellular membranes (Schaffran et al., 2009). As an example, the 1-octyl-3-methylimidazolium cation was demonstrated to cause severe disruptions in a supported phospholipid bilayer (Evans, 2008). Molecular dynamics simulation studies also tried to demonstrate the interactions of ILs with artificial lipid bilayers (Klein et al., 2013; Ruegg et al., 2014). These elegant models are, however, unable to reproduce the complexity of the heterogeneous lipid bilayers of biological membranes. Therefore, real biological systems, with all their intrinsic complexity, are required to evaluate the true effects of ILs.

The lack of systematic studies on membrane permeabilization as a mechanism for toxicity of ILs was the driving force of our work. We pioneered the use of filamentous fungi ecotoxicological assessment of ILs (Petkovic et al., 2009). They were shown to tolerate, in general, very high concentrations of ILs. Moreover, their high catabolic capacities render them as great candidates for biodegradation studies. We have successfully established all the methodologies necessary to evaluate the effects of ILs against filamentous fungi, excellent model organisms for moving beyond the state-of-the-art on ILs toxicity.

The first ILs subjected to in-depth investigation of their mechanisms of toxicity was a series of highly toxic alkyltributylphosphonium chlorides, $[P_{4\,4\,4\,n}]Cl$, where $n = 1$, 4–8, 10, 12, or 14. A fluorescence microscopy assessment demonstrated that these ILs were able to permeabilize the plasma membrane of the spores of the model filamentous fungus *Aspergillus nidulans* (Fig. 13.2) (Petkovic et al., 2012). Their toxicity was observed to increase with systematic elongation of one alkyl substituent, consistent with the increase in their 1-octanol/water partition coefficient. The data clearly suggested that the toxicity of $[P_{4\,4\,4\,n}]Cl$, where $n \geq 4$, is ruled by direct interaction with the plasma membrane. Further confirmation of these effects was attained by analyzing IL effects at a molecular level, in particular by measuring the expression levels of genes involved in plasma membrane biosynthesis in *A. nidulans* (Hartmann and Silva Pereira, 2013). The data suggested that the fungus alters the plasma membrane fluidity in response to the membrane permeabilization provoked by the ILs.

Both fluorescence microscopy and gene expression analysis were proven to be powerful and rapid tools to investigate in vivo the ability of ILs to permeabilize biological membranes. Accordingly, these methods were used to investigate the toxic effects of other important families of ILs. The imidazolium-based ILs are probably the most investigated group and one of the first to find application on an industrial scale (Plechkova and Seddon, 2008). The cholinium-based ILs are currently gaining great interest for the development of biocompatible ILs, mainly due to the benign nature of the cholinium cation (Klein et al., 2013; Pernak and Chwała, 2003; Pernak et al., 2007; Petkovic et al., 2010). We have chosen representative families of each of these IL groups, namely 1-alkyl-3-methylimidazolium chlorides ($[C_n\text{mim}]Cl$, where $n = 2$, 4, 6, 8, or 10) and cholinium alkanoates (anion = ethanoate, butanoate, hexanoate, octanoate, or decanoate), to perform a systematic investigation of their toxic effects and mechanisms of toxicity. As previously observed for the phosphonium-based ILs, both 1-alkyl-3-methylimidazolium chlorides and cholinium alkanoates present the same trend of toxicity against *A. nidulans*: their toxicity increased with the elongation of the alkyl

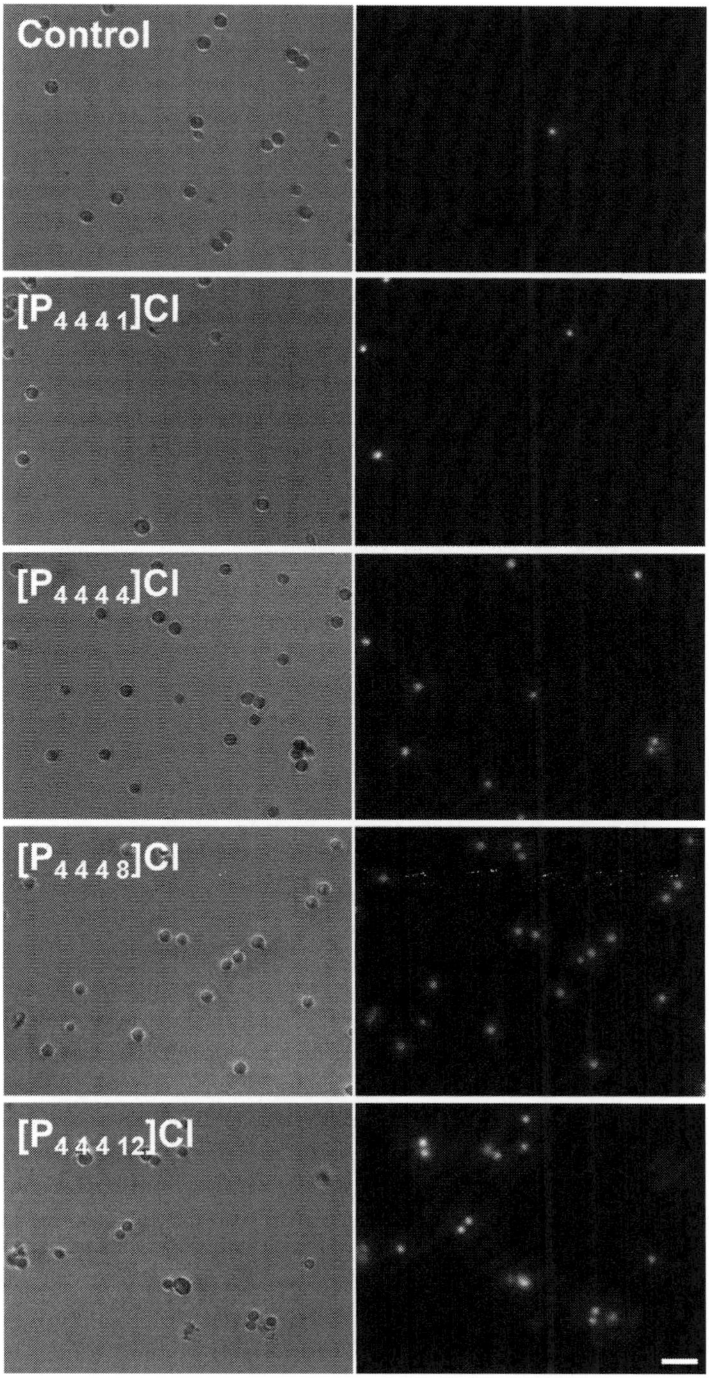

FIGURE 13.2 Fluorescence microscopy of asexual spores of *Aspergillus nidulans* exposed to alkyltributylphosphonium chlorides ($[P_{4\,4\,4\,n}]Cl$, $n = 1$, 4, 8, or 12) for 1 h and stained with propidium iodide, which only stains cells with permeabilized membranes. The left column shows the total number of spores and the right column shows membrane-damaged spores. The percentage of damaged cells increases with the elongation of the alkyl chain in the cation. Scale bar: 20 μm. *Adapted from Petkovic, M., Hartmann, D.O., Adamová, G., Seddon, K.R., Rebelo, L.P.N., Silva Pereira, C., 2012. Unravelling the mechanism of toxicity of alkyltributylphosphonium chlorides in* Aspergillus nidulans *conidia. New J. Chem. 36 (1), 56–63.*

chain either in the cation or the anion, respectively. However, while imidazolium-based ILs appear to be toxic and recalcitrant to degradation, cholinium alkanoates present low toxicity and are readily biodegradable (Petkovic et al., 2010).

Furthermore, 1-alkyl-3-methylimidazolium chlorides with long alkyl chains ([$C_n$mim]Cl, where $n \geq 6$) were able to permeabilize the fungal plasma membrane, leading to cell death, while cholinium alkanoates could not cause such an effect (Hartmann et al., 2015). These differences were most likely due to the nature of each IL family. The plasma membrane is charged along its surface, due to the different composition of neutral and negatively charged head groups of its composing phospholipids (Alberts et al., 2002). Toxicity of 1-alkyl-3-methylimidazolium chlorides seems to closely relate to the elongation of the alkyl chain in the cation, while for cholinium alkanoates, increasing number of carbons in the anion appears to rule their toxicity. Most probably, the negatively charged phospholipids assist the interaction of the imidazolium-based cations with the plasma membrane, allowing its permeabilization, while the anions from cholinium alkanoates are likely repelled. The cholinium cation, on the other hand, can interact with the surface of the plasma membrane, but its bulky nature does not allow membrane permeabilization. Elongating one of the alkyl chains of the cholinium cation, however, enables plasma membrane permeabilization. This could be observed for a series of alkyl-(2-hydroxyethyl)-dimethylammonium bromides ([$N_{1\ 1\ n\ 2OH}$]Br, where $n = 2$, 4, 6, 8, 10 or 12) (Hartmann et al., 2015). The analyses of the toxicity of these cholinium derivatives reveal that this family of ILs presented membrane-permeabilizing effects but maintains the low toxicity and the biodegradability of the archetype cholinium-based ILs. These observations constitute a unique asset for the development of new biocompatible ILs: linear elongation of alkyl chains should be preferentially introduced in the anions, in order to avoid permeabilization of biological membranes.

Interaction with the plasma membrane and its disruption is most likely a nonspecific mechanism of toxicity of ILs, probably observed for all cellular organisms. In addition to the plasma membrane the fungal cell envelope comprises another complex structure, the cell wall. The cell wall of *A. nidulans* presents an inner layer composed of polysaccharides, mainly chitin, 1,3-β-glucans, and 1,3-α-glucan (de Groot et al., 2009). The glucan network serves as a support for an outer layer of mannoproteins. The presence of these mannoproteins confers an overall negative charge to the cell wall that might also have an implication on the effects of ILs in fungi. Employing fluorescence microscopy, gene expression analysis, and scanning electron microscopy, we analyzed the ability of ILs to cause damage to the fungal cell wall and observed that alkyltributylphosphonium chlorides, in fact, damage this cellular structure, regardless of the alkyl chain length (Fig. 13.3) (Hartmann and Silva Pereira, 2013; Petkovic et al., 2012). Other ILs with long

FIGURE 13.3 Scanning electron microscopy analysis of asexual spores of *Aspergillus nidulans* exposed to alkyltributylphosphonium chlorides ($[P_{4\,4\,4\,n}]Cl$, $n = 1, 8$, or 12) for 2 h. Images reveal that ILs cause severe damage to the cell wall and surface of the cells. Micrographs on the left and on the right have 5000× and 15,000× magnification, respectively. Scale bars: 1 μm. *Adapted from Petkovic, M., Hartmann, D.O., Adamová, G., Seddon, K.R., Rebelo, L.P.N., Silva Pereira, C., 2012. Unravelling the mechanism of toxicity of alkyltributylphosphonium chlorides in* Aspergillus nidulans *conidia. New J. Chem. 36 (1), 56–63.*

alkyl chains in the cation, namely 1-alkyl-3-methylimidazolium chlorides and alkyl-(2-hydroxyethyl)-dimethylammonium bromides, also induce cell wall damage (unpublished data). Cholinium alkanoates, however, do not cause such effects. It seems possible that the negative charge of the cell wall also played a pivotal role in the effects of ILs upon this cellular structure.

Additional studies were undertaken by our group to further investigate the mechanisms used by filamentous fungi to tolerate ILs. We analyzed the mycelial proteome of two model fungi upon exposure to 1-ethyl-3-methylimidazolium chloride and cholinium chloride, which were selected since they present opposite toxicity and recalcitrance toward fungi (Martins et al., 2013). Data obtained showed that numerous stress-responsive proteins (eg, anti-ROS defense proteins and drug transporter proteins) as well as several critical biological processes and/or pathways were affected by either IL. Among other changes, these compounds altered developmental program in both fungi, for example, promoting the formation of Hülle cells (involved in sexual development of the fungus) or asexual sporulation, and led to accumulation of osmolytes, namely glycerol in *Neurospora crassa* and betaine in *A. nidulans*. This is consistent with the observation that addition of betaine to the bacterium *Enterobacter lignolyticus* growth media supplemented with 0.375 M of 1-ethyl-3-methylimidazolium chloride increased biomass production (Khudyakov et al., 2012). Interestingly, the proteomic study also shed light on the mechanism of toxicity of cholinium-based ILs. Major uptake of the cholinium cation led to the production of the toxic compound cyanide, which might be a cause of the observed growth inhibition provoked by this group of ILs (Martins et al., 2013). Another remarkable finding was the accumulation of a protein likely involved in the biosynthesis of unusual amino acids in *N. crassa* (unpublished data). In some fungi, these rare amino acids are involved in the formation of unusual metabolites, such as neoefrapeptins and acretocins, which usually show broad antimicrobial activity (Bruckner and Theis, 2010; Fredenhagen et al., 2006). Further tests are ongoing to confirm the presence of metabolites potentially carrying interesting biological activity. These results clearly demonstrate the importance of unraveling the mechanisms of toxicity of ILs at a molecular level, revealing that their full potential remains largely overlooked.

13.3 FUTURE: MOVING BEYOND TOXICITY TOWARD NEW POTENTIAL APPLICATIONS

From the knowledge obtained by the ecotoxicological assessment of ILs and their mechanisms of toxicity, two paths are possible. The first is the development of new green, safer ILs and their application in environmentally friendly processes. As an example, the biocompatible and biodegradable cholinium alkanoates developed by our group (Petkovic et al., 2010) have been observed to have a great solvent ability toward the plant polymer suberin

(Ferreira et al., 2012; Garcia et al., 2010; Petkovic et al., 2010). Cholinium hexanoate was successfully employed in the extraction of suberin from cork where it plays the dual roles of solvent and mild catalyst (Ferreira et al., 2014). This ensured the preservation of the native structure of the polymer, which allowed, for the first time, the formation of suberin films with antibacterial properties (Garcia et al., 2014). Moreover, deep knowledge on IL toxicity inspired the development of engineered bacteria carrying phenotypes with increased resistance to particular ILs in order to promote their efficient use in biorefineries and biofuel production (Khudyakov et al., 2012; Ruegg et al., 2014).

The second possible path is to explore ILs in biological applications, in which their unique toxic properties become useful. A good example of their early applications was the innovative use of 1-alkoxymethyl-3-methylimidazolium tetrafluoroborate in embalming and tissue preservation (Majewski et al., 2003), exploring their antimicrobial activity as a replacement for the highly volatile and toxic formalin, and as a wood preservative (Pernak et al., 2004a,b,c). The potential use of ILs as drugs for important etiological agents is also under current investigation. For example, imidazolium-based ILs have been explored for their potential in combating *Plasmodium falciparum* (Vlahakis et al., 2010) and *Trypanosoma cruzi* (Faral-Tello et al., 2014), the causative agents of malaria and Chagas disease, respectively. However, their cytotoxicity still hampers their clinical usage. A better example is choline-geranate that has been proposed as a coadjutant in the treatment of wound infections, due to its antimicrobial activity, minimal toxicity to epithelial cells as well as skin, and effective permeation enhancement for drug delivery (Zakrewsky et al., 2014).

Probably one of the most interesting and promising applications of ILs is their combination with active pharmaceutical ingredients (APIs) (Marrucho et al., 2014). These pure liquid salt forms of active pharmaceutical ingredients (API-ILs) are considered a novel strategy to overcome existing problems with solid-state APIs. This strategy potentially allows the liquefaction of the drug and when combined with adequate ionic counterparts can enhance its delivery. Several drugs have already been successfully synthesized in their API-IL form, such as the antiinflammatory ibuprofen, the antibiotic amoxicillin, and the anesthetic procaine (Marrucho et al., 2014). Progress in this field requires full compliance of the new API-ILs with the extant legislation. ILs have also been explored to increase the solubility of insoluble drugs and enhance their topical and transdermal delivery (Dobler et al., 2013; Shamshina et al., 2013). Our group has ongoing side projects that aim at using IL formulations to tune the solubility of well-established cytotoxic antifungal drugs, while establishing proof-of-concept that their mechanism of activity remains unaltered.

Rather unexplored is the potential of ILs to induce metabolic alterations and stress responses in living organisms that might be of great interest in biological sciences. Our studies on the mechanisms of toxicity of ILs revealed that some ILs not only cause plasma membrane permeabilization and cell wall damage, but also seem to induce alteration in lipid metabolism in filamentous fungi, more specifically in the synthesis of sphingolipids (unpublished data). Fungi respond to damage to the cell wall by activating several genes involved in its biosynthesis, creating conditions that allow them to reestablish its integrity. This response is regulated by a signaling cascade better understood in the yeast *Saccharomyces cerevisiae*, called cell wall integrity pathway (Levin, 2005). This pathway is also known to be present in a variety of filamentous fungi, but many knowledge gaps in their constitution and regulation call for further investigation. It is thought that *A. nidulans* possesses an alternative pathway for cell wall integrity regulation, still uncharacterized (Fujioka et al., 2007). Exposure to alkyltributylphosphonium chlorides, as an example, led to cell wall damage and was believed to induce the activation of an alternative cell wall integrity pathway (Hartmann and Silva Pereira, 2013). This scientific hypothesis is tremendously exciting, especially when considering the significance of this pathway as a potential pathogen-specific target for the development of new antifungal drugs. Importantly, there are only three major classes of antifungal drugs in clinical use and the current pace of drug development is inadequate to fight the threat of emerging fungal pathogens and multidrug resistance (Ostrosky-Zeichner et al., 2010).

Interfaces of the cell wall integrity pathway with other signaling pathways in *S. cerevisiae* have been observed, including the sphingolipid metabolism (Levin, 2005). This, however, has not yet been elucidated in the filamentous fungus *A. nidulans*. We have observed that some ILs that cause membrane permeabilization and cell wall damage also activate the sphingolipid biosynthetic pathway and induce accumulation of certain sphingolipid intermediates, such as sphingoid bases (unpublished data). It is possible that these molecules are acting as second messengers for the response to cell wall stress induced by ILs. Curiously, each family of ILs tested seemed to induce distinct profiles of accumulation of sphingoid bases, including unknown species (Fig. 13.4). ILs' unique stimuli could therefore be explored to activate cell wall integrity pathways in filamentous fungi and help establishing a connection of this signaling pathway with sphingolipid metabolism. Interestingly, cholinium alkanoates cannot permeabilize the plasma membrane nor cause cell wall damage, but still activate sphingolipid biosynthesis and lead to the accumulation of sphingoid bases, including unknown species. Whether these molecules are involved in the response of *A. nidulans* to the toxic effects of cholinium alkanoates remains unclear. The distinct profiles of accumulation of sphingoid bases displayed by each IL family further highlight the uniqueness of IL stimuli. It evidences that these compounds can be explored to further investigate the roles of sphingolipids and their intermediates as signaling molecules in filamentous fungi. Further investigations require the identification of the unknown intermediates seen to accumulate upon exposure to distinct ILs. Most of our studies focus on

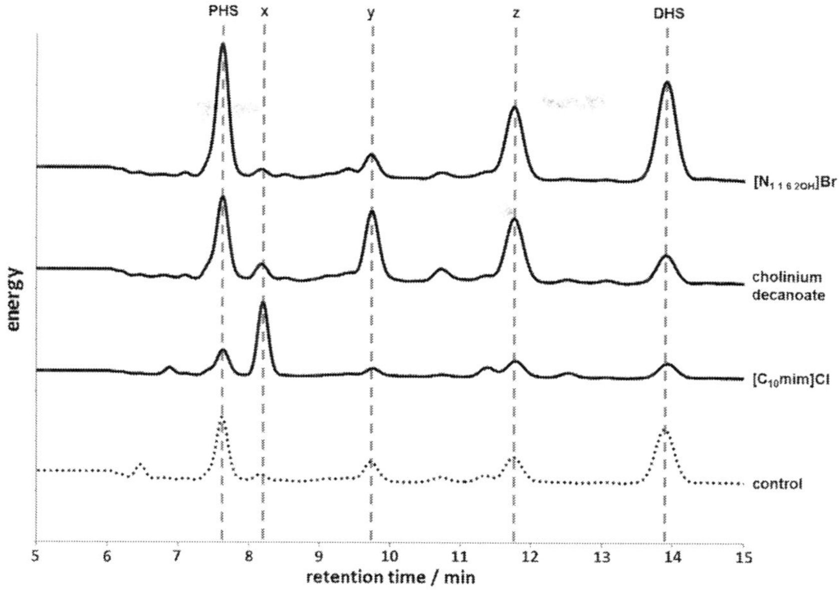

FIGURE 13.4 Chromatographic profile (high-performance liquid chromatography) of the accumulation of sphingoid bases in *Aspergillus nidulans* after exposure to three representative ILs: 1-decyl-methylimidazolium chloride ([C_{10}mim]Cl), cholinium decanoate, or hexyl-(2-hydroxyethyl)-dimethylammonium bromide ([$N_{1\ 1\ 6\ 2OH}$]Br). The control without ILs is indicated with a dashed line. Phytosphingosine (PHS) and dihydrosphingosine (DHS) are the sphingoid bases known to be produced by this fungus. **x**, **y**, and **z** represent unknown sphingoid bases with retention times at 8.2, 9.75, and 11.75 min, respectively (unpublished data).

A. nidulans, a model fungus that is phylogenetically closer to several emerging human-threatening fungal pathogens than *S. cerevisiae*.

The recent demonstration that ILs might occur in nature also opens unexpected paths of their impact in biological sciences (Chen et al., 2014). This recent study revealed that a protic IL is formed during the confrontation of two ant species, *Solenopsis invicta* and *Nylanderia fulva*, probably as a form of defense mechanism against the venom of the former ant species. In line with the previous, another good example of their unexpected impacts in biological systems is the demonstration by our group that ILs alter the metabolic footprint in filamentous fungi (Petkovic et al., 2009). This study suggested that fungal cultures respond to specific ILs by changing their cell biochemistry, resulting in an altered pattern of extracellular metabolites. The hypothesis that these newly formed compounds might play uncharacterized roles in fungi deserves further investigation (Petkovic and Silva Pereira, 2012).

Pathogenic and opportunistic filamentous fungi are responsible for numerous fungal infections and are emerging as causative agents in a broad diversity of clinical conditions (Brown et al., 2012). Furthermore, resistance to antifungal drugs is now an increasing risk to the lives of millions of individuals

(Anderson, 2005). Hence, the current challenge is to better understand the biology of filamentous fungi and identify novel targets for antifungal drugs. As our group gets closer to understanding the primary effects of each IL family and move toward the identification of their specific gene targets, the vision of developing antifungal ILs and materials by taking advantage of elegant progresses in this field might become a reality. The innovative frontier research in the cross-disciplinary field of ILs might provide us unforeseen means to address these global concerns.

All these new data can shed a new light on the studies of ILs toxicity. These compounds show a tremendous potential to induce unique stress responses in living organisms. Hence, ILs can in turn be explored as powerful tools in biological sciences, bringing a new boost of interest on IL research.

ACKNOWLEDGMENTS

D.O.H. is grateful to Fundação para a Ciência e a Tecnologia (FCT)—Portugal for the fellowship SFRH/BD/66396/2009. The work was partially supported by FCT Grants: PTDC/AAC-CLI/119100/2010 and PTDC/QUI—QUI/120982/2010.

REFERENCES

Abbott, A.P., Capper, G., Davies, D.L., Munro, H., Rasheed, R.K., Tambyrajah, V., 2003. Electrochemical studies of ambient temperature ionic liquids based on choline chloride. ACS Symp. Ser. 856, 439—452.

Abbott, A.P., Capper, G., Davies, D.L., Rasheed, R.H., Tambyrajah, V., 2002. Quaternary ammonium zinc- or tin-containing ionic liquids: water insensitive, recyclable catalysts for Diels—Alder reactions. Green Chem. 4 (1), 24—26.

Abbott, A.P., Capper, G., Davies, D.L., Rasheed, R.K., Tambyrajah, V., 2001. Novel ambient temperature ionic liquids for zinc and zinc alloy electrodeposition. Trans. Inst. Met. Finish 79, 204—206.

Abbott, A.P., Davies, D.L., 2000. Ionic Liquids Prepared as Low Melting Salts and Compounds of Quaternary Ammonium Halides with Metal Halides. World.

Abbott, A.P., Davies, D.L., Jenkins, P., 2004. Reactions in ionic liquids based on choline chloride. Spec. Chem. 24, 36—37.

Alberts, B., Johnson, J., Lewis, J., Raff, M., Roberts, K., Walter, P., 2002. Molecular Biology of the Cell, fourth ed. Garland Science, New York.

Anderson, J.B., 2005. Evolution of antifungal-drug resistance: mechanisms and pathogen fitness. Nat. Rev. Microbiol. 3 (7), 547—556.

Armand, M., Endres, F., MacFarlane, D.R., Ohno, H., Scrosati, B., 2009. Ionic-liquid materials for the electrochemical challenges of the future. Nat. Mat. 8 (8), 621—629.

Bernot, R.J., Kennedy, E.E., Lamberti, G.A., 2005. Effects of ionic liquids on the survival, movement, and feeding behavior of the freshwater snail, *Physa acuta*. Environ. Toxicol. Chem. 24 (7), 1759—1765.

Bradaric, C.J., Downard, A., Kennedy, C., Robertson, A.J., Zhou, Y.H., 2003. Industrial preparation of phosphonium ionic liquids. Green Chem. 5 (2), 143—152.

Brown, G.D., Denning, D.W., Gow, N.A., Levitz, S.M., Netea, M.G., White, T.C., 2012. Hidden killers: human fungal infections. Sci. Trans. Med. 4 (165), 165rv13.

Bruckner, H., Theis, C., 2010. Sequences of the polypeptide antibiotics (peptaibiotics) acretocins. J. Pept. Sci. 16, 141.

Busetti, A., Crawford, D.E., Earle, M.J., Gilea, M.A., Gilmore, B.F., Gorman, S.P., Laverty, G., Lowry, A.F., McLaughlin, M., Seddon, K.R., 2010. Antimicrobial and antibiofilm activities of 1-alkylquinolinium bromide ionic liquids. Green Chem. 12 (3), 420–425.

Chen, L., Mullen, G.E., Le Roch, M., Cassity, C.G., Gouault, N., Fadamiro, H.Y., Barletta, R.E., O'Brien, R.A., Sykora, R.E., Stenson, A.C., et al., 2014. On the formation of a protic ionic liquid in nature. Angew. Chem. Int. Ed. 53 (44), 11762–11765.

Cho, C.W., Jeon, Y.C., Pham, T.P.T., Vijayaraghavan, K., Yun, Y.S., 2008. The ecotoxicity of ionic liquids and traditional organic solvents on microalga *Selenastrum capricornutum*. Ecotoxicol. Environ. Saf. 71 (1), 166–171.

Cieniecka-Rosłonkiewicz, A., Pernak, J., Kubis-Feder, J., Ramani, A., Robertson, A.J., Seddon, K.R., 2005. Synthesis, anti-microbial activities and anti-electrostatic properties of phosphonium-based ionic liquids. Green Chem. 7 (12), 855–862.

Cole, A.C., Jensen, J.L., Ntai, I., Tran, K.L.T., Weaver, K.J., Forbes, D.C., Davis, J.H., 2002. Novel Brønsted acidic ionic liquids and their use as dual solvent-catalysts. J. Am. Chem. Soc. 124 (21), 5962–5963.

Costello, D.M., Brown, L.M., Lamberti, G.A., 2009. Acute toxic effects of ionic liquids on zebra mussel (*Dreissena polymorpha*) survival and feeding. Green Chem. 11 (4), 548–553.

Couling, D.J., Bernot, R.J., Docherty, K.M., Dixon, J.K., Maginn, E.J., 2006. Assessing the factors responsible for ionic liquid toxicity to aquatic organisms via quantitative structure–property relationship modeling. Green Chem. 8 (1), 82–90.

Deetlefs, M., Seddon, K.R., 2006. Ionic liquids: fact and fiction. Chim. Oggi Chem. Today 24 (2), 16–23.

Deetlefs, M., Seddon, K.R., 2010. Assessing the greeness of some typical laboratory ionic liquids preparations. Green Chem. 12, 17–30.

Demberelnyamba, D., Kim, K.S., Choi, S.J., Park, S.Y., Lee, H., Kim, C.J., Yoo, I.D., 2004. Synthesis and antimicrobial properties of imidazolium and pyrrolidinonium salts. Bioorg. Med. Chem. 12 (5), 853–857.

Dipeolu, O., Green, E., Stephens, G., 2009. Effects of water-miscible ionic liquids on cell growth and nitro reduction using *Clostridium sporogenes*. Green Chem. 11 (3), 397–401.

Dobler, D., Schmidts, T., Klingenhofer, I., Runkel, F., 2013. Ionic liquids as ingredients in topical drug delivery systems. Int. J. Pharm. 441 (1–2), 620–627.

Docherty, K.M., Kulpa, C.F., 2005. Toxicity and antimicrobial activity of imidazolium and pyridinium ionic liquids. Green Chem. 7 (4), 185–189.

Dong, Q., Muzny, C.D., Kazakov, A., Diky, V., Magee, J.W., Widegren, J.A., Chirico, R.D., Marsh, K.N., Frenkel, M., 2007. ILThermo: a free-access web database for thermodynamic properties of ionic liquids. J. Chem. Eng. Data 52 (4), 1151–1159.

Evans, K.O., 2008. Supported phospholipid bilayer interaction with components found in typical room-temperature ionic liquids—a QCM-D and AFM study. Int. J. Mol. Sci. 9 (4), 498–511.

Faral-Tello, P., Liang, M., Mahler, G., Wipf, P., Robello, C., 2014. Imidazolium compounds are active against all stages of *Trypanosoma cruzi*. Int. J. Antimicrob. Agents 43 (3), 262–268.

Ferreira, R., Garcia, H., Sousa, A.F., Petkovic, M., Lamosa, P., Freire, C.S.R., Silvestre, A.J.D., Rebelo, L.P.N., Pereira, C.S., 2012. Suberin isolation from cork using ionic liquids: characterisation of ensuing products. New J. Chem. 36 (10), 2014–2024.

Ferreira, R., Garcia, H., Sousa, A.F., Guerreiro, M., Duarte, F.J.S., Freire, C.S.R., Calhorda, M.J., Silvestre, A.J.D., Kunz, W., Rebelo, L.P.N., et al., 2014. Unveiling the dual role of the cholinium hexanoate ionic liquid as solvent and catalyst in suberin depolymerisation. RSC Adv. 4 (6), 2993–3002.

Fredenhagen, A., Molleyres, L.P., Bohlendorf, B., Laue, G., 2006. Structure determination of neoefrapeptins A to N: peptides with insecticidal activity produced by the fungus *Geotrichum candidum*. J. Antibiot. 59 (5), 267–280.

Fujioka, T., Mizutani, O., Furukawa, K., Sato, N., Yoshimi, A., Yamagata, Y., Nakajima, T., Abe, K., 2007. MpkA-dependent and -independent cell wall integrity signaling in *Aspergillus nidulans*. Eukaryot. Cell 6 (8), 1497–1510.

Fukaya, Y., Iizuka, Y., Sekikawa, K., Ohno, H., 2007. Bio ionic liquids: room temperature ionic liquids composed wholly of biomaterials. Green Chem. 9 (11), 1155–1157.

de Groot, P.W.J., Brandt, B.W., Horiuchi, H., Ram, A.F.J., de Koster, C.G., Klis, F.M., 2009. Comprehensive genomic analysis of cell wall genes in *Aspergillus nidulans*. Fungal Genet. Biol. 46, S72–S81.

Garcia, H., Ferreira, R., Martins, C., Sousa, A.F., Freire, C.S.R., Silvestre, A.J.D., Kunz, W., Rebelo, L.P.N., Silva Pereira, C., 2014. Ex situ reconstitution of the plant biopolyester suberin as a film. Biomacromolecules 15 (5), 1806–1813.

Garcia, H., Ferreira, R., Petkovic, M., Ferguson, J.L., Leitão, M.C., Gunaratne, H.Q.N., Seddon, K.R., Rebelo, L.P.N., Silva Pereira, C., 2010. Dissolution of cork biopolymers by biocompatible ionic liquids. Green Chem. 12 (3), 367–369.

Hartmann, D.O., Shimizu, K., Siopa, F., Leitão, M.C., Afonso, C.A.M., Canongia Lopes, J.N., Silva Pereira, C., 2015. Plasma membrane permeabilisation by ionic liquids: a matter of charge. Green Chem. 17 (9), 4587–4598.

Hartmann, D.O., Silva Pereira, C., 2013. A molecular analysis of the toxicity of alkyltributylphosphonium chlorides in *Aspergillus nidulans*. New J. Chem. 37, 1569–1577.

Holbrey, J.D., Seddon, K.R., 1999. Ionic liquids. Clean Prod. Process. 1, 223–236.

Jones, R.A., 2001. Quaternary Ammonium Salts: Their Use in Phase-Transfer Catalysis, second ed. Academic Press, New York.

Khudyakov, J.I., D'haeseleer, P., Borglin, S.E., DeAngelis, K.M., Woo, H., Lindquist, E.A., Hazen, T.C., Simmons, B.A., Thelen, M.P., 2012. Global transcriptome response to ionic liquid by a tropical rain forest soil bacterium, *Enterobacter lignolyticus*. Proc. Natl. Acad. Sci. U.S.A. 109 (32), E2173–E2182.

Klein, R., Muller, E., Kraus, B., Brunner, G., Estrine, B., Touraud, D., Heilmann, J., Kellermeier, M., Kunz, W., 2013. Biodegradability and cytotoxicity of choline soaps on human cell lines: effects of chain length and the cation. RSC Adv. 3 (45), 23347–23354.

Kumar, R.A., Papaïconomou, N., Lee, J.M., Salminen, J., Clark, D.S., Prausnitz, J.M., 2009. In vitro cytotoxicities of ionic liquids: effect of cation rings, functional groups, and anions. Environ. Toxicol. 24 (4), 388–395.

Levin, D.E., 2005. Cell wall integrity signaling in *Saccharomyces cerevisiae*. Microbiol. Mol. Biol. Rev. 69 (2), 262–291.

Mackay, D., Arnot, J.A., Petkova, E.P., Wallace, K.B., Call, D.J., Brooke, L.T., Veith, G.D., 2009. The physicochemical basis of QSARs for baseline toxicity. SAR QSAR Environ. Res. 20 (3), 393–414.

Majewski, P., Pernak, A., Grzymisławski, M., Iwanik, K., Pernak, J., 2003. Ionic liquids in embalming and tissue preservation. Can traditional formalin-fixation be replaced safely? Acta Histochem. (2), 135–142.

Marrucho, I.M., Branco, L.C., Rebelo, L.P., 2014. Ionic liquids in pharmaceutical applications. Annu. Rev. Chem. Biomol. Eng. 5, 527–546.

Martins, I., Hartmann, D.O., Alves, P.C., Planchon, S., Renaut, J., Leitao, C., Rebelo, L.P.N., Silva Pereira, C., 2013. Proteomic alterations induced by ionic liquids in *Aspergillus nidulans* and *Neurospora crassa*. J. Proteomics 94C, 262–278.

Morrissey, S., Pegot, B., Coleman, D., Garcia, M.T., Ferguson, D., Quilty, B., Gathergood, N., 2009. Biodegradable, non-bactericidal oxygen-functionalised imidazolium esters: a step towards "greener" ionic liquids. Green Chem. 11 (4), 475–483.

Nockemann, P., Thijs, B., Driesen, K., Janssen, C.R., Van Hecke, K., Van Meervelt, L., Kossmann, S., Kirchner, B., Binnemans, K., 2007. Choline saccharinate and choline acesulfamate: ionic liquids with low toxicities. J. Phys. Chem. B 111 (19), 5254–5263.

Ostrosky-Zeichner, L., Casadevall, A., Galgiani, J.N., Odds, F.C., Rex, J.H., 2010. An insight into the antifungal pipeline: selected new molecules and beyond. Nat. Rev. Drug Disc 9 (9), 719–727.

Pernak, J., Branicka, M., 2003. The properties of 1-alkoxymethyl-3-hydroxypyridinium and 1-alkoxymethyl-3-dimethylaminopyridinium chlorides. J. Surfactants Deterg. 6 (2), 119–123.

Pernak, J., Chwała, P., 2003. Synthesis and anti-microbial activities of choline-like quaternary ammonium chlorides. Eur. J. Med. Chem. 38 (11–12), 1035–1042.

Pernak, J., Feder-Kubis, J., 2005. Synthesis and properties of chiral ammonium-based ionic liquids. Chem. Eur. J. 11 (15), 4441–4449.

Pernak, J., Kalewska, J., Ksycińska, H., Cybulski, J., 2001a. Synthesis and anti-microbial activities of some pyridinium salts with alkoxymethyl hydrophobic group. Eur. J. Med. Chem. 36 (11–12), 899–907.

Pernak, J., Rogoża, J., Mirska, I., 2001b. Synthesis and antimicrobial activities of new pyridinium and benzimidazolium chlorides. Eur. J. Med. Chem. 36 (4), 313–320.

Pernak, J., Sobaszkiewicz, K., Mirskam, I., 2003. Anti-microbial activities of ionic liquids. Green Chem. 5 (1), 52–56.

Pernak, J., Goc, I., Mirska, I., 2004a. Anti-microbial activities of protic ionic liquids with lactate anion. Green Chem. 6 (7), 323–329.

Pernak, J., Sobaszkiewicz, K., Foksowicz-Flaczyk, J., 2004b. Ionic liquids with symmetrical dialkoxymethyl-substituted imidazolium cations. Chem. Eur. J. 10 (14), 3479–3485.

Pernak, J., Zabielska-Matejuk, J., Kropacz, A., Foksowicz-Flaczyk, J., 2004c. Ionic liquids in wood preservation. Holzforschung 58 (3), 286–291.

Pernak, J., Syguda, A., Mirska, I., Pernak, A., Nawrot, J., Prądzyńska, A., Griffin, S.T., Rogers, R.D., 2007. Choline-derivative-based ionic liquids. Chem. Eur. J. 13 (24), 6817–6827.

Petkovic, M., Ferguson, J.L., Bohn, A., Trindade, J., Martins, I., Carvalho, M.B., Leitão, M.C., Rodrigues, C., Garcia, H., Ferreira, R., et al., 2009. Exploring fungal activity in the presence of ionic liquids. Green Chem. 11 (6), 889–894.

Petkovic, M., Ferguson, J.L., Gunaratne, H.Q.N., Ferreira, R., Leitão, M.C., Seddon, K.R., Rebelo, L.P.N., Silva Pereira, C., 2010. Novel biocompatible cholinium-based ionic liquids-toxicity and biodegradability. Green Chem. 12 (4), 643–649.

Petkovic, M., Seddon, K.R., Rebelo, L.P.N., Silva Pereira, C., 2011. Ionic liquids: a pathway to environmental acceptability. Chem. Soc. Rev. 40 (3), 1383–1403.

Petkovic, M., Hartmann, D.O., Adamová, G., Seddon, K.R., Rebelo, L.P.N., Silva Pereira, C., 2012. Unravelling the mechanism of toxicity of alkyltributylphosphonium chlorides in *Aspergillus nidulans* conidia. New J. Chem. 36 (1), 56–63.

Petkovic, M., Silva Pereira, C., 2012. Pioneering biological processes in the presence of ionic liquids: the potential of filamentous fungi. In: Plechkova, N.V., Seddon, K.R. (Eds.), Ionic Liquids UnCOILed: Critical Expert Overviews. John Wiley & Sons, Inc., New Jersey, pp. 283–303.

Pham, T.P.T., Cho, C.W., Min, J., Yun, Y.S., 2008. Alkyl-chain length effects of imidazolium and pyridinium ionic liquids on photosynthetic response of *Pseudokirchneriella subcapitata*. J. Biosci. Bioeng. 105 (4), 425–428.

Plechkova, N.V., Seddon, K.R., 2007. Ionic liquids: "designer" solvents for green chemistry. In: Tundo, P., Perosa, A., Zecchini, F. (Eds.), Methods and Reagents for Green Chemistry: An Introduction. Wiley, New York, pp. 105–130.

Plechkova, N.V., Seddon, K.R., 2008. Applications of ionic liquids in the chemical industry. Chem. Soc. Rev. 37 (1), 123–150.

Pretti, C., Chiappe, C., Pieraccini, D., Gregori, M., Abramo, F., Monni, G., Intorre, L., 2006. Acute toxicity of ionic liquids to the zebrafish (*Danio rerio*). Green Chem. 8 (3), 238–240.

Pretti, C., Chiappe, C., Baldetti, I., Brunini, S., Monni, G., Intorre, L., 2009. Acute toxicity of ionic liquids for three freshwater organisms: *Pseudokirchneriella subcapitata*, *Daphnia magna* and *Danio rerio*. Ecotoxicol. Environ. Saf. 72 (4), 1170–1176.

Ranke, J., Mölter, K., Stock, F., Bottin-Weber, U., Poczobutt, J., Hoffmann, J., Ondruschka, B., Filser, J., Jastorff, B., 2004. Biological effects of imidazolium ionic liquids with varying chain lengths in acute *Vibrio fischeri* and WST-1 cell viability assays. Ecotoxicol. Environ. Saf. 58 (3), 396–404.

Ranke, J., Müller, A., Bottin-Weber, U., Stock, F., Stolte, S., Arning, J., Störmann, R., Jastorff, B., 2007a. Lipophilicity parameters for ionic liquid cations and their correlation to in vitro cytotoxicity. Ecotoxicol. Environ. Saf. 67 (3), 430–438.

Ranke, J., Stolte, S., Stormann, R., Arning, J., Jastorff, B., 2007b. Design of sustainable chemical products—the example of ionic liquids. Chem. Rev. 107 (6), 2183–2206.

Renshaw, R.R., 1909. Preparation of choline and some of its salts. Org. Biol. 128–130.

Ruegg, T.L., Kim, E.-M., Simmons, B.A., Keasling, J.D., Singer, S.W., Soon Lee, T., Thelen, M.P., 2014. An auto-inducible mechanism for ionic liquid resistance in microbial biofuel production. Nat. Commun. 5.

Sakaebe, H., Matsumoto, H., Tatsumi, K., 2007. Application of room temperature ionic liquids to Li batteries. Electrochim. Acta 53 (3), 1048–1054.

Schaffran, T., Justus, E., Elfert, M., Chen, T., Gabel, D., 2009. Toxicity of N,N,N-trialkylammoniododecaborates as new anions of ionic liquids in cellular, liposomal and enzymatic test systems. Green Chem. 11 (9), 1458–1464.

Shamshina, J.L., Barber, P.S., Rogers, R.D., 2013. Ionic liquids in drug delivery. Exp. Opin. Drug Del. 10 (10), 1367–1381.

Stark, A., Seddon, K.R., 2007. Ionic liquids. In: Seidel, A. (Ed.), Kirk–Othmer Encyclopaedia of Chemical Technology. John Wiley & Sons, Inc., New Jersey, pp. 836–920.

Stepnowski, P., Składanowski, A.C., Ludwiczak, A., Łaczyńska, E., 2004. Evaluating the cytotoxicity of ionic liquids using human cell line HeLa. Hum. Exp. Toxicol. 23 (11), 513–517.

Stolte, S., Arning, J., Bottin-Weber, U., Müller, A., Pitner, W.R., Welz-Biermann, U., Jastorff, B., Ranke, J., 2007a. Effects of different head groups and functionalised side chains on the cytotoxicity of ionic liquids. Green Chem. 9 (7), 760–767.

Stolte, S., Matzke, M., Arning, J., Böschen, A., Pitner, W.R., Welz-Biermann, U., Jastorff, B., Ranke, J., 2007b. Effects of different head groups and functionalised side chains on the aquatic toxicity of ionic liquids. Green Chem. 9, 1170–1179.

Swatloski, R.P., Holbrey, J.D., Memon, S.B., Caldwell, G.A., Caldwell, K.A., Rogers, R.D., 2004. Using *Caenorhabditis elegans* to probe toxicity of 1-alkyl-3-methylimidazolium chloride based ionic liquids. Chem. Commun. 668–669.

Vlahakis, J.Z., Lazar, C., Crandall, I.E., Szarek, W.A., 2010. Anti-Plasmodium activity of imidazolium and triazolium salts. Bioorg. Med. Chem. 18 (16), 6184–6196.

Wang, X.F., Ohlin, C.A., Lu, Q.H., Fei, Z.F., Hu, J., Dyson, P.J., 2007. Cytotoxicity of ionic liquids and precursor compounds towards human cell line HeLa. Green Chem. 9, 1191–1197.

Weaver, K.D., Kim, H.J., Sun, J.Z., MacFarlane, D.R., Elliott, G.D., 2010. Cyto-toxicity and biocompatibility of a family of choline phosphate ionic liquids designed for pharmaceutical applications. Green Chem. 12 (3), 507–513.

Wells, A.S., Coombe, V.T., 2006. On the freshwater ecotoxicity and biodegradation properties of some common ionic liquids. Org. Process Res. Dev. 10 (4), 794–798.

Ying, G.G., 2006. Fate, behavior and effects of surfactants and their degradation products in the environment. Environ. Int. 32 (3), 417–431.

Zakrewsky, M., Lovejoy, K.S., Kern, T.L., Miller, T.E., Le, V., Nagy, A., Goumas, A.M., Iyer, R.S., Del Sesto, R.E., Koppisch, A.T., et al., 2014. Ionic liquids as a class of materials for transdermal delivery and pathogen neutralization. Proc. Natl. Acad. Sci. U.S.A. 111 (37), 13313–13318.

Chapter 14

Ionic Liquids in Lipid Analysis

Jingbo Li
Aarhus University, Department of Engineering, Aarhus, Denmark

14.1 INTRODUCTION

Ionic liquids (ILs) are molten salts at <100°C, which are typically composed of a bulky heteroatom-containing dissymmetrical cation such as alkyl-imidazolium, pyrodinium, pyrrolidinium, ammonium or phosphonium, and so on; and an inorganic or organic anion such as chloride, triflate, dicyanamide, acetate, trifluoroacetate, trifluoromethylsulfate, or hexafluorophosphate, and so on. ILs are called designer solvents, because, in principle, the property of an IL can be tailor-made by judicious selection/synthesis of cation or anion with different substituents; up to 10^{18} possible combinations of cation and anion are theoretically available (Jolanta, 2007). Chemically ILs are regarded as ionic or nonmolecular compounds. This structural feature endows ILs with distinct properties different from either conventional molecule solvents or inorganic salts, such as a broad liquid range, extremely low vapor pressure, high thermal stability, viscosity, and electric conductivity, which make them very special in the chemical shelf (Pandey, 2006; Koel, 2005). These unique properties have resulted in applications in different fields (Fig. 14.1); among which the application in analytical chemistry is an important area (Koel, 2005; Soukup-Hein et al., 2009).

Fig. 14.2 showed the typical properties of ILs and the corresponding applications in analytical chemistry.

Properties of ILs, including high thermal stability, low volatility, strong wetting ability, and high viscosity made them good candidates for stationary phase for gas chromatography (Jolanta, 2007; Koel, 2005). In addition, ILs also have good solvation properties of organic and inorganic compounds, thus making them good additives or modifiers of mobile phase or stationary phase for liquid chromatography (LC) (Berthod et al., 2008). The excellent electrical conductivity of the molten salts (ILs) makes them a better option for capillary electrophoresis (CE) (Jolanta, 2007). Thus, ILs are suitable for use as stationary phase in gas chromatography (GC) and LC column, additives to mobile phase in high-performance liquid chromatography (HPLC), electrolytes

Ionic Liquids in Lipid Processing and Analysis. http://dx.doi.org/10.1016/B978-1-63067-047-4.00014-3

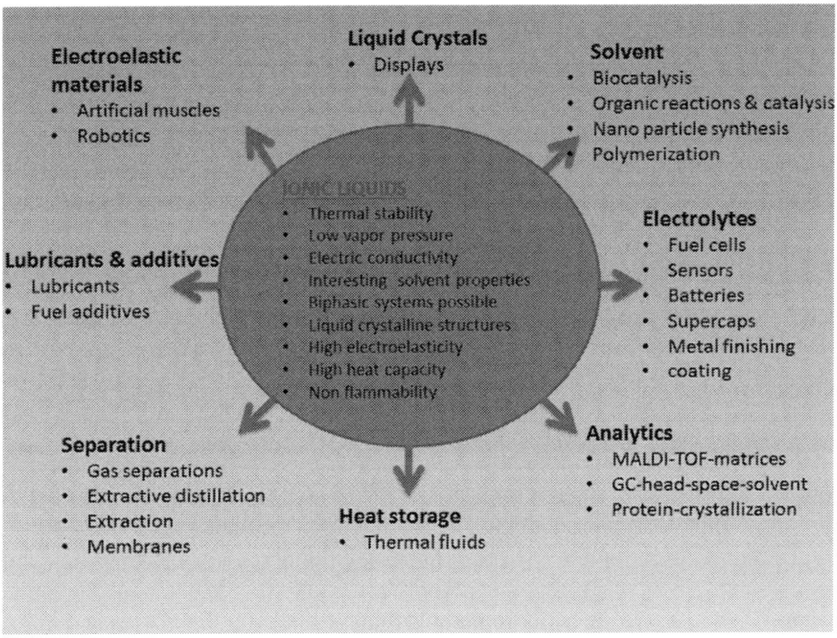

FIGURE 14.1 Main properties of ILs and some applications. *From http://www.sigmaaldrich.com/technical-documents/articles/chemfiles/ionic-liquids.html.*

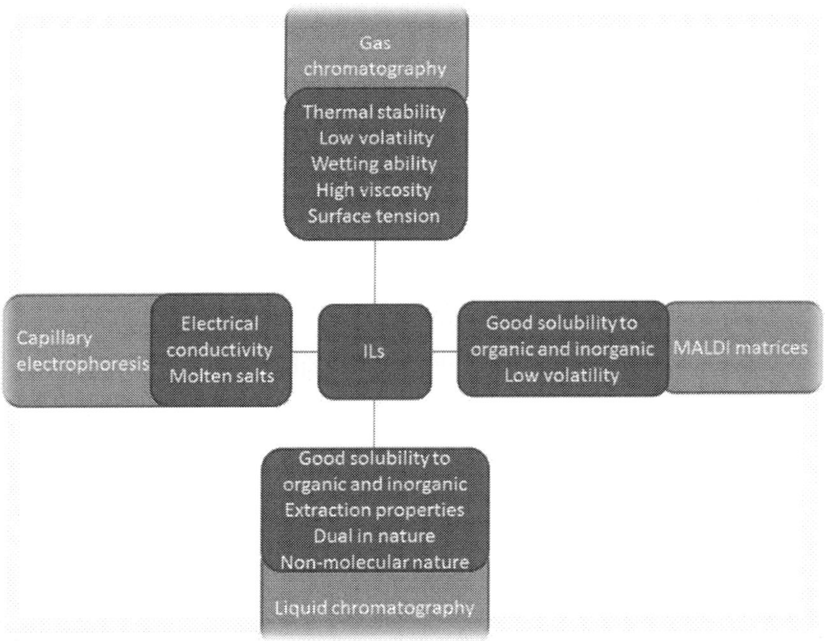

FIGURE 14.2 The properties and corresponding applications of ILs in analytical chemistry. Blue (dark gray in print versions) indicates properties and green (light gray in print versions) indicates applications.

in CE, or as Matrix-Assisted Laser Desorption Ionization matrices. Most interestingly, ILs consisting of imidazolium, pyridinium, and pyrrolidinium, and so on have aromatic property, which can form reversible $\pi-\pi$ complexation with polyunsaturated fatty acids (PUFAs). Hence these types of ILs can be used for enrichment of omega-3 PUFAs (Cheong et al., 2011; Li et al., 2009a,b).

The physicochemical properties and chemical structure of cations of some selected ILs, which have been investigated for applications in analytical chemistry, are listed in Table 14.1.

14.2 IONIC LIQUIDS IN GAS CHROMATOGRAPHY ANALYSIS

In gas chromatography, the choice of the proper stationary phase plays a crucial role in improvement or optimization of chromatographic separation of analytes. Common stationary phases in open tubular columns are cyanopropylphenyl dimethyl polysiloxane, carbowax polyethyleneglycol, biscyanopropyl cyanopropylphenyl polysiloxane, and diphenyl dimethyl polysiloxane.

Some reviews have been published on the application of ILs for column of gas chromatography (Koel, 2005; Soukup-Hein et al., 2009; Berthod et al., 2008). IL-based gas chromatography column showed a like-separates-like behavior in the separation of analytes (Berthod et al., 2005). Polar compounds, such as acids, bases, and compounds capable of forming hydrogen bonding are better separated if polar ILs are used as stationary phases; meanwhile, less polar compounds are better separated using less polar ILs as stationary phase. Some IL columns separated the nonpolar compounds in almost the same way and with similar retention times as a nonpolar commercial column under the same separation conditions (Wilkes and Zaworotko, 1992). Han and Armstrong (2007) highlighted the role of ILs in separation techniques, and emphasized that the high thermal stability of ILs is the most important factor to consider to compete with the conventional columns on the market. For example, ethylammonium nitrate employed as the stationary phase has a working temperature range from 20 to 120°C (Pacholec et al., 1982); while ethylpyridinium bromide can be operated no more than 160°C (Pacholec and Poole, 1983). In 2003, two new ILs, 1-benzyl-3-methylimidazolium trifluoromethanesulfonate and 1-(4-methoxyphenyl)-3-methylimidazolium trifluoromethanesulfonate, were engineered and synthesized to overcome the thermostability problem, with high thermal stability up to 260°C (Anderson and Armstrong, 2003). Capillary GC column using 1-butyl-3-methylimidazolium hexafluorophosphate and the analogous chloride salt as stationary phase have a working temperature up to 300°C (Armstrong et al., 1999). Recently, a new imidazolium IL polymer was synthesized and prepared as the stationary phase of GC, which exhibited excellent thermal stability from room temperature to 380°C (Kao et al., 2014).

TABLE 14.1 Physicochemical Properties of Selected Ionic Liquids (Berthod et al., 2008; Reichardt, 2005)

Chemical Formula	Anion	Abbreviation	m.w. (Da)	Melting Point (°C)	Density (25°C)	Refractory Index	Viscosity (25°C) (cp)	Solubility in Water	E_T^N (×100)
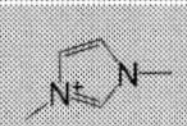	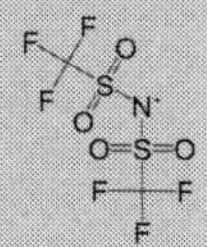	[MMIM][NTfO$_2$]	377	22	1.559	1.422	44	N/A	N/A
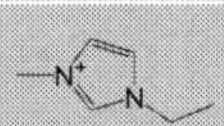	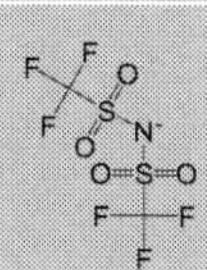	[EMIM][NTfO$_2$]	391	−17	1.52	1.4321	18	N	N/A
		[EMIM][DCA]	177	−21	1.06		21	N/A	N/A
	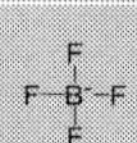	[EMIM][BF$_4$]	197.8	6	1.248		66	S	71
	Cl^-	[EMIM]Cl	146.5	89	1.12*	Solid	Solid	S	N/A
	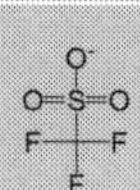	[EMIM][TfO]	260	−9	1.39	1.4332	45	S	N/A

	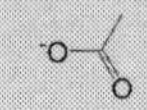	[EMIM][OAc]	170	−20	1.03		91	S	N/A
		[EMIM][SCN]	169	−6				N/A	N/A
	$[AlCl_4^-]$	[EMIM][$AlCl_4$]	280	−10	1.24		26	Decompose	N/A
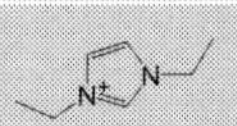	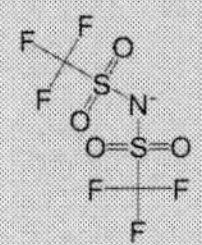	[EEIM][$NTfO_2$]	405	14	1.452	1.426	35	N	N/A
	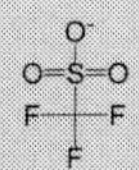	[EEIM][TfO]	274	23	1.33	1.4367	53	N/A	N/A
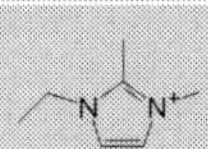	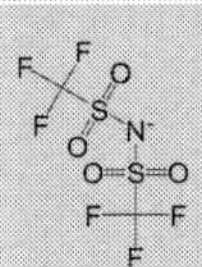	[EM2MIM] [$NTfO_2$]	405	20	1.495	1.4305	88	N/A	N/A
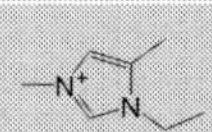	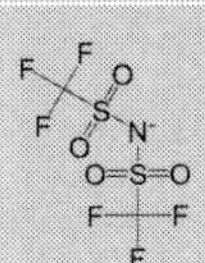	[EMM5IM] [$NTfO_2$]	405	−39	1.47	1.4275	37	N/A	N/A

Continued

TABLE 14.1 Physicochemical Properties of Selected Ionic Liquids (Berthod et al., 2008; Reichardt, 2005)—cont'd

Chemical Formula	Anion	Abbreviation	m.w. (Da)	Melting Point (°C)	Density (25°C)	Refractory Index	Viscosity (25°C) (cp)	Solubility in Water	E_T^N (×100)
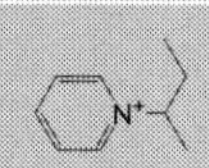	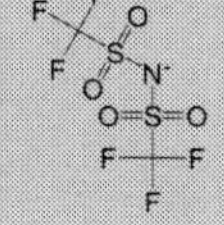	[MPPyr][NTfO$_2$]	416	0	1.44		39	N	N/A
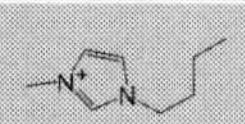	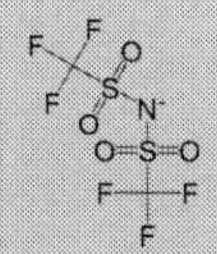	[BMIM][NTfO$_2$]	419	−4	1.429	1.4271	80	N/A	64
	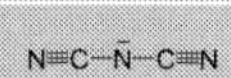	[BMIM][DCA]	205	−6	1.06		37	S	N/A
	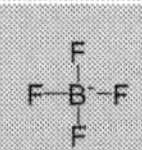	[BMIM][BF$_4$]	225.8	−82	1.208	1.429	233	S	67
	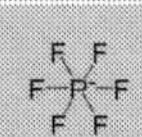	[BMIM][PF$_6$]	284	10	1.373	1.411	400	18 g/L	67
	Cl^-	[BMIM]Cl	174.5	65	1.10*	Solid	Solid	S	N/A
	Br^-	[BMIM]Br	218.9					S	N/A

	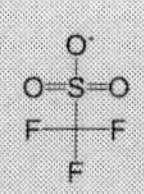	[BMIM][TfO]	288	16	1.29	1.438	90	S	67
	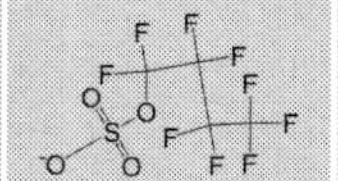	[BMIM][PFBS]	438	20	1.473	1.4052	373	N/A	N/A
	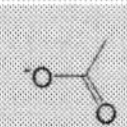	[BMIM][Act]	198	−20	1.06		525	S	N/A
	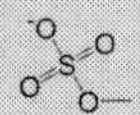	[BMIM][MS]	250	−20	1.2		180	S	N/A
	$[AlCl_4^-]$	[BMIM][$AlCl_4$]	308	−10	1.24		26	Decompose	N/A
	I^-	[BMIM][I]	265.9	−72			1110	N/A	N/A
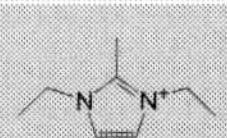	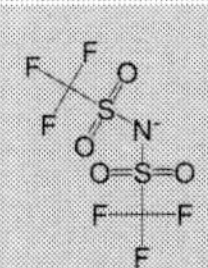	[EEMIM][$NTfO_2$]	419	−22	1.432	1.43	36	N/A	N/A

Continued

TABLE 14.1 Physicochemical Properties of Selected Ionic Liquids (Berthod et al., 2008; Reichardt, 2005)—cont'd

Chemical Formula	Anion	Abbreviation	m.w. (Da)	Melting Point (°C)	Density (25°C)	Refractory Index	Viscosity (25°C) (cp)	Solubility in Water	E_T^N (×100)
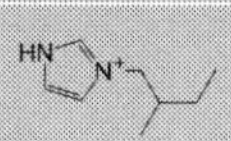	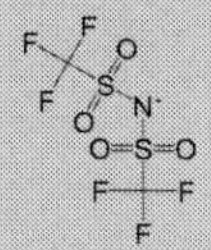	[MM2PMIM] [$NTfO_2$]	419	15	1.46		41	N	N/A
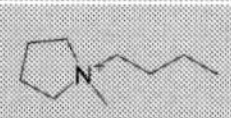	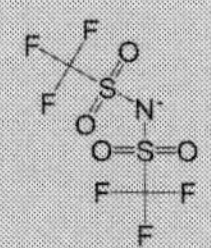	[BMPyrrol] [$NTfO_2$]	422	−50	1.4		71	N	N/A
		[BMPyrrol] [DCA]	208	−55	0.93		50	N/A	54
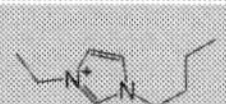	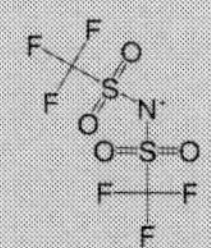	[BEIM][$NTfO_2$]	433	−8	1.404	1.4285	48	N/A	51
	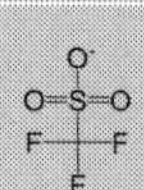	[BEIM][TfO]	302	2	1.18	1.441	90	S	67

	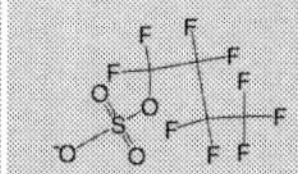	[BEIM][PFBS]	452	21	1.427	1.4025	320	N/A	N/A
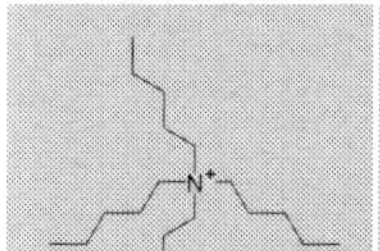	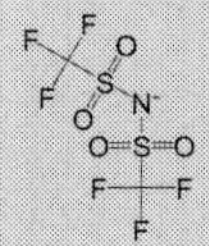	$[(C_5)_4N][NTfO_2]$	578	25.2	1.16		430	N/A	51
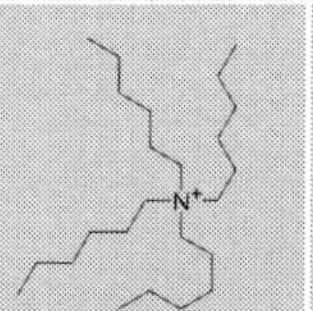	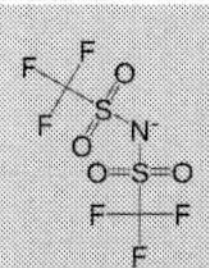	$[(C_6)_4N][NTfO_2]$	634	−7	1.11		435	N/A	N/A
	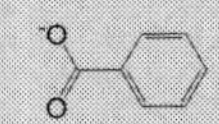	$[(C_6)_4N][Bzt]$	476	−50	0.938		895	N	42
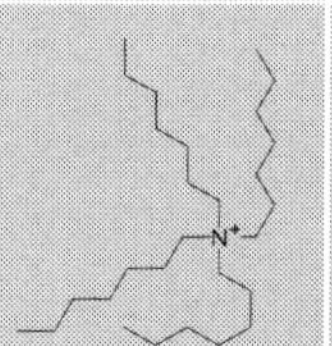	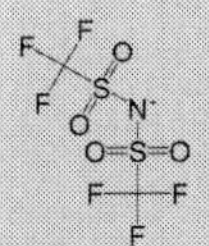	$[(C_7)_4N][NTfO_2]$	690	11.2	1.1		453	N/A	N/A
	Cl^-	$[(C_7)_4N][Cl]$	445.5	−9	0.882		598	S	N/A

Continued

TABLE 14.1 Physicochemical Properties of Selected Ionic Liquids (Berthod et al., 2008; Reichardt, 2005)—cont'd

Chemical Formula	Anion	Abbreviation	m.w. (Da)	Melting Point (°C)	Density (25°C)	Refractory Index	Viscosity (25°C) (cp)	Solubility in Water	E_T^N (×100)
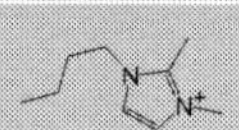		[BMMIM][BF_4]	239.8	37	1.2		780	S	57.6
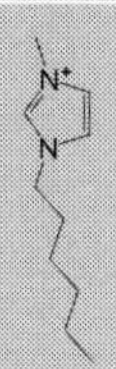	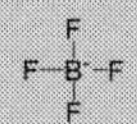	[HMIM][BF_4]	253.8	−82	1.208		310	P	70.7
	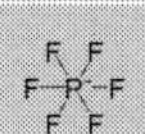	[HMIM][PF_6]	312	−61	1.304		800	N	65.7
	Cl^-	[HMIM][Cl]	202.5	−75	1.05		7500	S	56.2
	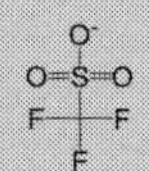	[HMIM][TfO]	316	29	1.20		160	N	N/A

	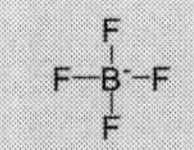	[OMIM][BF_4]	281.8	−79	1.11		440	P	54
	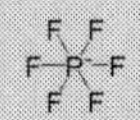	[OMIM][PF_6]	340	−40	1.2		810	N	59.6
	Cl^-	[OMIM][Cl]	230.5	0	1		16,000	S	54.9
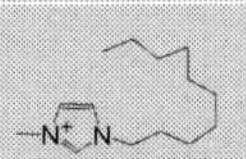	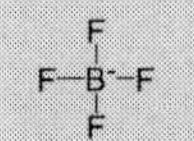	[DMIM][BF_4]	309.8	−25	1.072		930	N/A	63

E_T^N is the Reichardt polarity index (y index (E_T^N water = 100)). Density and viscosity values at 25°C except when an (*) indicates that the given value corresponds to the salt melting temperature. Solubility in water at room temperature: *N*, nonsoluble (two phases form); *S*, soluble; *P*, partly soluble; *TfO*, triflate anion or trifluoromethyl sulfate; *PFES*, perfluoroethyl sulfate; *PFBS*, perfluorobutyl sulfate; *TFAct*, trifluoroacetate or trifluoroethanoate anion; *Bzt*, benzoate; *N/A*, not available.

Poole and Poole (2011) reviewed the progress of IL as a stationary phase for GC, which emphasized the design of ILs with low melting point, high thermal stability, viscosity, and variable selectivity as the stationary phase for separations. Ragonese et al. (2012) published a review on the use of IL stationary phases in the field of GC-based hyphenated techniques as well as in the multidimensional GC field. They concluded that IL phases with flexible selectivities and high thermal stabilities opened a new, exciting, and stimulating era in the field of GC separations. In short, designing ILs for specific analytes as well as specific analytical equipment will be the future research area.

Utilization of a dicationic liquid [1,9-di(3-vinyl-imidazolium) nonane bis-(trifluoromethyl) sulfonyl imidate] stationary phase in the fast and conventional gas chromatographic analysis of cis/trans fatty acid methyl esters isomers was examined by Ragonese et al. (2009). Selectivity of this IL column was superior to other well-established capillary columns (Fig. 14.3). The test using a longer column (100 m) with the same IL stationary phase demonstrated that the retention time and peak area intraday precision ($n = 5$) were excellent, with RSD values lower than 0.07% and 6.6%, respectively. Limit of detection for C18:1Δ^{9tr} and C18:2$\Delta^{9tr,12tr}$ was 0.15 and 0.18 ng; but the corresponding limit of quantitation for was 0.49 and 0.60 ng, respectively. A linear correlation between the sample concentration and the response was also found in the 10–2000 ppm range for both *cis* and *trans* fatty acid methyl esters.

A new commercialized GC column employing 1,5-Di(2,3-dimethyl imidazolium)pentane bis(trifluoromethylsulfonyl)imide as stationary phase named SLB-IL111 from Supelco Inc. was used to separate fatty acid methyl esters to compare its performance with other commercial columns SP-2560 and CP-Sil 88 by Delmonte et al. (2011). The SLB-IL111 column achieved an improved separation of *cis*- and *trans*-18:1 and *cis*/*trans* conjugated linoleic acid isomers. It was declared that their result was the first direct GC separation of c9, t11-from t7,c9-conjugated linoleic acid, and t15-18:1 from c9-18:1; and the SLB-IL111 column also provided partial resolution of t13/t14-18:1, c8-from c6/c7-18:1, and t,t-conjugated linoleic acid isomer pairs.

The SLB-IL111 column was also used to separate *cis*-6 from *cis*-8 monounsaturated fatty acid position isomers such as *cis*-6 18:1 from *cis*-8 18:1 methyl esters since it is unable to achieve such separations using a cyanoalkyl-based highly polar capillary column (Destaillats et al., 2011). A baseline separation between *cis*-6 18:1 from *cis*-8 18:1 methyl esters was achieved, where an ambiguous identification was avoided when using mass spectrometry.

One-dimensional GC/mass spectrometry and comprehensive two-dimensional GC of fatty acid methyl esters from microalgae oils were tested by comparing polyethylene glycol and cyanopropyl substituted polar stationary phase with IL stationary phases SLB-IL 82 and SLB-IL100

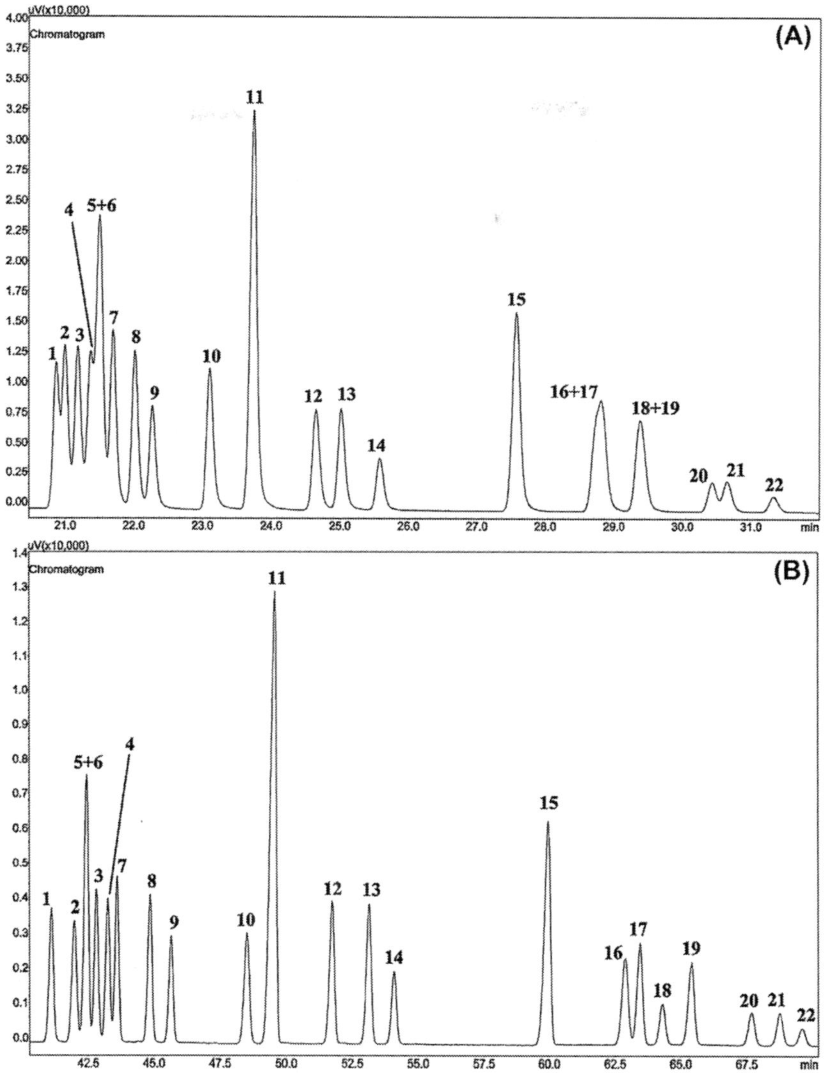

FIGURE 14.3 (A) SP-2560 column result and (B) 100 m IL column result for the FAMEs test mixture. *(Adapted from Ragonese, C., Tranchida, P.Q., Dugo, P., Dugo, G., Sidisky, L.M., Robillard, M.V., Mondello, L., 2009. Evaluation of use of a dicationic liquid stationary phase in the fast and conventional gas chromatographic analysis of health-hazardous C18 Cis/trans fatty acids. Anal. Chem. 81 (13), 5561–5568.)* Peak assignment (1) C18:1Δ^{6tr}; (2) C18:1Δ^{9tr}; (3) C18:1Δ^{11tr}; (4) C18:1Δ^{12tr}; (5) C18:1Δ^{6c}; (6) C18:1Δ^{7c}; (7) C18:1Δ^{9c}; (8) C18:1Δ^{11c}; (9) C18:1Δ^{12c}; (10) C18:1Δ^{15c}; (11) C18:2$\Delta^{9tr,12tr}$; (12) C18:2$\Delta^{9c,12tr}$; (13) C18:2$\Delta^{9tr,12c}$; (14) C18:2$\Delta^{9c,12c}$; (15) C18:3$\Delta^{9tr,12tr,15tr}$; (16) C18:3$\Delta^{9tr,12tr,15c}$; (17) C18:3$\Delta^{9tr,12c,15tr}$; (18) C18:3$\Delta^{9c,12tr,15tr}$; (19) C18:3$\Delta^{9c,12c,15tr}$; (20) C18:3$\Delta^{9c,12tr,15c}$; (21) C18:3$\Delta^{9tr,12c,15c}$; (22) C18:3$\Delta^{9c,12c,15c}$.

(Gu et al., 2011). The resolution between them was comparable but lower column bleeding was found for the IL-based column. The lower bleeding further enhanced the sensitivity of mass spectrometry. The established two-dimensional GC using a combination of a polar IL column and an apolar polydimethyl siloxane showed an excellent group-type separation of fatty acids with different carbon numbers and number of unsaturation. Thus, complex sample containing polar and nonpolar compounds can be better separated using a combination of IL-based and conventional column.

Delmonte et al. (2012) evaluated the SLB-IL111 column for determination of fatty acid composition in milk fat. The column was operated under combined temperature and eluent flow gradient conditions and successfully resolve most of the fatty acids in milk fat with a single run in 80 min. Under their attempted conditions, a good separation of short-chain (from 4:0), long-chain polyunsaturated fatty acids, and most of the unsaturated fatty acid positional/geometric isomers was observed.

SLB-IL111 seems to be one of the most favorable IL-based GC columns for fatty acid composition analysis. In 2012, Tyburczy et al. determined the content of *trans* fat in 32 representative fast food samples according to AOCS Official Method Ce 1j-07 using 100 m SP-2560 CPS column. For comparison, the fatty acid ethyl esters were also analyzed using the 200 m SLB-IL111 IL column. They found that comparable quantity of total, *trans*, saturated, and *cis* unsaturated fat were obtained using these two columns. Furthermore, the 200 m SLB-IL111 IL column showed the advantage in determination of individual *trans* 18:1, *trans* 18:2, and *trans* 18:3 fatty acid isomeric compositions under conditions suitable for routine sample analysis.

The maximum temperature of SP-2560 and SLB-IL11 columns is 250 and 270°C, respectively. The polarity of SLB-IL111 belongs to an extremely polar group while that of SP-2560 belongs to a high polar group (https://www.sigmaaldrich.com/content/dam/sigma-aldrich/docs/Supelco/General_Information/1/t411139.pdf). The comparison of separation performance of SP-2560 and SLB-IL111 is shown in Fig. 14.4. It might be suggested that development of more polar and more thermal stable IL as the stationary phase of GC column for analyzing lipids is one of the future directions.

14.3 IONIC LIQUIDS IN LIQUID CHROMATOGRAPHY ANALYSIS

Mobile phase used in LC can be either a pure solvent or a mixture of different solvents with or without dissolved salts. Being both liquid solvent and salt, ILs can, therefore, be used in LC analysis either as a mobile phase modifier in reversed-phase LC, mobile phase additives, or as LC stationary phases (Berthod et al., 2008).

Berthod et al. (2008) pointed out some observed problems when ILs are used as mobile phase modifiers in reversed-phase LC. The first one is the

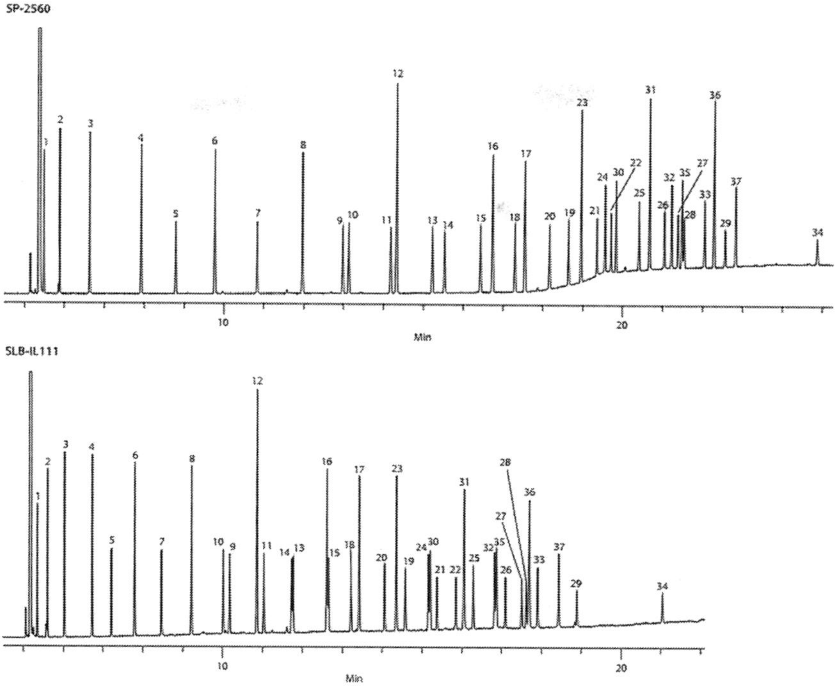

FIGURE 14.4 Comparison of performance between SP-2560 and SLB-IL111. *(Adapted from https://www.sigmaaldrich.com/content/dam/sigma-aldrich/docs/Supelco/General_Information/1/t411139.pdf.)* For each peak assignment: (1) Butyric Acid Methyl Ester (C4:0) at 4 wt%; (2) Caproic Acid Methyl Ester (C6:0) at 4 wt%; (3) Caprylic Acid Methyl Ester (C8:0) at 4 wt %; (4) Capric Acid Methyl Ester (C10:0) at 4 wt%; (5) Undecanoic Acid Methyl Ester (C11:0) at 2 wt%; (6) Lauric Acid Methyl Ester (C12:0) at 4 wt%; (7) Tridecanoic Acid Methyl Ester (C13:0) at 2 wt%; (8) Myristic Acid Methyl Ester (C14:0) at 4 wt%; (9) Myristoleic Acid Methyl Ester (C14:1) at 2 wt%; (10) Pentadecanoic Acid Methyl Ester (C15:0) at 2 wt%; (11) *cis*-10-Pentadecenoic Acid Methyl Ester (C15:1) at 2 wt%; (12) Palmitic Acid Methyl Ester (C16:0) at 6 wt%; (13) Palmitoleic Acid Methyl Ester (C16:1) at 2 wt%; (14) Heptadecanoic Acid Methyl Ester (C17:0) at 2 wt%; (15) *cis*-10-Heptadecenoic Acid Methyl Ester (C17:1) at 2 wt%; (16) Stearic Acid Methyl Ester (C18:0) at 4 wt%; (17) Oleic Acid Methyl Ester (C18:1n9c) at 4 wt%; (18) Elaidic Acid Methyl Ester (C18:1n9t) at 2 wt%; (19) Linoleic Acid Methyl Ester (C18:2n6c) at 2 wt%; (20) Linolelaidic Acid Methyl Ester (C18:2n6t) at 2 wt%; (21) γ-Linolenic Acid Methyl Ester (C18:3n6) at 2 wt%; (22) Linolenic Acid Methyl Ester (C18:3n3) at 2 wt%; (23) Arachidic Acid Methyl Ester (C20:0) at 4 wt%; (24) *cis*-11-Eicosenoic Acid Methyl Ester (C20:1n9) at 2 wt%; (25) *cis*-11,14-Eicosadienoic Acid Methyl Ester (C20:2) at 2 wt%; (26) *cis*-8,11,14-Eicosatrienoic Acid Methyl Ester (C20:3n6) at 2 wt%; (27) *cis*-11,14,17-Eicosatrienoic Acid Methyl Ester (C20:3n3) at 2 wt%; (28) Arachidonic Acid Methyl Ester (C20:4n6) at 2 wt%; (29) *cis*-5,8,11,14,17-Eicosapentaenoic Acid Methyl Ester (C20:5n3) at 2 wt%; (30) Heneicosanoic Acid Methyl Ester (C21:0) at 2 wt%; (31) Behenic Acid Methyl Ester (C22:0) at 4 wt%; (32) Erucic Acid Methyl Ester (C22:1n9) at 2 wt%; (33) *cis*-13,16-Docosadienoic Acid Methyl Ester (C22:2) at 2 wt%; (34) *cis*-4,7,10,13,16,19-Docosahexaenoic Acid Methyl Ester (C22:6n3) at 2 wt%; (35) Tricosanoic Acid Methyl Ester (C23:0) at 2 wt%; (36) Lignoceric Acid Methyl Ester (C24:0) at 4 wt%; (37) Nervonic Acid Methyl Ester (C24:1n9) at 2 wt%.

high viscosity of ILs. Less viscous ILs such as monoalkylammonium nitrate, formate, and acetate still have the higher viscosities of 10 cP at room temperature. The high viscosity of ILs causes very high back pressure of columns. Furthermore, the metallic parts of the HPLC are corroded when using thiocyanate ILs. The alkylammonium salts are inferior to the classical solvents methanol and acetonitrile in reversed-phase LC since ILs have lower UV transparency and inferior peak efficiencies. Another disadvantage of using ILs as mobile phase is their high cost. All these drawbacks limit the use of ILs as organic modifiers of mobile phases in routine reversed-phase LC analysis.

Basic compounds most often bear positively charged amine groups when added to low pH mobile phase. As a consequence, these compounds are retained by a combination of electrical (charge–charge) and hydrophobic interactions with the stationary phase and the ions of the mobile phase, because of the interactions between the cationic sites of the compounds and the anionic silanols of the stationary phase, including charge interactions, hydrophobic interactions, dispersion interactions, and electrostatic interactions. In aqueous mobile phases, the charge interactions are usually stronger and slower than the hydrophobic interactions, leading to peak tailing and long retention times. The cations in ILs can interact and compete with silanol groups on the alkyl silica base surface with the polar group of the analytes. At the same time, the nonpolar alkyl groups of the stationary phase can interact with different alkyl groups of the heterocyclic ring or quaternary cation. Thus, it can shield the residual silanols efficiently and improve the peak shapes, while also reducing the chromatographic retention times of the analytes (Han and Row, 2010). When ILs are added in a low concentration, they behave as ordinary salts. As a salt, ILs are composed of cations and anions with equal charges. Both species can affect the chromatographic results and work either synergistically or antagonistically.

Table 14.2 summarizes some recent applications of ILs either as mobile phase modifiers, additives, or stationary phase. The effect of 1-butyl-3-methylimidazolium tetrafluoroborate as mobile phase modifier on the retention performance of *N*-carbobenzyloxy-D-phenylalanine and D-tryptophan was examined by Polyakova and Row (2006). For mobile phases containing a relatively low concentration of [BMIM][BF4] (from 2 to 5 mmol/L) as a modifier, increases in the concentration of the modifier resulted in increased retention of *N*-CBZ-D-phenylalanine. This might be due to formation of a layer of IL on the surface of the stationary phase resulting in an electrostatic potential which causes additional retention of ionized analytes, and therefore, significant adsorption of the *N*-CBZ-D-phenylalanine. Under these conditions the chromatographic behavior of D-tryptophan was fairly stable. These results also suggested that not all the analytes respond to the IL addition (Polyakova and Row, 2006). Individual experiments should be carried out to examine the effects of ILs on analyte separation.

TABLE 14.2 Application of Ionic Liquid as Either Mobile Phase Additives or Modifiers or Stationary Phase

Target Analyte	Mobil Phase	Ionic Liquid	Role of Ionic Liquid	References
N-carbobenzyloxy-D-phenylalanine and D-tryptophan	Water-acetonitrile	1-butyl-3-methylimidazolium tetrafluoroborate	Modifier	Polyakova and Row (2006)
Aromatic compounds	Acetonitrile/water	1-hexyl-3-methylimidazolium tetrafluoroborate ([Hmim][BF_4]) and 1-methyl-3-octylimidazolium tetrafluoroborate ([Omim][BF_4])	Modifier	Tian et al. (2010)
Peptides		Butylimidazolium bromide	Stationary phase	Chitta and Van Meter (2010)
Biogenic amines	Acetonitrile-water	1-ethyl-3-methyl-imidazolium hexafluorophosphate (EMIM PF_6)	Additive	Flieger and Czajkowska-Żelazko (2011)
Alkaloids	Acetonitrile —acetate buffer	ILs 3-methyl-1-octylimidazolium tetrafluoroborate; 1-butyl-2,3-dimethylimidazolium tetrafluoroborate; *N*-butyl-3-methylpyridinium tetrafluoroborate; methyltrioctylammonium trifluoroacetate; 1-butyl-3-methylimidazolium tetrafluoroborate; 1-methyl-3-octy-loxymethylimidazolium tetrafluoroborate	Additive	Petruczynik (2012)
Fangchinoline and tetrandrine	Methanol-water or methanol-water	1-alkyl-3-methylimidazolium-based ionic liquids	Additive	Tang et al. (2013)
Naphazoline[a]	Acetonitrile—water	Imidazolium-class ionic liquid	Additive	Marszałł et al. (2013)
Mandelic acid and their derivative enantiomers	Methanol	Isoleucine ionic liquids	Additive	Zhou et al. (2014)

[a]*HPTLC.*

A comparison of chaotropic salt and IL as mobile phase additives in reversed-phase HPLC for separation of biogenic amines was conducted (Flieger and Czajkowska-Żelazko, 2011). IL, 1-ethyl-3-methyl-imidazolium hexafluorophosphate, and chaotropic salt $NaPF_6$ on Discovery HS C18 column at acidic conditions was used for the study. It turned out that both additives gave rise to comparable efficiency and selectivity toward biogenic amines. However, the limit of detection of chaotropic salt-involved eluent system (0.05−0.7 μg/mL) is much lower than that of an IL-modified system (1.5−5.8 μg/mL).

Effects of the addition of IL to mobile phase on separation and system efficiency for HPLC of selected alkaloids on different stationary phases was investigated (Petruczynik, 2012). The results confirmed that the application of ILs to HPLC was a simple and effective way to analyze and separate basic compounds. The addition of ILs to the mobile phases gave rise to decreased band tailing, increased retention, and improved system efficiency and resolution. The solvent system containing 1-butyl-2,3-dimethylimidazolium tetrafluoroborate proved to be most suitable in terms of selectivity, efficiency, and peak symmetry. An increase in IL concentration caused a decrease in alkaloid retention, improvement in peak symmetry, and increase of theoretical plate number in most cases. The best shape of peaks and the highest theoretical plate number for most investigated alkaloids were obtained in eluent solvent systems containing IL on pentafluorophenyl phase, when the double protection against disadvantageous interaction between cations of basic analytes and free silanol was used (Fig. 14.5).

Tang et al. (2013) investigated the effects of four types of 1-alkyl-3-methylimidazolium-based ILs as mobile phase additive to separate fangchinoline and tetrandrine; the possible separation mechanism with the presence of ILs is also discussed. The optimal analysis condition was found to be methanol/water (80/20, v/v) with 50 mM [HMIM][BF_4] at pH 3; and the possible mechanism of ILs as additive to mobile phase to improve separation was attributed to the role of the dual nature of ILs in reversed-phase HPLC. Several changes in a chromatographic system took place when ILs were used as the mobile phase additives. First, it leads to a competition between imidazolium cations and polar group of analytes for the free silanol groups on the surface of the stationary phase by the specific electrostatic interactions. Second, the hydrophobic stationary phase could adsorb the anion of ILs. An asymmetric bilayer could be created depending on the relative adsorption of the cation and anion. As a consequence, the retention of analyte changes are caused by a mixed mechanism such as ion-pairing, ion-exchange, and hydrophobic portioning. It seems that whether the separation is good or not depends on several factors including the properties of analytes, mobile phase composition, stationary phase, and even detector.

The previously listed publications mainly focus on improvement of addition of ILs on separation and resolution of small molecular analytes.

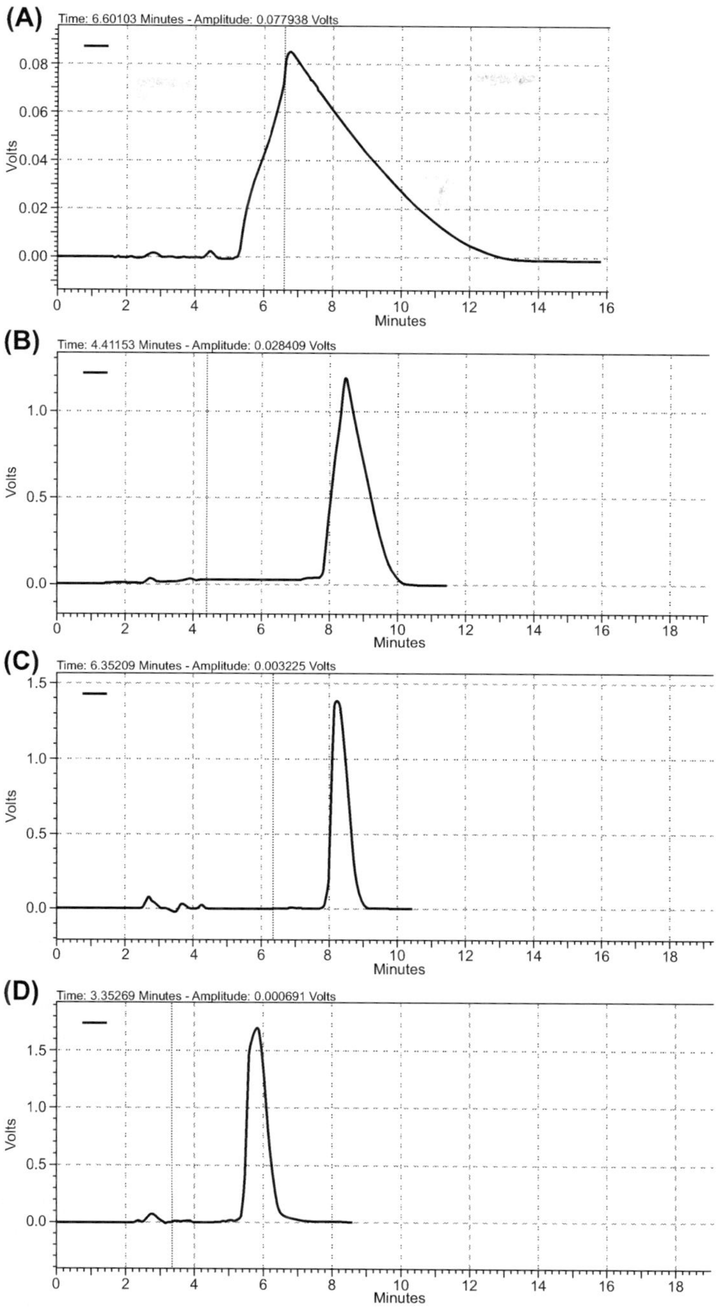

FIGURE 14.5 Chromatograms obtained for boldine on pentafluorophenyl column with different mobile phases: (A) 15% MeCN and 20% acetate buffer at pH 3.5; (B) 15% MeCN, 20% acetate buffer at pH 3.5%, and 0.5% IL2; (C) 15% MeCN, 20% acetate buffer at pH 3.5%, and 2% IL2; (D) 15% MeCN, 20% acetate buffer at pH 3.5%, and 5% IL2. *Adapted from Petruczynik, A. 2012. Effect of ionic liquid additives to mobile phase on separation and system efficiency for HPLC of selected alkaloids on different stationary phases. J. Chromatogr. Sci. 21, 287–293.*

Zhou and Danielson (2013) used isopropylammonium formate as a mobile phase modifier to improve protein stability during reversed phase LC. They examined three types of protein including nonsubunit, two subunit, and four subunit proteins. The results showed that a significant enhancement in protein stability was observed when isopropylammonium formate was used as mobile phase modifier. Most of the proteins of study could maintain the native form within up to 50% of isopropylammonium formate. As observed in a comparison study (Zhou and Danielson, 2013), even 10% of acetonitrile caused protein denaturation. This study broadens the application of ILs in LC to some extent. This demonstrated that application of ILs as mobile phase modifier or additive may improve both the separation and resolution, as well as enhance the stability of bioactive macromolecules.

Another attractive application of ILs is chiral ILs, which represents a new type of chiral recognition agent. Chiral ILs have been employed as chiral selectors in capillary electrophoresis and HPLC. Since most amino acids have chiral carbon atoms, they are selected as highly promising candidates for preparing chiral ILs, and the resultant ILs are named amino acid ILs. Zhou et al. (2014) recently used isoleucine ILs as additives to separate mandelic acid and their derivative enantiomers by HPLC. They successfully separated mandelic acid, 2-chloromandelic acid, 4-methoxymandelic acid, and 3,4,5-trimethoxymandelic acid by using isoleucine ILs and copper ion as the chiral selector. This application greatly expanded the application of ILs in the analysis field.

ILs have also been introduced into thin-layer chromatography. Kaliszan et al. (2004) added the imidazolium tetrafluoroborate class of ILs at concentration of 0.5−1.5% to block silanols, and the excellent thin-layer chromatographic separations of strongly basic drugs was obtained. In 2005, other researchers (Baczek et al., 2005) investigated the effect of IL addition into the mobile phase for separation of peptides by using thin-layer chromatography. In thin-layer chromatography without ILs, R_m can be described well with a quadratic function. With the presence of ILs, the retention behavior is better delineated with a third-degree polynomial function. Most current HPLC and TLC stationary phases use silica gel as the solid support. This brings up a common problem: the silica surface has residual acidic silanol groups, which have deleterious effects on separations. Therefore, amine additives are always added to block the acidic surface and ameliorate these effects. ILs, due to their proton acceptor properties, provide the potential for this application (Sun and Armstrong, 2010). This may explain that why addition of ILs improves performance of separation.

TLC is used widely in analytical chemistry for several reasons: (1) it is convenient and simple with commercialized plates available, (2) the required equipment is easily available, (3) several samples can be analyzed simultaneously, (4) it is good for analysis of suspicious compounds that might damage HPLC column, and (5) it consumes less sample than HPLC

(Fuchs et al., 2011). It can be used for analyses of most lipids including neutral glycerolipids, glycerophospholipids, glycosphingolipids, phosphosphingolipids, cholesteryl esters, gallic acids, steroid hormones, and lipid vitamins (Fuchs et al., 2011). In the review by Fuchs et al. (2011), they almost summarized all the applications of TLC on lipids analysis. However, no application of ILs either as additives or as modifiers was referred to. It seems that applying ILs in LC for analyzing lipids has not yet been explored. It may be because of the properties of lipid-relevant compounds that lead to the restriction of application of ILs as mobile phase modifier, additive, and stationary phase. However, it should be noted that compatibility of the selected detector with ILs should be taken into account when ILs are used to assist in the separation of analytes.

14.4 IONIC LIQUIDS AS MALDI MATRICES

MALDI (matrix-assisted laser desorption ionization) is a soft ionization technique for mass spectrometry, which allows detection of intact, large biomolecules (eg, proteins and oligonucleotides) and synthetic polymers. The analyte is dissolved in a volatile solvent containing the matrix and then spotted on the MALDI plate. The role of the matrix in this technique is to prevent analyte molecules from being destroyed by direct laser energy absorption and facilitating their volatilization and ionization (Sun and Armstrong, 2010) (Fig. 14.6). Thus, the selection of matrix is important in MALDI-MS analysis. An ideal matrix should possess several properties: (1) absorption at the laser wavelength, (2) capabilities of dissolving or cocrystalizing with the sample, (3) low volatility, (4) suppressing analyte decompositions, and (5) promoting the ionization of analytes. The problem of utilization of the solid matrix is that heterogeneity of the sample makes the MS signal greatly affected by the position of the laser spot. However, the liquid matrixes, which could provide a homogeneous distribution, led to peak broadening, adduct formation, and volatility problems (Sun and Armstrong, 2010).

In 1999, Schiller et al. first applied matrix-assisted laser desorption ionization time-of-flight mass spectrometry (MALDI-TOF-MS) for analysis of diacylglycerols, phosphatidylcholines, and (poly) phosphoinositides. With 2,5-dihydroxybenzoic acid as matrix, they found that the molecular ions (M+1) of phosphatidylcholines as well as the corresponding adducts of different phosphoinositides could be easily detected even in complex mixtures. MALDI-TOF-MS is a fast and convenient method: one sample can be analyzed in less than 1 min. MALDI-TOF targets (sample plates) allow simultaneous (even often automated) analysis of many samples. The amount of lipid samples in the range of a few nanograms are sufficient for MS analysis. The lipid and the matrix are readily soluble in organic solvents; all manipulations can be performed in a single organic phase. This results in extremely homogeneous matrix/analyte cocrystals and an excellent reproducibility in

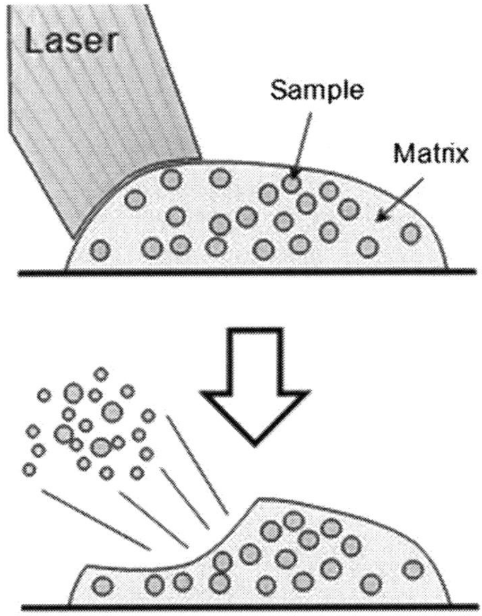

FIGURE 14.6 Schematic illustration of MALDI. *Adapted from Kuwabata, S., 2013. Use of ionic liquid under vacuum conditions. In: Kadokawa, J. (Ed.), Ionic Liquids – New Aspects for the Future. InTech.*

comparison to more polar molecules that require solvent mixtures. MALDI-TOF mass spectrometers are widespread and available at comparably low prices. Such devices can be used for a number of analytical problems and for a variety of biomolecules ranging from, for example, lipids to proteins and oligosaccharides. There is no need to buy an additional device exclusively for lipid analysis (Schiller et al., 2004).

ILs merit further study as matrixes of MALDI because of the low volatility of ILs, which makes them useful as solvents working their more amorphous solid analogs in high vacuum. ILs also have excellent solubilizing properties and vacuum stability as compared to other commonly used liquid and solid matrixes. However, certain ionic matrices produce homogeneous solutions of higher ion peak intensity and equivalent or lower detection limits than currently used solid matrixes (Koel, 2005).

The first application of IL matrices was compared to conventional solid matrices such as α-cyano-hydroxycinnamic acid on the basis of their ability to ionize traditional analytes such as insulin, bradykinin, and PEG. The Armstrong group further tested IL matrices for ionizing oligonucleotides (Soukup-Hein et al., 2009). However, IL was found not suitable for the oligonucleotides since it is liquid (Soukup-Hein et al., 2009). IL matrices made from sinapinic, α-cyano-hydroxycinnamic, and 2,5-dihydroxybenzoic acids as well as equimolar amounts of organic bases such as pyridine, 1-methylimidazole, and tributylamine were used to test small molecule ionization. Compared to conventional solid matrices, the IL matrices yielded better linearity, smaller standard deviations, and decreased analysis time. Especially, the use of IL matrices resulted in the suppression of fragmentation for phospholipids (Soukup-Hein et al., 2009).

Li et al. (2005) investigated several kinds of IL matrices with 2,5-dihydroxybenzoic acid as a comparison (Table 14.3). The results showed that IL matrices have advantages over the crystalline 2,5-dihydroxybenzoic acid for MALDI analysis of phospholipids. The use of IL matrices for phospholipids affords higher signal intensity, smaller spot size, improved spot homogeneity, better signal reproducibility, and comparable or better detection limits compared to that of the crystalline matrix 2,5-dihydroxybenzoic acid. 2,4,6-Trihydroxyacetophenone (THAP) was evaluated for the analysis of various lipid classes including neutral storage lipids (triacylglycerols), polar membrane lipids (glycerophospho- and sphingolipids), and glycosphingolipids. The results showed that THAP was a versatile matrix for the routine analysis of various lipids from biological samples (Stübiger and Belgacem, 2007). IL matrix was also used for MALDI MS imaging of gangliosides in mouse brain (Chan et al., 2009). The employed IL matrix offered excellent sensitivity for detection of gangliosides without significant loss of sialic acid residues. If the IL matrix was not used, the sialic acid residues can be dissociated in the ionization process. It means that IL matrices are somewhat a protection for sensitive molecules.

ILs have been used for extraction or pretreatment of lipids relevant for further determination. For example, Calvano et al. (2012) used ILs as extraction solvent to extract and enrich phospholipids in samples and then analyzed the composition by using MALDI-TOF-MS. They found that the IL tributylamine-α-cyano-4-hydroxycinnamic acid as MALDI matrix was very suitable for phospholipids analysis. It increased the phospholipids signals and at the same time decreased the signals from triacylglycerols and diacylglycerols according to the mass spectra obtained. The whole method is applicable to separate phospholipids in oil.

Lipids are of high interest in different respects. Neutral lipids such as triacylglycerols are of immense commercial interest. Lipids are also of high relevance for different in vivo processes such as diagnostics of changes of the lipoproteins of human blood for identifying vascular diseases. Phospholipids are crucial for the functions of the cellular membrane and thus some selected phospholipids can be used as biomarker to identify diseases (Fuchs and Schiller, 2009). Therefore, like protomics and genomics, lipidomics is now proposed and studied extensively. Lipidomics is a rapidly growing field that provides an insight on how specific lipids play roles in normal physiological and disease states. MALDI-TOF-MS is one of the methods to determine and/or identify phospholipids in tissues or complex samples. Compared to the other determination methods, HPLC-MS is the most versatile tool for lipid analysis with or without prior separation steps.

TABLE 14.3 Comparison of 2,5-DHB versus Ionic Liquid Matrices (ILMs) of α-cyano-4-hydroxycinnamic Acid Used as Matrices for MALDI-MS of Six Classes of PLs: PC, PE, PS, PA, PG, and PI[a] in Both Positive (Upper Row) and Negative (Lower Row) Ion Modes (Signal Intensities Are Divided by 1000)

Structure of Used ILMs	Physical State at 25°C	pK_a of Cation	Intensity of $[M+H]^+$ or $[M+Na]^+$ Signal[b]					
			PC	PE	PS	PA	PG	PI
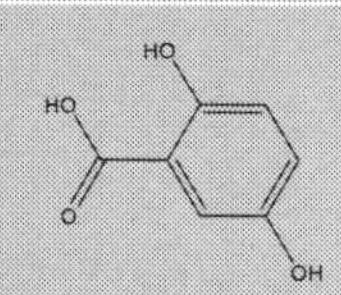	Solid	~2.5	6.0	3.2	15	7.0	9.0	3.0
			N/A	N/A	12	N/A	1.0	20
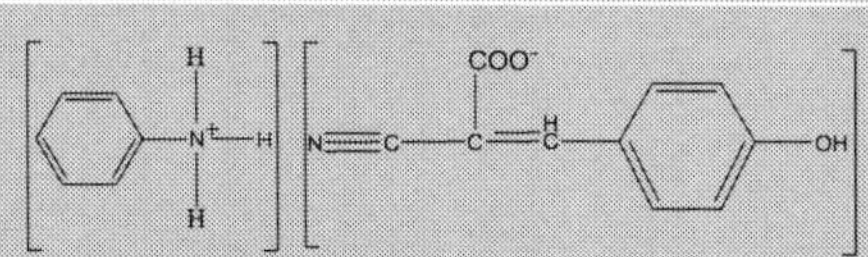	Solid	4.60 (25°C)	18	30	34	22	27	54
			N/A	N/A	56	4.0	17	47
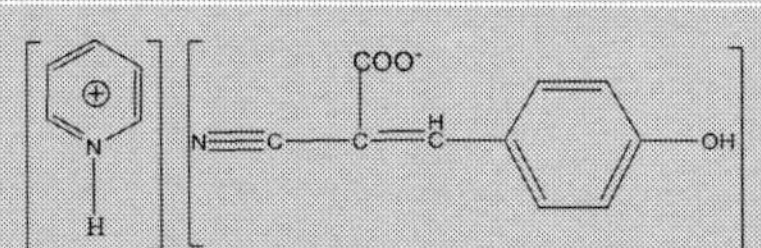	Solid	5.25 (25°C)	14	18	49	14	15	18
			N/A	N/A	27	6.0	15	45
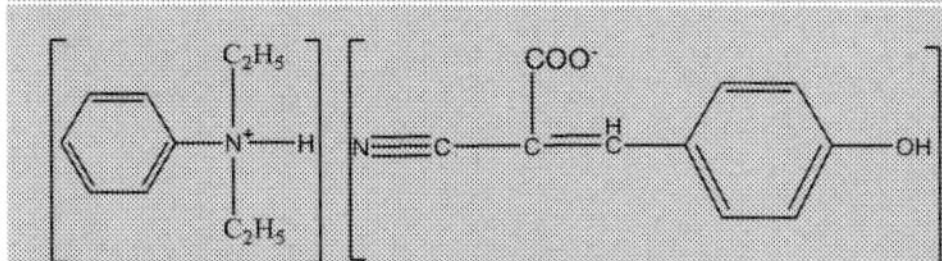	Liquid	6.61 (22°C)	17	10	50	30	6.0	14
			N/A	N/A	48	N/A	31	30

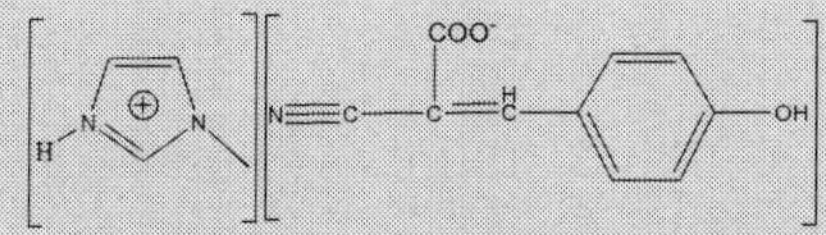	Liquid	6.95 (25°C)	25	10	29	24	8.0	5.0
			N/A	1.0	28	14	5.0	4.0
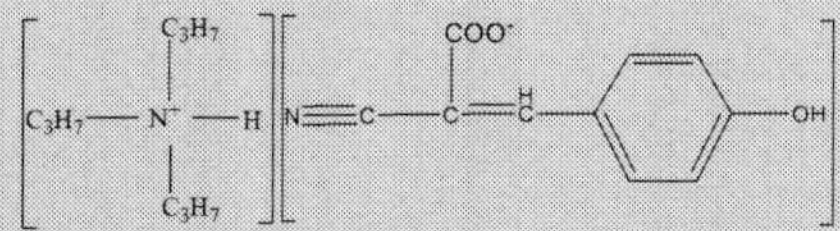	Liquid	10.65 (25°C)	21	9.0	15	8.0	10	3.0
			N/A	4.0	27	10	14	13
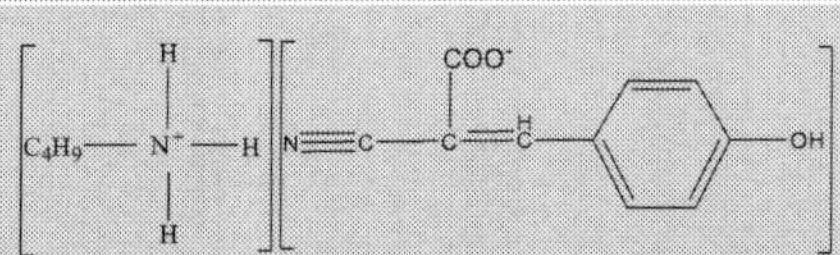	Liquid	10.78 (20°C)	18	12	20	42	30	18
			N/A	4.0	58	16	8.0	32
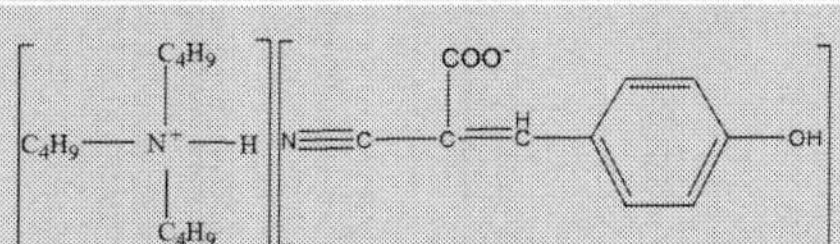	Liquid	10.89 (25°C)	20	N/A	13	3.0	N/A	1.0
			N/A	N/A	56	20	10	41

[a]*Phosphatidylcholine (PC, 16:0/18:1, monoisotopic MW = 759.58); phosphatidylethanolamine (PE, 16:0/18:1, monoisotopic MW = 717.53); phosphatidylserine (PS, 16:0/18:1, monoisotopic MW = 761.52); phosphatidic acid (PA, 16:0/18:1, monoisotopic MW = 674.45); phosphatidylglycerol (PG, 16:0/18:1, monoisotopic MW = 748.53); phosphatidylinositol (PI, bovine liver, monoisotopic MW for main product, 18:0/20:4 = 885.55). All measurements are from spotting lipid solutions at concentration of 1 μg/μL.*

[b]*The first row for each phospholipid species indicated signal intensity detected in the positive-ion mode, the second row indicated signal intensity in the negative-ion mode. When no signal was detected, the space is N/A.*

Adapted from Li, Y.L., Gross, M.L., Hsu, F.-F., 2005. Ionic-liquid matrices for improved analysis of phospholipids by MALDI-TOF mass spectrometry. J. Am. Soc. Mass Spectrom. 16 (5), 679–682.

Among several different ionization techniques that are able to produce ions from the lipids of interest, MALDI-TOF-MS is the simplest and most convenient method (Fuchs and Schiller, 2009). ILs as matrices in MALDI-TOF-MS has shown many advantages in lipids analysis as discussed earlier. Therefore, with the expanding of lipidomics, the application of ILs in this field will be growing steadily.

14.5 IONIC LIQUIDS IN CAPILLARY ELECTROPHORESIS ANALYSIS TECHNIQUE

Capillary electrophoresis (CE) has become a powerful separation technique for its low sample loading and low reagent consumption, high efficiency, and simplicity. The principle of CE is shown in Fig. 14.7. This method applies to the molecules that are either positively or negatively charged, which migrate in an electrical field with different velocities. The time-dependent separations of the different constituents in mixture depend on two principle factors called electrophoretic mobility and electroosmotic flow. As shown in Fig. 14.7, the apparent mobility is the vector addition of effective electrophoretic mobility and electroosmotic flow. Fused-silica capillaries are generally used for CE, and silanol groups on the inner surface are normally negatively charged, resulting in the formation of an electroosmotic flow. In addition, the inner capillary

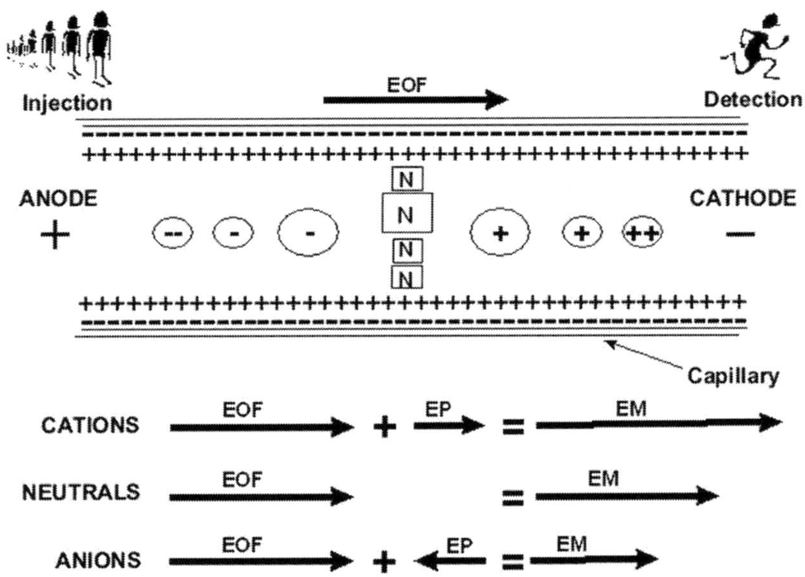

FIGURE 14.7 Principle of capillary electrophoresis. *EOF*, electroosmotic flow; *EP*, effective electrophoretic; *EM*, apparent mobility. *From http://www.doping.chuv.ch/en/lad_home/lad-prestations-laboratoire/lad-prestations-laboratoire-appareils/lad-prestations-laboratoire-appareils-ec.htm.*

surface could participate in interacting with analytes and further affecting separations (Sun and Armstrong, 2010). CE makes use of support electrolytes that are salts. Since ILs are viscous conductive salts, it may be difficult to use them as solvent in capillaries directly because the electric current would be high and/or the applied voltage and electric field would be low. The main disadvantage of CE is an electroosmotic flow provoked by silanol groups on the inner surface of silica capillaries. To reverse this process, ILs could be used as capillary wall coating materials. They are also possible running electrolytes in aqueous solutions or in polar organic solvents. Modified ILs were used to perform chiral and micellar electrokinetic chromatography separations (Berthod et al., 2008).

First, ILs can be used as capillary wall covalent coating materials to improve the separation performance. Qin and Li (2002) covalently bonded propyl methyl imidazolium chloride onto the silica capillary surface and the electroosmotic flow was reversed over pH range of 3.5−7. The adsorption of the analytes onto the bare capillary wall was eliminated by the IL coating and the drugs were baseline-separated within 14 min with detection limits of 14 and 17 ng/mL for sildenafil and itsmetabolite UK-103,320. The imidazolium-based IL was covalently coated to capillary for separating DNA (Qin and Li, 2003). The electroosmotic flow of the capillary was reversed between pH 4.5 and 9. The results indicated that the IL-coated capillary could work stably in the run buffer at least 96 h without notable deterioration in performance. The electroosmotic flow of bare silica capillary which was reversed by covalently bonded 1-ethyl-3-methylimidazolium coating was used to separate positively charged inorganic ions NH_4^+ and K^+ in human urine. Additionally, K^+, Na^+, Li^+, Ca^+, Mg^+, and Ba^{2+} in red wine were baseline separated in the modified capillary with run buffer of 1-ethyl-3-methylimidazolium OH-acetic acid at pH 5 (Qin et al., 2003). The main advantage of these capillaries appears to be increased in the working pH range of the running buffer with good efficiency and compatibility with mass spectrometry.

The second application of ILs on capillary electrophoresis is used as background electrolyte and dynamic wall coating in aqueous capillary electrophoresis, due to their electrolytes properties. The hydrophobic character of their cationic part plays the key role for a significant adsorption on the capillary wall. The immobilized adsorbed IL cations generate a change in the electroosmotic flow direction that can be used for separation. As mentioned previously, the hydrophobicity of the ILs cation strongly associated with the alkyl chain length. Therefore, the dynamic wall coating can be adjusted at will playing with the nature of IL cations and with its concentration (Berthod et al., 2008). Utilization of 1-alkyl-3-methylimidazolium-based ILs as running electrolytes in capillary electrophoretic for resolving phenolic compounds found in grape seed extracts was studied (Yanes et al., 2001). The electrophoretic method proved to be simple and reliable, providing good reproducibility in terms of migration times. The separation mechanism relies on the association of the polyphenols with the imidazoliun cations either coating the capillary wall or electrophoretically migrating in the bulk solution. The mechanism of polyphenols separation using 1-alkyl-3-methylimidazolium-based ILs is shown in Fig. 14.8. The delay in the migration time of the polyphenols may be attributed to two factors. The polyphenols may associate either with the positively charged imidazolium groups coating the capillary wall or with the free imidazolium ions in the bulk solution. This association could be partially driven by hydrophobic, hydrogen bonding, or ion-dipole/ion-induced-dipole interactions between the polyphenols and the imidazolium cations.

The same IL was used as the running electrolytes in capillary electrophoresis for separating and analysis of basic proteins including lysozyme, cytochrome c, trypsinogen, and α-chymotrypinogen A (Jiang et al., 2003). The use of ILs as the running electrolytes led to a surface charge reversal on the capillary wall. By using 90 mM 1-alkyl-3-methylimidazolium tetrafluoroborate solution, baseline separation, high efficiencies, and symmetrical peaks of four proteins were obtained. The low RSD values of $<0.68\%$ and $<0.3\%$ for migration times and peak areas were achieved, respectively. It is also concluded the separation mechanism seems to involve association between the imidazolium cations and the proteins. It seems that 1-alkyl-3-methylimidazolium-based ILs are very good candidates as background electrolytes in capillary electrophoresis. Tian et al. (2007) employed only 1-alkyl-3-methylimidazolium-based ILs as background electrolytes for simultaneous determination of five anthroquinone derivatives including aloe-emodin, emodin, chrysophanol, physcion, and rhein in Rhubarb species. The optimal analysis conditions were 90 mM 1-alkyl-3-methylimidazolium tetrafluoroborate (pH 11.0) with an applied voltage of 20 kV. By employing this optimized condition, the determination of anthraquinones in extracts

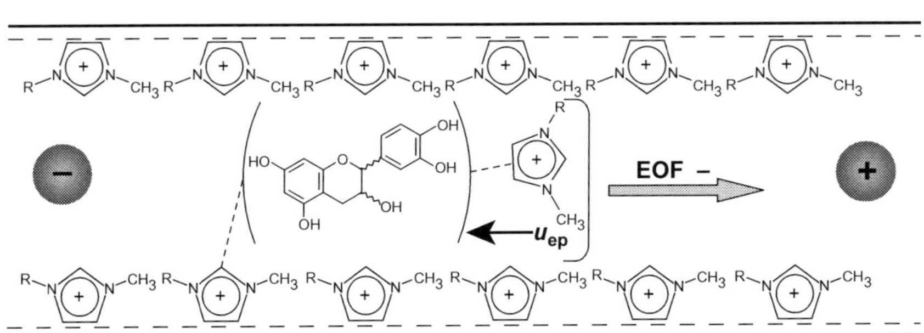

FIGURE 14.8 Mechanism of polyphenol separation using 1-alkyl-3-methylimidazolium-based ionic liquids. *Adapted from Yanes, E.G., Gratz, S.R., Baldwin, M.J., Robison, S.E., Stalcup, A.M., 2001. Capillary electrophoretic application of 1-Alkyl-3-Methylimidazolium-based ionic liquids. Anal. Chem. 73 (16), 3838–3844.*

was successful and completed within 12 min. Separation of inorganic anions in CE is often a challenging task because the electrophoretic mobilities of inorganic anions are comparable to or even greater than the electroosmotic flow mobility. In 2008, Mo et al. used 1-alkyl-3-methylimidazolium-based ILs as background electrolytes in capillary electrophoresis for the analysis of inorganic anions. This allows the anions to comigrate with the electroosmotic flow and yield a rapid separation. Increasing alkyl chain length of the ILs and background electrolytes concentration showed significant improvement on the separation resolution. Good separation of five model anions including Br^-, I^-, NO_2^-, NO_3^-, and SCN^- was achieved with 40 mM 1-butyl-3-methylimidazolium tetrafluoroborate as background electrolytes. The separation efficiency was as high as 34,600−155,000 and the RSDs of migration times were less than 0.8%, suggesting very good reproducibility.

ILs could be also used as background electrolytes in nonaqueous capillary electrophoresis (NACE). In NACE, the CE liquid phase is much less polar than water. The dynamic coating of the capillary wall is greatly decreased and/or nonexistent. The electroosmotic flow direction is not changed by IL addition and the solute/IL interaction is the main mechanism responsible for enhanced separation (Berthod et al., 2008). The first publication referring to use of ILs in NACE was authored by Vaher et al. (2001). The dialkylimidazolium-based ILs were used as buffer electrolytes in their work to separate water-insoluble dyes in acetonitrile. Anionic part of IL changes the general electrophoretic mobility of the system as well as the concentration of IL. Separation of analytes is achieved because of dissociation of analytes in the presence of ILs in the separating medium. The migration order of analytes is based on the differences in effective charge-hydrodynamic radius ratio of the analytes. 1-alkyl-3-methylimidazolium-based ILs were applied in NACE for separating Brønsted acids and polyphenolic compounds (Vaher and Koel, 2003). But the focus of Vaher et al.'s work was to assess the interactions between the analytes and the ionic additive presented to find the influences of the type and concentration of the ionic additives, as well as the nature of the nonaqueous medium employed. Meanwhile, different organic solvents including acetonitrile and methanol were assessed from their different contributions to the conversion of analytes into a charged form (Vaher et al., 2002a). The results indicated that acetonitrile is a good solvent to study the use of alkylimidazolium-based salts—ILs—as background electrolyte components. The IL electrolytes were helpful to achieve the separation of acetonitrile-soluble phenolic compounds. But the ILs studied as components of background electrolytes are not suitable for separation of polyphenolic compounds in acetonitrile. It is possible to change the mobilities of the analytes and the separation by changing the composition of a solvent mixture and the salt concentration. The dialkylimidazolium-based ILs were used as electrolytes to investigate the function of the anionic part as well as the concentration of an IL for affecting the general electrophoretic mobility of the buffer system (Vaher et al., 2002b). They found that the separation of different analytes could be achieved because they became charged in the presence of ILs in separation media. In addition, another possibility for complex formation between the solute and the electrolyte which may alter the mobility of the solute also existed. NACE has offered suitable electrolyte systems for a number of analytical problems, but there is more room for other studies with different kinds of salts offering other kinds of interactions in nonaqueous media.

14.6 APPLICATION OF IONIC LIQUIDS IN ENRICHMENT OF *n*-3 POLYUNSATURATED FATTY ACIDS/ESTERS

Marine omega-3 polyunsaturated fatty acids (*n*-3 PUFA) such as eicosapentaenoic acid (EPA) and docosahexaenoic acid (DHA) have long been recognized for their role in reducing the risk of heart disease and inflammatory disease and ameliorating brain function and mental health (Ruxton et al., 2004). It is always recommendable to consume essential *n*-3 PUFAs but not intake other fats to add weight. There are a number of different methods to produce *n*-3 PUFA concentrates, namely, enzymatic concentration (Halldorsson et al., 2004), urea inclusion complexation (Ackman et al., 1988), low-temperature fractional crystallization (Yokochi et al., 1990), and silver salt-based liquid/liquid extraction (Teramoto et al., 1994).

ILs, particularly those with cationic aromatic rings, contain delocalized π electrons and hence are strong π electron acceptors. They form $\pi-\pi$ complexation with electron donors such as double-bond-containing compounds, resulting in a charge transfer complex. The $\pi-\pi$ complexation is weak and thus can be easily reversed to release the initial compounds. This concept can be employed in selective extraction and enrichment of *n*-3 PUFAs from a bulk lipidic phase (Cheong et al., 2011). When added to a bulk lipidic phase, ILs with aromatic rings may selectively attract *n*-3 PUFAs into the IL phase by forming π bonds with the *n*-3 PUFA. Following that, a stripping solvent (*n*-hexene) can be used to break such bondings to release the *n*-3 PUFAs. This working principle can be demonstrated in Fig. 14.9. In fact, this concept has been employed for selective extraction and enrichment of *n*-3 polyunsaturated fatty acid methyl esters (*n*-3 PUFAME) (Li et al., 2009a,b; Pham et al., 2009) and also polyunsaturated triacylglycerols (TAG) (Pham et al., 2009). In these studies, silver salt is added to strengthen π-complexing capacity to improve the extraction and enrichment capacities of *n*-3 PUFAs.

Cheong et al. (2011) examined 11 kinds of ILs with different anions, belonging to aromatic imidazolium, pyridinium, and nonaromatic ammonium and pyrrolidinium, respectively, for their efficiency and selectivity in extraction and enrichment of *n*-3 PUFAs. ILs containing aromatic rings such as the *N,N′*-dialkylimidazolium and *N*-alkylpyridinium are found to selectively extract and enrich *n*-3 polyunsaturated compounds. The IL extraction process reached equilibrium relatively quickly (<30 min), and the extraction

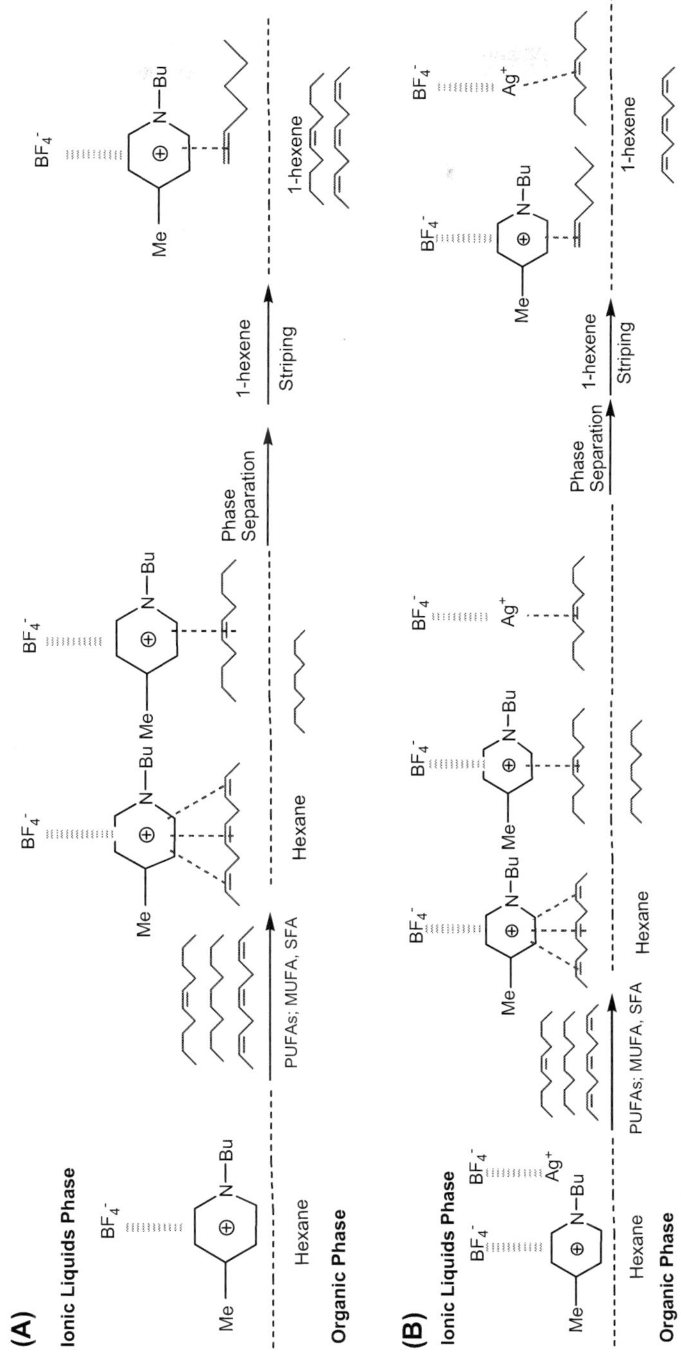

FIGURE 14.9 Schematic representation of aromatic IL (A) and IL/Ag+ based liquid–liquid extraction for enrichment of *n*-3 PUFAs (B).

efficiency can be further enhanced by increasing the volume ratio of IL to solvent, addition of silver tetrafluoroborate, and usage of double-bond-containing solvents such as 1-hexene as stripping solvent. It turned out 1-butyl-3-methylpyridinium dicyanamide is the most efficient IL for *n*-3 enrichment with highest selectivity (3.2/1, *n*-3 PUFAEE to non-*n*-3 FAEE); moreover, the addition of silver tetrafluoroborate further increased selectivity for *n*-3 PUFAEE up to 4.4/1. Through multistep extractions, the purity of *n*-3 PUFA can be enriched from 15.88% to 45%; and for *n*-3 PUFAEE from initial 72.56−89%.

Li's group has carried out intensively studies in ILs-based enrichment of *n*-3 PUFAs (Li et al., 2009a,b; Pham et al., 2009). For example, they developed mesoporous SBA-15-supported IL compounds with silver salts. They achieved omega-3 PUFAMEs (EPA + DHA) enrichment from 18% in the original cod liver oil to 90.5% by using reusable $AgBF_4$/SBA-15-IL-PF_6 π-complexing sorbents (Li et al., 2009a). Li and his coworkers (Pham et al., 2009) also examined Mesoporous silica (SBA-15) supported ILs with immobilized silver salt ($AgBF_4$/SBA15_HPSiOEtIM_PF_6) as HPLC adsorbent for selective absorption of polyunsaturated triacylglycerols. It is found that in mobile phase the difference of the concentration of linolenin and tripalmitin up to 1:10, where linolenin is trapped by the adsorbent.

14.7 PROSPECTS

The unusual characteristics and potential environment-friendly behavior of ILs make them usable for more and more applications in a variety of analytical purposes. As a stationary phase modifier, the stability of the stationary phase is an emerging handicap for commercialization. The excellent theromostability and low melting point are contradictory to achieve in IL design; however this is just the property of the ILs needed to be an ideal stationary phase. Therefore, to develop ILs with low melting point but high boiling point and high theromostability is definitely a target worth pursuing. Synthesis of ILs with chiral properties is another direction that may be explored for chiral analytes separation. Functional ILs made from natural chiral compounds such as amino acids and sugars are the new directions to both reach analysis target and improve the sustainability of IL technique. The range of separation capability needs to be broadened from normal chemicals to bioactive compounds as well.

The limitations of the application of ILs in LC are more obvious than that in GC. The high viscosity IL gives an extreme high back pressure of LC when used as mobile phase modifier. Some ILs could corrode the metallic parts of the HPLC system. Significant UV absorbance by ILs limits the use of UV detector and nonvolatile nature precludes the use of evaporative light-scattering detector. Either new ILs without these limitations or a newly developed detector or HPLC system that could be compatible with the properties of ILs may be developed in the future.

Almost all the works select some ILs and then study their functions either as stationary phase, mobile phase, modifier/additive, matrix, capillary coating materials, or background electrolytes. It still lacks effective methods to design unique ILs for a specific application. Therefore, the interactions between ILs and analyst should be further investigated to achieve improved understanding, which could serve more specific design of ILs for target use.

Although ILs have some excellent performance in analytical chemistry, the cost of using ILs should be taken into account. Recovery method of ILs could be developed from the perspective of economy.

REFERENCES

Anderson, J., Armstrong, D., 2003. High-stability ionic liquids. A new class of stationary phases for gas chromatography. Anal. Chem. 75 (18), 4851–4858.

Armstrong, D.W., He, L., Liu, Y.-S., 1999. Examination of ionic liquids and their interaction with molecules, when used as stationary phases in gas chromatography. Anal. Chem. 71 (17), 3873–3876.

Ackman, R.G., Ratnayake, W.M.N., Olsson, B., 1988. The "basic" fatty acid composition of Atlantic fish oils: potential similarities useful for enrichment of polyunsaturated fatty acids by urea complexation. J. Am. Oil Chem. Soc. 65 (1), 136–138.

Berthod, A., Ruiz-Angel, M.J., Carda-Broch, S., 2008. Ionic liquids in separation techniques. J. Chromatogr. A 1184 (1–2), 6–18.

Berthod, A., Ruiz-Angel, M.J., Huguet, S., 2005. Nonmolecular solvents in separation methods: dual nature of room temperature ionic liquids. Anal. Chem. 77 (13), 4071–4080.

Baczek, T., Marszałł, M.P., Kaliszan, R., Walijewski, L., Makowiecka, W., Sparzak, B., Grzonka, Z., Wiśniewska, K., Juszczyk, P., 2005. Behavior of peptides and computer-assisted optimization of peptides separations in a normal-phase thin-layer chromatography system with and without the addition of ionic liquid in the eluent. Biomed. Chromatogr. 19 (1), 1–8.

Cheong, L.-Z., Guo, Z., Yang, Z., Chua, S.-C., Xu, X., 2011. Extraction and enrichment of N-3 polyunsaturated fatty acids and ethyl esters through reversible π-π complexation with aromatic rings containing ionic liquids. J. Agric. Food Chem. 59 (16), 8961–8967.

Chitta, K.R., Van Meter, D.S., Stalcup, A.M., 2010. Separation of peptides by HPLC using a surface-confined ionic liquid stationary phase. Anal. Bioanal. Chem. 396 (2), 775–781.

Chan, K., Lanthier, P., Liu, X., Sandhu, J.K., Stanimirovic, D., Li, J., 2009. MALDI mass spectrometry imaging of gangliosides in mouse brain using ionic liquid matrix. Anal. Chim. Acta 639 (1–2), 57–61.

Calvano, C.D., De Ceglie, C., D'Accolti, L., Zambonin, C.G., 2012. MALDI-TOF mass spectrometry detection of extra-virgin olive oil adulteration with hazelnut oil by analysis of phospholipids using an ionic liquid as matrix and extraction solvent. Food Chem. 134 (2), 1192–1198.

Delmonte, P., Fardin Kia, A.-R., Kramer, J.K.G., Mossoba, M.M., Sidisky, L., Rader, J.I., 2011. Separation characteristics of fatty acid methyl esters using SLB-IL111, a new ionic liquid coated capillary gas chromatographic column. J. Chromatogr. A 1218 (3), 545–554.

Destaillats, F., Guitard, M., Cruz-Hernandez, C., 2011. Identification of Δ6-monounsaturated fatty acids in human hair and nail samples by gas-chromatography-mass-spectrometry using ionic-liquid coated capillary column. J. Chromatogr. A 1218 (52), 9384–9389.

Delmonte, P., Fardin-Kia, A.R., Kramer, J.K.G., Mossoba, M.M., Sidisky, L., Tyburczy, C., Rader, J.I., 2012. Evaluation of highly polar ionic liquid gas chromatographic column for the determination of the fatty acids in milk fat. J. Chromatogr. A 1233, 137–146.

Flieger, J., Czajkowska-Żelazko, A., 2011. Comparison of chaotropic salt and ionic liquid as mobile phase additives in reversed-phase high-performance liquid chromatography of biogenic amines. J. Sep. Sci. 34 (7), 733–739.

Fuchs, B., Süss, R., Teuber, K., Eibisch, M., Schiller, J., 2011. Lipid analysis by thin-layer chromatography—a review of the current state. J. Chromatogr. A 1218 (19), 2754–2774.

Fuchs, B., Schiller, J., 2009. Application of MALDI-TOF mass spectrometry in lipidomics. Eur. J. Lipid Sci. Technol. 111 (1), 83–98.

Gu, Q., David, F., Lynen, F., Vanormelingen, P., Vyverman, W., Rumpel, K., Xu, G., Sandra, P., 2011. Evaluation of ionic liquid stationary phases for one dimensional gas chromatography-mass spectrometry and comprehensive two dimensional gas chromatographic analyses of fatty acids in marine biota. J. Chromatogr. A 1218 (20), 3056–3063.

Han, X., Armstrong, D., 2007. Ionic liquids in separations. Acc. Chem. Res. (1977), 1079–1086.

Han, D., Row, K.H., 2010. Recent applications of ionic liquids in separation technology. Molecules 15 (4), 2405–2426.

Halldorsson, A., Kristinsson, B., Haraldsson, G.G., 2004. Lipase selectivity toward fatty acids commonly found in fish oil. Eur. J. Lipid Sci. Technol. 106 (2), 79–87.

Jolanta, F., 2007. Application of ionic liquids in liquid chromatography. In: Ionic Liquids: Applications and Perspectives, pp. 243–272.

Jiang, T.-F., Gu, Y.-L., Liang, B., Li, J.-B., Shi, Y.-P., Ou, Q.-Y., 2003. Dynamically coating the capillary with 1-alkyl-3-methylimidazolium-based ionic liquids for separation of basic proteins by capillary electrophoresis. Anal. Chim. Acta 479 (2), 249–254.

Koel, M., 2005. Ionic liquids in chemical analysis. Crit. Rev. Anal. Chem. 35 (3), 177–192.

Kao, C.-L., Lin, S.-Y., Hsieh, Y.-N., Ho, W.-Y., Chen, Y.-F., Huang, P.-C., Kuei, C.-H., 2014. A new imidazolium ionic polymer as a gas chromatography stationary phase for separation of high and wide temperature range complex samples. RSC Adv. 4 (96), 53828–53832.

Kaliszan, R., Marszałł, M.P., Jan Markuszewski, M., Bączek, T., Pernak, J., 2004. Suppression of deleterious effects of free silanols in liquid chromatography by imidazolium tetrafluoroborate ionic liquids. J. Chromatogr. A 1030 (1–2), 263–271.

Kuwabata, S., 2013. Use of ionic liquid under vacuum conditions. In: Kadokawa, J. (Ed.), Ionic Liquids – New Aspects for the Future. InTech.

Li, M., Pham, P.J., Pittman, C.U., Li, T., 2009a. SBA-15-Supported ionic liquid compounds containing silver salts: novel mesoporous π-complexing sorbents for separating polyunsaturated fatty acid methyl esters. Microporous Mesoporous Mater. 117 (1–2), 436–443.

Li, M., Pham, P.J., Wang, T., Pittman, C.U., Li, T., 2009b. Solid phase extraction and enrichment of essential fatty acid methyl esters from soy-derived biodiesel by novel pi-complexing sorbents. Bioresour. Technol. 100 (24), 6385–6390.

Li, Y.L., Gross, M.L., Hsu, F.-F., 2005. Ionic-liquid matrices for improved analysis of phospholipids by MALDI-TOF mass spectrometry. J. Am. Soc. Mass Spectrom. 16 (5), 679–682.

Marszałł, M.P., Sroka, W.D., Balinowska, A., Mieszkowski, D., Koba, M., Kaliszan, R., 2013. Ionic liquids as Mobile phase additives for feasible assay of naphazoline in pharmaceutical formulation by HPTLC-UV-densitometric method. J. Chromatogr. Sci. 51 (6), 560–565.

Mo, H., Zhu, L., Xu, W., 2008. Use of 1-alkyl-3-methylimidazolium-based ionic liquids as background electrolytes in capillary electrophoresis for the analysis of inorganic anions. J. Sep. Sci. 31 (13), 2470–2475.

Pandey, S., 2006. Analytical applications of room-temperature ionic liquids: a review of recent efforts. Anal. Chim. Acta 556 (1), 38–45.

Pacholec, F., Butler, H., Poole, C., 1982. Molten organic salt phase for gas-liquid chromatography. Anal. Chem. 1941 (18), 1938–1941.

Pacholec, F., Poole, C., 1983. Stationary phase properties of the organic molten salt ethylpyridinium bromide in gas chromatography. Chromatographia 17 (7).

Poole, C.F., Poole, S.K., 2011. Ionic liquid stationary phases for Gas chromatography. J. Sep. Sci. 34 (8), 888–900.

Polyakova, Y., Row, K., 2006. Retention behaviour of n-CBZ-D-phenylalanine and D-tryptophan: effect of ionic liquid as mobile-phase modifier. Acta Chromatogr. (17), 210–221.

Petruczynik, A., 2012. Effect of ionic liquid additives to mobile phase on separation and system efficiency for HPLC of selected alkaloids on different stationary phases. J. Chromatogr. Sci. (21), 287–293.

Pham, P.J., Pittman, C.U., Li, T., Li, M., 2009. Selective extraction of polyunsaturated triacylglycerols using a novel ionic liquid precursor immobilized on a mesoporous complexing adsorbent. Biotechnol. Prog. 25 (5), 1419–1426.

Qin, W., Li, S., 2002. An ionic liquid coating for determination of sildenafil and UK-103,320 in human serum by capillary zone electrophoresis-ion trap mass spectrometry. Electrophoresis (23), 4110–4116.

Qin, W., Li, S.F.Y., 2003. Electrophoresis of DNA in ionic liquid coated capillary. Analyst 128 (1), 37–41.

Qin, W., Wei, H., Li, S., 2003. 1,3-Dialkylimidazolium-based room-temperature ionic liquids as background electrolyte and coating material in aqueous capillary electrophoresis. J. Chromatogr. A 985, 447–454.

Reichardt, C., 2005. Polarity of ionic liquids determined empirically by means of solvatochromic pyridinium N-phenolate betaine dyes. Green Chem. 7 (5), 339.

Ragonese, C., Sciarrone, D., Tranchida, P.Q., Dugo, P., Mondello, L., 2012. Use of ionic liquids as stationary phases in hyphenated gas chromatography techniques. J. Chromatogr. A 1255, 130–144.

Ragonese, C., Tranchida, P.Q., Dugo, P., Dugo, G., Sidisky, L.M., Robillard, M.V., Mondello, L., 2009. Evaluation of use of a dicationic liquid stationary phase in the fast and conventional gas chromatographic analysis of health-hazardous C18 Cis/trans fatty acids. Anal. Chem. 81 (13), 5561–5568.

Ruxton, C.H.S., Reed, S.C., Simpson, M.J.A., Millington, K.J., 2004. The health benefits of omega-3 polyunsaturated fatty acids: a review of the evidence. J. Hum. Nutr. Diet. 17 (5), 449–459.

Soukup-Hein, R.J., Warnke, M.M., Armstrong, D.W., 2009. Ionic liquids in analytical chemistry. Annu. Rev. Anal. Chem. (Palo Alto, Calif.) 2, 145–168.

Sun, P., Armstrong, D.W., 2010. Ionic liquids in analytical chemistry. Anal. Chim. Acta 661 (1), 1–16.

Schiller, J., Arnhold, J., Benard, S., Müller, M., 1999. Lipid analysis by matrix-assisted laser desorption and ionization mass spectrometry: a methodological approach. Anal. Biochem. 56, 46–56.

Schiller, J., Süss, R., Arnhold, J., Fuchs, B., Lessig, J., Müller, M., Petković, M., Spalteholz, H., Zschörnig, O., Arnold, K., 2004. Matrix-assisted laser desorption and ionization time-of-flight (MALDI-TOF) mass spectrometry in lipid and phospholipid research. Prog. Lipid Res. 43 (5), 449–488.

Stübiger, G., Belgacem, O., 2007. Analysis of lipids using 2, 4, 6−trihydroxyacetophenone as a matrix for MALDI mass spectrometry. Anal. Chem. 79 (8), 3206−3213.

Tyburczy, C., Delmonte, P., Fardin-Kia, A.R., Mossoba, M.M., Kramer, J.K.G., Rader, J.I., 2012. Profile of trans fatty acids (FAs) including trans polyunsaturated FAs in representative fast food samples. J. Agric. Food Chem. 60 (18), 4567−4577.

Tian, M., Li, S., Row, K.H., 2010. Retention mechanism of some solutes using ionic liquids as mobile phase modifier in RP-high-performance liquid chromatography (HPLC). Korean J. Chem. Eng. 28 (2), 357−363.

Tang, Y., Sun, A., Liu, R., Zhang, Y., 2013. Simultaneous determination of fangchinoline and tetrandrine in *Stephania tetrandra* S. Moore by using 1-alkyl-3-methylimidazolium-based ionic liquids as the RP-HPLC Mobile phase additives. Anal. Chim. Acta 767, 148−154.

Tian, K., Wang, Y., Chen, Y., Chen, X., Hu, Z., 2007. Application of 1-alkyl-3-methylimidazolium-based ionic liquids as background electrolyte in capillary zone electrophoresis for the simultaneous determination of five anthraquinones in rhubarb. Talanta 72 (2), 587−593.

Teramoto, M., Matsuyama, H., Ohnishi, N., Uwagawa, S., Nakai, K., 1994. Extraction of ethyl and methyl esters of polyunsaturated fatty acids with aqueous silver nitrate solutions. Ind. Eng. Chem. Res. 33 (2), 341−345.

Vaher, M., Koel, M., Kaljurand, M., 2001. Non-aqueous capillary electrophoresis in acetonitrile using ionic-liquid buffer electrolytes. Chromatographia 53, 302−306.

Vaher, M., Koel, M., 2003. Separation of polyphenolic compounds extracted from plant matrices using capillary electrophoresis. J. Chromatogr. A 990, 225−230.

Vaher, M., Koel, M., Kaljurand, M., 2002a. Application of 1-alkyl-3-methylimidazolium-based ionic liquids in non-aqueous capillary electrophoresis. J. Chromatogr. A 979, 27−32.

Vaher, M., Koel, M., Kaljurand, M., 2002b. Ionic liquids as electrolytes for nonaqueous capillary electrophoresis. Electrophoresis 23 (3), 426−430.

Wilkes, J., Zaworotko, M., 1992. Air and water stable 1-ethyl-3-methylimidazolium based ionic liquids. J. Chem. Soc. Chem. Commun. 965−967.

Yanes, E.G., Gratz, S.R., Baldwin, M.J., Robison, S.E., Stalcup, A.M., 2001. Capillary electrophoretic application of 1-alkyl-3-methylimidazolium-based ionic liquids. Anal. Chem. 73 (16), 3838−3844.

Yokochi, T., Usita, M.T., Kamisaka, Y., Nakahara, T., Suzuki, O., 1990. Increase in the γ-linolenic acid content by solvent winterization of fungal oil extracted from *Mortierella* genus. J. Am. Oil Chem. Soc. 67 (11), 846−851.

Zhou, J., Zhao, S., Fu, G., Zhang, Z., 2014. Isoleucine ionic liquids as additives to separate mandelic acid and their derivative enantiomers by HPLC. Anal. Methods 6 (15), 5627.

Zhou, L., Danielson, N.D., 2013. The ionic liquid isopropylammonium formate as a mobile phase modifier to improve protein stability during reversed phase liquid chromatography. J. Chromatogr. B. Analyt. Technol. Biomed. Life Sci. 940, 112−120.

Index

'*Note*: Page numbers followed by "f" indicate figures and "t" indicate tables.'

C

D

E

F

G

M

N

O

P

Q

R

S

T

U

V

W

FSC
www.fsc.org
MIX
Papier | Fördert gute Waldnutzung
FSC® C083411

Kolibri 360 GmbH
Ferdinand-Jühlke-Straße 7
99095 Erfurt, Deutschland
produktsicherheit@kolibri360.de